Dust in the Galactic Environment

Second Edition

D0322107

UNIVERSITY OF NOTTINGHAM
WITHDRAWN
FROM THE LIBRARY

DATE DUE FOR RETURN

Series in Astronomy and Astrophysics

Series Editors: **M Birkinshaw**, University of Bristol, UK
M Elvis, Harvard–Smithsonian Center for Astrophysics, USA
J Silk, University of Oxford, UK

The Series in Astronomy and Astrophysics includes books on all aspects of theoretical and experimental astronomy and astrophysics. Books in the series range in level from textbooks and handbooks to more advanced expositions of current research.

Other books in the series

An Introduction to the Science of Cosmology
D J Raine and E G Thomas

The Origin and Evolution of the Solar System
M M Woolfson

The Physics of the Interstellar Medium
J E Dyson and D A Williams

Dust and Chemistry in Astronomy
T J Millar and D A Williams (eds)

Observational Astrophysics
R E White (ed)

Stellar Astrophysics
R J Tayler (ed)

Forthcoming titles

The Physics of Interstellar Dust
E Krügel

Very High Energy Gamma Ray Astronomy
T Weekes

Dark Sky, Dark Matter
P Wesson and J Overduin

Series in Astronomy and Astrophysics

Dust in the Galactic Environment

Second Edition

D C B Whittet

Professor of Physics, Rensselaer Polytechnic Institute,
Troy, New York, USA

Institute of Physics Publishing
Bristol and Philadelphia

1004591387

© IOP Publishing Ltd 2003

All rights reserved. No part of this publication may be reproduced, stored in a retrieval system or transmitted in any form or by any means, electronic, mechanical, photocopying, recording or otherwise, without the prior permission of the publisher. Multiple copying is permitted in accordance with the terms of licences issued by the Copyright Licensing Agency under the terms of its agreement with Universities UK (UUK).

British Library Cataloguing-in-Publication Data

A catalogue record for this book is available from the British Library.

ISBN 0 7503 0624 6

Library of Congress Cataloging-in-Publication Data are available

First Edition published 1992

Series Editors: **M Birkinshaw**, University of Bristol, UK
M Elvis, Harvard–Smithsonian Center for Astrophysics
J Silk, University of Oxford, UK

Commissioning Editor: John Navas
Production Editor: Simon Laurenson
Production Control: Sarah Plenty
Cover Design: Victoria Le Billon
Marketing: Nicola Newey and Verity Cooke

Published by Institute of Physics Publishing, wholly owned by The Institute of Physics, London

Institute of Physics Publishing, Dirac House, Temple Back, Bristol BS1 6BE, UK

US Office: Institute of Physics Publishing, The Public Ledger Building, Suite 929, 150 South Independence Mall West, Philadelphia, PA 19106, USA

Typeset in LaTeX 2ε by Text 2 Text, Torquay, Devon
Printed in the UK by J W Arrowsmith Ltd, Bristol

Ever the dim beginning,
Ever the growth, the rounding of the circle,
Ever the summit and the merge at last, (to surely start again,)

All space, all time,
The stars, the terrible perturbations of the suns,
Swelling, collapsing, ending, serving their longer, shorter use,

Ever the mutable,
Ever materials, changing, crumbling, recohering....

Walt Whitman

Image copyright WIYN Consortium, Inc., courtesy of Christopher Howk (Johns Hopkins University), Blair Savage (University of Wisconsin-Madison) and Nigel Sharp (NOAO).

Contents

Preface to the second edition

Dust is a ubiquitous feature of the cosmos, impinging directly or indirectly on most fields of modern astronomy. Dust grains composed of small (submicron-sized) solid particles pervade interstellar space in the Milky Way and other galaxies: they occur in a wide variety of astrophysical environments, ranging from comets to giant molecular clouds, from circumstellar shells to galactic nuclei. The study of this phenomenon is a highly active and topical area of current research. This book aims to provide an overview of the subject, covering general concepts, methods of investigation, important results and their significance, relevant literature and some suggestions for promising avenues of future research. It is aimed at a level suitable for those embarking upon postgraduate research but will also be of more general interest to researchers, teachers and students as a review of a significant area of astrophysics. As a formal text for taught courses, it will be particularly useful to advanced undergraduate and beginning postgraduate students studying the interstellar medium. My aim throughout is to create a compact, coherent text that will stimulate the reader to investigate the subject further.

Our concept of interstellar space has changed over the years, from a passive, static 'medium' to an active 'environment'. For this reason, the underlying theme of the book is the significance of dust in interstellar astrophysics, with particular reference to the interaction of the solid particles with their environment. The discussion is focused on interstellar dust in the solar neighbourhood of our own Galaxy, the Milky Way: our Galaxy is both the environment of planetary systems and the most accessible example of the building blocks of the Universe. If we can better understand the nature and evolution of dust in our local Galaxy, this will greatly aid us in our quest to comprehend both its role in the origins of stars and planetary systems such as our own and its influence on the observed properties of distant galaxies.

Many important new discoveries have been made in the field of cosmic dust since the first edition of *Dust in the Galactic Environment* was completed in mid-1991. The *Astrophysical Journal* alone typically publishes a hundred or more research papers per year on interstellar dust and related topics. Major advances have been made in fields as diverse as meteoritics, infrared astronomy and fractal grain theory. A new edition thus seems timely for two primary reasons:

(1) To bring the text up to date. This is especially urgent in the light of exciting new results from space missions such as the Hubble Space Telescope, the Cosmic Background Explorer and the Infrared Space Observatory, together with the latest developments in ground-based observational astronomy, laboratory astrophysics and theoretical modelling.
(2) To expand the scope of the text to provide a context for future research opportunities. In the first decade of the new millennium, we can anticipate discoveries linked to missions such as SIRTF (the Space Infrared Telescope Facility) and STARDUST (a cometary/interstellar dust collection and return mission). Key goals for these missions include the study of dust both 'near' and 'far', in our own Solar System, in protoplanetary discs around other nearby stars, and in distant galaxies.

The new edition places greater emphasis on these topics and has increased in overall length by more than 30%. The text is divided into ten chapters. The first provides a historical perspective for current research, together with an overview of interstellar environments and the role of dust in astrophysical processes. Chapter 2 discusses the cosmic history of the chemical elements expected to be present in dust and examines the effect of gas–dust interactions on gas phase abundances. Chapters 3–6 describe the observed properties of interstellar grains, i.e. their extinction, polarization, absorption and emission characteristics, respectively. In chapters 7–9, we discuss the origin and evolution of the dust, tracing its lifecycle in a sequence of environments from the circumstellar envelopes of old stars to diffuse interstellar clouds, molecular clouds, protostars and protoplanetary discs. The final chapter summarizes progress toward a unified model for galactic dust. Dust in other galaxies is discussed as an integral part of the text rather than as a distinct topic requiring separate chapters.

It is assumed throughout that the reader is familiar with basic concepts in stellar and galactic astronomy, such as stellar magnitude and distance scales and the spectral classification sequence and has a qualitative familiarity with galactic structure and stellar evolution according to current models. The reader with little or no background in astronomy will find many suitable introductory texts available: *Foundations of Astronomy* by Michael A Seeds (Wadsworth 1997) would be an excellent choice.

Système Internationale (SI) units are used in addition to the units of astronomy but the unsuspecting reader should be aware that the cgs system is still widespread in the astronomical literature. There are a few isolated exceptions to SI in the present text. For example, it is convenient to use microgauss ($1\ \mu G = 10^{-10}$ T) to specify interstellar magnetic flux densities and Ångstroms ($1\ Å = 10^{-10}$ m) to denote the wavelengths of spectral features in the visible and ultraviolet regions of the spectrum.

The author has found astrophysical dust to be a challenging and rewarding topic of study. An important reason for this is the wide variety of techniques involved, embracing observational astronomy over much of the electromagnetic spectrum, laboratory astrophysics and theoretical modelling. Interpretation and modelling of observational data may lead the investigator into such diverse fields as solid state physics, scattering theory, mineralogy, organic chemistry, surface chemistry and small-particle magnetism. Moreover, despite much activity and considerable progress in recent years, there is no shortage of challenging problems. If this book attracts students of physical sciences to study cosmic dust, it will have succeeded in its primary aim. Physicists with interest and expertise in small-particle systems may also be encouraged to consider grains in the laboratory of space. As Huffman (1977) remarked,

> *"it is a difficult experimental task to produce particles a few hundred Ångstroms in size, keep them completely isolated from one another and all other solids, maintain them in ultra-high vacuum at low temperature and study photon interactions with the particles at remote wavelengths ranging from the far infrared to the extreme ultraviolet. This is the opportunity we have in the case of interstellar dust."*

Acknowledgments are due, first and foremost, to my family: my parents for nurturing my educational development and encouraging my childhood interest in astronomy; my children Clair and James for everything they have been, are and will be; and Polly, my soulmate, partner and dearest friend, for her love and support. This book is dedicated to them. Many colleagues and friends have contributed over the years to the development of my knowledge and ideas on interstellar dust. My research has benefited immeasurably from interactions with others attracted to this strangely fascinating topic. I wish to record my thanks, especially, to the late Kashi Nandy for stimulating my early interest in the topic; to Andy Adamson for a collaboration that has thrived for more than a decade; to Walt Duley and Peter Martin for providing hospitality, intellectual stimulus and practical support during a period of sabbatical leave in Toronto, at a time when my research career had seemed in danger of suffocating under the weight of other responsibilities; to Thijs de Graauw for inviting and encouraging my participation in his guaranteed-time observations with the Infrared Space Observatory; to Rensselaer Polytechnic Institute for a new career opportunity; to my Rensselaer colleague Wayne Roberge and my recent doctoral students, Jean Chiar, Perry Gerakines, Kristen Larson, Erika Gibb and Sachin Shenoy, for all their hard work and dedication to the task of understanding dust in the galactic environment; and to the National Aeronautics and Space Administration (NASA) for financial support of our endeavours. I am especially grateful to John Mathis for his thorough reading of the entire manuscript and for his many insightful and constructive suggestions. Thanks are due also to Paul Abell, Eli Dwek, Roger Hildebrand, James Hough, Alex Lazarian, Mike Sitko, Paul Wesselius, Adolf Witt and Nicolle Zellner for helpful comments and ideas, and to John Navas, Simon Laurenson and their colleagues at IoP Publishing for their encouragement, support and (above all) patience. Finally, I am grateful to those who provided me with illustrations: these are acknowledged in the appropriate figure captions.

The lines from Whitman's poem 'Eidólons' prefacing this book were a source of inspiration. Reciting them quietly to myself seemed to get me through those times when I thought the book would never be finished. They first came to my attention in an entirely different context, by virtue of the fact that they were inscribed by Danish composer Vagn Holmboe (1909–96) on the score of his Tenth Symphony. This symphony, completed in 1971, is based on the principle of metamorphosis, pioneered by Jean Sibelius, in which musical themes undergo continuous evolution – sometimes slow, almost imperceptible, sometimes abrupt and dramatic. The analogy with cosmic evolution is apt. The Swedish company BIS has issued a complete cycle of the Holmboe symphonies, thus helping to rescue from obscurity one of the greatest composers of the 20th century. The music of these and other composers – Wolfgang Amadeus Mozart, Antonin Dvorak and Wilhelm Stenhammar, to name but three – was a source of solace and relaxation after long nights at the word-processor.

The text was produced by the author using the Institute of Physics macro package for the TEX typesetting system, 'intended for the creation of beautiful books – and especially for books that contain a lot of mathematics' (Knuth 1986). I leave the reader to judge the irrelevance of this quotation.

Readers are welcome to send comments or questions on the text to the author via electronic mail to whittd@rpi.edu.

D C B Whittet
Rensselaer Polytechnic Institute
June 2002

Chapter 1

Dust in the Galaxy: Our view from within

> *"The discovery of spiral arms and – later – of molecular clouds in our Galaxy, combined with a rapidly growing understanding of the birth and decay processes of stars, changed interstellar space from a stationary 'medium' into an 'environment' with great variations in space and in time."*
>
> H C van de Hulst (1989)

1.1 Introduction

Interstellar space is, by terrestrial standards, a near-perfect vacuum: the average particle density in the solar neighbourhood of our Galaxy is approximately 10^6 m^{-3} (one atom per cubic centimetre), a factor of about 10^{19} less than in the terrestrial atmosphere at sea level. However, dense objects such as stars and planets occupy a tiny fraction of the total volume of the Galaxy and the tenuous interstellar medium[1] contributes roughly a fifth of the mass of the galactic disc. The stellar and interstellar components are continually interacting and exchanging material. New stars condense from interstellar clouds and, as they evolve, they bathe the surrounding ISM with radiation; ultimately, many stars return a substantial fraction of their mass to the ISM, which is thus continuously enriched with heavier elements fused from the primordial hydrogen and helium by nuclear processes occurring in stars. A major proportion of these heavier atoms are locked up in submicron-sized solid particles (dust grains), which account for roughly 1% of the mass of the ISM and are almost exclusively responsible for its obscuring effect at visible wavelengths. Despite their relatively small contribution to the total mass, the remarkable efficiency with which such particles scatter, absorb and re-radiate starlight ensures that they have a very significant impact on our view of the Universe. For example, the attenuation between us and the

[1] For convenience, the term 'interstellar medium' (ISM) is used to refer, collectively, to interstellar matter over all levels of density, embracing a wide range of environments (section 1.4).

centre of the Galaxy is such that, at visible wavelengths, only one photon in every 10^{12} reaches our telescopes. The energy absorbed by the grains is re-emitted in the infrared, accounting for some 20% of the total bolometric luminosity of the Galaxy.

The influence of interstellar dust may be discerned with the unaided eye on a dark, moonless night at a time of year when the Milky Way is well placed for observation. In the Northern hemisphere, the background light from our Galaxy splits into two sections in Aquila and Cygnus. Southern observers are best placed to view such irregularities: the dark patches and rifts were seen by Aborigine observers as a 'dark constellation' resembling an emu, with the Coal Sack as its head, the dark lane passing through Centaurus, Ara and Norma as its long, slender neck and the complex system of dark clouds toward Sagittarius as its body and wings. Discoveries in the 20th century enabled us to recognize the Milky Way in Sagittarius as the nuclear bulge of a dusty spiral galaxy, seen from a vantage point within its disc at a distance of a few kiloparsecs from the centre. Our view of our home Galaxy is impressively illustrated by wide-angle, long-exposure photographs of the night sky, such as that shown in figure 1.1. The Milky Way is a fairly typical spiral, with a nucleus and disc surrounded by a spheroidal halo containing globular clusters (see Mihalas and Binney (1981) for a wide-ranging review of the structure and dynamics of the Galaxy). There is a striking resemblance between figure 1.1 and photographs of external spiral galaxies of similar morphological type seen edge-on, such as NGC 891, illustrated in figure 1.2. The visual appearance of such galaxies tends to be dominated by the equatorial dark lane that bisects the nuclear bulge. Obscuration is less evident (but invariably present) in spirals inclined by more than a few degrees to the line of sight. These results indicate that dark, absorbing material is a common characteristic of such galaxies and that this matter is concentrated into discs that are thin in comparison to their radii.

This chapter aims to provide a broad overview of the phenomenon of galactic dust and its role in astrophysical processes. We first review the early development of knowledge on interstellar dust (section 1.2) and assess the impact of its obscuring properties and spatial distribution on our view of the Universe (section 1.3), whilst simultaneously introducing some basic concepts and definitions. We then examine the environments to which the grains are exposed (section 1.4) and discuss the importance of dust as a significant chemical and physical constituent of interstellar matter (section 1.5). A summary of current models for interstellar dust grains appears in the final section (section 1.6).

1.2 Historical perspective: Discovery and assimilation

The study of extinction by interstellar dust can perhaps be said to have begun with Wilhelm Struve's analysis of star counts (Struve 1847). Struve demonstrated that the apparent number of stars per unit volume of space declines in all directions

Figure 1.1. A wide-angle photograph of the sky, illustrating the Milky Way from Vulpecula (left) to Carina (right). The nuclear bulge in Sagittarius is below centre. Photograph courtesy of W Schlosser and Th Schmidt-Kaler, Ruhr Universität, Bochum, taken with the Bochum super wide-angle camera at the European Southern Observatory, La Silla, Chile. The secondary mirror of the camera system and its support are seen in silhouette.

with distance from the Sun (see Batten (1988) for a modern account of this work). This led him to hypothesize that starlight suffers absorption in proportion to the distance travelled and, on this basis, he deduced a value for its amplitude in remarkably good agreement with current estimates. This proposal did not gain acceptance, however, and no further progress was made until the beginning of the 20th century, when Kapteyn (1909) recognized the potential significance of extinction:

> "*Undoubtedly one of the greatest difficulties, if not the greatest of all,*

Figure 1.2. An optical CCD image of the edge-on spiral galaxy NGC 891 (Howk and Savage 1997). Light from stars in the disc and nuclear bulge of the galaxy is absorbed and scattered by dust concentrated in the mid-plane, with filamentary structures extending above and below. The image was taken with the 3.5 m WIYN Telescope at Kitt Peak National Observatory, Arizona, USA, operated by the National Optical Astronomy Observatory and the Association of Universities for Research in Astronomy, with support from the National Science Foundation. Image copyright WIYN Consortium, Inc., courtesy of Christopher Howk (Johns Hopkins University), Blair Savage (University of Wisconsin-Madison) and Nigel Sharp (NOAO).

> *in the way of obtaining an understanding of the real distribution of the stars in space, lies in our uncertainty about the amount of loss suffered by the light on its way to the observer.*"

Both Struve and Kapteyn envisaged uniform absorption but Barnard's photographic survey of dark 'nebulae' provided evidence for spatial variations

(Barnard 1910, 1913, 1919, 1927). The existence of dark regions in the Milky Way had been known for many years: William Herschel regarded them as true voids in the distribution of stars ('holes in the sky'), a view that still prevailed in the early 20th century. However, detailed morphological studies convinced Barnard that at least some of the 'holes' contain interstellar matter that absorbs and scatters starlight. For example, the association of dark and bright nebulosities in the well studied complex near ρ Ophiuchi strongly supports this view (e.g. Barnard 1919; see Seeley and Berendzen 1972a, b and Sheehan 1995 for in-depth historical reviews). It was also suggested at about this time (Slipher 1912) that the diffuse radiation surrounding the Pleiades cluster might be explained in terms of scattering by particulate matter.

Confirmation that the interstellar extinction hypothesis is correct came some years later as the result of two distinct lines of investigation by the Lick Observatory astronomer R J Trumpler (1930a, b, c). If dust is present in the interstellar medium, its obscuring effect will clearly influence stellar distance determinations, introducing another degree of freedom in addition to apparent brightness and intrinsic luminosity. Trumpler sought to determine the distances of open clusters by means of photometry and spectroscopy of individual member stars. Spectral classification provides an estimate of the luminosity and the distance modulus is obtained by comparing apparent and absolute magnitudes. In the Johnson (1963) notation[2], the standard distance equation may be written:

$$V - M_V = 5 \log d' - 5 \tag{1.1}$$

where V and M_V are the apparent and absolute visual magnitudes, respectively and d' is the apparent mean cluster distance in parsecs. Having evaluated d', Trumpler then deduced the linear diameter of each cluster geometrically from the measured angular diameter. When this had been done for many clusters, a remarkable trend became apparent: the deduced cluster diameters appeared to increase with distance from the Solar System. From this, Trumpler inferred the presence of a systematic error in his results due to obscuration in the interstellar medium and concluded that a distance-dependent correction must be applied to the left-hand side of equation (1.1) in order to render the cluster diameters independent of distance:

$$V - M_V - A_V = 5 \log d - 5 \tag{1.2}$$

where d is now the true distance. The quantity A_V represents interstellar 'absorption' at visual wavelengths in the early literature but should correctly be termed 'extinction' (the combined effect of absorption and scattering). A_V tends to increase linearly with distance in directions close to the galactic plane; for the open clusters, a mean rate of $\sim$1 mag kpc^{-1} is required.

Trumpler then considered the implications of his discovery for the colours of stars. A problem that had puzzled stellar astronomers in the 1920s was the fact

[2] Trumpler used an early magnitude system but we adopt modern usage.

that many stars close to the galactic plane appear *redder* than expected on the basis of their spectral types. In essence, there appeared to be a discrepancy in stellar temperature deduced by spectroscopy and photometry. Spectral classification gives an estimate of temperature based on the presence and relative intensities of spectral lines in the stellar photosphere, whereas colour indices such as $(B - V)$ are indicators of temperature based on the continuum slope and its equivalent blackbody temperature. Many stars that show spectral characteristics indicative of high surface temperature (the 'early-type' stars) have colour indices more appropriate to much cooler ('late-type') stars. This anomaly is easily explained if they are reddened by foreground interstellar dust along the line of sight. By comparing the apparent brightnesses over a range of wavelengths of intrinsically similar stars with different degrees of reddening, Trumpler showed that interstellar extinction is a roughly linear function of wavenumber ($A_\lambda \propto \lambda^{-1}$) in the visible region of the spectrum. This important result, subsequently verified by more detailed studies (e.g. Stebbins *et al* 1939), implies the presence of solid particles with dimensions comparable to the wavelength of visible light. Such particles may be expected to contain $\sim 10^9$ atoms if their densities are comparable with those of terrestrial solids[3].

The process by which interstellar dust reddens starlight is exactly analogous to the reddening of the Sun at sunset by particles in the terrestrial atmosphere. A photon encountering a dust grain is either absorbed or scattered (chapter 3). An absorbed photon is completely removed from the beam and its energy converted into internal energy of the particle, whereas a scattered photon is deflected from the line of sight. Reddening occurs because absorption and scattering are, in general, more efficient at shorter wavelengths in the visible: thus red light is less extinguished than blue light in the transmitted beam, whereas the scattered component is predominantly blue. The appearance of a stellar spectrum over a limited spectral range is not drastically altered by moderate degrees of such reddening, in the sense that the wavelengths and relative strengths of characteristic lines are essentially unchanged: spectral classification therefore gives a good indication of the temperature of a star independent of foreground reddening. However, colour indices depend on both temperature and reddening, information that can be separated only if the spectral type of the star is known.

The degree of reddening or 'selective extinction' of a star is quantified as

$$E_{B-V} = (B - V) - (B - V)_0 \tag{1.3}$$

in the Johnson photometric system, where $(B - V)$ and $(B - V)_0$ are observed and 'intrinsic' values of the colour index and E_{B-V} is the 'colour excess'. As the extinction is always greater in the B filter (central wavelength 0.44 μm) than in V

[3] The term 'smoke' was often used to describe these particles in the early literature. 'Smoke' implies the product of combustion, whereas 'dust' implies finely powdered matter resulting from the abrasion of solids. The former is arguably more appropriate as a description of the particles condensing in stellar atmospheres, now regarded as an important source of interstellar grains. However, 'dust' has become firmly established in modern usage.

(0.55 μm), E_{B-V} is a positive quantity for reddened stars and zero (to within observational error) for unreddened stars. Intrinsic colours are determined as a function of spectral type by studying nearby stars and stars at high galactic latitudes that have little or no reddening. Colour excesses may be defined for any chosen pair of photometric passbands by analogy with equation (1.3): another commonly used measure of reddening in the blue–yellow region is the colour excess E_{b-y} based on the Strömgren (1966) intermediate passband system ($E_{b-y} \approx 0.74 E_{B-V}$; Crawford 1975). The relationship between total extinction at a given wavelength and a corresponding colour excess depends on the wavelength-dependence of extinction, or extinction curve. In the Johnson system, the extinction in the visual passband may be related to E_{B-V} by

$$A_V = R_V E_{B-V} \tag{1.4}$$

where R_V is termed the ratio of total to selective visual extinction. The quantity E_{B-V} is directly measurable, whereas A_V is generally much harder to quantify: often, for individual stars, the only viable method of evaluating A_V is to determine E_{B-V} and assume a plausible value of R_V. If the assumed value of R_V is wrong, then the inferred distance to the star will also be in error (equation (1.2)). Following the discovery of interstellar extinction, much effort was devoted in subsequent years to the empirical evaluation of R_V (e.g. Whitford 1958, Johnson 1968 and references therein). Theoretically, R_V is expected to depend on the composition and size distribution of the grains. However, in the low-density ISM, R_V has been shown to be virtually constant and a value of

$$R_V \approx 3.05 \pm 0.15 \tag{1.5}$$

may be assumed for most lines of sight. The origin of this result and its limits of applicability are discussed in chapter 3. For researchers whose primary interest is the determination of reliable distances, equation (1.5) is perhaps the most important in this book.

Extinction by dust renders interstellar space a polarizing as well as an attenuating medium. This was first demonstrated by Hall (1949) and Hiltner (1949), who showed that the light of reddened stars is partially plane polarized, typically at the 1–5% level. The origin of this effect is widely accepted to be the directional extinction of flattened or elongated grains that are aligned in some way, i.e. their long axes have some preferred direction. A model that produces alignment by means of an interaction between the spin of the particles and the galactic magnetic field was proposed by Davis and Greenstein (1951). These authors assumed that the grains are paramagnetic and are set spinning by collisions with atoms in the interstellar gas. Paramagnetic relaxation then results in the grains tending to be orientated with their angular momenta parallel (and hence their long axes perpendicular) to the magnetic field lines. Although it has since been shown that alignment cannot occur in precisely the manner suggested by Davis and Greenstein, nevertheless it seems highly probable that an analogous process is occurring in the interstellar medium (section 4.5).

1.3 The distribution of dust and gas

1.3.1 Overview

Studies of other galaxies give us a qualitative picture of the large-scale distribution of dust in typical spirals like the Milky Way: dust is most evident in galactic discs, producing conspicuous equatorial dark lanes in edge-on spirals such as NGC 891 (figure 1.2). In contrast, there is a general sparsity of dust in elliptical galaxies. As a general rule, dust in spiral galaxies is most closely associated with relatively young stars of the 'disc' population, whereas the older 'halo' population formed out of matter deficient in the chemical elements needed to make dust (see section 2.3). Within the disc, most of the material (both stars and interstellar matter) is confined to the spiral arms.

Quantitative investigations of the variation of reddening (E_{B-V}) with direction and distance in the solar neighbourhood of our Galaxy (out to a few kiloparsecs) have been carried out by several authors (FitzGerald 1968, Lucke 1978, Neckel and Klare 1980, Perry and Johnston 1982). These studies are based on photometry and spectral classifications for large numbers of stars; the method makes use of equations (1.2)–(1.4), or equivalent forms, together with the absolute magnitude versus spectral type calibration (e.g. Schmidt-Kaler 1982), to determine E_{B-V} (or A_V) and d from observed quantities. Data on the total extinction A_V in individual dark clouds may also be obtained statistically by means of star counts: in this method, the number of stars per unit area of sky toward the cloud is compared with that of the background population, as measured in unobscured adjacent fields (Bok 1956). The distribution of dark clouds as a function of their opacity has been studied by Feitzinger and Stüwe (1986). Analogous techniques have also been used to study the foreground reddening and extinction of extragalactic objects by dust in our Galaxy (e.g. Burstein and Heiles 1982, de Vaucouleurs and Buta 1983). Results from all of these investigations confirm that the particles responsible for reddening are quite closely constrained to the plane of the Milky Way (see figure 1.3), essentially within a layer no more than $\sim$200 pc thick in the solar neighbourhood. For example, FitzGerald (1968) determined the scale height of reddening material, measured from the mid-plane and averaged for different longitude zones, to be in the range 40–100 pc.

Comparisons between the distribution of optical extinction and atomic or molecular emissions show that dust and gas are generally well mixed in the ISM, as illustrated in figure 1.3. Visible extinction determined from dark clouds in the solar neighbourhood (Feitzinger and Stüwe 1986) is plotted in the upper frame. This is compared with two tracers of interstellar gas (radio CO-line emission from molecular gas; γ-ray emission from the interaction of atomic nuclei with cosmic rays) and another tracer of the dust (far infrared continuum emission; see section 1.3.4). Some differences occur (e.g. the γ-ray map includes bright point sources identified with supernova remnants; and the extinction map lacks

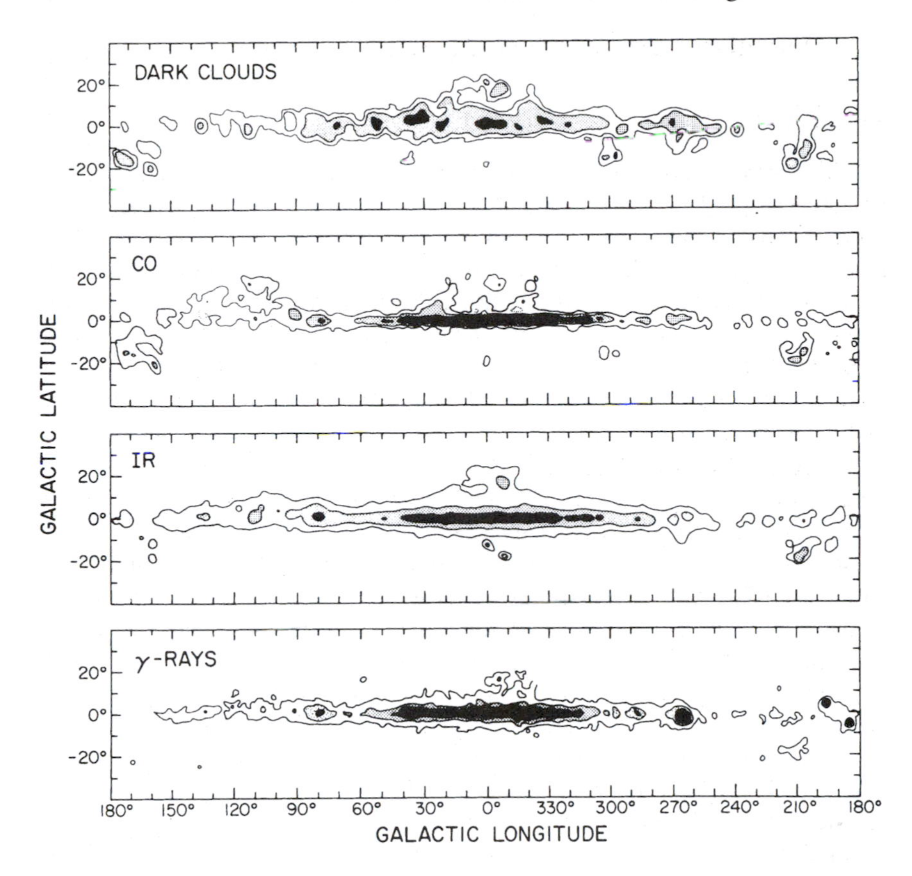

Figure 1.3. Maps comparing the distributions of dust and gas in the Milky Way. The galactic nucleus is at the centre of each frame. From the top: visual extinction due to dust, as determined from studies of dark clouds in the solar neighbourhood; line emission at 2.6 mm wavelength from CO gas; infrared emission from dust at 100 μm wavelength, measured by IRAS; and γ-ray emission in the energy range 70 MeV–5 GeV, measured by the COS B satellite, arising from the interaction of interstellar gas with cosmic rays. The resolution of each map is $\sim$2.5°. Several individual clouds and complexes may be discerned, including those in Taurus-Auriga ($\ell \approx 170°, b \approx -13°$), Ophiuchus ($\ell \approx 353°$, $b \approx 17°$) and Orion ($\ell \approx 209°$, $b \approx -19°$). Prominent sources in the γ-ray map (lower frame) include the Crab and Vela supernova remnants, which lie close to the galactic plane at $\ell \approx 184°$ and $\ell \approx 263°$, respectively. (Data from Dame *et al* 1987 and references therein.)

the intense central ridge because it is dominated by material somewhat closer to the Sun than the other tracers). However, the overall general similarity is striking.

1.3.2 The galactic disc

Although it is often convenient to visualize the macroscopic distribution of interstellar matter in the disc of the Galaxy as a continuous layer 100–200 pc thick, the distribution is, in reality, extremely uneven. Inhomogeneities occur on all size scales from 10^{-4} pc (the dimensions of solar systems) to 10^3 pc (the dimensions of spiral arms). Clumps of above-average density with sizes typically in the range 1–50 pc are traditionally termed 'clouds' (section 1.4.3). The general tendency for extinction to increase with pathlength arises stochastically, dependent on the number of clouds that happen to lie along a given line of sight. Currently, our Solar System happens to reside in a relatively transparent ('intercloud') region of the Galaxy near the edge of a spiral arm, with little or no reddening ($E_{B-V} < 0.03$) for stars within 50–100 pc in any direction. On average, a column $L = 1$ kpc long in the galactic disc intersects several ($\sim$5) diffuse clouds that produce a combined reddening typically of $E_{B-V} \approx 0.6$. Making use of equations (1.4) and (1.5) to express this in terms of total extinction, the mean ratio of visual extinction to pathlength (known as the 'rate of extinction') is

$$\left\langle \frac{A_V}{L} \right\rangle \approx 1.8 \text{ mag kpc}^{-1}. \tag{1.6}$$

This result is applicable only as a general average for lines of sight close to the plane of the Milky Way and for distances up to a few kiloparsecs from the Sun. At greater distances, $\langle A_V/L \rangle$ is difficult to estimate, as even luminous OB stars and supergiants become too faint to observe at visible wavelengths. The visual magnitude of a typical supergiant may exceed 20 for distances greater than 6.5 kpc and average reddening. Photometry at infrared wavelengths may be used to penetrate to greater distances if a sufficiently luminous background source is available; assumptions regarding the wavelength-dependence of extinction and its spatial uniformity then allow visual extinctions to be calculated. The extinction toward the infrared cluster at the galactic centre is estimated to be $A_V \sim 30$ mag over the $\sim$8 kpc path (Roche 1988), a result that implies an increase in the rate of extinction per unit distance, compared with the solar neighbourhood, as we approach the nucleus.

The concentration of dust in the galactic disc seriously hinders investigation of the structure and dynamics of our Galaxy using visually luminous spiral-arm tracers such as early-type stars and supergiants. Observations that extend beyond about 3 kpc are based almost entirely on long-wavelength astronomy (radio and infrared), although a few 'windows' in the dust distribution, where the rate of extinction is unusually low, allow studies at visible wavelengths to distances $\sim$10 kpc. However, in general, the morphological structure of our own Galaxy is less well explored than that of our nearest neighbours. Another implication of some significance is that it is extremely difficult to detect novae and supernovae in the disc of the Milky Way. Studies of external galaxies suggest that the expected mean supernova rate in spirals of similar Hubble type to our own is approximately

1 per 50 years (van den Bergh and Tammann 1991), with an uncertainty of about a factor two; but historical records suggest that only five visible supernovae have been seen in our Galaxy in the past 1000 years, the last of which was 'Kepler's star' in 1604 (Clark and Stephenson 1977). The apparent discrepancy is attributed to the presence of extinction: supernova explosions presumably occur in our Galaxy at approximately the expected rate but many are hidden by foreground dust; for external systems, our viewing angle is generally more favourable.

The correlation of dust with gas in the galactic disc has been studied using ultraviolet absorption-line spectroscopy of reddened stars within ~1 kpc of the Sun (Savage *et al* 1977, Bohlin *et al* 1978). The spectroscopic technique provides a measure of the hydrogen column density N_{H} (representing the number of hydrogen nucleons in an imaginary column of unit cross-sectional area, extending from the observer to the star, in units of m^{-2}). Separate measurements for atomic and molecular hydrogen (H I and H_2) are summed to give N_{H}:

$$N_{\mathrm{H}} = N(\mathrm{H\ I}) + 2N(\mathrm{H}_2) \tag{1.7}$$

where the factor two allows for the fact that H_2 contains two protons. Strictly speaking, equation (1.7) should include an additional term to allow for an ionized component of the gas (section 1.4.2) but this contributes only a tiny fraction of the total mass of interstellar material in the disc of the Galaxy and may be neglected here. Bohlin *et al* (1978) demonstrated that N_{H} and E_{B-V} are well correlated, confirming that gas and dust are generally well mixed in the ISM. The mean ratio of hydrogen column density to reddening is

$$\left\langle \frac{N_{\mathrm{H}}}{E_{B-V}} \right\rangle \approx 5.8 \times 10^{25}\ \mathrm{m}^{-2}\ \mathrm{mag}^{-1} \tag{1.8}$$

with scatter for individual stars typically less than 50%. Converting reddening in equation (1.8) to extinction via equations (1.4) and (1.5), we have

$$\left\langle \frac{N_{\mathrm{H}}}{A_V} \right\rangle \approx 1.9 \times 10^{25}\ \mathrm{m}^{-2}\ \mathrm{mag}^{-1}. \tag{1.9}$$

As we shall show in chapter 3, this result implies a dust-to-gas mass ratio of a little under 1%.

Equations (1.9) and (1.6) may be combined to eliminate A_V, giving a value for the mean hydrogen number density:

$$\langle n_{\mathrm{H}} \rangle = \left\langle \frac{N_{\mathrm{H}}}{L} \right\rangle \approx 1.1 \times 10^{6}\ \mathrm{m}^{-3} \tag{1.10}$$

or about one atom per cm^3. This is a good macroscopic average for the number density of the ISM in the solar neighbourhood of the galactic plane. Note, however, that individual regions may show orders-of-magnitude deviations from average behaviour (section 1.4.2), as matter tends to be distributed into clumps ('clouds') with $n_H \gg \langle n_H \rangle$ and interclump gas with $n_H \ll \langle n_H \rangle$. Equation (1.10) may be expressed in terms of mass density:

$$\langle \rho_H \rangle = m_H \langle n_H \rangle \approx 1.8 \times 10^{-21}\ \mathrm{kg\ m^{-3}}. \tag{1.11}$$

However, a more convenient measure of the contribution of hydrogen gas to the mass of the galactic disc is the surface mass density

$$\langle \sigma_H \rangle = 2h \langle \rho_H \rangle \approx 5.3\ \mathrm{M_\odot pc^{-2}} \tag{1.12}$$

(with attention to units), where $h \approx 100$ pc is the mean scale height for the ISM (Mihalas and Binney 1981). If the usual mean cosmic abundances are applicable to the ISM (see section 2.2), the result in equation (1.12) should be multiplied by a factor of about 1.4 to obtain the average surface density summed over all chemical elements:

$$\langle \sigma_{ISM} \rangle \approx 7.4\ \mathrm{M_\odot pc^{-2}}. \tag{1.13}$$

For comparison, the observed surface density of matter in stars in the disc of the Galaxy is

$$\langle \sigma_{stars} \rangle \approx 35\ \mathrm{M_\odot pc^{-2}} \tag{1.14}$$

(Kuijken and Gilmore 1989) and so the ISM contributes roughly 20% of the observed mass. The total surface density, σ_T, including all forms of mass in the disc of the Galaxy, may be estimated independently by investigating the motions of stars perpendicular to the galactic plane (z-motions). This technique, pioneered by Oort (1932), has been applied by Kuijken and Gilmore (1989) to obtain the value

$$\sigma_T = 46 \pm 9\ \mathrm{M_\odot pc^{-2}}. \tag{1.15}$$

Comparing the results in equations (1.13), (1.14) and (1.15), we see that the total surface density of *observed* mass, $\langle \sigma_{stars} \rangle + \langle \sigma_{ISM} \rangle \approx 42\ \mathrm{M_\odot pc^{-2}}$, is consistent with the dynamic value to within the uncertainty: there is no evidence for 'missing mass' in the solar neighbourhood of the galactic disc.

1.3.3 High galactic latitudes

The extinction in directions away from the galactic disc, although generally small, is of considerable significance as evaluation of its effect is a prerequisite for determining the intrinsic properties of external galaxies. Corrections for the dimming of primary distance indicators (such as Cepheids, novae and supernovae) in external systems by dust in our Galaxy influence the extragalactic distance scale. The reddening of high-latitude stars ($|b| > 20°$) is almost independent of distance beyond a few hundred parsecs, because of the general sparsity of dust in the halo of our Galaxy. If the disc is treated as a flat, uniform slab with the Sun in the central plane, a systematic dependence of extinction on latitude, b, is expected; it may easily be shown that this takes the form of a cosecant law[4]:

$$A_V(b) = A_P \operatorname{cosec}|b| \tag{1.16}$$

where A_P is the visual extinction at the galactic poles. The appropriate value of A_P has been disputed: some authors (e.g. McClure and Crawford 1971) argue in favour of polar 'windows', with $A_V(b) \leq 0.05$ for $b > 50°$, whereas de Vaucouleurs and Buta (1983) deduce $A_P \approx 0.15$, on the basis of galaxy counts and reddenings. In any case, this formulation should be used with the utmost caution, not so much because of uncertainties in A_P but, more crucially, because the distribution of dust is uneven and not well represented on small scales by any smoothly varying function.

The detection and study of high-latitude interstellar clouds was a major development in ISM research in the final decades of the 20th century, stimulated by the discovery in 1983 of infrared 'cirrus' by the Infrared Astronomical Satellite (section 1.3.4). Some high-latitude clouds are dense enough to contain a molecular phase (Magnani *et al* 1985, Reach *et al* 1995a) and to produce significant extinction ($A_V \sim 1$ or more; Penprase 1992). Many of the densest high-latitude clouds appear to be extensions of local dark-cloud complexes, such as those in Chamaeleon, Ophiuchus and Taurus; others appear to be isolated. Toward the cores of these clouds, the extinction will generally be much higher than predicted using equation (1.16) (see problem 3 at the end of this chapter for an example). The only reliable way to correct for the extinction of background objects is to evaluate A_V in each individual line of sight of interest. Burstein and Heiles (1982) used atomic hydrogen (H I) emission and galaxy counts to construct maps of galactic reddening that are helpful for this purpose, covering almost the entire sky for $|b| > 10°$ at a resolution of 0.6°. However, even this method can underestimate extinction in cloud cores, due to limited resolution and the effect of small-number statistics.

1.3.4 Diffuse galactic background radiation

The discussion so far has focused on the extinction properties of interstellar dust, i.e. on the attenuation and reddening of starlight. The energy removed from the transmitted beam when light passes through a dusty medium must reappear in another form: it is either scattered from the line of sight; or absorbed as heat (and subsequently re-emitted). The entire Galaxy is permeated by a diffuse interstellar radiation field (ISRF), representing the integrated light of all stars in the Galaxy. Interstellar grains effectively redistribute the spectrum of the ISRF: they absorb and scatter starlight most efficiently at ultraviolet and visible wavelengths; and emit in the infrared.

Direct observational evidence for scattered light in the ISM takes several forms: blue reflection nebulae surrounding individual dust-embedded stars or

[4] This relation is exactly equivalent to Bouguer's law for extinction in a plane-parallel planetary atmosphere, used to correct for telluric extinction in astronomical photometry.

clusters; bright filamentary nebulae and halos around externally heated dark clouds; and, on the macroscopic scale, weak ultraviolet background radiation from the disc of the Galaxy, termed the diffuse galactic light (DGL). It is interesting to note that the existence of faint reflection nebulosity at high galactic latitude (Sandage 1976) provided evidence for high-latitude clouds some years before they were studied in detail by other techniques. Observations of scattered light are important as they provide diagnostic tests for grain models, constraining the optical properties of the grains through determination of their albedo and phase function (section 3.3). They are also extremely difficult, however: because of its intrinsic weakness, the DGL component of the sky brightness cannot be easily separated from other diffuse emission, such as stellar background radiation, zodiacal light and airglow (Witt 1988).

An absorbing dust grain must re-emit a power equal to that absorbed to maintain thermal equilibrium. Grains that account for the visible extinction curve (often called 'classical' grains, with dimensions $\sim$0.1–0.5 μm) reach equilibrium at temperatures in the range 10–50 K under typical interstellar conditions (van de Hulst 1946). At such temperatures, the grains emit primarily in the far infrared (wavelengths $\sim$50–300 μm; see section 6.1). This emission has been mapped in the Milky Way and other spiral galaxies with instruments raised above the Earth's atmosphere, including the Infrared Astronomical Satellite (IRAS) and the Cosmic Background Explorer (COBE) (e.g. Sodroski *et al* 1997 and references therein). Correspondence between the distributions of absorbing and emitting grains in our Galaxy is evident from a comparison of the first and third frames in figure 1.3: both are broadly confined to the galactic disc. The scale height for 100 μm emission is comparable with those of reddening and H I and somewhat greater than that characteristic of CO (Beichman 1987). At higher latitudes, the 100 μm emission can be represented in terms of a smooth component that tends to follow a cosec $|b|$ law analogous to equation (1.16), upon which patchy emission (cirrus) associated with individual high-latitude clouds is superposed (Désert *et al* 1988).

In addition to far infrared emission attributed to classical dust grains with equilibrium temperatures $T_d < 50$ K, diffuse emission is also seen at shorter infrared wavelengths and attributed to the presence of a hotter ($T_d \sim$ 100–500 K) component of the dust. Classical grains in thermal equilibrium with their environment are expected to reach such high temperatures only in close proximity to individual stars or stellar associations, not in the ambient interstellar radiation field. However, smaller grains have much lower heat capacities and may undergo transient increases in temperature caused by absorption of individual energetic photons (section 6.1). A population of 'very small grains' (VSGs) with dimensions $<$0.01 μm may explain a number of the observed properties of interstellar dust (chapters 3–6), including not only continuum emission but also ultraviolet extinction and spectral absorption and emission features at various wavelengths.

1.4 Interstellar environments and physical processes

1.4.1 Overview

Interstellar gas is composed predominantly of hydrogen, which may be in one of three physical states or 'phases': molecular (H_2), atomic (H I) or ionized (H II). The properties of the gas are governed by the laws of thermodynamics and by its interaction with electromagnetic radiation, cosmic rays, shock waves and gravitational and magnetic fields. The physical processes involved lead to the presence of a vast range of environments, from tenuous, hot plasmas to cold, dense clouds where new solar systems are born. In this section, we discuss the factors that determine the physical state of the gas (section 1.4.2). We identify conditions of density, temperature and phase that characterize typical environments, including intercloud medium, diffuse and dense clouds (section 1.4.3) and H II regions (section 1.4.4). Finally, the current interstellar environment of our Solar System is considered (section 1.4.5). For more extensive reviews of physical processes in the interstellar medium, the reader is referred to Spitzer (1978) and Dyson and Williams (1997).

1.4.2 The physical state of the interstellar medium

The physical properties of interstellar gas are described by its number density (n) and temperature (T), together with its phase (molecular, atomic or ionized). For an ideal gas, the thermodynamic pressure is

$$p = nkT \tag{1.17}$$

where k is Boltzmann's constant. Thus, for pressure equilibrium between regions with different temperatures, the product nT should be constant (with a value typically $\sim 3 \times 10^9$ m^{-3} K; Mathis 2000). Such a general trend is, indeed, observed for many regions of the ISM (see figure 1.4). Note, however, that the ISM is a turbulent, not a static, medium (Elmegreen 1997), in which pressure equilibrium is not generally well established. Moreover, the magnetic field contributes to the pressure in an ionized or partially ionized gas ($p_B \propto B^2$, where B is the magnetic flux density) and this may exceed the thermodynamic pressure (Boulares and Cox 1990).

The temperature in a given environment depends on the balance of heating and cooling mechanisms, as discussed in detail by Spitzer (1978: pp 131–49). The dust is heated primarily by the absorption of energetic photons from the interstellar radiation field; and cooled by thermal emission of photons at longer wavelengths. The gas is heated not only by photon absorption but also by collisions with photoelectrons ejected from small grains (Bakes and Tielens 1994); and is cooled principally by spectral line emission. Cosmic rays also contribute to the heating of both dust and gas and are important especially in molecular clouds where the radiation field is strongly attenuated. The transitions

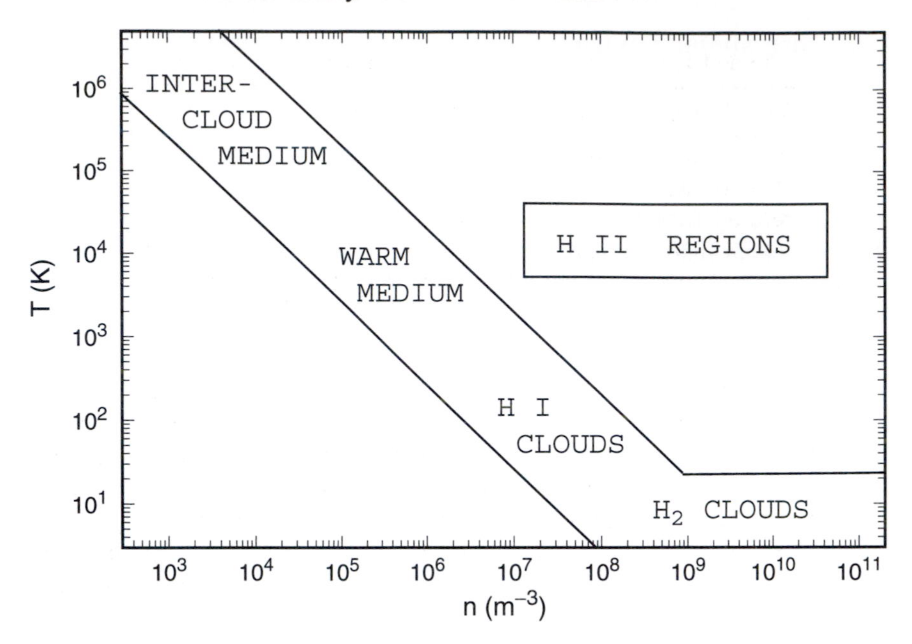

Figure 1.4. A schematic plot of kinetic temperature (T) versus number density (n) for interstellar gas, illustrating the loci of typical environments.

that cool molecular gas are generally rotational, driven by collisional excitation, such that the cooling rate varies as n^2. A small increase in density can therefore lead to a substantial increase in heat loss; this lowers the temperature and pressure, causing contraction of the region and further increase in density. The ISM thus tends naturally to separate into cool, dense regions (clouds) and hot, rarefied regions (the intercloud medium).

The physical state of the gas in a given region of the ISM is strongly influenced by the local intensity of the interstellar radiation field. Photons of the highest energies ($h\nu \geq 11.2$ eV) drive photoionization and photodissociation of hydrogen:

$$\mathrm{H} + h\nu \rightarrow \mathrm{p} + \mathrm{e} \qquad (h\nu \geq 13.6\ \mathrm{eV}) \tag{1.18}$$

and

$$\mathrm{H_2} + h\nu \rightarrow \mathrm{H} + \mathrm{H} \qquad (11.2 \leq h\nu \leq 13.6\ \mathrm{eV}). \tag{1.19}$$

Analogous reactions can be driven by other energy sources such as cosmic-ray impacts and shocks. Note, however, that dissociation of H_2 occurs at discrete energies in the range quoted. Interstellar H_2 normally exists in the ground state and there are no allowed transitions to higher states that result in dissociation. Reaction (1.19) is actually a two-step process, in which excitation by photon absorption to a higher level is followed by spontaneous transition to an excited vibrational level of the ground state from which dissociation can

Table 1.1. A four-component model for the interstellar medium.

Description	Phase	T (K)	n (m^{-3})	ϕ
Cold molecular	H_2	~ 15	$>1 \times 10^8$	<0.01
Cool atomic	H I	~ 80	$\sim 3 \times 10^7$	~ 0.03
Warm	H I, H II	~ 8000	$\sim 3 \times 10^5$	~ 0.2–0.5
Hot ionized	H II	$\sim 5 \times 10^5$	$\sim 5 \times 10^3$	~ 0.5–0.8

occur (see Spitzer 1978: p 124). Under conditions of equilibrium, ionization and dissociation reactions are balanced by recombination. Protons and electrons may recombine to form H I, with the release of kinetic and binding energy typically in the form of several lower-energy photons (recombination lines). Recombination of atomic H to form H_2 requires a third body (typically a dust grain). The equilibrium state is a strong function of density: high densities promote recombination and provide shielding against ionizing and dissociating radiation.

The phase structure of the ISM may be characterized conveniently in terms of four discrete components: cold (molecular), cool (atomic), warm (atomic or partially ionized) and hot (ionized) gas. Their average properties are listed in table 1.1 and their approximate distribution in the T versus n plane is shown in figure 1.4. This representation is loosely based on the so-called three-phase model of McKee and Ostriker (1977), with subdivision of the cold gas into atomic and molecular phases. In this model, the structure is regulated by supernova explosions. On average, a given point in the ISM is swept out by an expanding remnant once every few million years, leading to disruption of the existing cloud structure and the establishment of low-density bubbles of hot gas; thus, supernova explosions are primarily responsible for maintaining the intercloud medium in its hot, ionized state. The intercloud H II component should be distinguished from H II *regions*, which are generally much denser (figure 1.4) and result from ionization of the local interstellar gas by individual OB stars or associations (section 1.4.4).

In addition to phase, temperature and number density, table 1.1 lists the volume filling factor ϕ of each component, i.e. the average fractional volume of space that it occupies. Along a line of sight that intercepts many different regions, each typified by one of the four 'standard' environments, the average nucleon density is

$$\langle n \rangle = \sum \phi_i n_i \tag{1.20}$$

where ϕ_i and n_i are the volume filling factor and number density, respectively, of the ith component. The filling factors sum to unity ($\sum \phi_i = 1$) if all components of the ISM are accounted for. Their values are uncertain (Kulkarni and Heiles

1987), especially in terms of the relative importance of warm and hot components (Cox 1995). They are also spatially variable: as examples, molecular clouds are more common at galactocentric distances 3–7 kpc in the galactic disc than in the solar neighbourhood and hot ionized gas is pervasive in the galactic halo and in interarm regions of the disc. However, the contrast in density between the hotter and cooler components is sufficiently large that a general conclusion may be drawn on the basis of the crude estimates for ϕ in table 1.1: whilst the volume of the interstellar medium is mostly filled by tenuous, relatively hot plasma, the nucleon density (equation (1.20)) and hence the mass is dominated by the cooler, denser atomic and molecular phases, i.e. by *clouds*.

1.4.3 Interstellar clouds

We noted before (section 1.3.2) that the ISM is inhomogeneous, with a tendency to display structure over a wide range of size scales. The term 'cloud' has been adopted historically to describe visual features, such as Barnard's dark nebulae (section 1.2), and co-moving clumps of gas responsible for the Doppler components in the spectral line profiles (e.g. Routly and Spitzer 1952). This term may be somewhat misleading, however, in that the ISM now appears to be not only inhomogeneous but also hierarchical in structure ('clumps within clumps'). In the modern view, we may define clouds to be peaks in the density distribution on size scales that correspond to observed concentrations of interstellar gas and dust (Elmegreen 2002). Even clouds of similar size and mass may have quite different morphological structures.

It might seem futile to attempt characterization of 'representative' environments in the face of such diversity, yet the relatively simple phase structure of the ISM does allow this, to a degree. It is convenient to adopt the labels 'diffuse' and 'dense' to describe clouds in which the gas is predominantly atomic and molecular, respectively. An idealized representation of a cloud of each type is shown in figure 1.5. Both are assumed to lack internal sources of luminosity and to be immersed in a substrate of hot, ionized gas (the intercloud medium). They are externally heated by the ISRF, which permeates the intercloud medium virtually unattenuated (as it contains little dust and the plasma is optically thin to ionizing radiation). The properties of the two cloud types may be summarized as follows.

A *diffuse cloud* is a cloud of moderate density ($n_H \sim 10^7$–10^8 m^{-3}) and extinction ($0.1 < A_V < 1$) in which the dominant phase is H I. A typical example might have dimensions ~5 pc and mass ~30 $M_\odot$ (but with large scatter from one to another). A diffuse cloud is optically thick to radiation beyond the Lyman limit ($h\nu \geq 13.6$ eV) but remains relatively transparent to radiation of energy in the range 11.2–13.6 eV that can dissociate H_2 (equation (1.19)); nevertheless, some simple gas-phase molecules (e.g. CO, OH, CH and CH^+, as well as H_2) have detectable abundances. Cool H I gas is encased in a shell of warm gas, the outer 'halo' of which is partially ionized (equation (1.18)) by the hard ultraviolet

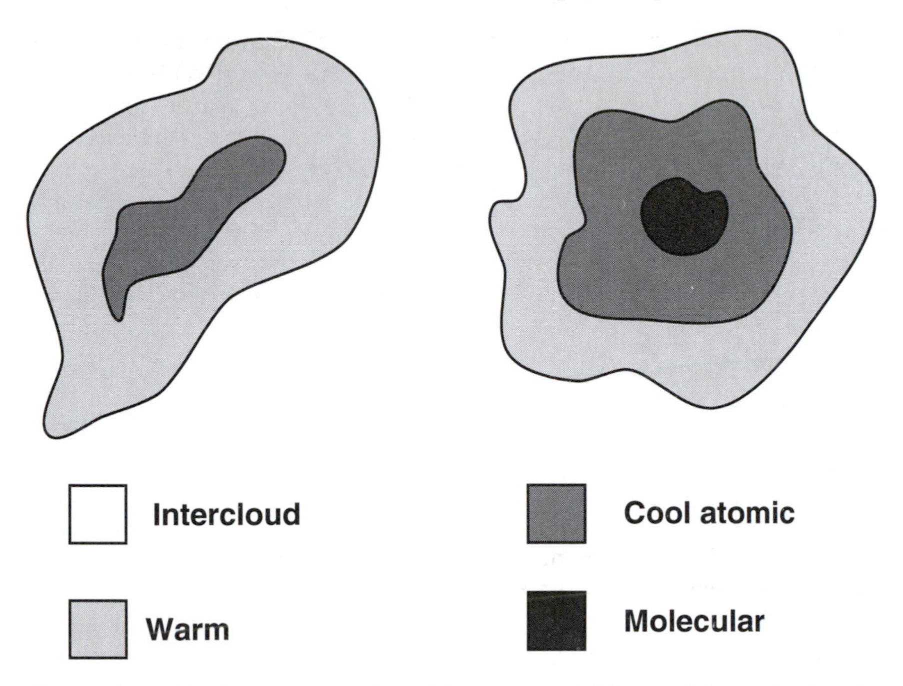

Figure 1.5. An idealized representation of the structure of diffuse and dense clouds in the ISM. Figure courtesy of Perry Gerakines.

photons in the ISRF, which it strongly absorbs. The warm gas is predominantly neutral, however, and heated by soft x-ray photons ($h\nu \sim 40$–120 eV) emitted by the hot intercloud gas. The least massive, most tenuous diffuse clouds may lack a cool phase entirely: these tend to be short lived due to evaporation into the surrounding hot medium.

A *dense cloud* contains regions sufficiently dense ($n > 10^8$ m^{-3}) that virtually all the H I is converted to H_2 by grain surface catalysis (section 8.1) on timescales of order a few million years, short compared with their expected lifetimes. Such conditions are found in regions ranging from small (<1 pc), low mass (<50 $M_\odot$) clumps within dark clouds to giant molecular clouds ranging up to $\sim$50 pc and $\sim 10^6$ $M_\odot$ in size and mass. Dense molecular gas is opaque to both ionizing and dissociating radiation. It is effectively self-shielded from the external ISRF by the outer layers of the cloud itself, which remain predominantly atomic and in which dissociating photons are attenuated by both gas and dust. The transition zone between the atomic and molecular gas is termed a photodissociation region (Hollenbach and Tielens 1997). The occurrence of this transition is demonstrated by observations of interstellar absorption lines in stellar spectra: figure 1.6 plots the column density of molecular hydrogen against the total column density (atomic plus molecular) for a large number of stars. Individual lines of sight may contain several clouds of different mean

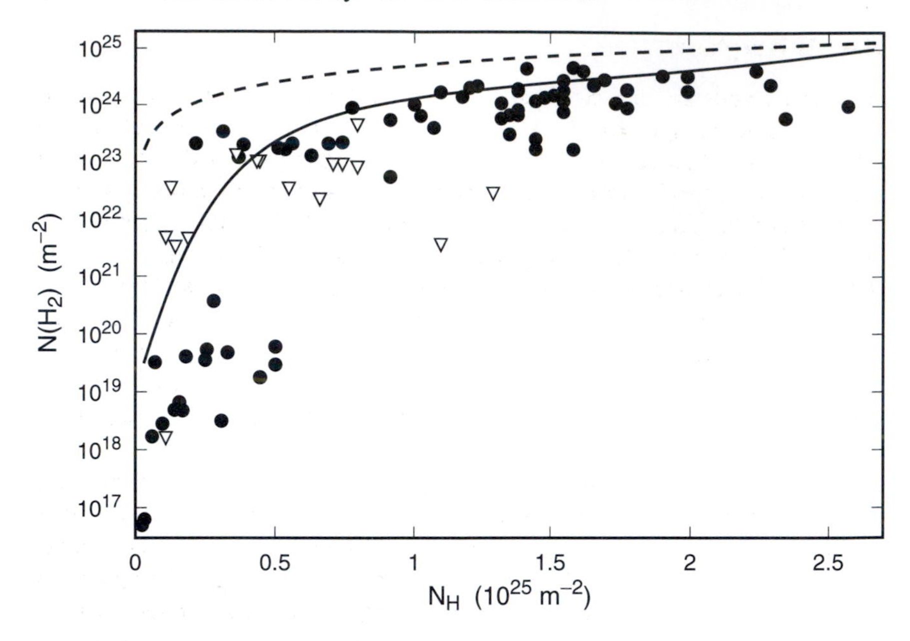

Figure 1.6. Plot of the column density of molecular hydrogen against that of total (atomic + molecular) hydrogen (log $N(H_2)$ versus N_H) for early-type stars (Savage *et al* 1977). Open triangles indicate upper limits for $N(H_2)$. The full curve represents a model in which the H_2 arises in clouds of internal density $\langle n_H \rangle = 6 \times 10^7$ m^{-3} and differing mean thickness (Spitzer 1982). The broken curve indicates the locus of total atomic to molecular conversion ($N_H = 2N(H_2)$).

density, which introduces scatter into the diagram, but there is a clear tendency for $N(H_2)$ to increase steeply with N_H initially and then to flatten off. The full curve represents a model for clouds with mean internal density $\langle n_H \rangle = 6 \times 10^7$ m^{-3} and differing mean thickness. At large column densities, the conversion of H I to H_2 is almost complete.

Although the energetic component of the external ISRF is strongly attenuated, it is important to note that appreciable ionization can occur deep within dense clouds due to penetration of relativistic particles (cosmic rays). Indeed, cosmic-ray ionization appears to play important roles in both the chemical and the physical evolution of dense clouds. In terms of chemistry, being more reactive than neutral species, the ions generally participate much more readily in chemical processes (section 8.2). In terms of physics, the magnetic field within a cloud exerts a force only on the ions but it may control the motions of neutral species through collisions between ions and neutrals. The efficiency with which a gravitationally bound cloud collapses to form stars (section 9.1) is strongly influenced by this effect.

1.4.4 H II regions

Interstellar clouds are in a continuous state of evolution. With the onset of star formation, internal sources of energy are created which disrupt and ultimately dissipate the natal cloud. The degree of disruption it imparts is directly related to the mass of a newly formed star. The most massive stars are both the most luminous and the quickest to evolve: for example, a 15 $M_{\odot}$ star (spectral type B0) has a luminosity of 13 000 $L_{\odot}$ and a main-sequence life-expectancy of only 10 Myr, small compared with typical cloud lifetimes of 10–100 Myr. Such massive stars emit most of their radiant energy in the ultraviolet and this leads to photodissociation and photoionization of the surrounding gas, forming H II regions. Examples of nebulae generated by this process are readily observable and are to be found in the colour illustrations of astronomy texts and glossy magazines throughout the world.

H II regions occur whenever a star of spectral type O or B is immersed in a medium of hydrogen. Of course, other species will be ionized as well but H II is always the dominant phase. Ions are formed by reaction (1.18): any excess photon energy on the left-hand side is added to the kinetic energy of the products on the right and the plasma is thus heated. Strong Coulomb attraction between protons and electrons leads to recombination, with the emission of recombination lines; in equilibrium, there is a balance between the rates of photoionization and recombination.

The physics of H II regions has been discussed by Spitzer (1978) and Dyson and Williams (1997) and only a few important results will be given here. The volume of gas a star can maintain in an ionized state is limited by recombination, such that the recombination rate is just equal to the rate (S_*) at which the star emits ionizing photons. In an idealized medium of uniform density, this volume will be a sphere (known as the Strömgren sphere) and will have radius

$$R_S = \left(\frac{3S_*}{4\pi n^2 \beta_2}\right)^{\frac{1}{3}} \tag{1.21}$$

where β_2 is the appropriate temperature-dependent recombination rate coefficient:

$$\beta_2 = 2 \times 10^{-16} T_e^{-\frac{3}{4}} \qquad (\mathrm{m^3\ s^{-1}}) \tag{1.22}$$

(see Dyson and Williams 1997: pp 67–70). Observed electron temperatures (T_e) are typically 10 000 K with little variation from one H II region to another, so $\beta_2 \approx 2 \times 10^{-19}$ m^3 s^{-1}. For a star of spectral type O5, $S_* \approx 5 \times 10^{49}$ s^{-1}, from which we deduce a Strömgren radius $R_S \approx 3.1 \times 10^{16}$ m ≈ 1 pc in a dense medium with $n = 10^9$ m^{-3}. For a B0 star, S_* is a factor of 100 less and hence R_S is a factor of $(100)^{\frac{1}{3}} \approx 5$ less. For stars later than B0 in spectral type, S_* declines rapidly and hence so does R_S (e.g. Spitzer 1978). Note that $R_S \propto n^{-\frac{2}{3}}$, thus the Strömgren sphere becomes larger in a more rarefied medium. These calculations are based on the assumption that H I gas is the only significant absorber of ionizing radiation: if the plasma contains dust, the effective size of the H II region will be reduced.

A newly formed OB star embedded in placental material generally forms a dense, compact H II region, the dimensions of which are limited by the availability of ionizing photons, as discussed earlier. Photons penetrating just beyond the ionization front will have energies < 13.6 eV and may be capable of dissociating H_2 (equation (1.19)) but not of ionizing H I. Thus, the H II region is encased in a photodissociation region, the dimensions of which are limited by the availability of photons in the energy range 11.2–13.6 eV. Beyond the photodissociation region lies molecular gas subject to irradiation only by photons of lower energy.

When an OB star emerges into lower density phases of the ISM, its sphere of influence increases dramatically. Ionization of ambient interstellar gas creates diffuse H II regions which can be orders of magnitude larger and more rarefied than the compact H II regions seen in molecular clouds. Diffuse H II regions may be density limited rather than photon limited, i.e. their boundaries may be determined by the physical limits of the cloud rather than the supply of ionizing photons. At intercloud densities, the nominal Strömgren radii for the most luminous O-type stars approach galaxian dimensions; such stars will contribute greatly to the high-energy flux of the general interstellar radiation field.

1.4.5 The interstellar environment of the Solar System

The term 'solar neighbourhood' has already appeared in earlier sections of this chapter: it is used to refer to regions of the Milky Way within a few kiloparsecs of the Sun. Such a volume of space may be reasonably homogeneous in terms of element abundances (sections 2.2–2.3) but will encompass a vast range of physical environments, including both 'arm' and 'interarm' regions of the Galaxy (e.g. Mihalas and Binney 1981: p 248). The Sun appears to lie near the inner edge of the Orion–Cygnus spiral arm. On scales of a few hundred parsecs, however, the distribution of matter is dominated less by spiral structure than by a local feature known as Gould's Belt, a disclike system of young stars and interstellar matter tilted at about 18° to the plane of the Milky Way. Gould's Belt contains the closest known regions of active star formation to the Solar System, including cloud complexes in Ophiuchus and Taurus, located at distances $\sim$150 pc in directions almost symmetrical about the Sun (Turon and Mennessier 1975). As previously noted in section 1.3.2, there is little evidence for reddening much closer than about 50 pc, indicating that the Solar System does not currently reside in the vicinity of an interstellar cloud of appreciable opacity: the closest dense cloud appears to be Lynds 1457, situated at a distance of about 65 pc in the constellation of Aries (Hobbs *et al* 1986). This result is not especially surprising: on the basis of estimates for the filling factors (table 1.1), the probability of finding the Solar System in an interstellar H I or H_2 cloud at any given time is no more than a few per cent.

There is, nevertheless, no lack of material in the current local interstellar medium (LISM). The gaseous component is detectable via observations of soft x-ray emission and of interstellar absorption lines at visible and UV wavelengths (see Ferlet 1999 for a review). On a scale of several tens of parsecs, the Sun is surrounded by an irregularly shaped, low-density region known as the local bubble (Cox and Reynolds 1987), with density $n \sim 4 \times 10^3$ m^{-3} and temperature $T \sim 10^6$ K, values typical of the intercloud medium (table 1.1). Embedded within this tenuous region are pockets of partially ionized gas of considerably higher density and lower temperature ($n \sim 10^5$ m^{-3}, $T \sim 7000$ K, ionization $\sim$50%; Cowie and Songaila 1986), typical of the warm phase of the ISM. The Sun itself appears to lie in one such pocket of warm gas, $\sim$3 pc in extent, named the local interstellar cloud (Linsky *et al* 1993, Lallement *et al* 1994, Ferlet 1999). Is the LISM a 'typical' interstellar environment? There seems to be little difficulty in understanding its physical characteristics in terms of ambient intercloud gas with embedded cloudlets. However, it is also possible that the LISM has been disturbed by a specific, local event such as a recent supernova explosion or the integrated stellar wind of the Scorpio–Centaurus association in Gould's Belt (Cox and Reynolds 1987, Ferlet 1999). The local bubble might represent the cavity blown by a single, active supernova remnant.

The kinematics of the local interstellar cloud (LIC) have been determined from spectral line studies of nearby stars such as Capella and Sirius. It is sweeping past us at a heliocentric speed of about 26 km s^{-1}, in a direction consistent with an origin in the Scorpio–Centaurus association. The Sun has already passed through the inner regions of the LIC and is currently located near its following edge. Independent confirmation that the Sun is embedded in the LIC was provided by the Ulysses space mission: neutral helium atoms were found to be entering the Solar System, with a mean heliocentric velocity consistent with that of the LIC as determined by stellar spectroscopy (Witte *et al* 1993).

The Ulysses and Galileo missions detected micrometre-sized dust grains near the orbit of Jupiter, also shown on the basis of their heliocentric velocities to be of interstellar origin (Grün *et al* 1993, 1994, Frisch *et al* 1999). Note that the presence of interstellar dust in the Solar System does not conflict with the fact that the LIC is essentially translucent, producing negligible extinction and reddening. Grün *et al* (1994) estimate the mass density of dust in the LIC to be about 20% of the interstellar mean, which implies that the rate of extinction per unit distance should be a factor $\sim$5 less than the value given in equation (1.6). Using this result and the estimated dimensions of the LIC ($\sim$3 pc), we may easily show that the visual extinction suffered by radiation passing through the LIC amounts to only $A_V \sim 0.001$ magnitudes, which is negligible. The grains detected by Ulysses and Galileo have radii in the range $0.1 < a < 4$ μm and are thus generally larger than typical interstellar grains. The apparent lack of a small-grain component is easily explained by consideration of the forces exerted on the particles by solar radiation pressure and the interplanetary magnetic field, which selectively block smaller particles from entering the inner Solar System (Grün *et al* 1994, Frisch *et al* 1999). Only the large end of the size distribution penetrates into the region sampled. Nevertheless, the detection of interstellar grains with $a > 1$ μm was unexpected, as it was not predicted by observations of interstellar extinction and may conflict with cosmic abundance constraints. Large grains are inefficiently destroyed by shocks: if the LIC is part of a shock-driven outflow from the Scorpio–Centaurus association, it might naturally contain a size distribution biased toward large grains (Frisch *et al* 1999).

1.5 The significance of dust in modern astrophysics

1.5.1 From Cinderella to the search for origins

Early studies of interstellar dust (section 1.2) were motivated primarily by the desire to correct photometric data for its presence rather than by an intrinsic interest in the dust itself or an appreciation of its true significance. According to Gaustad (1971) it was once the case, as far as a typical chauvinistic (male) astronomer was concerned, that "if you simply tell him the reddening law, particularly the ratio of total to selective extinction, he can unredden his clusters, correct the distance moduli, find the turnoff points, determine the age of the Galaxy and be happy!". Times have changed. Far beyond having annoyance value, dust is now recognized as a vital ingredient of the cosmos, a revelation which came about largely through exploration of new regions of the electromagnetic spectrum (ultraviolet, infrared, radio). Some of these developments are discussed in this section. Perhaps of greatest significance, however, is simply the realization that dust is the primary repository in the ISM for the chemical elements needed to make terrestrial planets (and life). "We are stardust", wrote Joni Mitchell[5], and this is literally true at the atomic level. The quest to understand the origin and evolution of interstellar dust is thus part of the quest for our origins.

1.5.2 Interstellar processes and chemistry

Evidence for a concentration of heavy elements into dust in the ISM came first by indirect means, via spectroscopic studies of interstellar gas-phase absorption lines at visible and ultraviolet wavelengths (e.g. Spitzer and Jenkins 1975)[6]. Abundances determined from line strengths were compared with standard reference values such as those in the Sun. Results indicated a dramatic shortfall in the abundances of many of the heavier elements, most readily explicable if the missing atoms are tied up in solid particles. This 'depletion' is almost total for the most refractory elements, such as Si, Ca, Fe, Ni and Ti (section 2.4). The

[5] From 'Woodstock' (*Ladies of the Canyon*), Reprise Records, 1970.

[6] The term 'heavy elements' is used in this book to mean chemical elements with atomic weight ≥ 12, i.e. carbon and above, that commonly condense to form solids. These elements are often referred to collectively as 'metals' in the astronomical literature.

depletion levels of several elements correlate with physical conditions, providing evidence for exchange of material between gas and dust: for example, depletions are enhanced in denser clouds where atoms are more likely to collide with (and stick to) grains and reduced in high-velocity clouds, where grains may be vaporized by shocks.

For many years, the only molecules known to exist in interstellar space were the radicals CH, CH^+ and CN, identified by their characteristic absorption lines in stellar spectra at blue-visible wavelengths (McKellar 1940). Since the mid-1960s, a host of new identifications have been made, primarily in the radio/microwave region of the spectrum. Snyder (1997) lists over 100 known interstellar molecules. Many polyatomic and organic species are included, some (e.g. HCN, H_2CO, HCOOH, CH_3NH_2) of potential prebiotic significance. These findings demonstrate the complexity of interstellar chemistry and the importance of chemical and physical interactions between gas and dust. Successful models must explain the formation and relative abundances of all the observed molecules. A fundamental theoretical problem concerned the production of H_2, the most abundant molecule of all (Hollenbach and Salpeter 1971): formation of H_2 by association of two H atoms in the gas phase cannot occur because the molecule has no means of releasing binding energy in the absence of a third particle. Formation on grain surfaces is now the accepted mechanism: H atoms attaching themselves to grains become trapped at surface defects in the grain structure; when they subsequently recombine, the binding energy is partly absorbed into the grain lattice and the resulting molecule is ejected from the surface and returned to the gas. Gas-phase reaction schemes, in which H_2 molecules released from grains react with heavier species, may explain the production of a number of other common interstellar molecules such as CH and OH (section 8.2). However, grains must be included in any complete model for interstellar chemistry (e.g. Williams 1993). As well as playing a catalytic role, they influence molecular abundances by reducing the intensity of photodissociative radiation. It follows that physical processes involving the dust, such as coagulation leading to a dearth of the small grains responsible for ultraviolet opacity, can have dramatic implications for the chemical evolution of a cloud (Cecci-Pestellina *et al* 1995).

When species such as O and OH attach themselves to grains, they tend to become hydrogenated via surface reactions to form H_2O. Unlike H_2, 'heavy' molecules such as H_2O will not generally be ejected from the grain when the binding energy is released. Similar arguments apply to other polyatomic molecules that may form by hydrogenation or oxidation reactions on grain surfaces, including CH_4, CH_3OH, CO_2 and NH_3. The rate at which these reactions occur increases rapidly with the density of the cloud. Thus, in dense clouds, the grains become nucleation centres for the growth of *icy mantles* and these mantles may become the dominant repository for molecular material (section 8.4). Conversely, mantled grains exposed to shocks or the unattenuated ISRF become new sources of gas-phase atoms and molecules. At all levels of density, the exchange of material between interstellar gas and grains is essential to models for the chemical composition and evolution of the ISM as a whole.

The detection of polycyclic aromatic hydrocarbons (PAHs) in the ISM, by means of characteristic vibrational modes observed in the infrared, adds another dimension to the problem of understanding interstellar chemistry (Omont 1986). PAHs are planar molecules composed of benzene rings, with sizes typically $\sim$1 nm. PAHs and PAH clusters may be taken to represent the transition zone between large molecules and small solid particles. The observations suggest that PAHs account for $\sim$5% of the available interstellar carbon: although they do not contain a large fraction of the mass, they may dominate in terms of *surface area* available for catalytic reactions. The origin of these particles is still debated but is presumably linked closely with that of other forms of solid carbon in the ISM.

1.5.3 Stars, nebulae and galaxies

Dust is a vital ingredient in models for many classes of infrared source, including both compact objects (young stars, evolved stars and some galactic nuclei) and extended objects (planetary nebulae, H II regions, 'starburst' regions, globules, cirrus clouds and entire galaxies). Models for these diverse phenomena have in common the presence of dust heated by the absorption of ambient starlight. In many cases, emission from the dust dominates the emergent spectral energy distribution in the infrared and constitutes an important cooling mechanism. Modelling such sources is often difficult and may involve detailed radiative transfer calculations as, for example, in the case of an optically thick circumstellar shell. Results depend critically on the optical properties assumed for the grains at far-infrared wavelengths and on their size distribution. Near- and mid-infrared emission in excess of that expected for classical grains will arise when very small particles are subject to transient heating.

A fundamental problem exists in determining the mass of a cloud opaque to starlight because the principal ingredient (H_2) cannot be observed directly: ultraviolet spectroscopy yields meaningful data (e.g. figure 1.6) only in relatively unobscured regions (diffuse clouds and the outer layers of dense clouds). It is usual to evaluate cloud masses from millimetre-wave observations of gas-phase CO, assuming a constant ratio of integrated CO emission intensity to H_2 column density (van Dishoeck and Black 1987). In reality, this ratio is likely to be sensitive to environmental factors: in cold, quiescent clouds, for example, a significant fraction of the CO may be frozen onto grains where, in solid form, it is undetected by the millimetre observations. However, the infrared and sub-millimetre continuum emission from a cloud is proportional to the mass of emitting dust (section 6.1) and observations of this emission thus provide an independent estimate of total cloud mass, subject to an assumed dust-to-gas ratio.

Dust is intimately involved in the formation and evolution of stars and planetary systems. Stars are born within cocoons of dust and gas and theoretical considerations suggest that the presence of dust may catalyse the gravitational collapse of protostars, providing an efficient radiator of excess heat energy. As a protostar contracts, rotation flattens the circumstellar envelope into a disclike structure. Protoplanetary discs have been detected around young stars by virtue of extinction and emission arising in the dust they contain (section 9.2). Planetary systems may be common products of accretion in discs around single stars; but in a number of cases, of which Vega is the best-known example, circumstellar debris discs persist into the main-sequence phase. The effect of radiation pressure on grains may strongly influence the dynamics of stellar envelopes surrounding both young and evolved stars. Grains condensing in the envelopes of late-type giants and supergiants (section 7.1) may drive the outflows from such objects, resulting in mass-loss rates sufficient to influence the ultimate fate of a star during the crucial late phases of its evolution. The dusty outflows of evolved stars contribute significantly to the enrichment of the ISM with heavy elements.

A major discovery in extragalactic astronomy was the identification of luminous infrared galaxies (LIGs), i.e. galaxies which emit more energy in the mid/far infrared ($\sim$5–500 μm) than all other wavelengths combined (Soifer *et al* 1987, Sanders and Mirabel 1996). For comparison, normal spiral galaxies such as the Milky Way emit no more than $\sim$30% of their total energy output at these wavelengths. Soifer *et al* identify three mechanisms thought to contribute to infrared emission in galaxies: photospheric emission from stars, synchrotron emission from relativistic electrons and emission from dust heated by other sources of luminosity. They conclude that *emission from dust* is the dominant mechanism in the vast majority of galaxies detected by IRAS. In a typical spiral galaxy, diffuse emission from dust dispersed throughout its disc and heated by the integrated ISRF contributes to the total infrared flux. However, in the majority of LIGs, the emission is dominated by regions of intense star formation (the starburst phenomenon) in which dust is heated by local concentrations of massive young stars. The trigger for a starburst appears to be the interaction or merger of young spirals with massive interstellar media. A starburst must be a transient phase in galaxian evolution, as the star formation rate is such that all the available material will be consumed on timescales of a few 100 Myr. Many of the most luminous LIGs have active nuclei in which circumnuclear dust mixed with gas funnelled into the nucleus is the dominant source of infrared flux. It is possible that such galaxies are the precursors of quasars and radio galaxies (Sanders and Mirabel 1996).

1.5.4 Back to basics

Finally, the original reason for the study of interstellar dust – to correct for its effect on astronomical observations – remains vitally important. Although the observed properties (extinction, reddening, etc) of the dust are now reasonably well known in general terms, the sheer complexity of the ISM often precludes indiscriminate use of general results. The *degree* of extinction varies stochastically with line of sight, even in directions away from the plane of the Milky Way (section 1.3). The *spectral dependence* of extinction varies with the size, shape and composition of the grains, factors that depend on both their history and the prevailing environment. In many cases, recovering the intrinsic spectrum of an object obscured by dust from observed fluxes is not a simple task. The safest approach would be to determine the detailed extinction curve applicable to the specific line of sight over all wavelengths of interest, but this is not always practical or even possible. A less direct approach is to assume an extinction curve that seems applicable to the type of environment, as determined by other measurements. An important step toward 'standardization' of extinction corrections was taken by Cardelli *et al* (1989), who showed that some of the environmental effects that lead to changes in the extinction curve are systematic and may be characterized by a single parameter, the ratio of total-to-selective extinction (R_V).

In view of the significance of the parameter R_V, it is important to establish its degree of variability and its sensitivity to environment. Pronounced, systematic and widespread variations in the value of R_V would have serious consequences for our understanding of galactic structure and the extragalactic distance scale (Johnson 1968) but no such variations have been substantiated. Adoption of the standard average value of $R_V \approx 3.1$ (equation (1.5)) is adequate in most situations, at least in the solar neighbourhood – the most important deviations are observed in dense clouds and H II regions, where values as high as $\sim$5.0 may occur. Dust associated with Gould's Belt appears to have a mean R_V value somewhat higher than the galactic average, and this results in a small systematic dependence on galactic longitude (Whittet 1977, 1979).

On the wider stage, the question of whether cosmologically distant objects are obscured by dust is vital to our understanding of the structure and evolution of the Universe (Ostriker and Heisler 1984). Significant extinction by dust either within or between intervening galaxies could hinder detection of objects at very high redshift and obscure our view of galaxy formation (Disney 1990). Mirabel *et al* (1998) show that dust could profoundly bias morphological classification of colliding/merging galaxies, especially if only visible wavelengths (equivalent to ultraviolet in the rest-frame of the galaxy) are considered. Evidence for dust out to redshifts of $z \approx 5$ has been found (Soifer *et al* 1998, Bertoldi and Cox 2002), suggesting that the chemical elements needed to make dust became available quite soon after the birth of the Universe in the big bang.

1.6 A brief history of models for interstellar dust

Observations imply the existence of dust particles in interstellar space that range in size from $\sim$0.001 to $\sim$1 μm. In view of the level of research activity devoted to these particles since Trumpler's pioneering work (section 1.2), it is perhaps surprising that their chemical composition has proved difficult to establish and is, indeed, a matter of continuing debate. The reason for this is not hard to find: a number of the observed properties of interstellar dust do not provide very specific compositional information and there is consequently a lack of uniqueness in the modelling. The 'continuum' effects of extinction, scattering, polarization and emission have much to tell us about various properties of the dust (e.g. size, shape, alignment properties) but do not place decisive constraints on its composition. Conversely, this uncertainty does not preclude some important investigations: for example, one can deduce empirical corrections for interstellar reddening by careful observation, without making any assumptions concerning the nature of the grains responsible.

A fundamental constraint on grain models is provided by information on the relative abundances of the chemical elements in the ISM. Of the most abundant elements, helium can obviously be disregarded owing to its extreme volatility and chemical inertia, whilst significant quantities of hydrogen can be incorporated into solids only if the H atoms are chemically bonded to heavier elements. The bulk of the grain mass is undoubtedly provided by elements with atomic weight $\geq$12, of which the most abundant are C, N, O, Mg, Si and Fe. The depletions of these elements in the gas phase, most pronounced for the metals, provide indirect but compelling evidence for their inclusion in grains (chapter 2). The combined mass of all the condensible elements is $\sim$2% of the total mass, setting an upper limit on the dust-to-gas ratio.

The key to the problem of dust composition lies in detailed study of dust-related absorption and emission features in astronomical spectra and comparison of observed spectra with appropriate laboratory data. In this section, we trace the development of ideas and summarize the main theories. Modelling strategies leading to the formulation of grain theories are discussed in greater depth in subsequent chapters of this book.

1.6.1 Dirty ices, metals and Platt particles

Amongst the first materials to be proposed for interstellar grains were ices (Lindblad 1935) and metals such as iron (Schalen 1936). Ices are particularly favoured by the high cosmic abundances of the constituent atoms[7]. Metals alone, on the other hand, are not sufficiently abundant to account for interstellar extinction and neither do they explain the scattering properties of interstellar dust and the presence of reflection nebulae (Greenberg 1968). Lindblad suggested that particles composed of saturated molecules (H_2O, NH_3 and CH_4) may form in interstellar clouds by random accretion, for which indirect support is provided by the presence of radicals such as CH, CN and OH that could be the precursors or destruction products of ice particles. This proposal was developed and extended by Oort and van de Hulst (1946) into a detailed model for ice nucleation and growth in interstellar clouds. Incorporation of metallic elements leads to the concept of 'dirty ices'. Extinction calculations for such particles, assuming

[7] The generic term 'ices' is used in this book to describe any volatile molecular solid composed primarily of the CHON group of elements. Typically, interstellar ices appear to be $\sim$70% H_2O, the remainder being made up of species such as CO, CO_2, CH_3OH, CH_4, NH_3, N_2 and O_2.

grains with sizes ranging up to approximately the wavelength of visible light, give an excellent fit to the observed extinction curve from the infrared to the near ultraviolet (Greenberg 1968). The simultaneous development of the 'dirty snowball' model for comets in our Solar System (Whipple 1950, 1951) was suggestive of the possibility that these bodies may be accumulations of interstellar grains.

Although the ice model gained widespread acceptance in the 1950s and early 1960s, Platt (1956) argued that random accretion would tend to produce small particles composed of unsaturated molecules with unfilled energy bands, rather than classical ice grains. These 'Platt particles' were postulated to have radii no more than $a \sim 1$ nm. They are better described as macromolecules than solid particles, in the sense that their optical properties are determined largely by quantum mechanics rather than solid-state band theory. Although little attention was paid to them at the time, Donn (1968), in a prophetic paper, drew attention to the possible astrophysical significance of very small grains, linking Platt particles with polycyclic aromatic hydrocarbons. These became the subject of renewed interest and intensive research in the 1980s, as possible carriers of a number of observed spectral features and diffuse continuum emission in the infrared, discussed in detail in chapter 6.

1.6.2 Graphite and silicates

It seemed by the early 1960s that the nature of interstellar dust was well established and that subsequent research would merely clarify some of the details rather than radically change the picture. The ice model of Oort and van de Hulst could explain the observed properties of the particles in the spectral range then accessible and it presented a logical and self-consistent picture of grain formation and evolution. The information explosion that accompanied technological advances in observational astronomy during the latter part of the 20th century had two dramatic effects on research related to interstellar dust: it greatly stimulated interest in the subject, for reasons discussed in section 1.5, and simultaneously demonstrated that the extant model was deficient or at best incomplete. It is no longer possible to explain all of the relevant observations in terms of a single substance or class of substances condensing under similar conditions; mixtures of materials with diverse origins must be invoked, in which different components account for different aspects of the data.

Vital compositional clues are provided by spectroscopic observations in the ultraviolet and the infrared. Ices have absorption edges in the ultraviolet (Field *et al* 1967), occurring, for example, in the case of H_2O-ice, at $\lambda \sim 1600$ Å. The spectra of reddened stars show no evidence for structure near this wavelength but do show a very strong absorption feature centred at 2175 Å, widely attributed to the presence of small graphitic grains with sizes <0.02 μm (Stecher 1965, Stecher and Donn 1965; section 3.5 of this book). A continuous rise in extinction observed at far-ultraviolet wavelengths ($\lambda < 1500$ Å) also supports the presence of very small particles which may be distinct from those that absorb at 2175 Å.

In the infrared region, vibrational transitions can be used to test for the presence of grain signatures in regions that are too heavily obscured to permit observation in the ultraviolet. For example, stretching vibrations of O–H bonds in H_2O-ice produce a strong feature at $\lambda \sim 3$ μm (chapter 5) and this allows a sensitive diagnostic for the presence of interstellar ice. First attempts to observe this feature in the spectra of reddened stars were unsuccessful (Knacke *et al* 1969), confirming that ice is not a major constituent of dust in diffuse clouds. Subsequent investigations have shown that the presence or absence of H_2O and other ices in the ISM is critically dependent on physical conditions along the particular line of sight observed. Stars reddened by long pathlengths of low density material do not generally show appreciable ice absorption, whereas those obscured by dense clouds invariably do (Gillett *et al* 1975a, Willner *et al* 1982, Whittet 1993). Ices require a shielded environment to condense and survive; only in dense molecular clouds do they appear to be ubiquitous.

Spectroscopy at longer infrared wavelengths demonstrated the presence of a quite different class of grain material. A broad feature centred at approximately 9.7 μm is observed in the spectra of many different types of object, including reddened stars (Gillett *et al* 1975a, Roche and Aitken 1984a), H II regions (Gillett *et al* 1975b), O-rich circumstellar envelopes (Merrill and Stein 1976a), comets (Bregman *et al* 1987) and some galactic nuclei (Roche *et al* 1991). It may be seen in either absorption or emission, depending on the nature of the object and the degree of foreground extinction (whereas the 3 μm ice feature is seen only in absorption). Absorption at 9.7 μm occurs in objects observed through either diffuse or dense regions of the ISM, with a strength that correlates with the visual extinction. The corresponding emission feature is frequently observed in the spectra of H II regions and stars (young or old) that are embedded in dust. Objects with 9.7 μm emission or absorption generally also show a weaker feature centred near 19 μm. The carrier must be robust, capable of surviving in a wide variety of physical conditions, and O rich, as it is typically observed in circumstellar shells only when the abundance of O exceeds that of C (chapter 7). Both 9.7 and 19 μm features are widely attributed to silicate dust, which has the required refractory properties and spectral activity at the appropriate wavelengths, produced by Si–O stretching and O–Si–O bending vibrations, respectively. Indirect support for this interpretation is provided by the ubiquity of such materials in meteorites and interplanetary dust.

The term 'silicate' is generic, covering (in the geological context) a wide range of possible chemical compositions and mineral structures based on SiO_4 units (figure 1.7). These units are tetrahedral in shape and bear negative charge. Oxygen atoms may be shared between adjacent units, joining them together, or they may bond to positively charged cations (typically Mg^{2+} or Fe^{2+}), forming chains or three-dimensional lattices. In silicates of the olivine group (e.g. Mg_2SiO_4), the SiO_4 units are isolated from each other by cations, whereas in other groups such as pyroxenes (e.g. $MgSiO_3$) they are linked by bridging

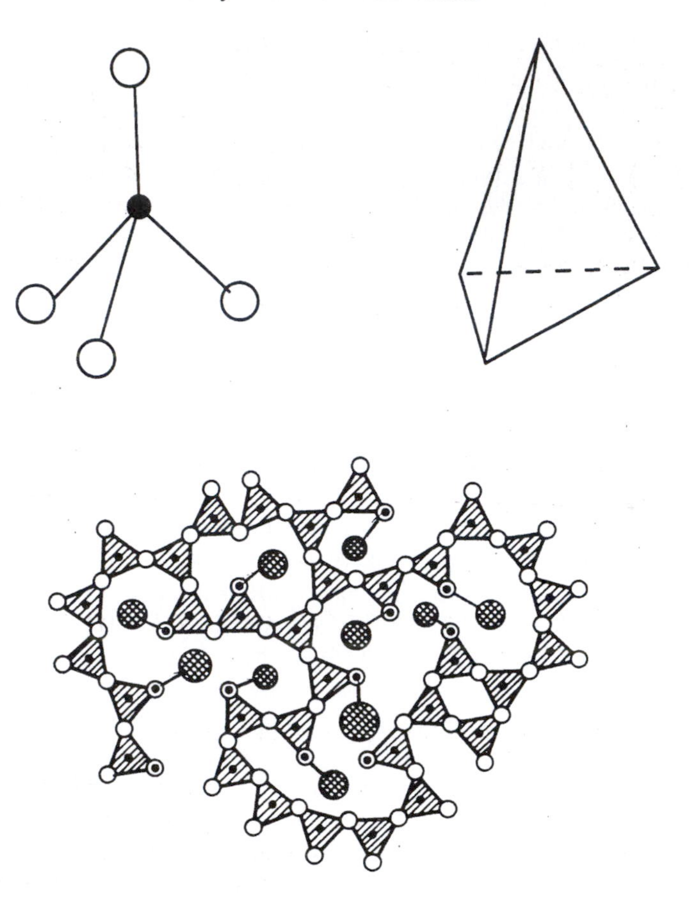

Figure 1.7. A schematic illustration of the structure of silicates. Top: the SiO_4 unit (left) and its tetrahedral structure (right). The Si atom (black circle) is located at the centre of the tetrahedron and O atoms (open circles) are located at its corners. Bottom: a two-dimensional representation of a segment of amorphous silicate. Anions (SiO_4^{2-}) and cations (e.g. Mg^{2+} or Fe^{2+}) are denoted by triangles and hatched circles, respectively. Some O atoms bridge adjacent tetrahedra (open circles) whilst others are bonded to cations (dotted circles).

O atoms. Layered structures can also occur, as in talc ($Mg_3Si_4O_{10}(OH)_2$, a hydrated silicate). The degree of long-range order (crystallinity) in the structure affects the spectroscopic properties of the silicate: highly ordered forms tend to produce sharp, structured features, whereas disordered forms tend to produce broader, smoother features. The profile of the 9.7 μm feature observed in astronomical spectra is generally broad and smooth (section 5.2.2), indicating that interstellar silicates are predominantly amorphous rather than highly crystalline. Abundance considerations suggest that they are likely to be rich in both Mg and

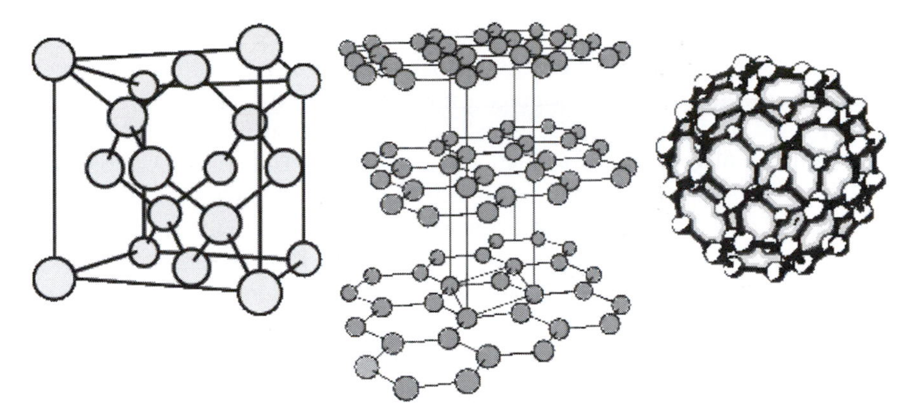

Figure 1.8. Schematic representation of the three basic ordered allotropes of carbon: diamond (left), graphite (centre) and fullerene (C_{60}, right). Figure courtesy of Mark Rawlings.

Fe (section 2.5) but their detailed composition is not uniquely specified by the observations.

1.6.3 Unmantled refractory and core/mantle models

Spectroscopic detection of silicates and, more tentatively, graphite, led to the formulation of a model in which the diffuse-cloud extinction is produced by a mixture of refractory grains rather than volatile ices (see Wickramasinghe and Nandy (1971) and references therein for a description of early work in this area). Such particles may originate in the atmospheres of evolved stars of differing C/O ratio and so it is natural to assume the existence of distinct O-rich and C-rich populations. Mathis *et al* (1977) showed in a key paper, widely referred to as 'MRN', that the extinction curve from the near infrared to the far ultraviolet can be explained by separate populations of unmantled graphite and silicate grains. An important feature of the MRN model is the use of a power-law size distribution, subsequently adopted by other investigators, in which the relative numbers of particles with respect to their size (assumed spherical of radius a) within a given range is $n(a) \propto a^{-3.5}$ (see chapter 3). This is the form of size distribution expected when particles are subject to collisional abrasion, as may occur in red-giant winds (Biermann and Harwit 1980). The particles are assumed to remain uncoated in diffuse interstellar clouds and to provide nucleation centres for the growth of ices when immersed in molecular clouds.

Solid carbon can assume a variety of forms, ranging from highly ordered (graphite, diamond, fullerene: see figure 1.8) to amorphous (sootlike) material: see Duley (1993) for a review. C atoms naturally tend to become arranged into planar groups of hexagonal rings and individual groups containing typically

~5–30 hydrogenated rings are essentially PAH molecules (see Robertson and O'Reilly 1987). In amorphous carbon, such groups or 'islands' are assembled randomly and there is no long-range order. In contrast, graphite (figure 1.8) is composed of regular stacks of platelets formed from planar groups of rings. Intermediate polycrystalline forms may also occur (e.g. Tielens and Allamandola 1987a). The MRN model assumes that graphite particles contribute significantly to the extinction at all wavelengths. However, whilst some graphite is required to explain the 2175 Å feature, it is reasonable to suppose that the bulk of the carbon is in a less ordered state. This is consistent with observations of C-rich late-type stars and with models for grain nucleation in their atmospheres (see chapter 7), indicating that the dust in their ejecta is predominantly amorphous rather than graphitic. It is difficult to understand how the carbon in interstellar grains can become highly ordered, yet no extant model has explained the ultraviolet extinction curve convincingly without graphite.

Developments of the MRN model have considered extinction produced by porous aggregates of smaller grains rather than by idealized, monolithic, chemically homogeneous particles (Mathis and Whiffen 1989, Wolff *et al* 1994, Dominik and Tielens 1997, Fogel and Leung 1998). This scenario is physically reasonable in terms of grain physics (particles are expected to form aggregates as the result of low-velocity collisions in dense interstellar clouds and circumstellar envelopes); and it gains indirect support from the known structure of interplanetary dust grains in the Solar System (e.g. Brownlee 1978). Sophisticated techniques have been developed for calculating the optical properties of such 'fractal' structures of arbitrary size, shape and porosity, allowing a more realistic representation of dust in theoretical models.

The main rival to the unmantled-refractory grain model postulated by MRN is the core/mantle model. Two distinct scenarios have been discussed. Greenberg and co-workers (e.g. Li and Greenberg 1997) propose, on the one hand, that the visual extinction is produced primarily by silicate grains with 'organic refractory' mantles. Justification for this structure is argued on the basis of mantle growth and subsequent evolution of the mantles: when hot stars form in molecular clouds, irradiation of ices present on mineral grains in the surrounding molecular cloud may convert them into complex organic polymers. This process has been demonstrated in the laboratory under simulated interstellar conditions. Duley *et al* (1989a), on the other hand, propose that silicates acquire mantles of hydrogenated amorphous carbon in interstellar clouds, as the result of direct depletion of C atoms onto the grain surface. As irradiation over cosmic timescales will tend to convert organic mantles to amorphous carbon, these scenarios may be indistinguishable in terms of their predictive power: they both attribute visual extinction to silicates coated with non-graphitic C-rich mantles but the mantles are assumed to originate in different ways. Both predict significant absorption at 3.4 μm, as a result of C–H bond vibrations, but the properties of the feature observed at this wavelength (Butchart *et al* 1986, Sandford *et al* 1991) suggest an origin in very small grains rather than classical core/mantle grains (Adamson *et al* 1999, Schnaiter *et al* 1999). There is no direct evidence to indicate that grain mantles survive the dissipation of a molecular cloud. Both the Greenberg and Duley proposals require separate populations of small grains to account for the ultraviolet extinction. A distinct feature of the Duley model is its interpretation of the mid-ultraviolet (2175 Å) absorption in terms of OH^- sites on the surfaces of silicates (rather than to graphite) but the existence of such an absorption has not been substantiated.

1.6.4 Biota

In contrast to models based on conventional materials such as minerals and ices, Hoyle and Wickramasinghe (1986) proposed that interstellar dust is composed of materials of biological origin including bacteria, viruses, diatoms, proteins and polysaccharides. Calculated extinction curves for dry bacteria are fitted to the observed extinction curve in the visible and identifications for various spectral features are claimed. Graphite spheres, assumed to be the irradiation products of organisms, provide the 2175 Å ultraviolet extinction bump. There are a number of fatal objections to this model, however, which are briefly summarized here. For further discussion, see Duley (1984) and Davies (1986, 1988).

The spectra of biological materials are, in general, complex, frequently exhibiting features that have no counterpart in astronomical spectra (Davies *et al* 1984). Strong absorptions are predicted in the ultraviolet, notably in the wavelength range 2500–3000 Å, but these are not observed in the ISM (McLachlan and Nandy 1984, Savage and Sitko 1984, Whittet 1984a). Similarly, the infrared spectra of bio-organics show complex features that do not match the observed interstellar features (Butchart and Whittet 1983, Pendleton and Allamandola 2002). A further test is provided by the cosmic availability of phosphorus, the least abundant of the chemical elements essential to biology. Phosphates are integral to the structure of nucleic acids. The ratio of P to C by number of atoms is approximately 1:50 in biota, whereas the equivalent cosmic abundance ratio is 1:1000. It follows that no more than a few per cent of the interstellar grain mass can be in 'biological' grains (Whittet 1984b). Of course, these results do not exclude the possibility that biological organisms exist in interstellar space; they simply indicate that they do not constitute a viable model for interstellar dust.

Recommended reading

- *Galactic Astronomy* (second edition), by Dimitri Mihalas and James Binney (W H Freeman and Company, San Francisco, 1981).
- *Atlas of Selected Regions of the Milky Way*, by Edward Emerson Barnard (ed E B Frost and M R Calvert, Carnegie Institute of Washington, 1927).
- *The Immortal Fire Within: The Life and Work of Edward Emerson Barnard*, by William Sheehan (Cambridge University Press, 1995), pp 369–83.

- *Physical Processes in the Interstellar Medium*, by Lyman Spitzer (John Wiley and Sons, New York, 1978).
- *The Physics of the Interstellar Medium* (second edition), by John Dyson and David Williams (Institute of Physics Publishing, Bristol, 1997).

Problems

Most of the problems here may be solved with reference to information and concepts presented in this chapter. Reference to material in other chapters may also be helpful in some cases.

1. Describe *three* types of observation that would enable you to determine whether a void in an optical photograph of the Milky Way represents a true minimum in the distribution of background stars or evidence for a foreground interstellar cloud.
2. Explain why Trumpler found that the dimensions of open clusters appeared to increase systematically with distance from the Sun.
3. (a) The star HD 210121 is located behind a high-latitude cloud at galactic coordinates $\ell \approx 57°$, $b \approx -45°$. Calculate the expected visual extinction toward HD 210121, assuming that dust in the Galaxy is uniformly distributed in a flat disc and that the mean visual extinction at the galactic poles is 0.15 ± 0.05.
   (b) A study of its *reddening* determined that the visual extinction toward HD 210121 is $A_V = 0.80 \pm 0.07$. Compare this with the value you calculated in (a). What can be deduced from the agreement or lack of agreement?
4. (a) Suppose that a type II supernova with an absolute visual magnitude of -18 at maximum luminosity were to explode at the centre of our Galaxy (a distance 8 kpc from the Solar System). Estimate the apparent visual magnitude (V_{max}) of the supernova at maximum luminosity, assuming that the mean rate of visual extinction toward the galactic centre is 4 mag kpc^{-1}.
   (b) Estimate and compare V_{max} for an identical supernova located at the same distance but in a direction sufficiently displaced from the galactic disc to suffer negligible extinction. Discuss the implications of your calculations for detection and study of supernovae in our Galaxy. (*Note*: the condition for naked-eye visibility is $V < 6$; objects detectable with large telescopes have $V < 26$.)
5. Explain why H_2 is photodissociated only by photons with energies >11.2 eV, although its binding energy is 4.48 eV.
6. An interstellar cloud of uniform gaseous number density 3×10^7 m^{-3} and temperature 80 K is in pressure equilibrium with the intercloud medium, which has a mean number density of 5×10^3 m^{-3}. Calculate the temperature of the intercloud medium.

7. Suppose that the solar neighbourhood were to pass through a giant molecular cloud of mean density 10^9 m^{-3}, dimensions ~50 pc and normal dust-to-gas ratio. Estimate the effect of the resulting extinction on the apparent visual magnitudes of nearby bright stars such as Sirius ($V = -1.5$, $d = 2.6$ pc) and Vega ($V = 0.0$, $d = 7.5$ pc) when the Sun is near the centre of the cloud. What would be the implications of such a situation on our view of the Universe?
8. Summarize the principal similarities and differences, comparing the 'unmantled refractory' and 'organic refractory mantle' models for dust in the diffuse ISM.

Chapter 2

Abundances and depletions

"Figuratively, when we study depletions, it is as if we were looking at the crumbs left on the plate after the grains have eaten their dinner."

E B Jenkins (1989)

It is well known that hydrogen and helium together comprise some 98% by mass of all observed matter in the Universe. As inhabitants of a terrestrial planet, we are accustomed to an anomalous situation in which the rarer, heavier elements predominate. The most abundant elements that commonly condense to form solids in some form or other are C, N, O, Mg, Si and Fe and these six elements are presumed to contribute most of the mass of interstellar grains. Observational evidence for their presence in grains is inferred from gas-phase abundances, which are generally much less (e.g. by a factor $\sim$100 in the case of Fe) in interstellar clouds, compared with stellar atmospheres where essentially all material is in the gas. The depletions of heavy elements in interstellar gas thus provide qualitative evidence for their inclusion in the dust and, more tentatively, quantitative information on the probable distribution of element abundances in the solid material. The sensitivity of the depletions to environment gives important clues on the nature of the formation and destruction mechanisms for dust grains that operate in the interstellar medium.

In this chapter, we begin (section 2.1) by outlining the origin and evolution of the chemical elements, with emphasis on the production of the key heavy elements. In section 2.2, we discuss abundance measurements for the Solar System, commonly adopted as a standard for the ISM. Galactic abundance trends are examined in section 2.3 and the depletions observed in the ISM are reviewed in detail in section 2.4. The implications of these results for the nature and evolution of interstellar grains are discussed in the final section (section 2.5).

2.1 The origins of the condensible elements

A detailed account of the synthesis of the chemical elements is beyond the scope of this book but an outline of the processes that lead to the presence of 'condensible' elements – those likely to form solids in the ISM – is pertinent. For extensive discussion and relevant literature, the reader is referred to reviews by Tayler (1975), Trimble (1975, 1991, 1997), Boesgaard and Steigman (1985) and Arnett (1996).

2.1.1 The cosmic cycle: an overview

A flow chart illustrating the cosmic 'astration' cycle is shown in figure 2.1. According to modern cosmological theory, the Universe was created in the primordial fireball or big bang (BB) some 15 billion years (15 Gyr) ago. The standard model predicts that matter emerging from the big bang was composed of hydrogen and helium, approximately in the ratio 10:1 by number[1]. This matter dispersed and subsequently condensed into galaxies. The first generations of stars presumably formed from material with element abundances essentially unchanged from the primordial values and nuclear reactions in stars led to subsequent synthesis of heavier elements. As stars evolve and die, processed material is injected into the ISM via stellar winds and cataclysmic events such as supernova (SN) explosions. This mass-loss leads to a steady increase in the heavy-element endowment of successive generations of stars, up to the present value of $\sim$2% by mass.

The temperatures and densities required for nucleosynthesis in the post-big-bang era are normally reached only in stellar cores. The yield of heavy elements from stars to the ISM is much lower than the rate of production. Due to the dominance of gravity, much material remains trapped within compact cores that ultimately collapse to become white dwarfs, neutron stars or black holes, whilst the ejected material often tends to remain relatively hydrogen rich. The majority of stars show little or no direct evidence for the presence of intrinsic heavy-element enrichment in their photospheric spectra. Exceptions are to be found most commonly amongst post-main-sequence stars such as red giants, where episodic mixing of core and envelope material may produce outflows enriched in carbon, for example. Subsequent core–envelope separation may lead to ejection of processed remnants (planetary nebulae, supernova remnants) that ultimately merge with the ISM. Supernovae are also primary sources of cosmic-ray particles and the interaction of cosmic rays with low-energy interstellar matter leads to further evolution of the elements, notably the production of Li, Be and B (Viola and Mathews 1987).

[1] It is conventional to denote the total mass fraction contained in H, He and all other elements combined by the symbols X, Y and Z, respectively. Z is often referred to as the metallicity, although it includes all elements heavier than He. Models predict initial (primordial) values of $X_{\rm P} \approx 0.76$, $Y_{\rm P} \approx 0.24$, $Z_{\rm P} \approx 0.00$ and observations indicate present-day values of $X \approx 0.73$, $Y \approx 0.25$, $Z \approx 0.02$. Note that $Z_{\rm P}$ is not identically zero as tiny amounts of Li were produced in the big bang.

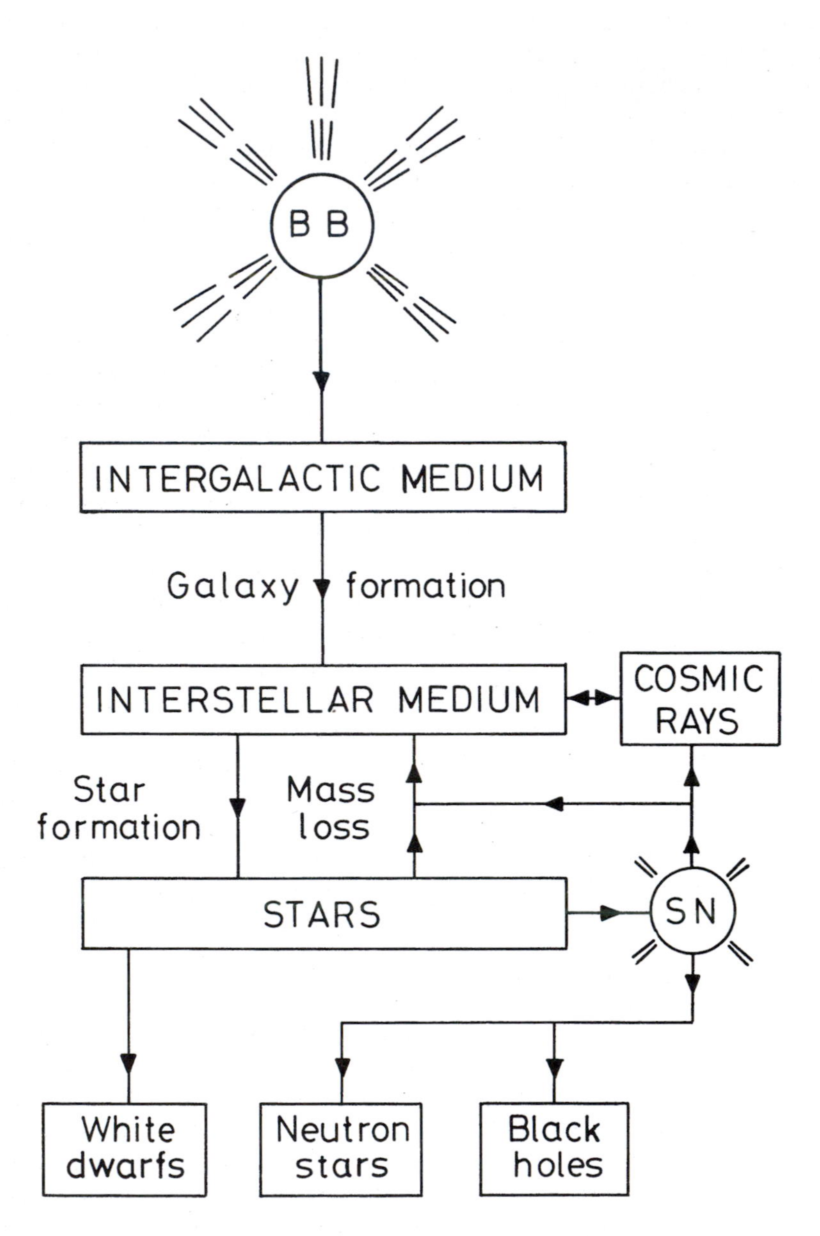

Figure 2.1. A schematic representation of cosmic evolution.

2.1.2 Nucleogenesis

According to standard models for the hot big bang, nuclear reactions occurring at temperatures $\sim 10^9$ K during the first few minutes of the expansion of the Universe led to the synthesis of substantial quantities of helium. The nucleosynthesis phase of the big bang began at time $t \sim 20$ s after the birth of the Universe, when deuterons (formed by fusion of protons and neutrons) became stable against photodestruction. Further fusion events led to the production of ^{3}H, ^{3}He and, ultimately, ^{4}He. The primordial abundance of helium relative to hydrogen was fixed by the relative numbers of neutrons and protons at the commencement of nucleosynthesis (approximately 1/6), as free neutrons are captured and processed to ^{4}He on a timescale short compared with their decay half-life of about 10 min. Models predict values of the helium mass fraction Y_{P} in the range 0.22–0.26 (dependent on nucleon density). Values determined from observations of emission nebulae in our Galaxy ($Y_{\mathrm{P}} \approx 0.23$; Peimbert 1992) and other galaxies ($Y_{\mathrm{P}} \approx 0.245$; Boesgaard and Steigman 1985) are in good agreement. Appreciable abundances of primordial deuterium (D/H $\sim 2 \times 10^{-5}$) are also predicted and observed. This consistency between theory and observations is a corner stone of the big-bang model for the origin of the Universe.

Heavier elements are not produced in the big bang (except for trace amounts of ^{7}Li with abundance Li/H $\sim 10^{-10}$). This conclusion is insensitive to the assumed density during the nucleogenesis phase and thus to whether the Universe is predicted to be open or closed. It arises because there are no stable elements of atomic mass 5 or 8: ^{4}He nuclei (α-particles) cannot form stable products by capture of protons or other α-particles. The production of carbon requires the fusion of three α-particles and densities much higher than those occurring during the nucleosynthesis phase of the big bang are required for this to occur at a significant rate. The rates of all fusion reactions drop rapidly to negligible levels as the density of the Universe falls due to expansion and, after $t \sim 10$ min, no further synthesis is likely to occur until the first generations of stars are born.

2.1.3 Stellar nucleosynthesis

The interaction of atomic nuclei is governed by two opposing forces: the strong nuclear interaction, which is attractive but has a very short range; and electrostatic repulsion caused by the fact that they bear positive charge. It is the former that leads to nuclear fusion and that binds nuclei together. For two nuclei to become sufficiently close for fusion to occur, they must approach each other at high speed. The rate at which a given fusion reaction proceeds is critically dependent on both temperature and density. As we consider successive phases of nucleosynthesis in stars and the production of progressively more massive nuclei, the required temperature increases with the nuclear charge of the 'target' element.

The first nuclear reactions to occur in a newly formed star involve deuterium and the rare lithium group of elements (Li, Be and B), which are converted to ^{4}He at temperatures $T \sim 10^6$ K. Hydrogen-burning, the staple source of energy for stars during their main-sequence lifetime, commences at $T \geq 10^7$ K. This process may be summarized by the equation

$$4\,^1\mathrm{H} \rightarrow \,^4\mathrm{He} + 2\mathrm{e}^+ + 2\nu \qquad (2.1)$$

i.e. four protons are converted into an α-particle, two positrons and two neutrinos. The chains of reactions that may lead to this conversion are diverse and include two distinct processes (e.g. Tayler 1975). Of these, the proton–proton (pp) chain requires no raw materials heavier than H and can thus operate in stars condensed from primordial material. The CNO cycle requires the pre-existence of ^{12}C (from helium-burning) and can therefore only operate in stars condensed from matter that has been enriched by earlier generations of stars. The complete cycle consists of six steps:

$$\begin{aligned} ^{12}\mathrm{C} + \mathrm{p} &\rightarrow \,^{13}\mathrm{N} + \gamma \\ ^{13}\mathrm{N} &\rightarrow \,^{13}\mathrm{C} + \mathrm{e}^+ + \nu \\ ^{13}\mathrm{C} + \mathrm{p} &\rightarrow \,^{14}\mathrm{N} + \gamma \\ ^{14}\mathrm{N} + \mathrm{p} &\rightarrow \,^{15}\mathrm{O} + \gamma \\ ^{15}\mathrm{O} &\rightarrow \,^{15}\mathrm{N} + \mathrm{e}^+ + \nu \\ ^{15}\mathrm{N} + \mathrm{p} &\rightarrow \,^{4}\mathrm{He} + \,^{12}\mathrm{C}. \end{aligned} \qquad (2.2)$$

As ^{12}C is both consumed and produced, it can be regarded as a catalyst.

Helium-burning is the first crucial stage in the production of elements heavier than those already available from the big bang. This occurs in the cores of stars in which the central supply of H has been exhausted; the central temperature of the star rises due to contraction caused by the resulting imbalance of gravitational and thermodynamic forces, whilst the envelope expands and cools as the star evolves away from the main sequence. The core typically contains $\sim$10% of the star's mass. When a core temperature $T \geq 2 \times 10^8$ K is reached, helium nuclei may fuse:

$$^4\mathrm{He} + \,^4\mathrm{He} \rightleftharpoons \,^8\mathrm{Be} + \gamma \qquad (2.3)$$

but ^{8}Be is unstable and rapidly dissociates back to helium. Carbon production requires the quasi-simultaneous fusion of three He nuclei (the triple-α process):

$$3\,^4\mathrm{He} \rightarrow \,^{12}\mathrm{C} + \gamma. \qquad (2.4)$$

This reaction occurs primarily through an excited level of carbon with quantum mechanical properties that enhance its rate (i.e. a resonance). But for this enhancement, helium could not burn and no heavier elements could form (e.g. Trimble 1997). Reaction (2.4) may be followed by the capture of a further α-particle to produce oxygen:

$$^{12}\mathrm{C} + \,^4\mathrm{He} \rightarrow \,^{16}\mathrm{O} + \gamma. \qquad (2.5)$$

Thus, the net result of helium-burning is a mixture of carbon and oxygen, their relative abundance depending on the core temperature of the star.

Nitrogen, the fifth most abundant element, is not the end-product of any of the principal burning reaction sequences. The main source of nitrogen in the Universe appears to be the CNO cycle: a hydrogen-burning star that contains ^{12}C will undergo rapid $C \rightarrow N$ conversion by means of the first three reactions in the sequence (2.2). Although nitrogen is subsequently consumed, its abundance under equilibrium remains high because of its low proton capture cross section and hence the slow rate at which the fourth reaction in the cycle proceeds.

As the helium-burning phase of an evolving star draws to a close, exhaustion of the He fuel results in a further increase in core temperature as the star again attempts to establish equilibrium between gravitational and thermodynamic forces. Its ultimate fate depends critically on its mass: if it is sufficiently massive, temperatures are reached at which carbon (5×10^8 K) and oxygen (10^9 K) are ignited and a new phase of nucleosynthesis begins. The critical stellar mass for C and O ignition is essentially the same as the critical mass that separates stars destined to become supernovae from those destined to become white dwarfs (Trimble 1991). C and O burning leads to the production of elements with atomic weights in the range 20–32. The primary C-burning reactions are:

$$
\begin{aligned}
{}^{12}\mathrm{C} + {}^{12}\mathrm{C} &\rightarrow {}^{24}\mathrm{Mg} + \gamma \\
&\rightarrow {}^{23}\mathrm{Na} + \mathrm{p} \\
&\rightarrow {}^{20}\mathrm{Ne} + {}^{4}\mathrm{He}.
\end{aligned} \tag{2.6}
$$

These may be followed by

$$
\begin{aligned}
{}^{20}\mathrm{Ne} + {}^{4}\mathrm{He} &\rightarrow {}^{24}\mathrm{Mg} + \gamma \\
{}^{24}\mathrm{Mg} + {}^{4}\mathrm{He} &\rightarrow {}^{28}\mathrm{Si} + \gamma
\end{aligned} \tag{2.7}
$$

a sequence sometimes referred to as neon-burning. The primary oxygen-burning reactions are:

$$
\begin{aligned}
{}^{16}\mathrm{O} + {}^{16}\mathrm{O} &\rightarrow {}^{32}\mathrm{S} + \gamma \\
&\rightarrow {}^{31}\mathrm{P} + \mathrm{p} \\
&\rightarrow {}^{31}\mathrm{S} + \mathrm{n} \\
&\rightarrow {}^{28}\mathrm{Si} + {}^{4}\mathrm{He}.
\end{aligned} \tag{2.8}
$$

The existence of a number of competing reactions is a common feature of nucleosynthesis involving relatively massive nuclei. The protons, neutrons and α-particles released by reactions (2.6) and (2.8) rapidly undergo further processing and, typically, the main product of carbon- and oxygen-burning is ^{28}Si, a particularly strongly bound nucleus.

Following the production of nuclei in the vicinity of silicon in the periodic table, the final phase of the stellar nucleosynthesis chain is the production of elements in the vicinity of iron and nickel. A logical route would be the direct association of two ^{28}Si nuclei to form ^{56}Ni (which will subsequently β-decay to ^{56}Fe); but in reality the situation is more complex. When temperatures $T > 2 \times 10^9$ K are reached, ambient thermal photons have sufficient energy ($h\nu > 2 \times 10^5$ eV) to remove protons, neutrons and α-particles from heavy nuclei and these are rapidly captured by other nuclei to form a range of products. Destructive and constructive reactions thus operate in parallel and the equilibrium abundance of any given element will depend on its binding energy. The binding energy per nucleon is greatest for elements in the region of Fe and so their abundances build up in the core of the star.

The fusion reactions discussed so far are *exothermic*: there is a net release of energy, as each successive compound nucleus is more tightly bound than its parent nuclei. Because Fe has the greatest binding energy per nucleon, there are no exothermic reactions that can utilize it to form still heavier elements. The production of elements beyond the Fe–Ni group in the Periodic Table is thought to depend on neutron capture reactions. Free neutrons are by-products of some fusion processes (e.g. the third reaction in (2.8)) and they are readily produced by photodissociation of nuclei at the highest core temperatures reached in massive stars. Capture of a neutron results in a unit increase in atomic mass. As the neutron has no charge, there is no electrostatic potential barrier to overcome. The resultant nucleus is generally unstable to β-decay, leading to a unit increase in atomic number. A specific example is the production of ^{59}Co from ^{58}Fe:

$$\begin{aligned} {}^{58}\mathrm{Fe} + \mathrm{n} &\rightarrow {}^{59}\mathrm{Fe} \\ {}^{59}\mathrm{Fe} &\rightarrow {}^{59}\mathrm{Co} + \mathrm{e}^- + \bar{\nu}. \end{aligned} \tag{2.9}$$

Neutron capture is usually a slow process as the number density of free neutrons is normally low, but in some circumstances, such as a supernova explosion, neutrons are generated very rapidly. In this situation, the mean free time between n-captures may be similar to, or less than, the decay half-life and a nucleus may undergo several captures before decaying to a stable form. Isotopes produced by slow neutron capture (the 's-process') tend to have relatively large numbers of protons in their nuclei, whereas rapid neutron capture (the 'r-process') leads to isotopes rich in neutrons. See Trimble (1991) for more detailed discussion.

2.1.4 Enrichment of the interstellar medium

The sequential production of heavy elements by the exothermic fusion reactions (2.1)–(2.8) discussed earlier proceeds in massive ($M > 8$ M$_\odot$) stars until an iron-rich core is produced. The structure of such a star is onion-like, with the core surrounded by successive shells bearing the products of previous burning cycles. Despite the internal ferment, the outermost layers may remain hydrogen-rich. Such a star is destined to undergo core collapse and become a type II supernova. As no further energetically favourable nuclear reactions can occur in the core, its temperature rises as it contracts until the ambient photon field is sufficiently energetic to cause photodestruction of the Fe nuclei to α-particles and neutrons, *absorbing* energy and leading to catastrophic implosion of the core to form a neutron star. The gravitational energy thus released ejects the outer layers in a supernova explosion. Ironically, the immediate prelude to the collapse of the core is thus a reversal of the previous cycle of energy-releasing nuclear reactions, for which the debt is paid by gravity. Supernovae are quintessential sources of heavy elements in the ISM, their expanding remnants containing both the ashes of previous burning cycles and the products of r-process evolution in the supernovae themselves.

A major contribution to the elemental enrichment of the ISM comes also from stars of intermediate mass ($1 < M < 8\,M_\odot$), which are more numerous than high-mass stars, evolve more slowly and lose mass copiously during the red giant and asymptotic giant branch phases of their evolution. Nucleosynthesis in such stars does not progress as far as the silicon-burning phase: they do not develop iron-rich cores and become supernovae but evolve into white dwarfs, often with the ejection of their outer layers to form a planetary nebula. Red-giant winds and planetary nebulae are not only important contributors to the element enrichment of the ISM but also likely sources of dust, as discussed in detail in chapter 7.

Heat is transported to the surface of a normal star by convective currents. In a main-sequence star like the Sun, the convective layer is relatively shallow (no more than $\sim$30% of its radius). The products of nucleosynthesis reach the surface only if enriched material leaks into the convective zone or if the outer layers are stripped off. The structure of an evolving star is determined primarily by its age and mass (although it may also be influenced by a number of other factors, including pulsational instability and, in the case of close binaries, tidal effects and mass exchange). A single star on the asymptotic giant branch has a very compact, degenerate C-rich or O-rich core, surrounded by thin He-burning and H-burning shells and a deep, fully convective envelope (see, for example, Shu 1982). Temporary instabilities may lead to the episodic establishment of convection in the shell zone, resulting in transport of C, the product of He-burning, to the surface, a process referred to as 'dredge-up'. Many red-giant atmospheres appear to be enriched in this way. If the C abundance is enhanced to the extent that it exceeds that of O, this has a profound effect on the chemistry of the stellar atmosphere and on the composition of solid condensates in the stellar wind. Whereas O-rich stars produce silicate dust, C-rich stars produce silicon carbide and amorphous carbon (see chapter 7).

2.2 The Solar System abundances

2.2.1 Significance and methodology

The element abundances in the Solar System provide a reference set which is invaluable in astrophysics, both as a test for models of nucleosynthesis and as a basis for comparison with other regions of the Universe[2]. The Solar System is the natural choice for this purpose because abundances may be determined more accurately for more elements than is the case for any other sample.

Information is obtained in two ways: by spectroscopic analysis of the solar photosphere; and by laboratory analysis of meteorites. The solar atmosphere is likely to contain a representative cross section of virtually all the elements present at the time of its formation. Hydrogen-burning in the core will have no effect on the abundances available to measurement, for reasons discussed in the previous section. Results are thus expected to reflect abundances in the original cloud (the solar nebula) from which the sun and planets formed (section 9.2.3). The crusts of accessible planetary bodies (the Earth and Moon) have been modified by processes such as gravitational fractionation and loss of volatiles and they do not therefore provide reliable constraints. The most appropriate solids available for laboratory analysis are the C-type meteorites (carbonaceous chondrites; see Cronin and Chang 1993 for a review of their properties). These objects are thought to be fragments of primitive asteroids: they have a granular structure, suggestive of formation by an accretion process, and are rich in hydrous minerals, organic molecules and carbon. Isotopic abundance anomalies show that they contain some grains of pre-solar origin (section 7.2.4). It is, therefore, reasonable to regard the C-type meteorites as relatively pristine samples of protoplanetary material from the early Solar System. The time elapsed since the epoch of condensation is determined rather precisely by radiometric dating techniques to be 4.57 ± 0.03 Gyr (Kirsten 1978), which is the generally accepted value for the age of the Solar System.

2.2.2 Results

The Solar System abundances are thus based on two types of measurement: remote sensing of the solar atmosphere; and laboratory studies of selected meteorites. Results discussed in this chapter are taken from the compilation of Anders and Grevesse (1989), updated for some elements by Grevesse and Noels (1993) and Grevesse and Sauval (1998). Note that solar abundances are generally expressed relative to hydrogen, whereas meteoritic abundances are more naturally expressed relative to a condensible element such as silicon. Calibration of solar and meteoritic results is achieved by taking the average of the meteoritic-to-solar abundance ratios for refractory elements believed to be fully condensed in the meteorites (Anders and Grevesse 1989). The correlation of solar and meteoritic abundances, shown in figure 2.2, is good for most elements. The lightest elements tend to be less abundant in meteorites compared with the Sun: this is the case for the HCNO group (figure 2.2) and for noble gases such as He and Ne, which are only trace constituents of meteorites. These relatively volatile elements naturally tend to remain in the gas unless chemically bonded into condensible

[2] The Solar System abundances are often described in the literature as 'cosmic' abundances.

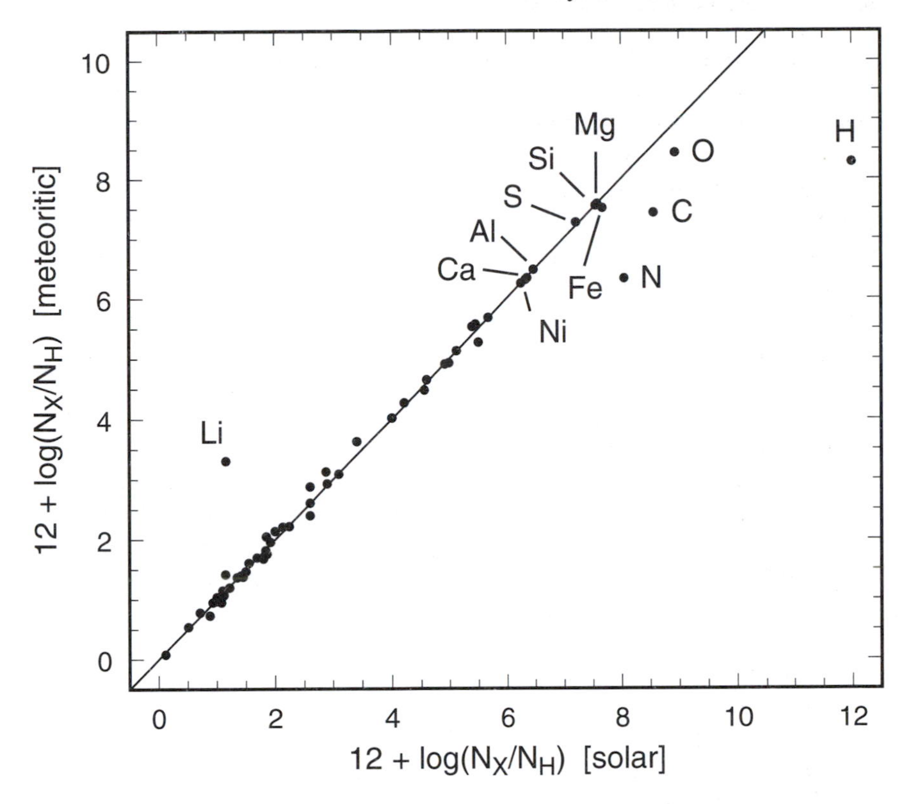

Figure 2.2. Plot of element abundances by number in the solar atmosphere versus those in carbonaceous chondrites (based on data from Anders and Grevesse 1989 and references therein). Meteoritic abundances have been renormalized to $\log N(\mathrm{H}) = 12$. Noble gases are excluded. The straight line represents exact agreement. Several elements discussed in the text are labelled.

compounds. This cannot, of course, occur for the noble gases but elements from the HCNO group are partially condensed into, for example, hydrated silicate and carbonate minerals, organic matter and solid carbon. The solar abundance is taken to be appropriate for these elements. One element plotted in figure 2.2 that is significantly anomalous in the opposite sense, i.e. less abundant in the Sun compared with the meteoritic value, is lithium; as discussed in section 2.1, Li is easily destroyed in stars and its general rarity in the photospheres of the Sun and other main sequence stars implies that this process is operating in material which is being transported by convection currents to and from the surface. In this case, the meteoritic value is more likely to represent the true initial abundance.

Figure 2.3 plots mean Solar System abundances against atomic number z for elements in the range $1 \leq z \leq 83$. The general trend is a steady decline from the very high abundances of the lightest elements, H and He, to the low

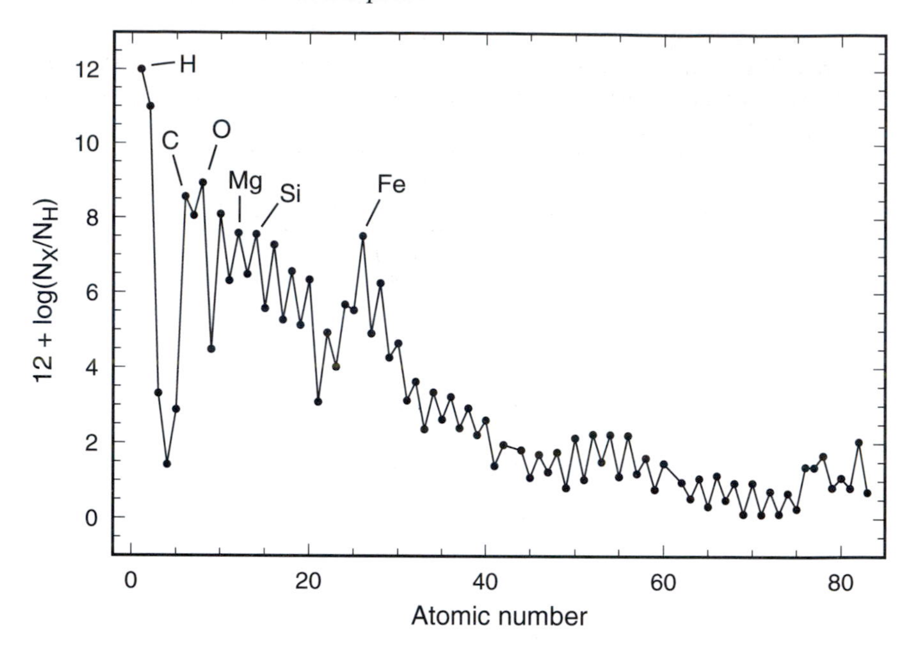

Figure 2.3. Plot of mean abundances by number in the Solar System against atomic number.

abundances of the elements with $z > 30$, with a total range of ~12 dex. Structure in the curve supports the paradigm that the heavy elements present in the Solar System are the products of nucleosynthesis within previous generations of stars (section 2.1.3). The trough at z-values 3–5 (lithium group) reflects the intrinsic fragility of these elements. Peaks occur for nuclei composed of integral numbers of α-particles (^{12}C, ^{16}O, ^{20}Ne, ^{24}Mg, ^{28}Si, etc) and the prominent iron (Fe) peak centred at $z = 26$ represents the build-up of elements at the end-point of exothermic nucleosynthesis.

Astrophysical abundances are traditionally expressed on a logarithmic scale relative to $N_{\rm H} = 10^{12}$, as in figures 2.2 and 2.3: i.e. for element X,

$$\log A(\mathrm{X}) = 12 + \log\left\{\frac{N_{\rm X}}{N_{\rm H}}\right\}. \tag{2.10}$$

The arbitrary constant 12 in equation (2.10) is merely a mathematical convenience, as the logarithmic abundance is then a positive number for even the rarest chemical elements (see figure 2.3). However, it is often more convenient to express the abundances of the more common heavy elements on a linear scale relative to $N_{\rm H} = 10^6$, i.e. in parts per million (ppm) relative to hydrogen:

$$A(\mathrm{X}) = 10^6\left\{\frac{N_{\rm X}}{N_{\rm H}}\right\}. \tag{2.11}$$

Table 2.1. The Solar System abundances of the 14 most abundant chemical elements likely to be present in interstellar dust. Values are listed by number relative to hydrogen in logarithmic and linear form.

Element	z	m $(\mathrm{g\,mol^{-1}})$	$\log A_\odot$ $(N_\mathrm{H} = 10^{12})$	$A_\odot$ (ppm)
H	1	1.01	12.00	10^6
C	6	12.01	8.56	360
N	7	14.01	7.97	93
O	8	16.00	8.83	676
Na	11	22.99	6.31	2
Mg	12	24.31	7.59	39
Al	13	26.98	6.48	3
Si	14	28.09	7.55	35
P	15	30.97	5.57	0.4
S	16	32.06	7.27	19
Ca	20	40.08	6.34	2
Cr	24	52.00	5.68	0.5
Fe	26	55.85	7.51	32
Ni	28	58.71	6.25	2

Table 2.1 lists solar abundances[3] in both formats for the most common elements (excluding noble gases), together with atomic number (z) and atomic weight (m). All isotopes are summed for individual elements. The experimental uncertainties are typically ± 0.04 dex in $\log A$. Note, however, that appreciable systematic errors may exist for some important elements, notably oxygen. The O abundance of 8.83 adopted here (from Grevesse and Sauval 1998) may be compared with 8.93 (Anders and Grevesse 1989) and 8.74 (Holweger 2001).

Interstellar dust is expected to be made up almost entirely of the elements listed in table 2.1, in one form or another. The mass fraction of heavy elements available to make dust may be estimated from the data by calculating

$$Z_\odot = 0.71 \sum \left\{ \frac{m_\mathrm{X} N_\mathrm{X}}{m_\mathrm{H} N_\mathrm{H}} \right\}_\odot \simeq 0.016 \qquad (2.12)$$

where the summation has been carried out over all elements from C to Ni in table 2.1 and the factor 0.71 allows for the contributions of the noble gases (primarily He and Ne) to the total mass. Although 13 elements heavier than H are included, the sum is dominated by only a few (principally O and C, followed by N, Mg, Si, Fe and S). The contributions of metals such as Al, Ca and Na are rather small ($\sim 1\%$) and those of all the rarer elements omitted from table 2.1 are

[3] For convenience, from this point on, the term 'solar abundances' will be used to mean 'Solar System abundances'.

entirely negligible. Note that the calculation of $Z_\odot$ in equation (2.12) assumes that H remains in the gas; however, H can make a minor contribution to the dust mass, e.g. by its presence in ices, organics or hydrated minerals. A mean hydrogenation factor of 2 (i.e. an average of two H atoms bonded to each heavier atom) would increase $Z_\odot$ to about 0.018. This is effectively an upper limit on the dust-to-gas ratio for solar abundances.

2.3 Abundance trends in the Galaxy

It was noted in section 2.2.1 that abundances measured in the Sun's photosphere are likely to be representative of those in the original cloud (the solar nebula) from which it condensed. Similarly, when we consider other stars, it is generally assumed that abundances measured in photospheric spectra broadly reflect initial composition: based on our knowledge of stellar structure and evolution, the products of internal nucleosynthesis are not usually mixed with the outer layers of the star. This axiom provides a basis for the investigation of systematic abundance trends in our Galaxy. There are obvious exceptions to the general rule, such as carbon stars, helium stars, peculiar metal-rich stars, Wolf–Rayet stars, etc, which are easily recognized. Note, however, that more subtle intrinsic variations can occur and these may be difficult to detect. For example, abundances in the atmosphere of a newly formed star may not match those in the local ISM if gas and dust are somehow segregated during the condensation process, e.g. as the result of differential drag or radiation pressure effects (chapter 9).

When comparing solar abundances with those measured in other stars, it is important to distinguish between temporal and spatial effects. Heavy-element abundances are expected to increase steadily with time as the Galaxy evolves; spatial variations will also occur if some regions of the Galaxy evolve more rapidly than others.

2.3.1 Temporal variation

The Galaxy has become progressively enriched (or contaminated!) with heavy elements over time, as the astration process (figure 2.1) has cycled through successive generations of stars. Each generation forms from material with a slightly higher average metal content than the previous one. As the age of the Sun ($\sim$4.6 Gyr) is roughly a third of the age of the Galaxy, we might expect to see clear differences when we compare abundances in the Sun with those in much older or much younger stars. This prediction is readily confirmed for the oldest stars, such as members of globular clusters formed some 15 Gyr ago, where metallicities are typically below solar values by factors of up to 100.

Early-type stars (spectral types O and B) have condensed from interstellar clouds recently, in astrophysical terms, i.e. within the past 100 Myr, and their atmospheres should thus provide a reliable guide to 'current' abundance levels in the ISM. This expectation is supported by the fact that abundances in OB stars are broadly consistent with those in H II regions (Gies and Lambert 1992). However, a detailed comparison with solar abundances (Savage and Sembach 1996, Snow and Witt 1996, Gummersbach *et al* 1998) leads to an unexpected result: the abundances of several heavy elements in OB stars appear to be *subsolar* by ~0.2 dex (i.e. ~63% of solar). If this discrepancy is real, it challenges our understanding of galactic evolution. Perhaps OB stars are systematically metal-poor compared with the present-day ISM (Sofia and Meyer 2001) or perhaps the Sun is unusually metal rich for a star of its age?

Whereas OB stars ideally provide a 'snap-shot' of abundances in the recent ISM, stars of lower mass (and longer main-sequence lifetimes) may be used to trace the development of heavy-element enrichment over galactic history (Twarog 1980, Edvardsson *et al* 1993, Gonzalez 1999). Metallicities for such stars are conveniently determined using the Strömgren (1966, 1987) technique based on narrow-band photometry and their ages are estimated with reference to theoretical isochrones. The logarithmic abundance of element X relative to its solar abundance is given by

$$\left[\frac{\mathrm{X}}{\mathrm{H}}\right] = \log\left\{\frac{N_{\mathrm{X}}}{N_{\mathrm{H}}}\right\} - \log\left\{\frac{N_{\mathrm{X}}}{N_{\mathrm{H}}}\right\}_{\odot} \tag{2.13}$$

and we may represent the metallicity by [Fe/H], determined from the Strömgren metallicity parameter[4]. Figure 2.4 plots [Fe/H] against age for several groups of stars (binned according to age). The expected trend of increasing stellar metallicity with decreasing stellar age is evident. Linear extrapolation to age zero predicts a present-day value for the metallicity of the ISM that is consistent with solar to within considerable uncertainty. The Sun's metallicity is enhanced relative to the average for main-sequence stars of similar age by about 0.2 dex.

2.3.2 Spatial variation

Spatial abundance variations may be investigated by studying objects of similar age and different location in the Galaxy. Large-scale variations are most manifest as a trend of decreasing heavy-element abundances with increasing galactocentric distance (R_{G}). This trend is most clearly seen in young objects from the galactic-disc population (OB stars and their H II regions; e.g. Gummersbach *et al* 1998, Shaver *et al* 1983). To a first approximation, logarithmic abundances are found to scale linearly with R_{G}: as an example, the case of Si is shown in figure 2.5. Similar correlations are seen for other elements, including C, N, O, Mg and Al, with gradients typically of order -0.08 dex kpc^{-1}. Extrapolation to $R_{\mathrm{G}} = 0$ suggests a mean heavy-element enrichment of 0.6 dex at the galactic centre compared with the current location of the Sun ($R_{\mathrm{G}} \approx 7.7$ kpc: Reid 1989), which corresponds to a fourfold linear increase. The implication of this result is that

[4] Fe is merely taken as a representative metal here; the Strömgren parameter is a photometric measure of average absorption line strengths for several metals.

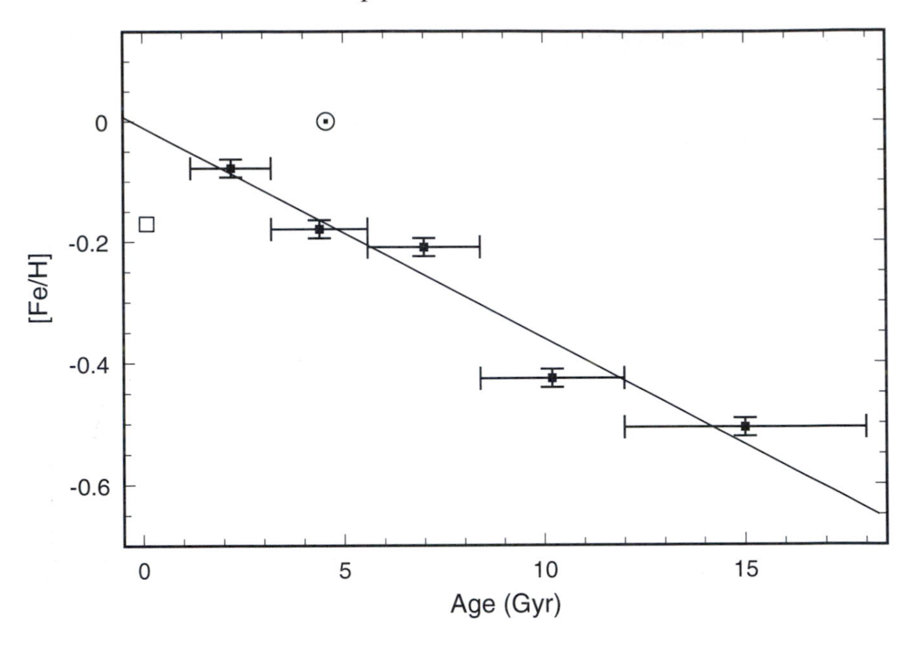

Figure 2.4. Temporal variation in heavy-element enrichment of the interstellar medium. Metallicity [Fe/H] is plotted against age for a total of 174 stars of intermediate spectral type (F, G, K) in the solar neighbourhood of the Galaxy, binned into five age ranges represented by the horizontal error bars. Data are from Gonzalez (1999) and references therein. The vertical error bars are standard errors in the mean. A linear least-squares fit to these points is shown. The Sun (⊙) and the mean for OB stars (square; Savage and Sembach 1996) are also plotted for comparison.

the nuclear region of the Galaxy has reached greater maturity compared with the outer arms (Wannier 1989) because of a more rapid turnover of material through the cosmic cycle (figure 2.1). Observations of external systems suggest that this is a common characteristic of spiral galaxies (Pagel and Edmunds 1981).

An increase in metallicity is naturally expected to lead to an increase in the dust-to-gas ratio, as more heavy elements are available to condense into solid particles in stellar outflows and to attach themselves to existing grain surfaces in the ISM itself. The observed trend of increasing metallicity towards the galactic centre is in qualitative agreement with the increase in the rate of extinction compared with the solar neighbourhood (section 1.3.2). Issa *et al* (1990) compared the spatial distributions of metallicity and dust-to-gas ratio in the Milky Way and in nearby external galaxies and showed that these quantities are, indeed, correlated. This is illustrated in figure 2.6.

Systematic changes in heavy-element abundances may lead to corresponding changes in the quality as well as the quantity of dust in the interstellar medium. The relative number densities of C-rich and O-rich red giants are sensitive to

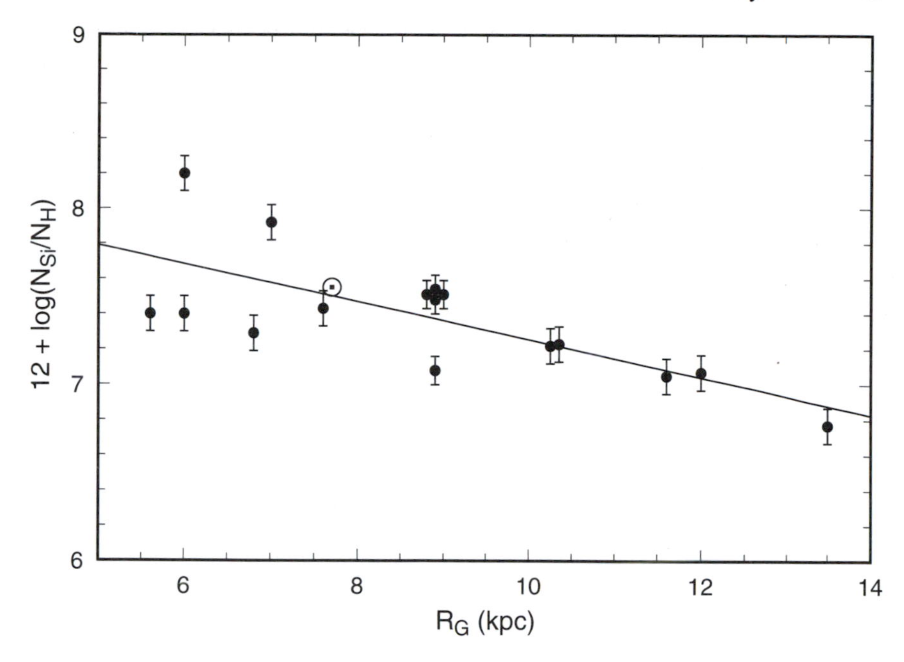

Figure 2.5. Spatial variation in heavy-element enrichment of the interstellar medium. The silicon abundance is plotted against galactocentric distance for B-type stars, using data from Gummersbach *et al* (1998) (points with error bars). The position of the Sun ($\odot$) is also shown. The diagonal line is the linear least-squares fit to the B-type stars.

their initial metallicities: as the natural excess of O over C is enhanced at high metallicity, a greater quantity of C must be dredged up to the surface to produce a C star with $N(\mathrm{C}) > N(\mathrm{O})$. Observational estimates of space densities confirm this (Thronson *et al* 1987). Thus, the ejection rates for carbonaceous and O-rich dust are predicted to vary with R_{G}, the latter dominating near the galactic centre.

2.3.3 Solar abundances in space and time

Do Solar System abundances provide an appropriate model for the composition of the local ISM? In considering this question, we should bear in mind that the currently available solar values are not definitive and may be revised. Nevertheless, results discussed earlier (section 2.3.1) strongly suggest that the Sun has an enhanced heavy-element endowment compared with both stars of its age group and young OB stars. This remarkable result might have anthropic significance, as it seems to be a common characteristic of stars with planets (Gonzalez 1999). Proposed explanations include the possibility that the Sun has migrated from a birth site some 2 kpc closer to the galactic centre than its present location or that the solar nebula was enriched by ejecta from a supernova explosion.

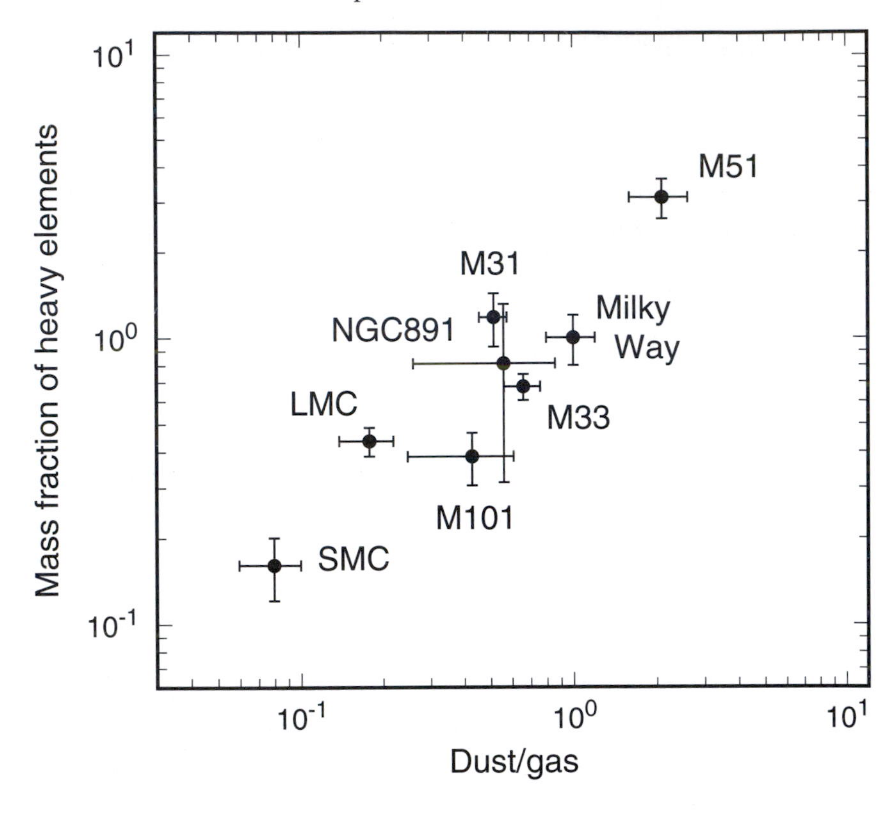

Figure 2.6. Plot of the mass fraction of heavy elements against dust-to-gas ratio for nearby galaxies. Each quantity is evaluated at a galactocentric distance equivalent to that of the Sun and normalized to our Galaxy. (Data from Issa *et al* 1990 and Alton *et al* 2000a.)

On the basis of the correlation shown in figure 2.4, the solar excess appears to be of the same order as the degree of galactic enrichment occurring over time in the past 4.6 Gyr, suggesting that the Sun might – by pure chance – be a reasonable standard for the current ISM. However, if this is so, an explanation must be sought for the lower metallicities of OB stars. The discrepancy between solar and OB-type stellar abundances has led some investigators to assume that the current ISM has subsolar metallicity. See Sofia and Meyer (2001) for further discussion of this important topic.

2.4 The observed depletions

2.4.1 Methods

The term depletion refers to the underabundance of a gas-phase element with respect to its standard reference abundance, resulting from its assumed presence in dust. The depletion index of element X is defined by analogy with equation (2.13):

$$D(\mathrm{X}) = \log\left\{\frac{N_\mathrm{X}}{N_\mathrm{H}}\right\} - \log\left\{\frac{N_\mathrm{X}}{N_\mathrm{H}}\right\}_\mathrm{ISM} \tag{2.14}$$

where the term on the right represents the standard reference abundance for the ISM (e.g. solar). Note that $D(\mathrm{X})$ becomes more negative for greater depletion and should never be positive if the standard is well chosen. It is also useful to define the fractional depletion:

$$\delta(\mathrm{X}) = 1 - 10^{D(\mathrm{X})} \tag{2.15}$$

which is bound by the limits $\delta(\mathrm{X}) = 0$ (all atoms in the gas) and $\delta(\mathrm{X}) = 1$ (all atoms in the dust). It is often convenient to express $\delta(\mathrm{X})$ as a percentage. The abundance of element X in the dust relative to total hydrogen (equation (2.11)) is then

$$A_\mathrm{dust} = \delta(\mathrm{X}) A_\mathrm{ISM} \tag{2.16}$$

where

$$A_\mathrm{ISM} = A_\mathrm{gas} + A_\mathrm{dust}. \tag{2.17}$$

The observed column densities N_X and N_H needed to determine the depletion (equation (2.14), first term on the right-hand side) are evaluated from analysis of interstellar absorption lines in stellar spectra using the curve of growth technique (e.g. Spitzer 1978). The column density of the absorber producing a weak (unsaturated) absorption line is

$$N_\mathrm{X} = \left(\frac{4\epsilon_0 m_\mathrm{e} c}{e^2}\right)\frac{W_\nu}{f} \tag{2.18}$$

where f is the oscillator strength of the transition, W_ν is the equivalent width in frequency units[5] and the various physical constants have their usual meaning. Equation (2.18) describes the linear region of the curve of growth, where $N_\mathrm{X} \propto W_\nu$: each absorbing atom along the line of sight sees essentially the full continuum level and an increase in N_X, the number of absorbers, would lead to a proportionate increase in the strength of the line. However, for stronger absorption (optical depth $\tau_\nu \geq 1$ at the line centre), the line becomes saturated and equation (2.18) underestimates the true column density. Column densities may be estimated from the curve of growth for saturated lines but results are inherently less accurate (typically by a factor of around five) compared with those deduced from unsaturated lines. Hence, the depletions of some elements are known to considerably greater precision than others.

Carbon is the most problematic of the elements expected to contribute significantly to the grain mass. As its first ionization potential (11.3 eV) is less

[5] Note that equivalent widths are usually expressed in wavelength units in the astronomical literature, where $W_\lambda = -(\lambda^2/c)W_\nu$.

than that of H I (13.6 eV), most of the available gas-phase carbon is in C II in H I clouds and in C I or CO in H_2 clouds. The abundance of C I is relatively easy to measure, that of C II much more difficult because the strong permitted resonance lines at 1036 and 1335 Å are invariably saturated (Jenkins *et al* 1983). The best available data on interstellar C II abundances come from observations of a very weak, unsaturated semi-forbidden C II line at 2325 Å (Hobbs *et al* 1982, Cardelli *et al* 1996, Sofia *et al* 1997). In contrast to the situation for carbon, oxygen (13.6 eV) and nitrogen (14.5 eV) have higher first ionization potentials than hydrogen and only the neutral species need be considered. Most metals have values in the range 5–8 eV and are usually singly ionized in interstellar clouds.

The gas-phase atomic lines used to evaluate abundances for most elements occur in the ultraviolet region of the spectrum and are thus accessible to observation only from space. High-resolution spectrometers on board the Copernicus satellite and the Hubble Space Telescope have provided a wealth of observational data (see Spitzer and Jenkins 1975, Jenkins 1987 and Savage and Sembach 1996 for reviews). Interstellar extinction places practical constraints on the lines of sight in which depletions can be investigated: suitable spectra are most readily obtained for lightly and moderately reddened stars ($E_{B-V} < 0.5$) and the environments sampled thus tend to be predominantly diffuse clouds and intercloud medium (although some data exist for lines of sight that contain appreciable molecular material). Most pathlengths studied are relatively short ($L < 2$ kpc) and galactic trends in metallicity (section 2.3.2) are not normally important. The environment sampled by a given line of sight can be characterized by the mean number density of hydrogen:

$$\langle n_{\mathrm{H}} \rangle = \frac{N(\mathrm{H\ I}) + 2N(\mathrm{H_2})}{L} \tag{2.19}$$

(Spitzer 1985). A column of low mean density, $\langle n_{\mathrm{H}} \rangle < 0.2 \times 10^6\ \mathrm{m^{-3}}$, is unlikely to intercept a cloud containing a cool phase. The presence of H I or H_2 clouds of various densities and filling factors along a line of sight (see section 1.4.2) elevate $\langle n_{\mathrm{H}} \rangle$ to values typically $\geq 10^6\ \mathrm{m^{-3}}$.

2.4.2 Average depletions in diffuse clouds

Gas-phase abundances and inferred depletions for various elements observed in diffuse clouds are listed in table 2.2. These are taken from the reviews by Jenkins (1987, 1989) and references therein, with updates from Cardelli *et al* (1996), Fitzpatrick (1996), Meyer *et al* (1997, 1998b) and Sofia *et al* (1994, 1997). Elements that show a strong correlation between depletion and density (section 2.4.3) have been standardized to a density of $\langle n_{\mathrm{H}} \rangle = 3 \times 10^6\ \mathrm{m^{-3}}$ (Jenkins 1987). Results are given for two values of the reference abundances: solar and 63% solar ($\log A_{\odot} - 0.2$ dex). The latter would be appropriate if abundances in B stars better represent the composition of the current ISM. The true ISM abundances seem likely to lie somewhere between these two extremes.

Table 2.2. Mean gas-phase abundances and depletions in diffuse clouds. For each element, the values of the depletion index (D), the fractional depletion (δ) and dust-phase abundance ($A_{\rm dust}$, ppm) are calculated for two values of the standard reference abundances. Note that with the lower reference abundances, D becomes marginally positive for two elements (N and S); in these cases, the depletion has been set to zero (values in brackets).

Element	$A_{\rm gas}$ (ppm)	(Standard ≡ Solar) D	δ	$A_{\rm dust}$	(Standard ≡ 63% Solar) D	δ	$A_{\rm dust}$
C	140	−0.41	0.61	220	−0.21	0.38	87
N	75	−0.09	0.19	17	(0)	(0)	(0)
O	320	−0.32	0.52	356	−0.12	0.24	106
Na	0.6	−0.50	0.68	1	−0.30	0.50	0.7
Mg	3.1	−1.10	0.92	36	−0.90	0.87	21
Al	0.01	−2.50	1.00	3	−2.30	0.99	2
Si	0.9	−1.60	0.97	34	−1.40	0.96	21
P	0.07	−0.74	0.82	0.3	−0.54	0.71	0.2
S	19	0.00	0.00	0	(0)	(0)	(0)
Ca	0.0005	−3.60	1.00	2	−3.40	1.00	1
Cr	0.04	−2.10	0.99	0.5	−1.90	0.99	0.3
Fe	0.32	−2.00	0.99	32	−1.80	0.98	20
Ni	0.01	−2.30	1.00	2	−2.10	0.99	1

The observed depletions correlate with condensation temperature, $T_{\rm C}$, as was first shown by Field (1974). For a given element, $T_{\rm C}$ is defined as the temperature at which 50% of the atoms condense into the solid phase in some form under thermodynamic equilibrium, assuming solar abundances (e.g. Savage and Sembach 1996). The plot of depletion index D against $T_{\rm C}$ (figure 2.7), known as the depletion pattern, provides a convenient means of displaying the depletions for the various elements and may have physical significance: for example, a correlation would be expected if grain formation occurs under equilibrium conditions in circumstellar shells around cool stars. The behaviour of the more volatile elements ($T_{\rm C} < 1000$ K) is quite distinct from that of the more refractory elements. The former show low depletions with no dependence on $T_{\rm C}$; indeed the data for N and S are consistent with *no* depletion. Depletions for the more refractory elements show a strong tendency to increase (i.e. D becomes more negative) with increasing $T_{\rm C}$. In percentage terms, the degree of depletion is almost 100% for the most refractory metals such as Al, Ca, Fe and Ni (independent of the choice of reference abundances). The correlation of D(X) with $T_{\rm C}$, although pronounced, shows scatter in excess of observational error: for example, Fe is generally more depleted than Mg although their condensation temperatures are similar.

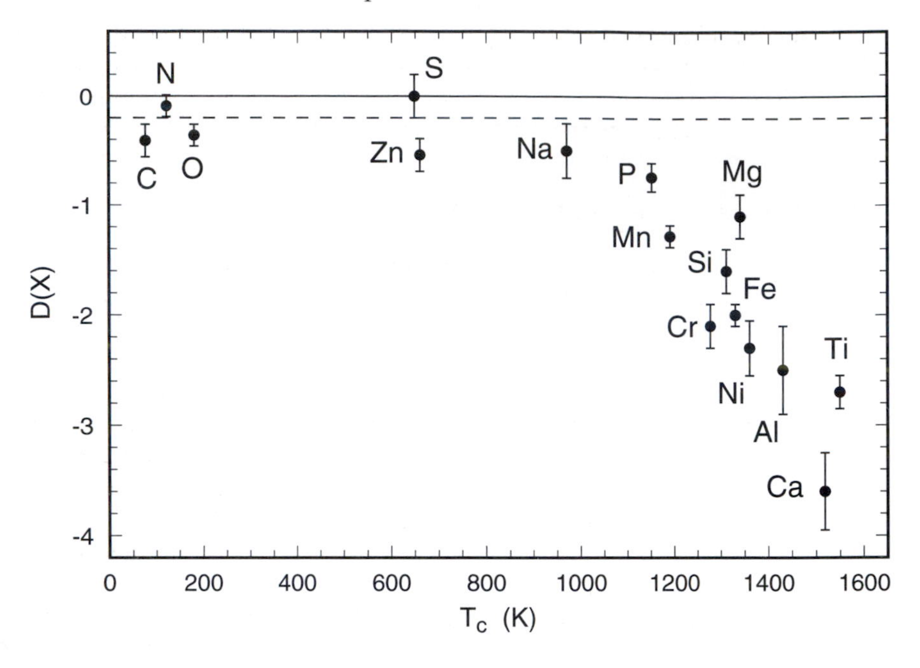

Figure 2.7. The depletion pattern for diffuse interstellar clouds. The mean depletion index $D(\mathrm{X})$ for solar reference abundances is plotted against condensation temperature T_C (K) for various elements. The full horizontal line at $D(\mathrm{X}) = 0$ represents solar abundances and the broken horizontal line at $D(\mathrm{X}) = -0.2$ represents mean OB star abundances.

The depletion results may be used to estimate the mass density, ρ_d, of material depleted into dust and the resulting dust-to-gas ratio, Z_d. The contribution of element X to ρ_d is

$$\rho_\mathrm{d}(\mathrm{X}) = 10^{-6} A_\mathrm{dust} \left(\frac{m_\mathrm{X}}{m_\mathrm{H}} \right) \rho_\mathrm{H} \tag{2.20}$$

where $\rho_\mathrm{H} \simeq 1.8 \times 10^{-21}$ kg m^{-3} (section 1.3.2) and A_dust (ppm) is the abundance of X in dust. Evaluating $\rho_\mathrm{d}(\mathrm{X})$ for each heavy element in table 2.2 and summing the results, we obtain

$$\rho_\mathrm{d} = \sum \rho_\mathrm{d}(\mathrm{X}) \approx 2.3 \times 10^{-23} \text{ kg m}^{-3} \tag{2.21}$$

and

$$Z_\mathrm{d} = 0.71 \frac{\rho_\mathrm{d}}{\rho_\mathrm{H}} \approx 0.009 \tag{2.22}$$

for solar standard abundances, where again the factor 0.71 allows for the contributions of the noble gases to the total mass of gas. Comparing the dust-to-gas ratio (equation (2.22)) with the total availability of condensible elements (equation (2.12)), the overall depletion is about 60% for solar abundances.

If the standard is set to 63% solar, the results in equations (2.21) and (2.22) are reduced substantially, by a factor of about 2.4 in each case, i.e. $\rho_d \approx 1.0 \times 10^{-23}$ kg m^{-3} and $Z_d \approx 0.004$. This difference arises primarily because of dramatically reduced contributions from O and C.

2.4.3 Dependence on environment

The results discussed in section 2.4.2 are representative of physical conditions in diffuse clouds. When depletions are considered over a range of physical conditions, a systematic trend of depletion with density is found for many elements: those exhibiting this trend include Mg, Si, P, Ca, Ti, Cr, Mn and Fe, whilst others (e.g. C, N, O, S, Zn) show little or no correlation (Harris *et al* 1984, Jenkins 1987, Sofia *et al* 1997). Figure 2.8 plots depletion index D versus mean density $\langle n_H \rangle$ for Ti, as an example of a highly correlated element. Although no more than a trace element in terms of its contribution to grain mass, Ti provides an excellent illustration of density-dependent depletion. The most likely interpretation of this result is that atoms are being exchanged between the gas and solid phases as a function of environment. Note, however, that the Ti depletion remains high even at the lowest densities for which data are available: $\sim$90% ($D \approx -1.0$) at $\langle n_H \rangle \sim 3 \times 10^4$ m^{-3} (figure 2.8). Other refractory elements (Cr, Mn, Fe) behave in a similar manner but for some (Mg, Si, P) the depletion becomes quite low (0–60%, dependent on the chosen standard) at the lowest densities (Jenkins 1987, Fitzpatrick 1996).

The correlation of depletion index with density may be stated mathematically for element X as

$$D(\mathrm{X}) = D_0(\mathrm{X}) + m \log(n/n_0) \tag{2.23}$$

where n represents $\langle n_H \rangle$, m is the slope of the correlation line and D_0 is the value of D at some reference density $n = n_0$. If the grains consist of separate 'refractory' and 'volatile' components, D_0 may be thought of as the 'base depletion' in the refractory material that remains after the more volatile fraction has evaporated at some sufficiently low density (e.g. $n_0 \simeq 3 \times 10^4$ m^{-3}; Jenkins 1987).

Although $\langle n_H \rangle$ is a reasonable parameter to use to delineate average physical conditions toward a given star, it has obvious limitations. For example, a line of sight in which most of the matter is in one or two compact, dense clouds and most of the volume is in the intercloud medium may have a similar value of $\langle n_H \rangle$ to another that is dominated by diffuse clouds. Fortunately, data from instruments such as the Goddard High-Resolution Spectrograph on board the Hubble Space Telescope are of sufficient quality to resolve individual cloud components (e.g. Savage and Sembach 1996). To take a well studied example, spectra of the star ζ Oph exhibit two Doppler components, at heliocentric velocities of -27 and -15 km s^{-1}, that contain warm neutral and cold molecular gas, respectively. The depletions measured in these two clouds show clear systematic differences, being

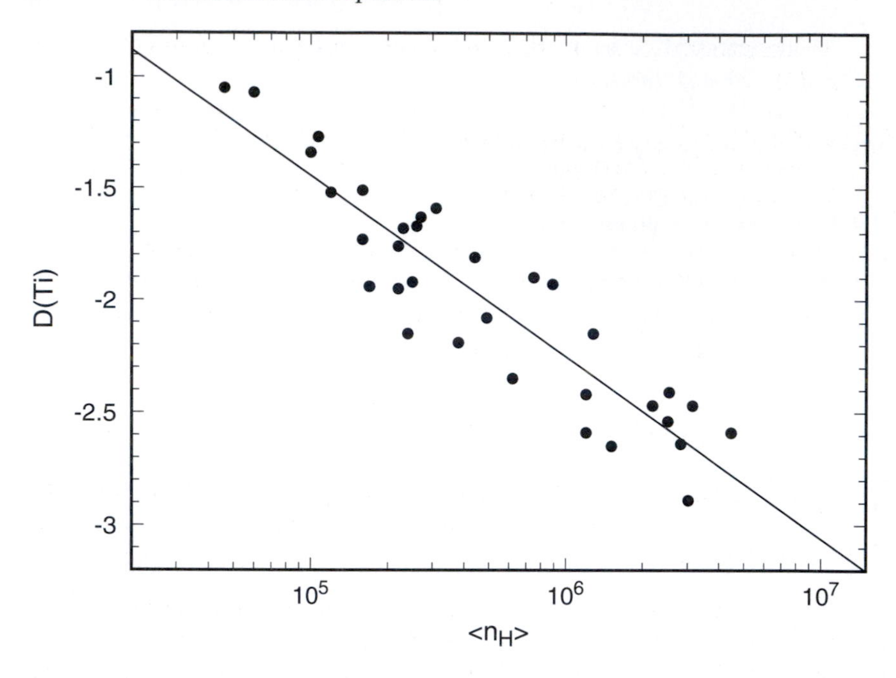

Figure 2.8. Correlation of the depletion index for titanium with mean hydrogen number density. The diagonal line is a least-squares fit of the form of equation (2.23); the correlation coefficient is 0.91. Data are from Stokes (1978); see also Harris *et al* (1984).

always higher in the cold, dense cloud. Moreover, the magnitude of the difference is greatest for the most depleted elements, such as Fe, Cr, Ni and Ti.

Element depletions are also sensitive to cloud velocity (Routly and Spitzer 1952, Shull *et al* 1977) and to vertical distance from the galactic plane (Spitzer and Fitzpatrick 1993, Sembach and Savage 1996, Fitzpatrick and Spitzer 1997). Diffuse high-velocity clouds, presumed to have been accelerated by shocks, display systematically lower depletions compared with their low-velocity counterparts. Hot, turbulent gas is ubiquitous in the outer disc and inner halo of the Galaxy and these regions also have depletions systematically lower than in a typical diffuse cloud in the galactic plane. The general pattern is clear: the harsher the environment, the higher the gas-phase abundances are for the most refractory elements.

2.4.4 Overview

The elements likely to be present in dust appear to fall into three distinct groups, in terms of their depletion properties:

Group I. These elements have low to moderate depletions (0–60%) that do not correlate strongly with physical conditions. This group includes C, N, O, S and Zn.

Group II. The depletions of these elements vary with density: they are high (80–100%) in diffuse clouds but can become quite low in the intercloud medium. This group includes Mg, Si and P.

Group III. The depletions of these elements also vary with density but they remain high (80–100%) even in the harshest environments. This group, which includes Fe, Ti, Ca, Cr, Mn and Ni, seems to represent an almost indestructible component of interstellar dust.

2.5 Implications for grain models

The observed depletions reviewed in the previous section provide constraints on models for the composition and evolution of interstellar dust grains. Important results to be considered include

(i) the correlation with condensation temperature for certain elements;
(ii) the strong dependence on environment for certain elements and
(iii) the general availability of depleted elements to explain other observational phenomena attributed to dust, such as interstellar extinction and spectral features.

The correlations with condensation temperature and with cloud density may be understood in terms of a general model for the origin, growth and destruction of interstellar dust (Field 1974, Dwek and Scalo 1980, Seab 1988, Tielens 1998, Dwek 1998). Such a model postulates that refractory grains (stardust) originating in stellar ejecta, such as supernova remnants and red-giant winds, are injected into the ISM. A range of compositions is expected, including silicates, metals, oxides and solid carbon (chapter 7). The grains subsequently cycle between the various phases of the ISM, where they may accumulate atoms from the gas or return atoms to the gas, depending on physical conditions. Adsorption increases with cloud density, whilst desorption is most rapid in the tenuous intercloud gas. One may thus envisage two types of grain material: an underlying component that remains in solid form and a variable component that migrates between gas and dust according to environment (see equation (2.23)). It is logical to associate these components with stardust cores and volatile mantles, respectively (Field 1974). However, energetic shocks will tend to destroy entire grains rather than merely resurface them. The underlying component is expected to contain only the most robust of grain materials.

How do the element groups (section 2.4.4) fit into this picture? Of the group I elements, one can discount N and S (along with Zn) as important constituents of the dust in the diffuse ISM. Interstellar grains do not seem to contain appreciable quantities of sulphates, sulphides, or N-bearing organic molecules. C and

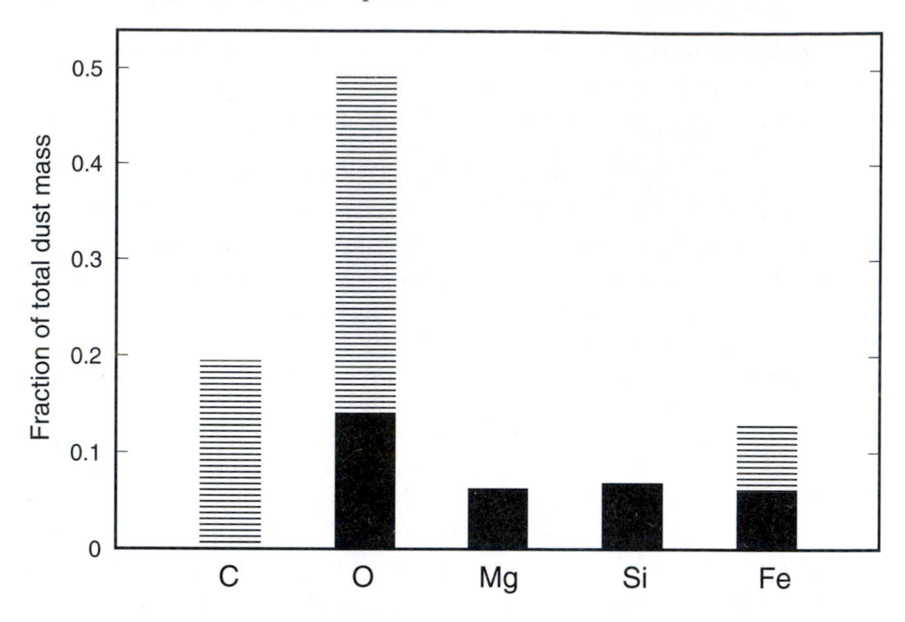

Figure 2.9. Bar chart showing the fractional masses of the five major elements depleted into dust in diffuse interstellar clouds, normalized to the total mass of depleted material. The ordinate represents the ratio $\rho_d(X)/\rho_d$ (equations (2.20) and (2.21)) for solar reference abundances (subsolar values tend to reduce the contributions of C and O relative to those of Mg, Si and Fe without affecting their rankings). Filled bars show the fraction of each element that may be tied up in olivine and pyroxene silicates with an Mg:Fe ratio of 5:2 (see text).

(especially) O, however, are so abundant that they dominate even when their depletions are not particularly high. As their gas-phase abundances do not correlate with density, any C and O atoms that reside in the migratory component of interstellar dust cannot be a large fraction of the total inventory of these elements. In contrast, the important group II elements, Mg and Si, clearly do reside primarily in the migratory component. Finally, the Group III elements, of which Fe is easily the most abundant, reside primarily in the underlying refractory component, with a minority in the migratory component. Fe atoms locked into refractory dust may originate in Type Ia supernovae (Tielens 1998). Some Fe is presumably tied up in silicates from red giants (see later) and some must enter the ISM via gaseous winds from hot stars that produce no dust at all (Jura 1987).

The fractional mass abundances of the 'big five' dust elements – C, O, Mg, Si and Fe – are displayed as a bar chart in figure 2.9. Between them, these elements account for ∼95% of the depleted mass. The optical properties of the dust, reviewed in the following chapters, give clear indications as to the likely chemical arrangement of these elements in the dust. Infrared spectroscopy (chapter 5) demonstrates that Si has a strong preference for O over C, i.e. it resides in silicates rather than silicon carbide. Silicates also naturally explain the requirement, arising from observations of scattering and polarization of starlight (chapters 3 and 4), that at least one component of the dust should be *dielectric* (non-absorbing) in character. Silicates that utilize the most abundant elements have the generic formulae $MSiO_3$ (pyroxene) and M_2SiO_4 (olivine), where M $\equiv$ Mg or Fe. Mg belongs to the same depletion group as Si, whereas Fe does not. It, therefore, seems reasonable to suppose that Mg is more abundant than Fe in interstellar silicates. A mixture of pyroxene and olivine with Mg:Fe $\approx$ 5:2 by number can accommodate all the depleted Mg and Si, as shown in figure 2.9. If this is a good model for interstellar silicates then a little under half of the available Fe resides in silicates. This conclusion is not sensitive to the choice of reference abundances as the relative masses of Mg, Si and Fe in dust are not much affected.

Oxygen accounts for $\sim$45% of the depleted mass and its dominance presents a dilemma. Only about a third of it is tied up in silicates (figure 2.9) for solar reference abundances. What form does the remaining oxygen take? Metal oxides are obvious candidates (Jones 1990, Sembach and Savage 1996), yet even if we assume that all the remaining metals are fully oxidized, we would still account for little more than half of the depleted oxygen. There seems little prospect of further grain constituents involving oxygen in the diffuse ISM: ices such as H_2O, CO_2, CO and O_2 cannot survive; and refractory organics that contain oxygen appear not to be common either (section 5.2.4). The problem is alleviated if subsolar reference abundances are assumed (Mathis 1996a); but it is hard to reconcile substantially subsolar abundances with the observed strength of the 9.7 μm silicate feature, requiring 80–100% of the solar Si abundance in silicates (section 5.2.2). The root of the problem may be a systematic error in the solar oxygen abundance, for which Holweger (2001) proposes a downward revision of about 0.1 dex.

Distinct populations of relatively big ('classical') grains and much smaller particles are needed to explain the extinction curve and other observed properties of interstellar dust (see section 1.6 for an overview and section 3.7 for specific examples). Although the big grains contain most of the mass, there are compelling reasons to suppose that it is the smallest grains that dominate the exchange of elements between gas and dust in the ISM. First, it is the small grains that provide most of the available *surface area*. The second reason arises from the electrostatic properties of the grains: whereas big grains tend to bear positive charge (due to electron loss via the photoelectric effect), small grains tend to be electrostatically neutral or weakly negative (due to electron capture). The accreting elements are predominantly singly ionized positive ions and will thus accrete preferentially onto negative or neutral substrates, i.e. the small grains. Weingartner and Draine (1999) consider the attachment of Fe to small C grains and conclude that as much as 60% of the available Fe might be accounted for in this way. Thus, it may be that much of the migratory population of Fe and other heavy metals is in the form of atoms that attach to and desorb from small C grains, rather than a component of Fe-rich silicate or oxide stardust that is destroyed and replenished.

Models for interstellar extinction (chapter 3) must explain the observed opacity per H atom and its wavelength dependence, without exceeding the quota of available elements set by the abundances and depletions. Mathis (1996a) describes a model that can accomplish this with $\sim$80% solar reference abundances, i.e. roughly intermediate between the two cases considered here, subject to certain assumptions regarding the composition and structure of the particles. Most of the mass is taken up by big grains composed of silicates, oxides and solid carbon, thus utilizing all of the major dust elements to some degree. The physical structure of these grains proves to be crucial to the model, as Mathis finds that the extinction per unit mass is optimized for porous aggregates containing $\sim$45% vacuum.

Recommended reading

- *Supernovae and Nucleosynthesis*, by David Arnett (Princeton University Press, 1996).
- *Abundances in the Interstellar Medium*, by T L Wilson and R T Rood, in Annual Reviews of Astronomy and Astrophysics, vol 32, pp 191–226 (1994).
- *Interstellar Abundances from Absorption-Line Observations with the Hubble Space Telescope*, by B D Savage and K R Sembach, in Annual Reviews of Astronomy and Astrophysics, vol 34, pp 279–329 (1996).
- *Dust Models with Tight Abundance Constraints*, by John S Mathis, in Astrophysical Journal, vol 472, pp 643–55 (1996).
- *Interstellar Abundance Standards Revisited*, by Ulysses J Sofia and David M Meyer, in Astrophysical Journal, vol 554, pp L221–4 (2001).

Problems

1. Nucleosynthesis in the big bang is initiated by a neutron capture reaction ($\mathrm{p} + \mathrm{n} \rightarrow \mathrm{d} + \gamma$), whereas nucleosynthesis in the first generation of stars is initiated by a proton capture reaction ($\mathrm{p} + \mathrm{p} \rightarrow \mathrm{d} + \mathrm{e}^{+} + \nu$). Give an explanation of this difference.
2. Why does nucleosynthesis leading to significant carbon production occur only in the cores of a helium-burning stars and not in the big bang?
3. Discuss the significance of the observed correlation between element abundances in the Sun's atmosphere and in carbonaceous chondrites (see figure 2.2). Explain why a few elements deviate from the general trend.
4. In studies of galactic chemical evolution based on spectral analysis of stellar atmospheres, it is generally assumed that the observed element abundances in a given star are closely similar to those in the interstellar medium from

which it originally formed. Discuss the arguments on which this assumption is based, noting any exceptional circumstances.

5. Explain what is meant by saturation in a spectral line. Comment on the effect on the estimated column density of wrongly assuming a saturated line to be unsaturated.
6. (a) The 2325 Å semi-forbidden line of C II has an equivalent width (in wavelength units) of 0.6 mÅ in the line of sight to the star δ Scorpii. Given that the oscillator strength of the transition is 6.7×10^{-8}, and assuming that the 2325 Å line is unsaturated, estimate the column density of C II towards this star.
   (b) Deduce the depletion index and the fractional depletion of carbon in the line of sight toward δ Sco, given that the total (atomic and molecular) hydrogen column density is observed to be $N_{\rm H} = 1.45 \times 10^{25}\ {\rm m}^{-2}$ and assuming that all of the available gas phase carbon is in C II. Compare your results with expected average values (e.g. figure 2.7 and table 2.2).
   (c) Comment briefly with reasoning on whether you think the assumption that all of the available gas-phase carbon is in C II is likely to be reasonable, given that the *molecular* hydrogen column density toward δ Sco is found to be $N({\rm H_2}) = 2.6 \times 10^{23}\ {\rm m}^{-2}$.
7. Biological organisms contain approximately 3% by mass of phosphorus (P). Assuming solar abundance data (table 2.2), estimate the contribution $\rho_{\rm d}$(P) of phosphorus to the mass density of dust (see equation (2.20)). Hence deduce the mass density of 'interstellar biota' if all depleted P were in this form. Express your answer as a fraction of the total dust density (equation (2.21)) in the solar neighbourhood.

Chapter 3

Extinction and scattering

"Further, there is the importance of getting an insight into the true spectrum of the stars, freed from the changes brought about by the medium traversed by light on its way to the observer."

J C Kapteyn (1909)

Extinction occurs whenever electromagnetic radiation propagates through a medium containing small particles. In general, the transmitted beam is reduced in intensity by two physical processes – absorption and scattering. The energy of an absorbed photon is converted into internal energy of the particle, which is thus heated, whilst a scattered photon is deflected from the line of sight. The spectrum as well as the total intensity of the radiation is modified. The spectral dependence of extinction, or extinction curve, is a function of the composition, structure and size distribution of the particles. Studies of extinction by interstellar dust are important because they provide information pertinent both to understanding the properties of the dust and to correcting for its presence.

In this chapter, we begin by outlining the theoretical basis for models of extinction and scattering (section 3.1) and the method for determining extinction curves from observational data (section 3.2). The relevant observations are described and discussed in sections 3.3–3.6, considering both continuum extinction and structure associated with discrete absorption features and their dependence on environment. Attempts to match observations with theory are reviewed in the final section (section 3.7). For the most part, this chapter is concerned with the spectral range from the ultraviolet to the near infrared (wavelengths 0.1 to 3 μm) over which interstellar extinction is well studied. Discussion of infrared absorption features is deferred to chapter 5.

3.1 Theoretical methods

3.1.1 Extinction by spherical particles

We begin by considering the optical properties of small spheres. As a representation of interstellar grains, this is obviously highly idealized: the polarization of starlight provides direct evidence that at least one component of the dust has anisotropic optical properties, a topic discussed in detail in chapter 4. However, spheres are a reasonable (and mathematically convenient) starting point, at least for situations where polarization is not being considered. Calculations indicate that the optical properties of spheroids are closely similar to those of spheres of equal volume when averaged over all orientation angles. Methods for calculating the extinction of more complex forms are reviewed in section 3.1.4.

Consider spheres of radius a, distributed with number density n_{d} per unit volume in a cylindrical column of length L and unit cross-sectional area along the line of sight from a distant star. The reduction in intensity of the starlight at a given wavelength resulting from the extinction produced in a discrete element of column with length $\mathrm{d}L$ is

$$\frac{\mathrm{d}I}{I} = -n_{\mathrm{d}} C_{\mathrm{ext}}\,\mathrm{d}L \tag{3.1}$$

where C_{ext} is the extinction cross section. Integrating equation (3.1) over the entire pathlength gives

$$I = I_0 \mathrm{e}^{-\tau} \tag{3.2}$$

where I_0 is the initial value of I (at $L = 0$) and

$$\begin{aligned} \tau &= \int n_{\mathrm{d}}\,\mathrm{d}L \cdot C_{\mathrm{ext}} \\ &= N_{\mathrm{d}} C_{\mathrm{ext}} \end{aligned} \tag{3.3}$$

is the optical depth of extinction caused by the dust. The quantity N_{d} in equation (3.3) is the column density of the dust, i.e. the total number of dust grains in the unit column. Expressing the intensity reduction in magnitudes, the total extinction at some wavelength λ is given by

$$\begin{aligned} A_\lambda &= -2.5 \log\left(\frac{I}{I_0}\right) \\ &= 1.086 N_{\mathrm{d}} C_{\mathrm{ext}} \end{aligned} \tag{3.4}$$

using equations (3.2) and (3.3). A_λ is more usually expressed in terms of the extinction efficiency factor Q_{ext}, given by the ratio of extinction cross section to geometric cross section:

$$Q_{\mathrm{ext}} = \frac{C_{\mathrm{ext}}}{\pi a^2}. \tag{3.5}$$

Hence,

$$A_\lambda = 1.086 N_{\mathrm{d}} \pi a^2 Q_{\mathrm{ext}}. \tag{3.6}$$

If, instead of grains of constant radius a, we have a size distribution such that $n(a)\,\mathrm{d}a$ is the number of grains per unit volume in the line of sight with radii in the range a to $a + \mathrm{d}a$, then equation (3.6) is replaced by

$$A_\lambda = 1.086\pi L \int a^2 Q_{\mathrm{ext}}(a) n(a)\,\mathrm{d}a. \tag{3.7}$$

The problem of evaluating the expected spectral dependence of extinction A_λ for a given grain model (with an assumed composition and size distribution) is essentially that of evaluating Q_{ext}. The extinction efficiency is the sum of corresponding efficiency factors for absorption and scattering,

$$Q_{\mathrm{ext}} = Q_{\mathrm{abs}} + Q_{\mathrm{sca}}. \tag{3.8}$$

These efficiencies are functions of two quantities, a dimensionless size parameter,

$$X = \frac{2\pi a}{\lambda} \tag{3.9}$$

and a composition parameter, the complex refractive index of the grain material,

$$m = n - \mathrm{i}k. \tag{3.10}$$

Q_{abs} and Q_{sca} may, in principle, be calculated for any assumed grain model and the resulting values of total extinction compared with observational data. The problem is that of solving Maxwell's equations with appropriate boundary conditions at the grain surface. A solution was first formulated by Mie (1908) and independently by Debye (1909), resulting in what is now known as the Mie theory. Excellent, detailed accounts of Mie theory and its applications appear in van de Hulst (1957) and Bohren and Huffman (1983), to which the reader is referred for further discussion.

To compute the extinction curve for an assumed grain constituent, the real and imaginary parts of the refractive index (equation (3.10)) must be specified. These quantities, n and k, somewhat misleadingly called the 'optical constants', are, in general, functions of wavelength. For pure dielectric materials ($k = 0$) the refractive index is represented empirically by the Cauchy formula

$$m = n \simeq c_1 + c_2\lambda^{-2} \tag{3.11}$$

where c_1 and c_2 are constants. In general, $c_1 \gg c_2$ and so n is only weakly dependent on λ for dielectrics. Ices and silicates are examples of astrophysically significant solids that behave approximately as dielectrics ($k < 0.1$) over much of the electromagnetic spectrum. For strongly absorbing materials such as metals, k is of the same order as n and both may vary strongly with wavelength.

Figures 3.1 and 3.2 illustrate the results of Mie theory calculations for weakly absorbing spherical grains of constant refractive index $1.5 - 0.05\mathrm{i}$. Values of efficiency factor Q_{ext} and its absorption and scattering components are plotted

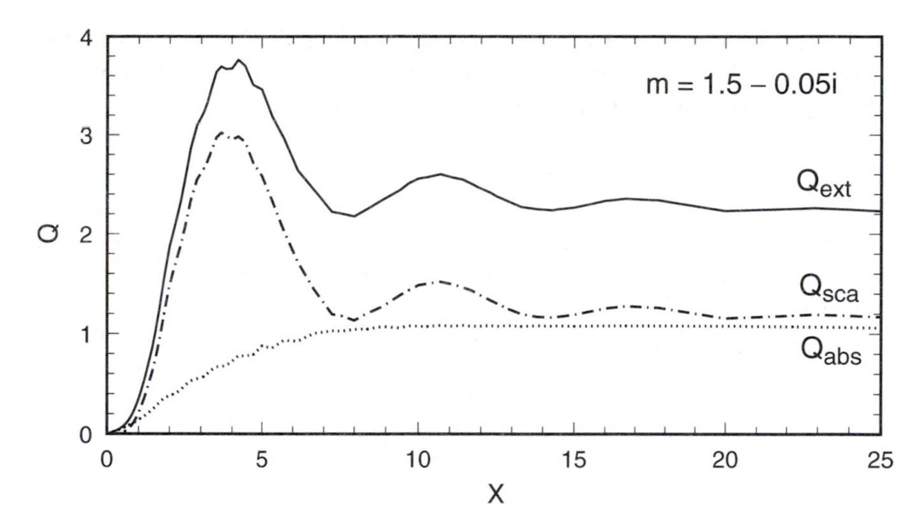

Figure 3.1. Results of Mie theory calculations for spherical grains of refractive index $m = 1.5 - 0.05\text{i}$. Efficiency factors Q_{ext}, Q_{sca} and Q_{abs} are plotted against the dimensionless size parameter $X = 2\pi a/\lambda$.

against $X = 2\pi a/\lambda$. Figure 3.2 is an enlargement of figure 3.1 near the origin. It is helpful to regard these plots in terms of the variation in extinction with λ^{-1} for constant grain radius (or with grain radius for constant wavelength). A few general features are evident. Q_{ext} increases monotonically with X for $0 < X < 4$. In this domain, extinction is dominated by scattering for the chosen refractive index and its magnitude is sensitive to the precise value of X. For $1 < X < 3$, Q_{ext} increases almost linearly with X. At higher values of X, peaks arise in the scattering component of Q_{ext} caused by resonances between wavelength and grain size and these will disappear when the contributions of grains with many different radii in a size distribution are summed (equation (3.7)). Q_{ext} becomes almost constant as X becomes large, indicating that the extinction is neutral (wavelength independent) for grains much larger than the wavelength.

3.1.2 Small-particle approximations

When $X \ll 1$ (i.e. the particles are small compared with the wavelength), useful approximations may be used to give simple expressions for the efficiency factors (see Bohren and Huffman 1983: chapter 5):

$$Q_{\text{sca}} \simeq \frac{8}{3}\left(\frac{2\pi a}{\lambda}\right)^4 \left|\frac{m^2 - 1}{m^2 + 2}\right|^2 \tag{3.12}$$

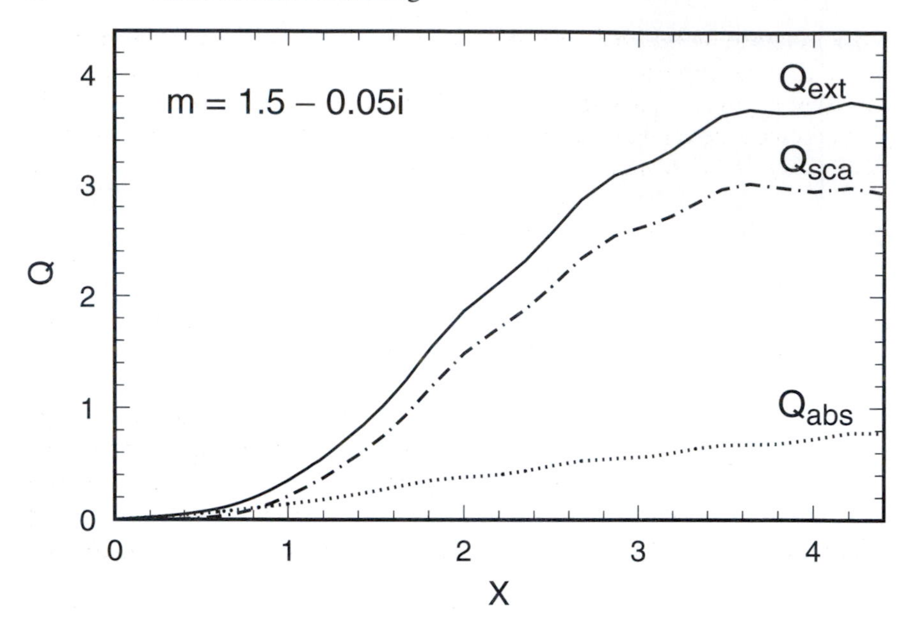

Figure 3.2. An enlargement of figure 3.1 showing the initial rise in extinction efficiency with X near the origin.

and

$$Q_{\mathrm{abs}} \simeq \frac{8\pi a}{\lambda} \operatorname{Im}\left\{\frac{m^2-1}{m^2+2}\right\}. \tag{3.13}$$

For pure dielectrics, m is real and almost constant with respect to wavelength, as discussed earlier. In this case, we have $Q_{\mathrm{sca}} \propto \lambda^{-4}$ and $Q_{\mathrm{abs}} = 0$, a situation termed Rayleigh scattering (Rayleigh 1871, Bohren and Huffman 1983). More generally, the quantity $(m^2-1)/(m^2+2)$ is often only weakly dependent on wavelength for materials that are not strongly absorbing, in which case $Q_{\mathrm{sca}} \propto \lambda^{-4}$ and $Q_{\mathrm{abs}} \propto \lambda^{-1}$ to good approximations. The wavelength dependence of extinction may thus be quite different for small particles in which either absorption or scattering is dominant.

3.1.3 Albedo, scattering function and asymmetry parameter

Quantities describing the scattering properties of the grains may be calculated from Mie theory. The albedo is defined

$$\alpha = \frac{Q_{\mathrm{sca}}}{Q_{\mathrm{ext}}} \tag{3.14}$$

and is bounded by the limits $0 \le \alpha \le 1$ (since $0 \le Q_{\rm sca} \le Q_{\rm ext}$), the extreme values representing perfect absorbers and pure dielectrics, respectively. More generally, a grain model based on a mix of absorbing and dielectric particles will predict some form for the dependence of α on wavelength. Note that the efficiency factors $Q_{\rm sca}$ and $Q_{\rm abs}$ may be written in terms of α:

$$\begin{aligned} Q_{\rm sca} &= \alpha Q_{\rm ext} \\ Q_{\rm abs} &= (1-\alpha) Q_{\rm ext}. \end{aligned} \tag{3.15}$$

It follows that if the extinction and albedo can be determined observationally over the same spectral range, the contributions of absorption and scattering may be easily calculated.

The scattering function $S(\theta)$ describes the angular redistribution of light upon scattering by a dust grain. It is defined such that, for light of incident intensity I_0, the intensity of light scattered into unit solid angle about the direction at angle θ to the direction of propagation of the incident beam is $I_0 S(\theta)$ (assuming axial symmetry). The scattering cross section, defined as $C_{\rm sca} = \pi a^2 Q_{\rm sca}$ by analogy with equation (3.5), is related to $S(\theta)$ by

$$C_{\rm sca} = 2\pi \int_0^\pi S(\theta) \sin\theta \, {\rm d}\theta. \tag{3.16}$$

The asymmetry parameter is defined as the mean value of $\cos\theta$ weighted with respect to $S(\theta)$:

$$\begin{aligned} g(\theta) &= \langle \cos\theta \rangle \\ &= \frac{\int_0^\pi S(\theta) \sin\theta \cos\theta \, {\rm d}\theta}{\int_0^\pi S(\theta) \sin\theta \, {\rm d}\theta} \\ &= \frac{2\pi}{C_{\rm sca}} \int_0^\pi S(\theta) \sin\theta \cos\theta \, {\rm d}\theta. \end{aligned} \tag{3.17}$$

Calculations for dielectric spheres show that $g(\theta) \simeq 0$ in the small particle limit, which corresponds to spherically symmetric scattering, whereas $0 < g(\theta) < 1$ for larger particles, indicating forward-directed scattering. As the ratio of grain diameter to wavelength increases from 0.3 to 1.0, a range of particular interest for studies of interstellar dust, the value of $g(\theta)$ increases from 0.15 to 0.75. Hence, the asymmetry parameter is a sensitive function of grain size (Witt 1989).

3.1.4 Composite grains

Spherical grains are considered in the preceding discussion but we noted at the outset that this is no more than a convenient generalization. Although sphericity might be a reasonable approximation in many situations, the growth and destruction processes that occur in the ISM (chapter 8) seem likely to result in grains with complex shapes and structures, such as porous aggregates composed of many subunits. How can the extinction produced by such particles be calculated? Classical Mie-type solutions are possible only for certain special cases, such as long cylinders, oblate or prolate spheroids and concentric core/mantle particles (e.g. Greenberg 1968). Two techniques have been developed. The discrete dipole approximation (DDA), first proposed by Purcell and Pennypacker (1973), represents a composite grain of arbitrary shape as an array of dipole elements: each dipole has an oscillating polarization in response to both incident radiation and the electric fields of the other dipoles in the array and the superposition of dipole polarizations leads to extinction and scattering cross sections. Examples of DDA calculations are described by (e.g.) Draine (1988), Bazell and Dwek (1990) and Fogel and Leung (1998); results generally agree well with Mie calculations in special cases where the two can be directly compared. An alternative approach, less rigorous but also less demanding on computer time, is effective medium theory (EMT), in which the optical properties of a collection of small particles are approximated by a single averaged optical constant and Mie-type calculations are then applied (e.g. Mathis and Whiffen 1989). The reader is referred to Wolff *et al* (1994, 1998) for further discussion of these techniques and their applicability to interstellar dust. A comparison between DDA calculations for composite grains and Mie calculations for spheres of the same composition and volume is reported by Fogel and Leung (1998). They find that the composite grains generally have larger extinction cross sections, thus requiring less grain material to produce a given opacity.

3.2 Observational technique

We next consider the problem of determining extinction curves observationally. The most reliable and widely used technique involves the 'pairing' of stars of identical spectral type and luminosity class but unequal reddening and determining their colour difference. The apparent magnitude of each star as a function of wavelength may be written:

$$\begin{aligned} m_1(\lambda) &= M_1(\lambda) + 5\log d_1 + A_1(\lambda) \\ m_2(\lambda) &= M_2(\lambda) + 5\log d_2 + A_2(\lambda) \end{aligned} \tag{3.18}$$

where M, d and A represent absolute magnitude, distance and total extinction, respectively (see section 1.2) and subscripts 1 and 2 denote 'reddened' and 'comparison' stars. The intrinsic spectral energy distribution, represented by $M(\lambda)$, is expected to be closely similar or identical for stars of the same spectral classification, thus we may assume $M_1(\lambda) = M_2(\lambda)$. If $A(\lambda) = A_1(\lambda) \gg A_2(\lambda)$, i.e. the extinction toward star 2 is negligible compared with that toward star 1, then the magnitude difference $\Delta m(\lambda) = m_1(\lambda) - m_2(\lambda)$ reduces to

$$\Delta m(\lambda) = 5\log\left(\frac{d_1}{d_2}\right) + A(\lambda). \tag{3.19}$$

The first term on the right-hand side of equation (3.19) is independent of wavelength and constant for a given pair of stars. Hence, the quantity $\Delta m(\lambda)$ may be used to represent $A(\lambda)$. The constant may be eliminated by means of normalization with respect to two standard wavelengths λ_1 and λ_2:

$$E_{\text{norm}} = \frac{\Delta m(\lambda) - \Delta m(\lambda_2)}{\Delta m(\lambda_1) - \Delta m(\lambda_2)} = \frac{A(\lambda) - A(\lambda_2)}{A(\lambda_1) - A(\lambda_2)} = \frac{E(\lambda - \lambda_2)}{E(\lambda_1 - \lambda_2)} \tag{3.20}$$

where $E(\lambda_1 - \lambda_2)$, the difference in extinction between the specified wavelengths, is equal to the colour excess (see section 1.2, equation (1.3)). The normalized extinction E_{norm} should be independent of stellar parameters and determined purely by the extinction properties of the interstellar medium. In practical terms, normalization is helpful because it allows extinction curves for different reddened stars to be superposed and compared: effectively, the degree of reddening in the line of sight is standardized (e.g. to $E_{B-V} = 1$; see section 3.3.1) and the slope of the curve between the standard wavelengths becomes a constant, irrespective of the number of dust grains in the line of sight. Theoretical extinction curves deduced from equation (3.7) for a given grain model may be normalized in the same way to allow direct comparison between observations and theory.

Observational data used to determine the interstellar extinction curve include broadband photometry and low-resolution spectrometry. Application of the pair method is particularly straightforward for broadband measurements in standard passbands such as the Johnson system, as unreddened (intrinsic) colours have been set up with respect to spectral type and there is no need to observe comparison stars. Extinction curves are commonly normalized with respect to the B and V passbands in the Johnson system, i.e. the normalized extinction (equation (3.20)) becomes $E_{\lambda-V}/E_{B-V}$. However, broadband photometry alone provides little information on structure in the curve. When spectrophotometry of resolution $\Delta\lambda < 50$ Å is used, the matching of individual stellar spectral lines in the reddened and comparison stars becomes important. Early-type stars (spectral classes O–A0) are generally selected for such investigations as their spectra are simpler and thus easier to match, compared with late-type stars; and their intrinsic luminosity and frequent spatial association with dusty regions also render them most suitable for probing interstellar extinction at optical and ultraviolet wavelengths. As an example of the application of the pair method, figure 3.3 plots ultraviolet spectra for a matching pair of stars observed by the International Ultraviolet Explorer satellite. Note the cancellation of stellar spectral lines to produce a relatively smooth extinction curve. A broad absorption feature centred near 4.6 μm^{-1} is conspicuous in the reddened star and weak or absent in the comparison star, resulting in a prominent peak in extinction.

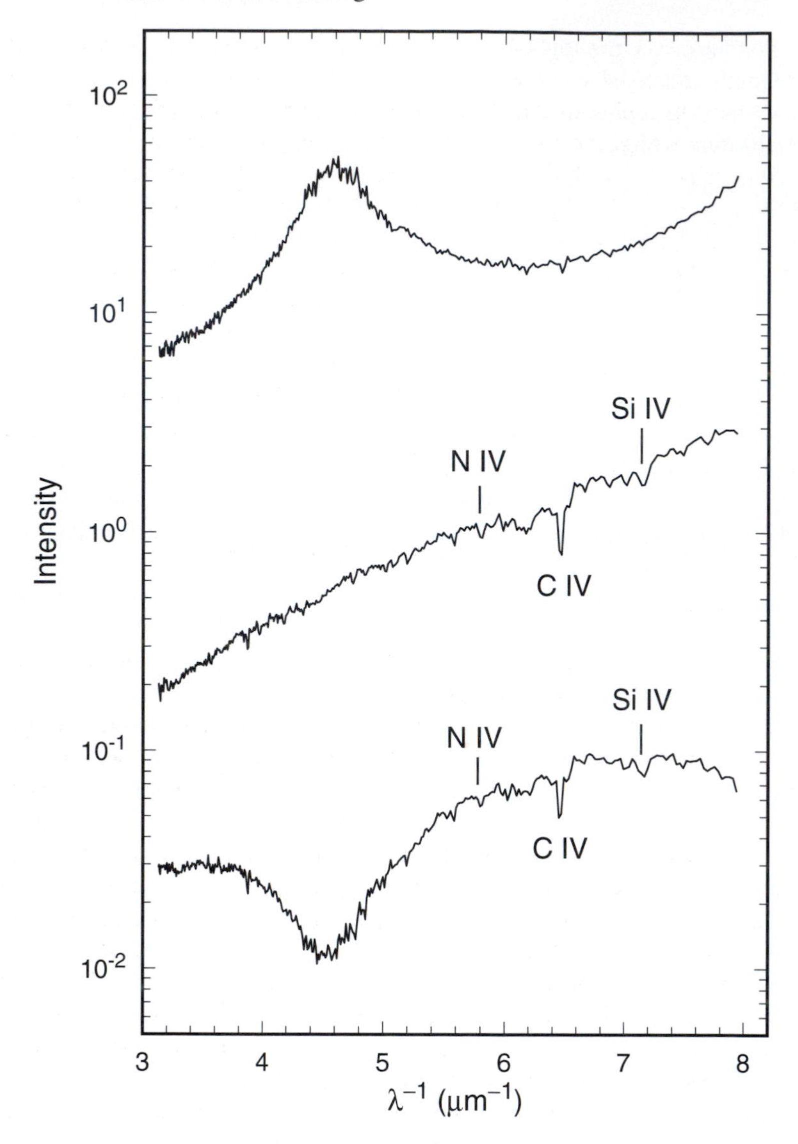

Figure 3.3. An illustration of the pair method for determining interstellar extinction curves. The lower curve is the ultraviolet spectrum of a reddened star (HD 34078, spectral type O9.5 V, $E_{B-V} = 0.54$) and the middle curve is the corresponding spectrum of an almost unreddened star of the same spectral type (HD 38666, $E_{B-V} = 0.03$). A few representative spectral lines are labelled. The vertical axis plots intensity in arbitrary units – note that the scale is logarithmic and hence equivalent to magnitude. The upper curve is the resulting extinction curve, obtained by taking the intensity ratio (equivalent to magnitude difference) of the two spectra. Based on data from the IUE *Atlas of Low-Dispersion Spectra* (Heck *et al* 1984).

Two difficulties associated with the pair method should be noted. First, there is a scarcity of suitable comparison stars for detailed spectrophotometry, as relatively few OB stars are close enough to the Sun or at high enough galactic latitude to have negligible reddening. The problem is most acute for supergiants and these are often excluded from studies of extinction in the ultraviolet, where mismatches in spectral line strengths can be particularly troublesome. Second, many early-type stars have infrared excess emission, due to thermal re-radiation from circumstellar dust or free–free emission from ionized gas. If one attempts to derive the extinction curve for such an object by comparing it with a normal star or with normal intrinsic colours, the derived extinction curve will be distorted in the spectral bands at which significant emission occurs. Stars with hot gaseous envelopes (shell stars) can usually be identified spectroscopically by the presence of optical emission lines, usually denoted by the suffix 'e' in the spectral classification; these should generally be avoided.

3.3 The average extinction curve and albedo

3.3.1 The average extinction curve

Reliable data on the wavelength dependence of extinction are available in the spectral region 0.1–5 μm. Studies of large samples of stars have shown that the extinction curve takes the same general form in many lines of sight. Regional variations are evident, particularly in the blue to ultraviolet, which will be discussed in section 3.4, but the average extinction curve for many stars provides a valuable benchmark for comparison with curves deduced for individual stars and regions and a basis for modelling. Table 3.1 lists values of mean normalized extinction at various wavelengths. The data represent a synthesis of previous literature, taken from reviews by Savage and Mathis (1979), Seaton (1979) and Whittet (1988). A correction has been applied to remove spurious structure near 1600 Å (6.3 μm^{-1}), which arose due to mismatched stellar C IV lines (Massa *et al* 1983) in part of the data set used by Savage and Mathis (1979). The stars included in the mean are reddened predominantly by diffuse clouds in the solar neighbourhood of the Milky Way, within 2–3 kpc of the Sun.

The values of extinction presented in table 3.1 make use of standard normalizations. The relative extinction (replacing the labels λ_1 and λ_2 in equation (3.20) with B and V) is

$$\frac{E_{\lambda-V}}{E_{B-V}} = \frac{A_\lambda - A_V}{E_{B-V}} = R_V \left\{ \frac{A_\lambda}{A_V} - 1 \right\}. \tag{3.21}$$

Thus, the *absolute* extinction A_λ / A_V may be deduced from the relative extinction if $R_V = A_V / E_{B-V}$, the ratio of total-to-selective extinction, is known. The

Table 3.1. The average interstellar extinction curve at various wavelengths in standard normalizations. Letters in square brackets denote standard photometric passbands. Values of the coefficients $a(x)$ and $b(x)$ are also listed (see section 3.4.3).

λ (μm)	λ^{-1} (μm^{-1})	$\frac{E_{\lambda-V}}{E_{B-V}}$	$\frac{A_\lambda}{A_V}$	$a(x)$	$b(x)$
∞	0	−3.05	0.00	0.000	0.000
4.8 [M]	0.21	−2.98	0.02	0.046	−0.043
3.5 [L]	0.29	−2.93	0.04	0.078	−0.072
2.22 [K]	0.45	−2.77	0.09	0.159	−0.146
1.65 [H]	0.61	−2.58	0.15	0.230	−0.243
1.25 [J]	0.80	−2.25	0.26	0.401	−0.398
0.90 [I]	1.11	−1.60	0.48	0.679	−0.623
0.70 [R]	1.43	−0.78	0.74	0.869	−0.366
0.55 [V]	1.82	0.00	1.00	1.000	0.000
0.44 [B]	2.27	1.00	1.33	1.000	1.000
0.40	2.50	1.30	1.43	0.978	1.480
0.36 [U]	2.78	1.60	1.52	0.953	1.909
0.344	2.91	1.80	1.59	0.870	2.333
0.303	3.30	2.36	1.77	0.646	3.639
0.274	3.65	3.10	2.02	0.457	4.873
0.25	4.00	4.19	2.37	0.278	6.388
0.24	4.17	4.90	2.61	0.201	7.370
0.23	4.35	5.77	2.89	0.122	8.439
0.219	4.57	6.47	3.12	0.012	9.793
0.21	4.76	6.23	3.04	−0.050	9.865
0.20	5.00	5.52	2.81	−0.059	8.995
0.19	5.26	4.90	2.61	−0.061	8.303
0.18	5.56	4.65	2.52	−0.096	8.109
0.17	5.88	4.57	2.50	−0.164	8.293
0.16	6.25	4.70	2.54	−0.250	8.714
0.149	6.71	5.00	2.64	−0.435	9.660
0.139	7.18	5.39	2.77	−0.655	10.810
0.125	8.00	6.55	3.15	−1.073	13.670
0.118	8.50	7.45	3.44	−1.362	15.740
0.111	9.00	8.45	3.77	−1.634	17.880
0.105	9.50	9.80	4.21	−1.943	20.370
0.100	10.00	11.30	4.70	−2.341	23.500

values of A_λ/A_V in table 3.1 are deduced for a value of $R_V = 3.05$ (see section 3.3.3).

The average extinction curve, plotted in figure 3.4, shows a number of distinctive features. It is almost linear in the visible from 1 to 2 μm^{-1}, with changes in slope in the blue near 2.2 μm^{-1} (the 'knee') and in the infrared

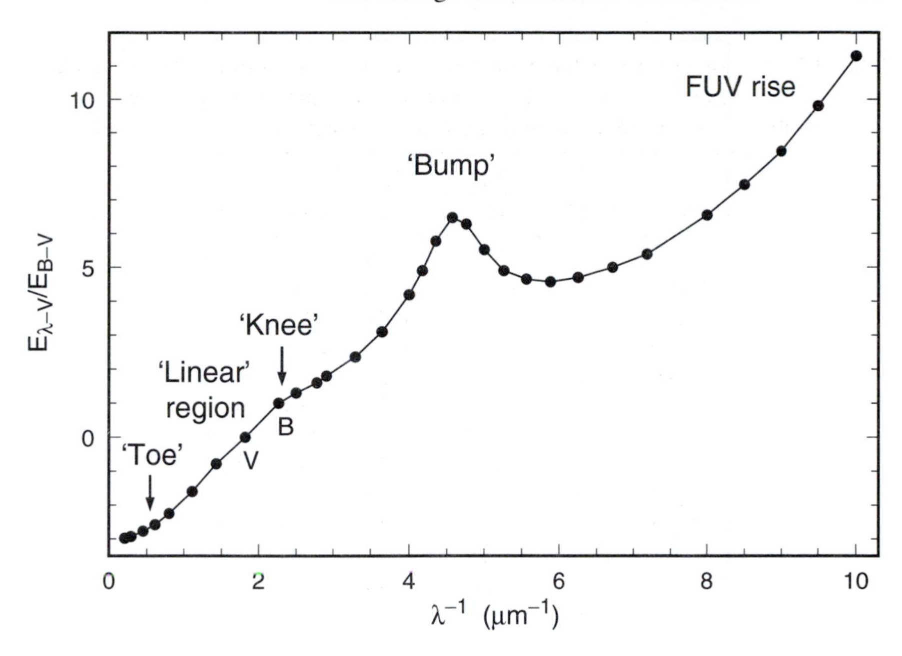

Figure 3.4. The average interstellar extinction curve ($E_{\lambda-V}/E_{B-V}$ versus λ^{-1}) in the spectral range 0.2–10 μm^{-1}. Data are from table 3.1. Various features of the curve discussed in the text are labelled. The positions of the *B* and *V* passbands selected for normalization are also indicated.

near 0.8 μm^{-1} (the 'toe'). This section of the curve resembles the dependence of Q_{ext} on λ^{-1} for a single grain size (figure 3.2) and for a refractive index $m \simeq 1.5 - 0.05i$, we deduce from equation (3.9) and figure 3.2 that Mie calculations for grains of radius $a \simeq 0.2$ μm would roughly reproduce its form. At shorter wavelengths, this comparison breaks down. The most prominent characteristic of the observed extinction curve is a broad, symmetric peak in the mid-ultraviolet centred at ~ 4.6 μm^{-1}: this is the 2175 Å 'bump', discussed in detail in section 3.5. Beyond the bump, a trough occurs near $\sim$6 μm^{-1}, followed by a steep rise into the far-ultraviolet (FUV, $\lambda^{-1} > 6$ μm^{-1}).

3.3.2 Scattering characteristics

The scattering properties of the grains may be investigated by observations of the diffuse galactic light (DGL), reflection nebulae and x-ray halos. Of all the phenomena that contribute to our understanding of interstellar grains, DGL is perhaps the most difficult to observe (because of its intrinsic faintness and the numerous sources of contamination) and also to analyse. The spectral dependence of the DGL in the satellite ultraviolet has been investigated in detail by Lillie and Witt (1976) and Morgan *et al* (1978) and additional optical data from the

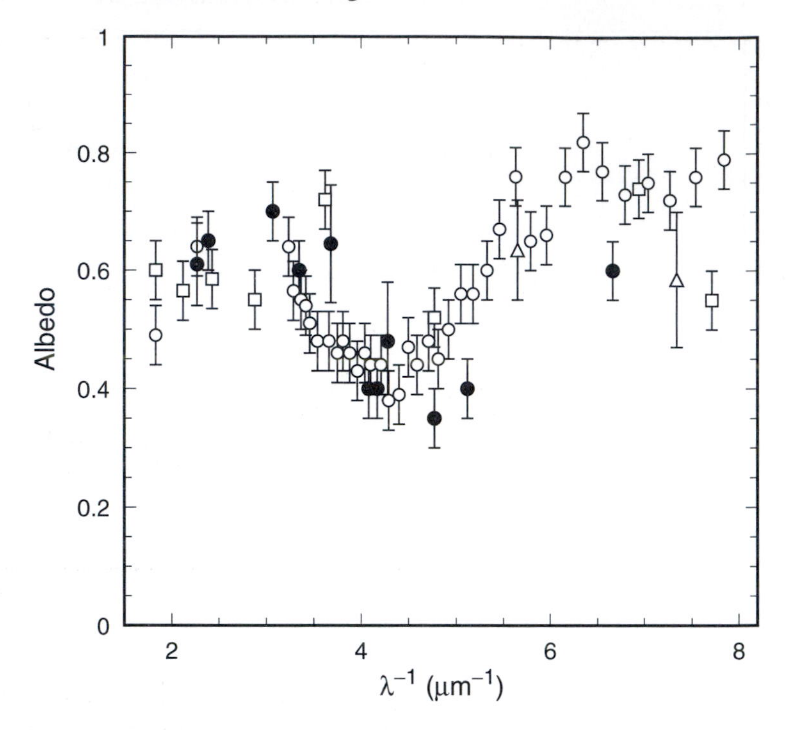

Figure 3.5. The spectral dependence of the grain albedo from observations of diffuse galactic light and reflection nebulae: full circles, diffuse galactic light (Lillie and Witt 1976, Morgan *et al* 1976, Toller 1981); open circles, IC 435 (Calzetti *et al* 1995); open squares, NGC 7023 (Witt *et al* 1982, 1992, 1993); and open triangles, Upper Scorpius (Gordon *et al* 1994).

Pioneer 10 spacecraft have been presented by Toller (1981); see also Witt (1988, 1989) for extensive reviews. Analysis depends on an idealized plane-parallel model for radiative transfer in the galactic disc and requires detailed knowledge of the spectral dependence of the illuminating radiation from the visual into the ultraviolet. Moreover, stars contributing to this radiation field at different wavelengths have different spatial distributions and its spectrum is thus dependent on geometry (Witt 1988). Observations of reflection nebulae are generally much easier to analyse, particularly in cases where the nebula is illuminated by a single embedded star: both the geometry and the spectrum of the illuminating radiation are then better constrained. However, such situations most commonly occur in relatively dense regions and the dust within the nebulae may not be typical of the ISM as a whole.

The observed spectral dependence of the albedo is plotted in figure 3.5, combining results from DGL and several reflection nebulae. The level of agreement between the two methods is reasonable and generally within estimated

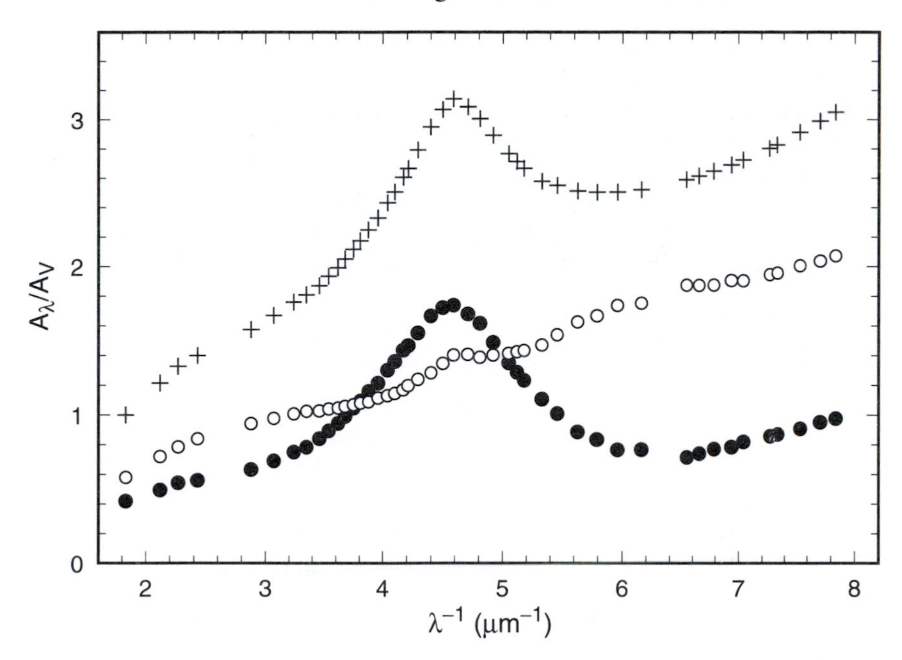

Figure 3.6. Absorption and scattering components of the total mean extinction curve. Absorption, scattering and extinction are denoted by full circles, open circles and crosses, respectively.

uncertainties. The albedo in the visual-blue region of the spectrum (1.8–2.3 μm^{-1}) is quite high ($\sim$0.6). In the ultraviolet, there is a clear trough near 4.5 μm^{-1}, beyond which α again becomes relatively high but with considerable scatter, in the FUV (6–8 μm^{-1}). This scatter is probably not attributable entirely to observational error but may reflect real differences in grain properties between different regions.

The contributions of absorption and scattering to the average extinction curve many be separated using the method described in section 3.1.3. To accomplish this, a mean albedo curve, α_λ, was found by fitting a smooth curve to the observational data plotted in figure 3.5. The mean extinction curve A_λ/A_V was then scaled by factors of α_λ and $1 - \alpha_\lambda$ to give scattering and absorption components, respectively (see equation (3.15)). Results are plotted in figure 3.6. Note that systematic errors may arise because extinction and albedo are not measured for identical samples: the FUV slope and bump strength are both sensitive to variation. Figure 3.6 should, however, provide a good indication of general behaviour.

This result leads to some important conclusions. The peak at 4.6 μm^{-1} in the extinction curve clearly corresponds quite closely to the trough in the albedo curve (figure 3.5), indicating the bump to be a pure absorption feature: its profile is revealed in figure 3.6. The relatively high albedo ($\alpha > 0.5$) over much of the spectrum indicates that at least one major component of the dust has optical properties that are predominantly dielectric. Indeed, scattering appears to be important at all wavelengths in the spectral range considered (figure 3.6).

The asymmetry factor has been evaluated over the same spectral range as the albedo and results used to constrain the size distribution of the particles responsible for scattering (Witt 1988 and references therein). In the visible, $g(\theta)$ is found to be quite high, typically 0.6–0.8, indicating predominantly forward-scattering by grains that are classical ($a \sim 0.1$–$0.3\ \mu$m) in size. In the FUV, $g(\theta)$ is generally lower than in the visible, indicating a trend toward more symmetric, less forward-biased scattering by much smaller grains. Since $g(\theta)$ is principally a function of the ratio of particle size to wavelength, $g(\theta)$ can decline systematically with wavelengths only in a situation where scattering is dominated by grains that decline in size faster than the wavelength itself (Witt 1988). This implies an upper limit $a < 0.04\ \mu$m on their radii.

Scattering halos around x-ray sources provide another potentially valuable diagnostic of dust properties. At x-ray wavelengths, all potential grain materials have refractive indices close to unity and so their optical properties are not sensitive to composition but they are sensitive to porosity and size distribution (Mathis and Lee 1991, Witt *et al* 2001). The best observed x-ray halo to date is that surrounding Nova Cygni 1992. Mathis *et al* (1995) show that the data are consistent with a high degree of porosity (>25% vacuum) for the large grains ($a > 0.1\ \mu$m). Calculations by Witt *et al* (2001) suggest that the size distribution extends to larger grains ($a \approx 2.0\ \mu$m) than are needed to fit the extinction curve (section 3.7) but if this is so then the implied value of R_V ($\approx$6.1) is much larger than is typical of the general diffuse ISM (section 3.3.3).

3.3.3 Long-wavelength extinction and evaluation of R_V

In the absence of neutral extinction (see section 3.3.4) R_V is related formally to the normalized relative extinction by the limit

$$R_V = -\left[\frac{E_{\lambda-V}}{E_{B-V}}\right]_{\lambda\to\infty} \tag{3.22}$$

and may thus be deduced by extrapolation of the observed extinction curve with reference to some model for its behaviour at wavelengths beyond the range for which data are available. Assuming that the small-particle approximation (section 3.1.2) applies at sufficiently long wavelengths in the infrared, an inverse power law of the form $Q_{\rm ext} \propto \lambda^{-\beta}$ appears to be a reasonable model. A limiting value of the index, $\beta = 4$, arises for scattering by pure dielectrics (equation (3.12)). Indices closer to unity are predicted for absorption-dominated extinction (see equation (3.13), which predicts $\beta = 1$ if m is independent of λ). The observed extinction in the infrared (figure 3.7) is, indeed, very well

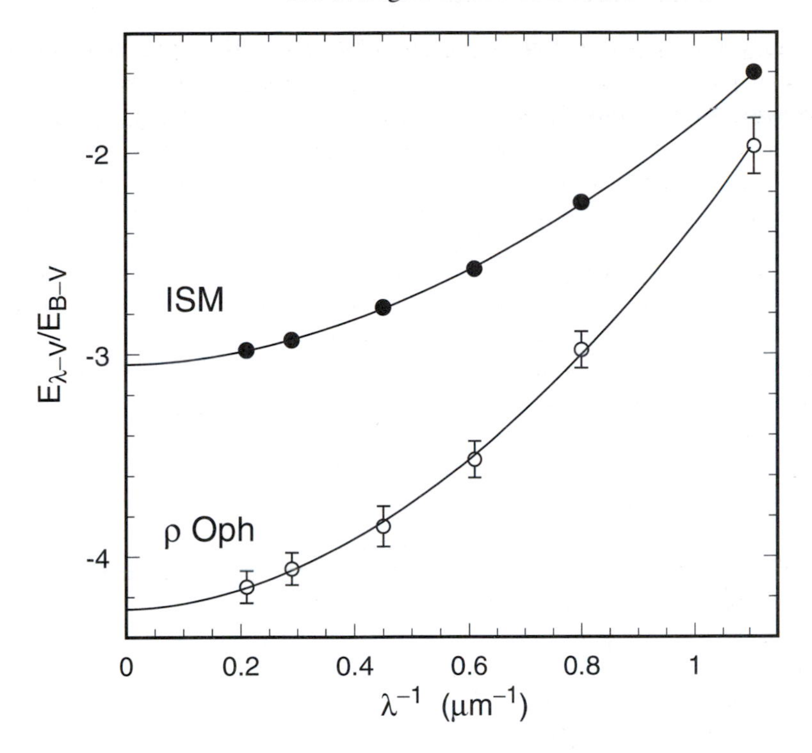

Figure 3.7. Mean infrared extinction curves in the range 0.2–1.1 μm^{-1}, comparing the general interstellar medium (ISM, full circles; table 3.1) with a representative dark cloud (ρ Oph, open circles with error bars; Martin and Whittet 1990). Standard deviations in the ISM points are comparable with the size of the plotting symbol. The curves are least-squares fits of an offset power law (equation (3.23)) to each dataset, which independently yield a consistent power law index $\beta = 1.84 \pm 0.02$ (see Martin and Whittet 1990). The intercepts yield the ratio of total-to-selective extinction, $R_V = 3.05$ (ISM) and 4.26 (ρ Oph).

represented by an inverse power law of the form

$$\frac{E_{\lambda-V}}{E_{B-V}} = \varepsilon\lambda^{-\beta} - R_V \tag{3.23}$$

where ε is a constant. A fit to the ISM data in figure 3.7 yields values of $\varepsilon = 1.19$, $\beta = 1.84$ and $R_V = 3.05$, with formal errors of about 1% (Martin and Whittet 1990). The measured index is thus consistent with predicted limits for idealized particles. We cannot assume, however, that this value of β is applicable at wavelengths beyond 5 μm. As $Q_{\rm sca}$ declines more rapidly than $Q_{\rm abs}$ with increasing λ, absorption should eventually dominate and the form of the extinction law will then depend on the nature of the grain material. However, values of R_V deduced by power-law extrapolation are not very sensitive to small changes in β and results are consistent with those obtained by fitting extinction curves calculated from Mie theory (e.g. Whittet *et al* 1976). Taking all these factors into consideration, we adopt

$$R_V = 3.05 \pm 0.15 \tag{3.24}$$

as the most likely average value of R_V in diffuse clouds. The quoted error in the mean represents the typical scatter in R_V and is higher than the formal error obtained from fitting the mean curve. Also shown in figure 3.7 are equivalent data for the ρ Oph dark cloud, which has higher R_V ($\approx$4.3) but is nevertheless fitted by a power law of identical index to within the uncertainty.

A useful approximation may be applied to relate R_V to the relative extinction in an infrared passband such as K:

$$R_V \simeq 1.1 \frac{E_{V-K}}{E_{B-V}}. \tag{3.25}$$

Effectively, we are setting λ in equation (3.22) to K (2.2 μm) and applying a scaling constant to represent extrapolation $\lambda \to \infty$. The value of the constant in equation (3.25) is consistent with theoretical extinction curves and applicable over a wide range of R_V (Whittet and van Breda 1978). Thus, photometry in three passbands (B, V and K) is sufficient to estimate R_V for a reddened star of known spectral type[1]. Note, however, that if 2.2 μm emission from a circumstellar shell is present, the colour excess ratio E_{V-K}/E_{B-V} will be anomalously large (because the K magnitude is numerically less) and this will lead to an overestimate of R_V.

3.3.4 Neutral extinction

Neutral (wavelength-independent) extinction occurs when particles large compared with the wavelength of observation are present – fog in the Earth's atmosphere is a good example. Any neutral extinction produced in the interstellar medium by 'giant grains' with dimensions $\gg$1 μm would be undetected by the pair method, yet its presence would affect distance determinations. Evaluation of R_V by extrapolation of the extinction curve (equation (3.22)) assumes implicitly that extinction $A_\lambda \to 0$ as the wavelength becomes very large; this is true only in the absence of grains large compared with the longest wavelengths at which extinction data are available, i.e. only if the size distribution $n(a) \to 0$ for $a \gg \lambda$. This assumption requires justification.

An independent method of evaluating R_V that includes any contribution from neutral extinction arises from Trumpler's method of determining open

[1] One could easily devise equivalent versions of equation (3.25) for different passbands and with different scaling constants. However, K is the preferred infrared passband for this purpose as it has the longest wavelength in the 'nonthermal' infrared: at longer wavelengths, thermal background radiation increases rapidly, limiting the accuracy of photometry with ground-based telescopes.

cluster diameters. The total visual (neutral plus wavelength-dependent) extinction averaged over a cluster is given by

$$\langle A_V \rangle = \langle V - M_V \rangle - \left\{ 5 \log \left(\frac{D}{\theta} \right) - 5 \right\} \tag{3.26}$$

where θ is the angular diameter of the cluster in radians and D the linear diameter in parsecs deduced from Trumpler's morphological classification technique. Thus, D/θ is the geometric distance (independent of extinction). $\langle V - M_V \rangle$ is determined by photometry and spectral classification of individual cluster members, from which the mean reddening $\langle E_{B-V} \rangle$ is also deduced. A plot of $\langle A_V \rangle$ against $\langle E_{B-V} \rangle$ for many clusters yields a linear correlation *passing through the origin* to within observational error and the slope gives $R_V = 3.15 \pm 0.20$ (Harris 1973), consistent with the value from the extinction curve (equation (3.24)). The consistency between the two methods of evaluating R_V suggests that any contribution to A_V from very large grains is likely to be negligible. Abundance considerations also argue against a substantial population of very large grains: if they existed, they would consume a major fraction of the available heavy elements (chapter 2) and exacerbate the problem of accounting for the wavelength-dependent component of the extinction (Mathis 1996a).

3.3.5 Dust density and dust-to-gas ratio

An estimate of the amount of grain material required to produce the observed mean rate of extinction with respect to distance in the galactic plane may be deduced from general principles described by Purcell (1969). The integral of Q_{ext} over all wavelengths can be obtained from the Kramers–Krönig relationship

$$\int_0^\infty Q_{\text{ext}}\, \mathrm{d}\lambda = 4\pi^2 a \left\{ \frac{m^2 - 1}{m^2 + 2} \right\} \tag{3.27}$$

for spherical grains of radius a and refractive index m. The mass density of dust in a column of length L is

$$\rho_{\text{d}} = \frac{N_{\text{d}} m_{\text{d}}}{L} \tag{3.28}$$

where N_{d} is the column density (equation (3.3)) and

$$m_{\text{d}} = \tfrac{4}{3}\pi a^3 s \tag{3.29}$$

is the mass of a spherical dust grain composed of material of specific density s. Using equation (3.6) to relate Q_{ext} to A_λ in equation (3.27) and substituting for N_{d} and m_{d} in equation (3.28), we have

$$\rho_{\text{d}} \propto s \left\{ \frac{m^2 + 2}{m^2 - 1} \right\} \int_0^\infty \frac{A_\lambda}{L}\, \mathrm{d}\lambda. \tag{3.30}$$

From a knowledge of the observed mean extinction curve, ρ_d may be expressed approximately in terms of $\langle A_V/L\rangle$ in mag kpc^{-1} (e.g. Spitzer 1978: p 153):

$$\rho_d \simeq 1.2 \times 10^{-27} s \left\{\frac{m^2+2}{m^2-1}\right\} \left\langle\frac{A_V}{L}\right\rangle. \tag{3.31}$$

From observations of reddened stars, $\langle A_V/L\rangle \sim 1.8$ mag kpc^{-1} in the diffuse ISM (section 1.3.2) and if we assume that $m = 1.50 - 0\text{i}$ and $s \simeq 2500$ kg m^{-3}, appropriate for low-density silicates, then equation (3.31) gives

$$\rho_d \simeq 18 \times 10^{-24} \text{ kg m}^{-3}. \tag{3.32}$$

This result is somewhat dependent on the assumed composition; for example, ice grains ($m = 1.33 - 0\text{i}$, $s = 1000$ kg m^{-3}) would yield a value ~40% less. It should also be noted that much larger values of A_V/L and hence of ρ_d, occur locally within individual clouds.

The dust-to-gas ratio, allowing for the presence of helium in the gas, is

$$Z_d = 0.71\frac{\rho_d}{\rho_H} \simeq 0.007 \tag{3.33}$$

where $\rho_H \simeq 1.8 \times 10^{-21}$ kg m^{-3} (section 1.3.2). This result is consistent with estimates of Z_d based on abundance and depletion data (section 2.4.2), which lie in the range 0.004–0.010 (dependent on choice of reference abundances). A similar calculation for reddened stars in the Large Magellanic Cloud (Koornneef 1982) yields $Z_d \sim 0.002$, a significant difference which is qualitatively consistent with the low metallicity of that galaxy compared with the Milky Way (see figure 2.6).

3.4 Spatial variations

3.4.1 The blue–ultraviolet

Regional variations in the optical properties of interstellar dust were first discussed by Baade and Minkowski (1937), who found that the extinction curves for stars in the Orion nebula (M42) differ from the mean curve for more typical regions in a manner consistent with the selective removal of small particles from the size distribution. Such an effect can be produced by a number of physical processes (see chapter 8), including grain growth by coagulation, size-dependent destruction and selective acceleration of small grains by radiation pressure in stellar winds. Star-to-star variations are most conspicuous at ultraviolet wavelengths, hinting that it is the smallest grains that are most subject to change. The largest deviations from average extinction are often observed in lines of sight that sample dense clouds associated with current or recent star formation.

It is convenient to characterize variations in the morphological appearance of the UV extinction curve in terms of a three-component empirical model:

$$\frac{E_{\lambda-V}}{E_{B-V}} = (c_1 + c_2x) + c_3D(x) + c_4F(x) \tag{3.34}$$

(Fitzpatrick and Massa 1986, 1988, 1990), where $x = \lambda^{-1}$ and c_1, c_2, c_3 and c_4 are constants for a given line of sight. The three components are:

(i) a linear term, $c_1 + c_2x$;
(ii) a 'bump' term, $c_3D(x)$, where $D(x)$ is a mathematical representation of the 2175 Å absorption profile (see section 3.5.1), and
(iii) a far-UV term, $c_4F(x)$, where $F(x)$ is defined by equation (3.35) below.

Each of these components can vary more or less independently from one line of sight to another.

Variations in the linear component are most easily seen as a change in slope in the 2–3 μm^{-1} segment of normalized extinction curves (e.g. Nandy and Wickramasinghe 1971, Whittet *et al* 1976), i.e. between the knee and the bump (figure 3.4). These changes are accompanied by variations in R_V and are associated with fluctuations in the size distribution of the grains responsible for continuum extinction in the visible to near ultraviolet: the slope declines as the mean grain size increases.

Star-to-star variations in the 2175 Å feature and the FUV rise are illustrated in figure 3.8. The sample includes stars that probe a range of environments, from dense clouds (HD 147701 and HD 147889 in the ρ Oph complex) and H II regions (Herschel 36 in M8, HD 37022 in M42) to more typical diffuse clouds. A variety of morphologies is evident. The 2175 Å feature shows variations in strength and width independent of changes in FUV extinction (compare, for example, the curves for HD 204827, HD 37367 and HD 37022 in figure 3.8). The behaviour of the bump is discussed further in section 3.5.1.

The FUV rise is represented in the empirical formula (equation (3.34)) by a polynomial:

$$F(x) = 0.5392(x - 5.9)^2 + 0.0564(x - 5.9)^3 \tag{3.35}$$

for $x > 5.9\ \mu\text{m}^{-1}$, with $F(x) = 0$ for $x \leq 5.9\ \mu\text{m}^{-1}$ (Fitzpatrick and Massa 1988). Figure 3.9 shows a fit based on this functional form to the average residual curve for 18 stars, with the linear background and 2175 Å feature removed. The parameter c_4 in equation (3.34) characterizes the *amplitude* of the FUV rise and this varies from star to star. However, the *shape*, $F(x)$, is essentially the same for all stars in the sample, regardless of environmental factors or the morphology of the extinction curve at longer wavelengths. This suggests that the FUV rise is not an artifact of the size distribution but a distinct optical property of an independent grain population (Fitzpatrick and Massa 1988). Variations in amplitude are then simply caused by variations in the abundance of the carrier grains.

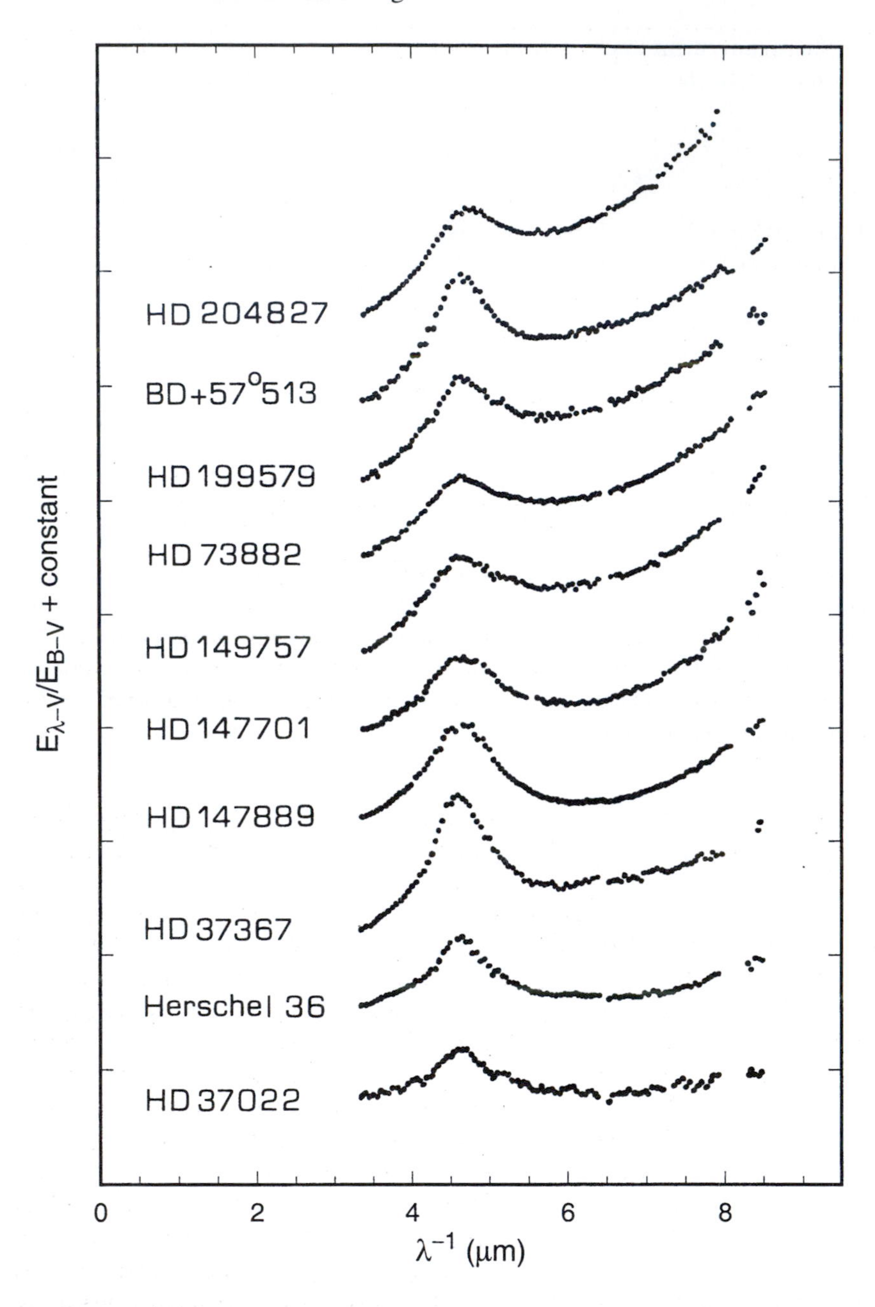

Figure 3.8. Comparison of the ultraviolet extinction curves of 10 stars observed with the International Ultraviolet Explorer satellite (Fitzpatrick and Massa 1986, 1988). The separation of tick marks on the vertical axis is 4 magnitudes, individual curves being displaced vertically for display.

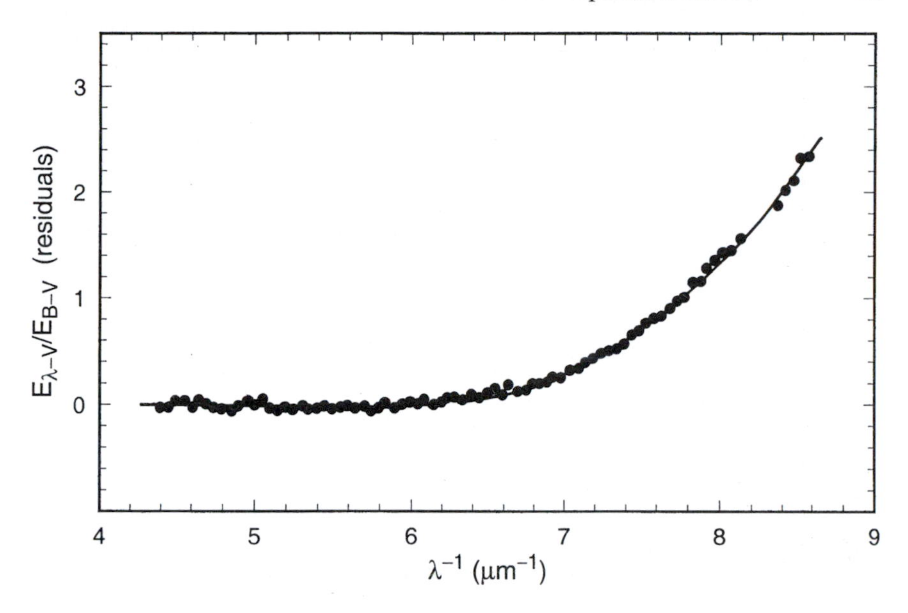

Figure 3.9. The shape of the far-ultraviolet rise in the extinction curve (Fitzpatrick and Massa 1988). Observational data for 18 stars are averaged and the residuals (points) are plotted after extraction of a linear background and a Drude profile representing the 2175 Å absorption. The smooth curve is a fit based on the polynomial function in equation (3.35).

It is informative to compare extinction curves for the Milky Way with those of other galaxies (Nandy 1984, Fitzpatrick 1989). Representative results are displayed in figures 3.10 and 3.11 for the two best studied cases, the Large and Small Magellanic Clouds. These dwarf galaxies are deficient in heavy elements by factors of about 2.5 (LMC) and 7 (SMC) compared with the solar standard (Westerlund 1997) and it seems likely that this could affect the quality as well as the quantity of dust in their interstellar media. In the case of the LMC, two distinct mean extinction curves have been found (figure 3.10). That for stars in the vicinity of the 30 Doradus complex differs from that for stars more widely distributed in the LMC, displaying a weaker bump and a stronger FUV rise (Koornneef 1982, Fitzpatrick 1985, 1986, Misselt *et al* 1999). This dispersion is no greater than is seen in the Milky Way, however (see Clayton *et al* 2000). What is more remarkable is that the form of the general extinction in the LMC is so *similar* to that of the Milky Way, resembling curves for certain individual stars (such as HD 204827 in figure 3.8). Clearly, the ingredients of interstellar dust that lead to bump absorption and FUV extinction in galactic extinction curves are also present in the LMC in broadly similar proportions. Dust in the SMC seems to be more radically different (Prévot *et al* 1984, Rodrigues *et al* 1997). The SMC curve (figure 3.11) displays a steep continuum and extremely weak or absent bump absorption. Empirical fits using equation (3.34) suggest that the continuum

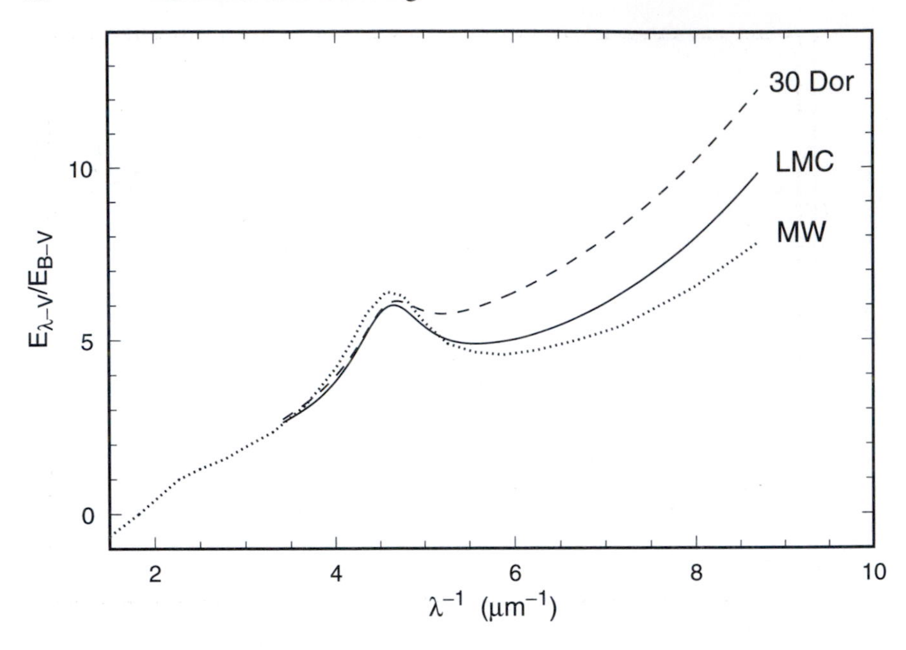

Figure 3.10. Ultraviolet extinction curves for the Large Magellanic Cloud (LMC), based on data from Fitzpatrick (1986). The average for stars widely distributed in the LMC (full curve) is compared with that for the 30 Doradus region of the LMC (broken curve) and the solar neighbourhood of the Milky Way (dotted curve; data from table 3.1).

extinction arises in an unusually steep linear component, with only a minor contribution from grains responsible for the FUV rise. These differences might be related to the low metallicity of the SMC and its effect on grain production. They could also be linked to differences in radiative environment; Gordon *et al* (1997) find that dust in starburst galaxies has similar optical properties to that in the SMC.

3.4.2 The red–infrared

Several investigations have demonstrated that the extinction law in the spectral range 0.7–5.0 μm is essentially invariant to within observational error (e.g. Koornneef 1982, Whittet 1988, Cardelli *et al* 1989, Martin and Whittet 1990). In contrast to the situation at shorter wavelengths, no significant differences are generally apparent between different regions in the solar neighbourhood, or between the Milky Way and the Magellanic Clouds. Differences in colour excess ratios $E_{V-\lambda}/E_{B-V}$, where λ represents an infrared wavelength or passband (see figure 3.7) are imposed by the change in slope of the extinction law in the blue–visible region, affecting the differential extinction between B and V. If R_V is determined and extinction curves renormalized to absolute extinction A_λ/A_V

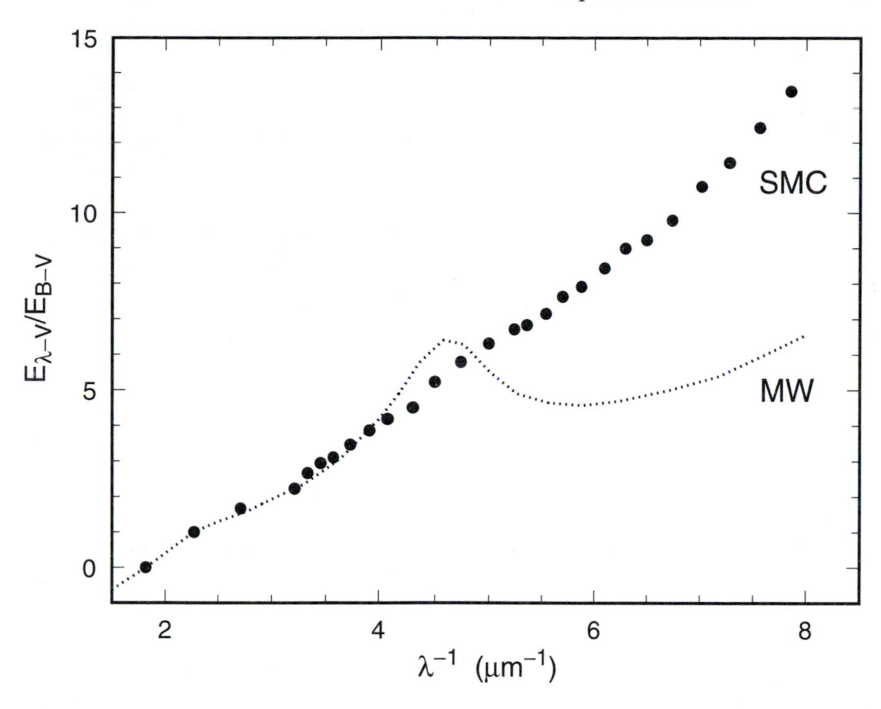

Figure 3.11. The ultraviolet extinction curve for the Small Magellanic Cloud (SMC, points; data from Prévot *et al* 1984). The average for the Milky Way is also shown for comparison (dotted curve; data from table 3.1).

(figure 3.12), these variations disappear. This convergence of extinction laws in the infrared may tell us something fundamental about the dust. It also has practical applications, as a simple mathematical form such as a power law (section 3.3.3) may be used to deduce the intrinsic spectral energy distributions of obscured infrared sources from observed fluxes.

The relative contributions of absorption and scattering in the near infrared are uncertain because the albedo is poorly constrained at these wavelengths (see Lehtinen and Mattila 1996). If scattering is dominant, then the wavelength dependences of extinction and albedo constrain the size distribution. If absorption is dominant, then the extinction law depends on the nature of the absorber. For a semiconductor such as amorphous carbon, the absorption spectrum is determined by intrinsic properties such as the band-gap energy and the distribution of additional energy states associated with impurities and structural disorder (Duley 1988, Duley and Whittet 1992). An invariant near-infrared extinction curve would require that the absorbing material has rather homogeneous properties that are not greatly affected by environment.

Objects with extremely high degrees of extinction, such as stars within or behind dense molecular-cloud cores, are often too faint to be observable at visible or ultraviolet wavelengths and all information on their extinction properties therefore comes from the infrared. If the spectral type of such an object can be estimated from infrared spectra, the extinction in the line of sight may be quantified in terms of an infrared colour excess such as E_{J-K}. More generally, diagnostic studies may involve use of infrared colour–colour diagrams such as $J-H$ versus $H-K$, which allow some discrimination between embedded stars with dust shells and reddened background stars with purely photospheric emission (see Itoh *et al* 1996 for an example). Fortuitously, because of the invariance of the infrared extinction law, the slope of the 'reddening vector' in such diagrams is also invariant, such that reddened stars are displaced from intrinsic colour lines in a predictable way; its value averaged over diverse environments is

$$\frac{E_{J-H}}{E_{H-K}} = 1.60 \pm 0.04 \tag{3.36}$$

(Whittet 1988, Kenyon *et al* 1998 and references therein).

It is often desirable to estimate the total *visual* extinction of an obscured infrared source from its infrared colours. This may be done by assuming a form for the extinction law from 0.55 μm to the spectral region of convergence. We may conveniently express A_V in terms of E_{J-K} thus:

$$A_V = rE_{J-K} \tag{3.37}$$

where r is the ratio of total visual to selective infrared extinction, analogous to R_V. The value of r has been determined empirically for different regions and found to vary from approximately 5.9 for average diffuse-cloud extinction to values in the range 4.6–5.4 for typical dense clouds (He *et al* 1995, Whittet *et al* 2001a). If one may characterize the environment along a given line of sight qualitatively as 'dense' or 'diffuse', the most appropriate value of r to use may be selected accordingly. A quantitative estimate of r is possible if the value of R_V for the region can be determined from other observations: r varies with R_V according to an empirical law

$$r = \frac{a}{b - R_V^{-1}} \tag{3.38}$$

(He *et al* 1995), where a and b are constants. Values $a \approx 2.38$ and $b \approx 0.73$ are consistent with available data for both diffuse and dense clouds. In lines of sight so reddened that even the J (1.25 μm) magnitude cannot be measured, an equivalent form of equation (3.37) may be used, based on E_{H-K}:

$$A_V = r'E_{H-K} \tag{3.39}$$

where $r' \approx 16$ for the diffuse ISM and 12–13 for dense clouds (Whittet *et al* 1996).

3.4.3 Order from chaos?

Interstellar extinction curves display a wide diversity in morphological structure (e.g. figures 3.8, 3.10, 3.11) but there are, nevertheless, common features that unify results for different environments. In the infrared, the situation is reasonably clear: the extinction behaves predictably and is well represented by a generic mathematical form (an inverse power law; section 3.3.3). In the ultraviolet, the 2175 Å bump feature is ubiquitous and seems quite stable in position and profile shape, whilst displaying variations in amplitude and width (section 3.5.1). The form of the FUV rise also appears to be well established (section 3.4.1). The ultraviolet segment of the extinction curve for a given line of sight may thus also be described in terms of a mathematical formula (e.g. equation (3.34)) with a manageable number of free parameters.

Progress toward a unique mathematical description of the extinction curve was made by Cardelli *et al* (1989), who proposed a relation between the general form of the curve and R_V. The entire extinction curve is represented by the equation

$$\frac{A_\lambda}{A_V} = a(x) + \frac{b(x)}{R_V} \tag{3.40}$$

where $a(x)$ and $b(x)$ are coefficients that have unique values at a given wavenumber $x = \lambda^{-1}$. Note that the extinction is expressed here in the absolute normalized form. The R_V-dependent extinction represented by equation (3.40) is often referred to as the 'CCM extinction law'. The coefficients are determined empirically from the slope and intercept of the correlation of A_λ/A_V with R_V^{-1} at selected wavelengths. Cardelli *et al* (1989) also list formulae that allow $a(x)$ and $b(x)$ to be calculated, assuming a power-law form for A_λ in the infrared, together with various polynomial forms in the visible and ultraviolet and a Drude profile (section 3.5.1) for the 2175 Å bump. Values of $a(x)$ and $b(x)$ at various wavenumbers are listed in table 3.1, taken from the results of Cardelli *et al* (1989) and O'Donnell (1994a).

R_V-dependent extinction is plotted for some representative values of R_V in figure 3.12. R_V tends to be higher in dense clouds, where grains grow by coagulation (section 8.3), compared with the average value of around 3.1 (section 3.3.3). Note that normalization sets all curves to unity at the V passband (1.8 μm^{-1}). The curves converge in the infrared ($\lambda < 1\ \mu$m^{-1}) and diverge in the ultraviolet ($\lambda > 2\ \mu$m^{-1}). The level of UV continuum extinction is a strong function of R_V.

3.5 The 2175 Å absorption feature

In this section, we take a closer look at the 2175 Å 'bump' feature, described by Draine (1989a) as *"a dramatic piece of spectroscopic evidence which should have much to tell us about at least a part of the interstellar grain population"*. We will review the evidence and attempt to interpret the message.

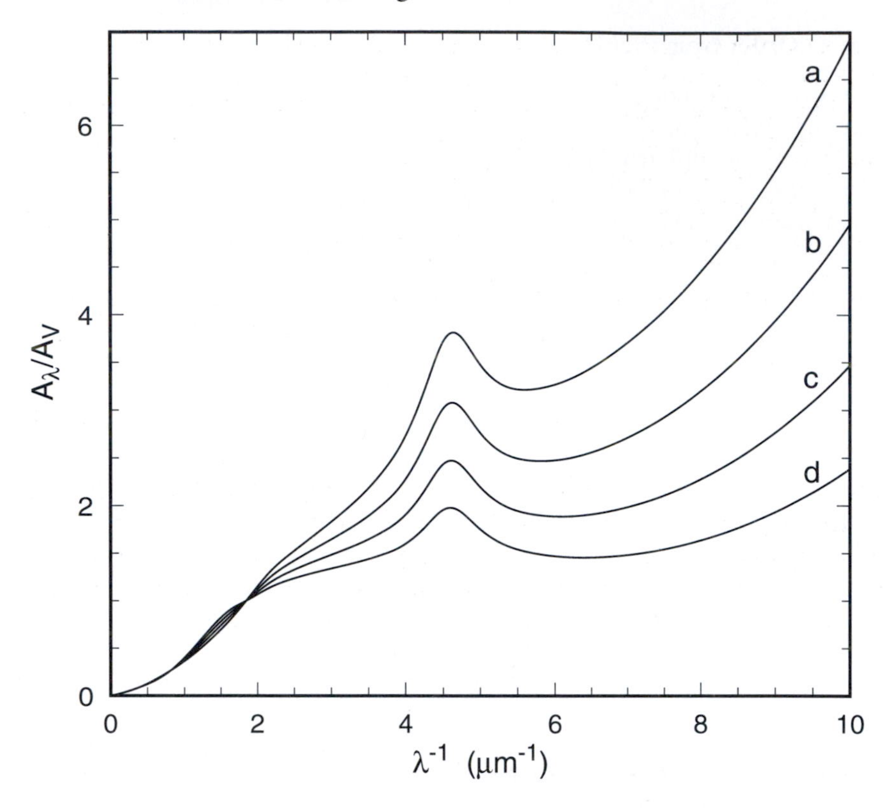

Figure 3.12. R_V-dependent variations in the extinction curve, based on empirical fits to data for many stars (originally proposed by Cardelli *et al* 1989 and shown here in the formulation of Fitzpatrick 1999): curve a, $R_V = 2.5$; curve b, $R_V = 3.1$; curve c, $R_V = 4.0$; and curve d, $R_V = 5.5$.

3.5.1 Observed properties

The most striking aspects of the bump are its ubiquity and strength, its stability of central wavelength and its uniformity of profile. It is almost invariably detectable in the spectra of stars with appreciable reddening ($E_{B-V} > 0.05$) and its strength is generally well correlated with E_{B-V} (Meyer and Savage 1981)[2]. Its central wavelength, λ_0, has a mean value

$$\langle \lambda_0 \rangle = 2175 \pm 10 \text{ Å} \tag{3.41}$$

[2] Whilst the correlation between bump strength and E_{B-V} confirms the interstellar nature of the feature, it does not imply that the absorbing agent necessarily resides in particles responsible for visual extinction and reddening. The various ingredients of the ISM (atoms, molecules, small grains, large grains) are generally well mixed, such that almost any unsaturated interstellar absorption feature will show a significant positive correlation with E_{B-V}.

the error representing a 2σ dispersion of only 0.46%. Comparing data for individual stars (see figure 3.8 for typical examples), λ_0 is generally constant to within observational error (Fitzpatrick and Massa 1986) with very few known exceptions (Cardelli and Savage 1988). In the most deviant cases the feature is shifted to shorter wavelength by $\sim$2.5%.

A mathematical representation of the 2175 Å profile shape is useful and may provide physical insight. To a good approximation, the feature is Lorentzian (Savage 1975, Seaton 1979) but Fitzpatrick and Massa (1986) show that an even better fit to the observations is obtained with a model based on the Drude theory of metals (Bohren and Huffman 1983). The Drude profile is defined in terms of $x = \lambda^{-1}$ by

$$D(x) = \frac{x^2}{(x^2 - x_0{}^2)^2 + \gamma^2 x^2} \tag{3.42}$$

where $x_0 = \lambda_0^{-1}$ and γ specify the position and width (FWHM) of the feature, respectively, in wavenumber units. Traditionally, the strength of the feature has been expressed in terms of its peak intensity in magnitudes, E_{bump}, relative to an assumed continuum level, such as the linear background. Setting $x = x_0$ in equation (3.42), we may show, with reference to equation (3.34), that

$$\frac{E_{\text{bump}}}{E_{B-V}} = \frac{c_3}{\gamma^2} \tag{3.43}$$

with respect to the linear background. However, a more appropriate measure of strength is the quantity

$$A_{\text{bump}} = c_3 \int_0^\infty D(x)\,\mathrm{d}x = \frac{\pi c_3}{2\gamma} \tag{3.44}$$

(Fitzpatrick and Massa 1986), effectively the equivalent width in wavenumber units: A_{bump} represents the area under the 2175 Å profile in the normalized extinction curve and is thus a measure of strength per unit E_{B-V}. Star-to-star variations in A_{bump} thus reflect scatter in the general correlation between bump strength and reddening.

The position, width and strength parameters λ_0, γ and A_{bump} have been evaluated for many lines of sight by fitting equations (3.34) and (3.42) to observational extinction curves in the range $3 < \lambda^{-1} < 6\ \mu\text{m}^{-1}$. Table 3.2 lists results for a selection of individual stars that sample diverse environments, together with mean values for the Milky Way and the Large Magellanic Cloud.

Perhaps the most remarkable observational property of the 2175 Å profile is the occurrence of variations in width that are unaccompanied by changes in position. Although two stars (HD 29647 and HD 62542) with exceptionally large values of x_0 also have unusually broad bumps, stars with essentially identical values of x_0 can have very different widths. Figure 3.13 compares the profiles with largest and smallest γ values in the group of 45 stars studied by Fitzpatrick

Table 3.2. Representative values of the 2175 Å bump parameters in units of μm^{-1} (Fitzpatrick and Massa 1986, Cardelli and Savage 1988, Welty and Fowler 1992). Abbreviations used to denote environments have the following meanings: DC (dark cloud), DISM (diffuse ISM), H II (compact H II region), HLC (high latitude cloud), OB (OB star cluster) and RN (reflection nebula).

Star	Environment	λ_0^{-1}	γ	A_{bump}
HD 29647	DC	4.70	1.62	3.35
θ^1 Ori C	H II	4.63	0.84	2.43
HD 37061	H II	4.57	1.00	2.69
HD 37367	DISM	4.60	0.91	7.04
HD 38087	RN	4.56	1.00	6.68
HD 62542	DC	4.74	1.29	3.11
HD 93028	OB	4.63	0.79	2.62
HD 93222	OB	4.58	0.81	3.33
ρ Oph	DC	4.60	0.99	5.57
HD 147889	DC	4.63	1.16	7.14
ζ Oph	DC	4.58	1.25	5.71
Herschel 36	H II	4.62	0.88	3.51
HD 197512	DISM	4.58	0.96	6.83
HD 204827	OB/DC	4.63	1.12	4.98
HD 210121	HLC	4.60	1.09	3.48
Mean ISM (45 stars)		4.60	0.99	5.17
General LMC (13 stars)		4.60	0.99	4.03
30 Dor LMC (12 stars)		4.61	0.89	2.62

and Massa (1986). This behaviour is hard to reconcile with any solid state model for the feature that involves particles with sizes comparable with the wavelength: calculations based on Mie theory would predict a strong correlation between x_0 and γ for such particles. The carrier cannot, therefore, be a component of the classical-sized grains responsible for visual extinction.

The width and strength of the bump display systematic dependences on environment (table 3.2). At least three trends can be discerned:

(i) stars that sample the diffuse ISM tend to have relatively strong bumps of average width;
(ii) stars associated with H II regions and/or OB star clusters generally have narrow, weak bumps and
(iii) stars that sample dense clouds tend to have broad bumps (but with wide dispersion in both width and strength).

Examples of stars in H II regions include θ^1 Ori and Herschel 36 (see figure 3.8). Bump widths for stars associated with dense clouds range from near-average

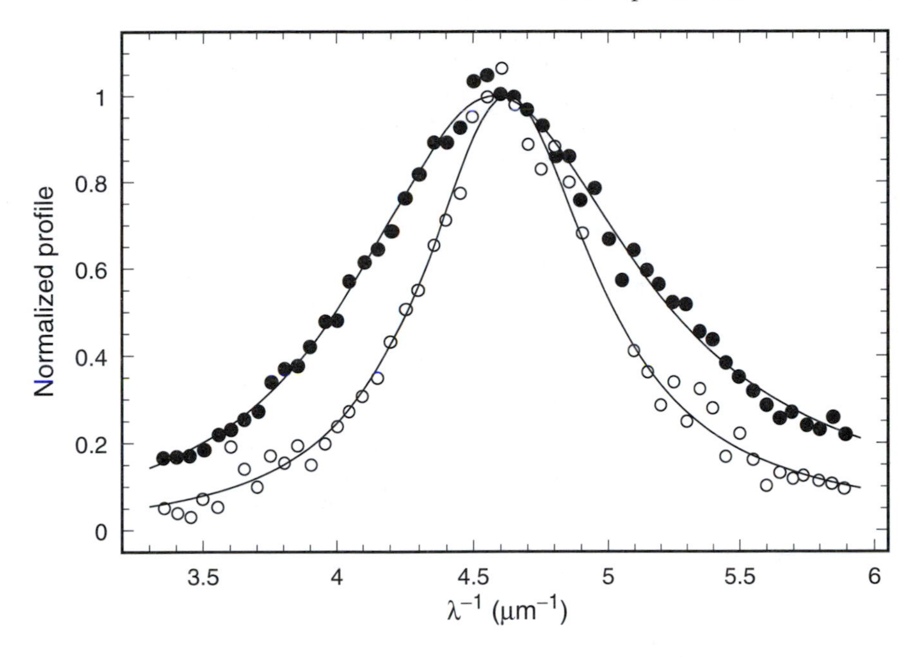

Figure 3.13. Normalized 2175 Å bump profiles toward two stars, illustrating variation in width: observational data for ζ Oph (filled circles) and HD 93028 (open circles) are fitted with the Drude function using values of the parameters listed in table 3.2. (Data from Fitzpatrick and Massa 1986.)

(ρ Oph) to the broadest known (HD 29647). Figure 3.14 compares bump profiles for HD 29647 and HD 62542 with the interstellar mean. In both cases, the feature is broadened by an apparent extension of the profile to longer wavenumber.

In addition to stars in our Galaxy, two independent groups of stars in the LMC are represented in table 3.2: the 30 Dor region and the general LMC. As previously noted, the bump in the general LMC is remarkably similar to the corresponding feature observed in the solar-neighbourhood ISM (figure 3.10): its position and width are identical to within observational error, whilst its strength is $\sim$20% less. In the 30 Dor region, the position and width are again similar but the strength is lower by $\sim$50%. For both LMC samples, the bump parameters are within the range of values observed toward individual stars in the Milky Way: the carrier is clearly a characteristic ingredient of dust in both galaxies. It is tempting to associate the weakness of the feature in the 30 Dor nebula with that observed in galactic H II regions (comparing results for 30 Dor, θ^1 Ori C and Herschel 36 in table 3.2) and thus to conclude that environmental influences are more important than metallicity effects (which would also apply to the general LMC, not just 30 Dor). The reason for the extreme weakness or absence of the feature in the SMC (figure 3.11) has yet to be discovered: this might provide an important clue

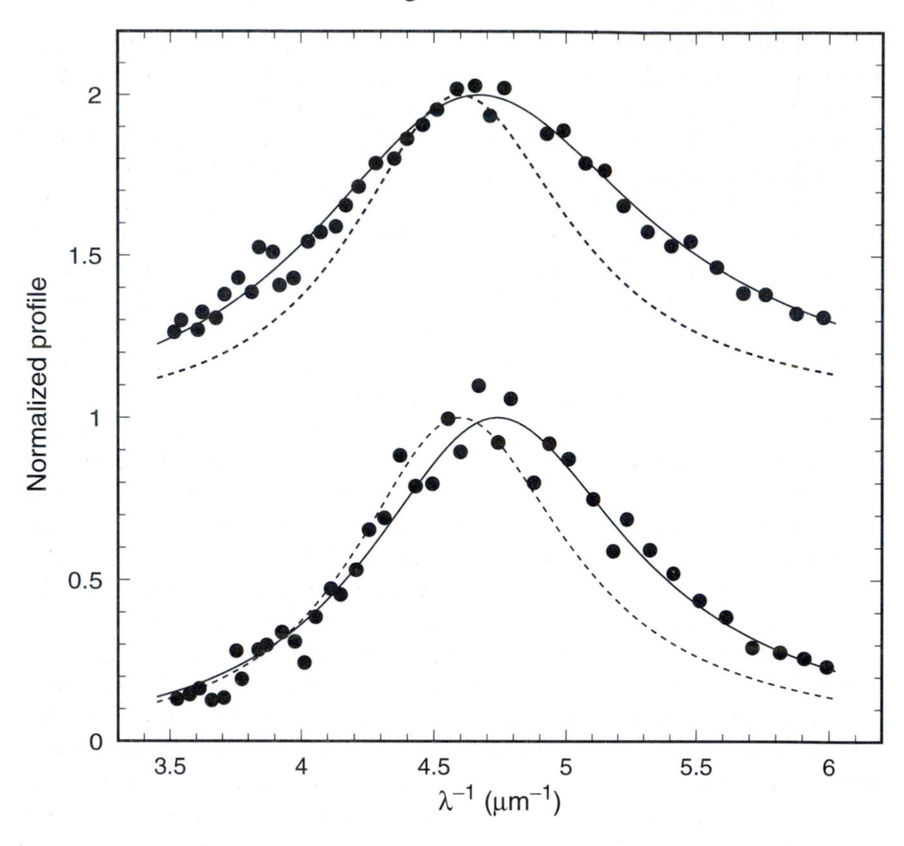

Figure 3.14. Normalized profiles for two stars with anomalous 2175 Å bumps: HD 29647 (top) and HD 62542 (bottom). In each case, observational data (points) are fitted with the appropriate Drude function (full curve) and compared with the average profile for the ISM (broken curve). The data for HD 29647 are displaced upward by one unit for display. Both stars are associated with relatively dense interstellar material: HD 29647 is located behind a dense clump (TMC-1) in the Taurus dark cloud, whilst HD 62542 is located behind material swept up by ionization fronts in the Gum nebula. (Data from Cardelli and Savage 1988.)

to the nature of the carrier.

Several studies have shown that the strength of the bump is generally uncorrelated with the amplitude of the FUV rise (Meyer and Savage 1981, Seab *et al* 1981, Witt *et al* 1984, Cardelli and Savage 1988, Jenniskens and Greenberg 1993). Toward HD 62542, for example, the bump is relatively weak and the FUV extinction is very strong, whereas ρ Oph has near-normal bump strength and weak FUV extinction. Stars embedded in H II regions show comparative weakness in both bump and FUV extinction. The obvious interpretation of this evidence is that the bump and the FUV rise originate in different grain populations, with environmental factors governing their relative contributions in a given line of sight. However, Fitzpatrick and Massa (1988) report a correlation between the *width* of the bump and the amplitude of FUV rise. This hints at a possible explanation of the latter: the FUV rise might perhaps be another absorption feature, centred at $\lambda < 1100$ Å in the extreme ultraviolet (EUV), i.e. beyond the spectral range of instruments commonly used to investigate interstellar extinction[3]. Only the long-wavelength wing of such a feature should thus be observed. If the bump and the EUV feature were produced by similar mechanisms, correlated changes in the widths of both might occur (see Rouleau *et al* 1997); in the case of the EUV feature, this behaviour would result in an apparent modulation of the FUV extinction amplitude.

It is of interest to consider whether other relevant features or structure might exist in the UV extinction. Searches have been generally unsuccessful or inconclusive, with no convincing evidence for structure greater in amplitude than 5% relative to that of the bump (Savage 1975, Seab and Snow 1985). A number of detections have been claimed but not confirmed. Discrepancies in stellar line strengths between reddened and comparison stars can lead to false-positive results: for example, a weak, broad hump near 6.3 μm^{-1} seen in extinction curves derived from TD-1 satellite data was caused by mismatch (section 3.2) in the λ1550 C IV lines (Massa *et al* 1983). Carnochan (1989) has argued that a broad, shallow absorption feature centred near 1700 Å (5.9 μm^{-1}) is present in TD-1 data but this is not seen in extinction curves from other satellite databases such as IUE (Fitzpatrick and Massa 1988). Apparent absorption at ~2700 Å, reported in IUE data and put forward as evidence for interstellar proteins, proved to be of instrumental origin (McLachlan and Nandy 1984, Savage and Sitko 1984). The 2175 Å bump remains unchallenged as the dominant feature in the extinction curve.

3.5.2 Implications for the identity of the carrier

Prime requirements of a 2175 Å absorber to be a viable candidate for the interstellar feature are that it is cosmically abundant, sufficiently robust to survive in a variety of interstellar environments and capable of matching closely the observed profile position, width and shape, without producing significant absorptions at other wavelengths where no feature is observed.

Let us first consider what constraints may be placed on the abundance of the carrier. The equivalent width is related to the strength parameter $A_{\rm bump}$ (equation (3.44)) by

$$W_{\nu} = cA_{\rm bump}E_{B-V}. \tag{3.45}$$

Using $A_{\rm bump} \simeq 5.2 \times 10^6$ m^{-1} (table 3.2) and $N_{\rm H}/E_{B-V} \simeq 5.8 \times 10^{25}$ m^{-2}

[3] Note that absorption by atomic hydrogen beyond the Lyman limit at 912 Å sets a fundamental constraint on EUV observations of extinction in diffuse clouds (see Longo *et al* 1989).

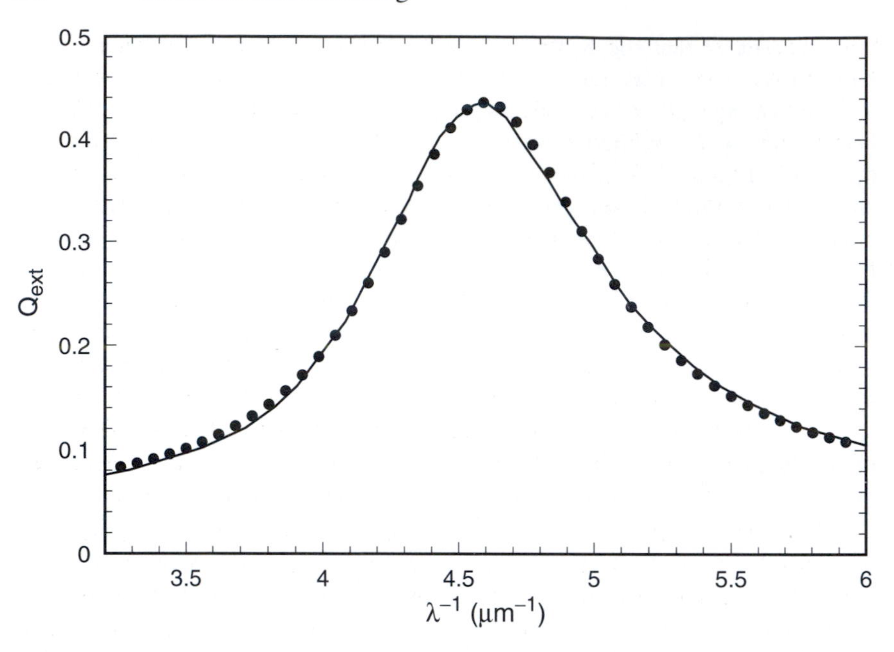

Figure 3.15. Plot of Q_{ext} against λ^{-1} for randomly orientated graphite spheroids with $a = 0.003$ μm and axial ratio $b/a = 1.6$, where $2a$ and $2b$ are the particle lengths parallel and perpendicular to the axis of symmetry (Draine 1989a, curve). This is compared with the mean observed 2175 Å profile (points) from Fitzpatrick and Massa (1986), normalized to the peak of the feature.

(section 1.3.2), the abundance relative to hydrogen of the carrier (X) is

$$\frac{N_{\mathrm{X}}}{N_{\mathrm{H}}} \simeq \frac{10^{-5}}{f} \tag{3.46}$$

where f is the oscillator strength per absorber associated with the feature (chapter 2, equation (2.18)). The strongest permitted transitions typically have $f \leq 1$ and so the abundance of the carrier must be $N_{\mathrm{X}}/N_{\mathrm{H}} \geq 10^{-5}$ (see Draine 1989a). With reference to the standard element abundances (section 2.2), X must therefore contain one or more from the set {C, N, O, Ne, Mg, Si, S, Fe}. This set can immediately be reduced from eight elements to four: Ne is a noble gas and can therefore be ruled out; S (abundance 1.8×10^{-5}) is only weakly depleted in the interstellar medium and can also be excluded; finally, N and O are rejected as they are electron acceptors. We may conclude that the carrier must contain one or more elements from the set {C, Mg, Si, Fe}.

Graphitic carbon is by far the most widely discussed material amongst candidates for the bump and this identification has gained a measure of acceptance. The average profile of the observed feature is well matched by theoretical models involving small graphite grains, as illustrated in figure 3.15. As a form of solid carbon, graphite easily satisfies abundance constraints: the oscillator strength per carbon atom is $f \sim 0.16$ in the small particle limit (Draine 1989a) and equation (3.46) gives $N_C/N_H \sim 6.3 \times 10^{-5}$, which is $\sim$ 20% of the solar abundance. Graphite is an optically anisotropic, uniaxial crystal. Once formed, it is sufficiently refractory to survive for long periods in the diffuse ISM. Absorption arises from excitation of electrons with respect to the positive ion background of the solid, to produce resonance peaks in the optical constants, a phenomenon referred to in the bulk material as plasma oscillations. Excitation of π electrons in graphite produces absorption in the mid-ultraviolet. Excitation of σ electrons produces a feature in the EUV, centred near 800 Å, the presence of which could have implications for the shape of the observed extinction from 912 to 1500 Å, as discussed in section 3.5.1. It should be noted that hydrogenation suppresses absorption by localizing the electrons (Hecht 1986) and the carrier grains must, therefore, be assumed to have low hydrogen content.

There are two specific problems with the graphite identification. First, the question of its origin is raised by observations of circumstellar dust, which suggest that the solid particles ejected into the ISM by carbon stars are predominantly non-graphitic, a topic reviewed in chapter 7. A plausible mechanism for the production of graphite in the ISM must, therefore, be formulated to strengthen the case for its inclusion in grain models[4]. Second, the observed properties of the bump place tight, and arguably unrealistic, constraints on the nature of the particles (their size and/or shape and the presence or absence of surface coatings). In the remainder of this section, we assess the feasibility of graphite as the carrier of the bump and examine the alternatives.

The observed properties of the bump strongly suggest that the 2175 Å absorbers are in the small particle limit, i.e. they have dimensions $a \ll \lambda$, which, in practical terms, means $a \leq 0.01$ μm for graphite. Above this limit, λ_0 increases systematically with particle size as a result of scattering, which contributes to extinction predominantly on the long-wavelength side of the absorption peak. Any spatial fluctuation in mean grain size would then lead to both star-to-star changes in λ_0 and a systematic trend in λ_0 with γ, neither of which have been found; and so it becomes necessary to adjust the size distribution artificially to some preferred value. Calculations based on Mie theory for graphite 'spheres' in the small particle limit produce a feature displaced to significantly shorter wavelength compared with the observed feature. To match the observations, a particle size distribution that is sharply peaked at a specific grain radius ($a = 0.018 \pm 0.002$ μm) would be required for spheres. The assumption of sphericity, is, in any case, highly artificial for grains composed of an anisotropic crystalline material. Non-spherical grains are much more reasonable physically, as graphite particles minimize their free energy when flattened (Draine 1989a).

[4] Graphitization of amorphous carbon requires an input of energy, perhaps in the form of ultraviolet photons. We return to this topic in chapter 8 (section 8.5.3).

Calculations for non-spherical graphite grains show that shape as well as size affects the position and profile of the absorption feature (Savage 1975, Draine 1989a). If the particles are assumed to be spheroidal, an excellent fit to the observed mean profile can be obtained by appropriate choice of axial ratio b/a, as illustrated in figure 3.15. An important point to note is that this fit is obtained *within the small particle limit*, in contrast to the situation for spheres (Draine 1989a). However, the goodness of fit depends on both b/a and the orientation of the crystal axis relative to the geometrical axis (Savage 1975) and so the properties of the particles remain highly constrained: a tight restriction on shape has replaced a tight restriction on size.

Given the crucial role played by Mie theory calculations in the identification of the bump, it is appropriate to review the reliability of the modelling process. There are two distinct issues – the propriety of the technique and the accuracy of the laboratory data that it uses. The reliability of calculations based on bulk optical constants must decline at very small particle sizes: it is intuitively obvious that if a solid is repeatedly subdivided, its optical properties must ultimately deviate from those of the bulk material as molecular orbital theory takes over from solid-state band theory. The grains invoked to explain the bump are probably at an intermediate point where the proximity of surfaces may significantly modify the behaviour of internal electrons (see Bohren and Huffman 1983: p 335). Errors of measurement for bulk optical constants are a more readily quantifiable source of uncertainty. In the case of graphite, data are required in two planes – parallel and perpendicular to the crystal axis – and in both cases there is substantial variation between datasets published by different authors (see Huffman 1977, 1989, Draine and Lee 1984 and Draine and Malhotra 1993 for further discussion and references). Extinction curves measured directly for laboratory-manufactured smokes provide a potentially valuable comparison for those generated from Mie theory but difficulties are encountered in determining and controlling the size distribution and degree of crystallinity of the samples. Day and Huffman (1973) demonstrated the presence of an absorption feature near 2200 Å in graphite smoke but noted discrepancies in the shape of the profile that could be due to clumping of particles in the sample. Stephens (1980) presented extinction measurements for amorphous carbon grains that are in reasonable agreement with the equivalent Mie calculations (but do not match the interstellar feature). In general, differences between the directly measured and calculated extinctions are as likely to be caused by saturation effects, uncertainties in the parameters of the laboratory analogues, or errors in the bulk optical constants used in the calculations, as by a breakdown in Mie theory.

One of the greatest challenges to the graphite (or, indeed, any) model for the bump feature is to explain the variations in its width whilst conserving the peak wavelength. Proposed mechanisms for width variation include particle clustering, mantle growth, compositional inhomogeneities, porosity variations and surface effects. Rouleau *et al* (1997) show that particle clustering can result in width changes of appropriate magnitude; however, the peak position, although conserved, falls at shorter wavelength than the observed feature in the small particle limit, in common with calculations for small spheres. Mathis (1994) discusses mantling; he shows that the observational properties of the feature can be met with graphite cores coated with absorptive mantles, provided that both the shape of the cores and the optical constants of the mantles are highly constrained. Mantles broaden the feature, so unmantled grains are associated with the narrowest bumps. The mantles may contain metals or PAHs but cannot be purely ices or amorphous carbon. Perrin and Sivan (1990) investigate the effects of impurities and porosity on unmantled grains. Although generally discouraging, their results for graphite with amorphous carbon inclusions do show that changes in width of $\sim$10% as a function of impurity concentration may be accompanied by changes in position of only $\sim$0.5% in the small particle limit. Draine and Malhotra (1993) argue that the width changes must be caused by some systematic modification of the optical properties of the graphitic material itself. Possibilities include varying hydrogenation, varying crystallinity and changes in electronic structure caused by adsorption and desorption of atoms from the surface.

Other carbon-based materials have been suggested as carriers of the bump. Sakata *et al* (1977, 1983) demonstrate the presence of absorption near 2200 Å in the spectra of carbonaceous extracts from the Murchison meteorite and synthetic quenched carbonaceous composites (QCCs). Polycyclic aromatic hydrocarbons (PAHs) and buckminsterfullerene (C_{60}) have also been widely discussed (Léger *et al* 1989, Krätschmer *et al* 1990, Joblin *et al* 1992, Duley and Seahra 1998, Arnoult *et al* 2000). PAHs, in particular, are known to exist in ISM from their infrared emission features (section 6.3.2) and must contribute to the UV extinction at some level. A common characteristic of QCCs and PAH clusters is that they produce absorptions near 2200 Å that are broader than the observed feature. In the case of PAHs, this arises when the electronic absorptions of several different species are superposed (see Léger *et al* 1989); it is possible to obtain more realistic simulations with specific molecules (Duley and Seahra 1998). However, PAHs also generally display absorptions in the 2400–4000 Å wavelength interval and no corresponding features have been observed in the ISM. Results reported by Krätschmer *et al* (1990) indicate a similar objection to C_{60}. Finally, synthetic carbon nanoparticles condensed in a hydrogen-rich atmosphere have been shown to absorb near 2175 Å (Schnaiter *et al* 1998). Although lacking long-range order, these particles are likely to contain a mix of hybridizaton states that include graphitic subunits. The profile of the absorption feature is Drude-like, displaying width variations that depend on the degree of clustering of the particles.

O-rich materials have also been proposed as the carrier of the bump but none are now thought to be viable. A dielectric such as a silicate will absorb continuously in the UV at sufficiently high photon energies, due to excitation of electrons to the conduction band. The rapid onset of absorption can coincide with a rapid decrease in scattering, such that the net effect on the extinction curve could be to simulate a broad peak near the absorption edge. Huffman and Stapp (1971) noted that this occurs near 2175 Å in enstatite ($MgSiO_3$) spheres of radius

$a = 0.06\ \mu$m. However, the feature position and shape are extremely sensitive to particle radius (even more so than in the case of graphite), requiring unreasonable fine-tuning of the size distribution. Steel and Duley (1987) inferred, on the basis of photoexcitation spectra, the possible presence of an absorption feature near 2175 Å associated with OH^- ions at low-coordination sites on the surfaces of silicate particles in the small particle limit, but as no laboratory measurements have been obtained to support this possibility, it cannot be subjected to critical analysis. Finally, absorptions near 2175 Å in small MgO and SiO_2 particles have been discussed by MacLean *et al* (1982) but these oxides produce stronger absorptions in the 1200–1600 Å region that have no counterparts in the observed extinction.

To summarize, graphite or partly graphitized carbon grains remain the most likely identification of the 2175 Å bump. As the feature is purely absorptive in character, the particles responsible must be small, with dimensions $\sim$0.01 μm or less and the position of the feature is then independent of size but dependent on shape: the mean profile may be fitted by small spheroids of appropriate axial ratio, requiring some 20% of the solar abundance of carbon to be in such particles. Critical observational tests of the graphite hypothesis might be provided by searches for other predicted spectral signatures. The EUV feature near 800 Å and its possible contribution to the FUV extinction is of great interest but such observations will be extremely difficult. Graphite also has an infrared feature at 11.5 μm (Draine 1984) but it is so weak that observational confirmation does not seem to be feasible. Further laboratory work and modelling calculations are needed, e.g. to test the various mechanisms proposed to explain the changes in width.

3.6 Structure in the visible

Having discussed at length the structure of the UV extinction curve, we now focus on the visible region. Sensitive studies have demonstrated the occurrence of discrete features or structure (e.g. Whiteoak 1966, York 1971, Herbig 1975) that vary greatly in scale: the broadest, termed 'very broadband structure' (VBS), may be 500–1000 Å or more in extent; narrower features include the 'diffuse interstellar bands' (DIBs), with typical widths in the range 1–30 Å.

The VBS may be illustrated by plotting residuals with respect to a linear baseline fit to the extinction curve in the visible against wavenumber, as illustrated in figure 3.16. The resulting profile shape may be described as a trough centred near 1.8 μm^{-1} and a peak centred near 2.05 μm^{-1}. Note that the precipitous drop in residuals with increasing wavenumber beyond 2.2 μm^{-1} is caused primarily by the change in the slope of the extinction curve in the blue–violet region (the 'knee' region of figure 3.4). Detailed discussion of the observed properties of the VBS may be found in Whiteoak (1966), van Breda and Whittet (1981), Reimann and Friedemann (1991) and Bastiaansen (1992).

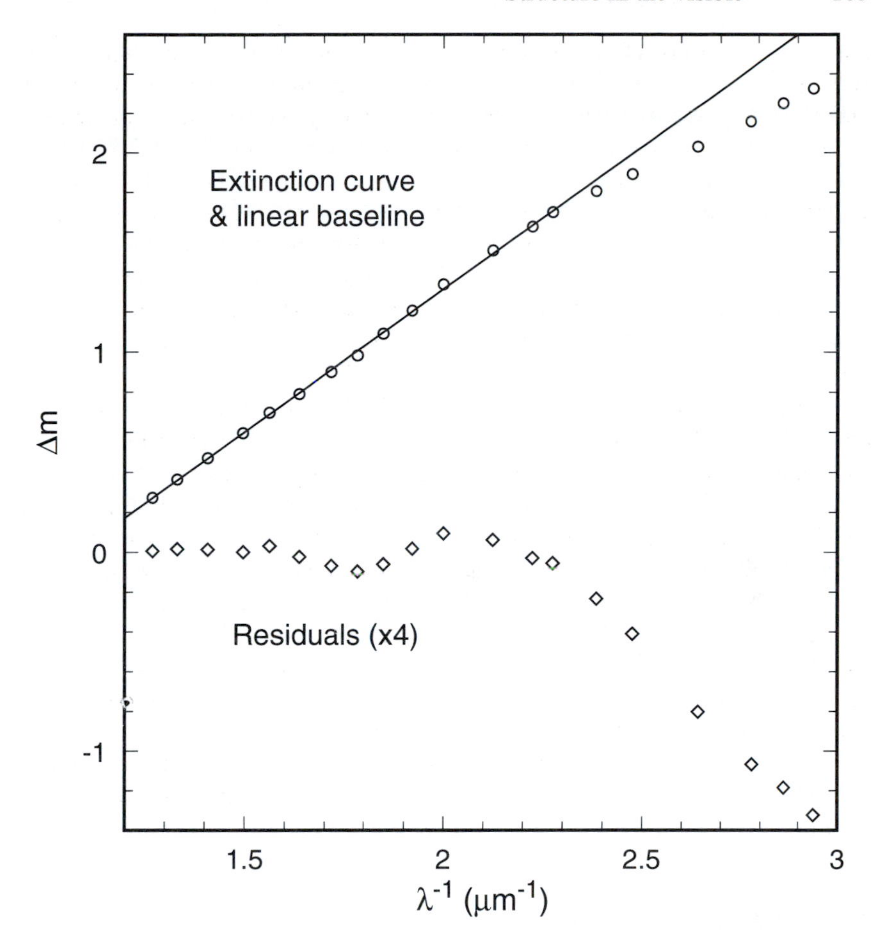

Figure 3.16. Very broadband structure (VBS) in the visible region of the extinction curve. The mean extinction curve for 19 reddened stars (circles) is from Bastiaansen (1992). A linear baseline is fit to the curve between 1.5 and 2.2 $\mu\mathrm{m}^{-1}$. Residuals obtained by subtracting the linear baseline from the extinction curve are plotted below. The VBS is characterized by the trough near 1.8 $\mu\mathrm{m}^{-1}$ and adjacent peak near 2.05 $\mu\mathrm{m}^{-1}$ in the residuals.

The origin of the VBS is unknown. Proposals include a broad absorption feature centred near 2.05 $\mu\mathrm{m}^{-1}$ or an emission (luminescence) feature centred near 1.8 $\mu\mathrm{m}^{-1}$ (Jenniskens 1994 and references therein). Structure in the optical constants of a continuous absorber such as magnetite (Fe_3O_4) have also been discussed (Huffman 1977) but attempts to model the profile are unconvincing (Millar 1982). The proximity of the knee is a complicating factor. The latter is attributed to a reduction in extinction efficiency for classical-sized dielectric grains as the wavelength becomes smaller than typical grain dimensions and this should lead to a smooth, continuous reduction in extinction slope. The knee may thus mask the true extent of the VBS; Jenniskens (1994) finds a correlation between VBS amplitude and the slope of the extinction curve *beyond* the knee, a result which suggests that the VBS extends into the ultraviolet. A linear baseline (figure 3.16) may not be the most appropriate choice. Jenniskens (1994) proposes the onset of continuous absorption at $\lambda^{-1} > 1.8\ \mu\text{m}^{-1}$ in small amorphous carbon grains as the most likely cause of the VBS.

The presence of diffuse absorption features in the optical spectra of reddened stars has been known for many years (see Herbig 1995 and Tielens and Snow 1995 for extensive reviews). Their interstellar origin is established on the basis of correlations between their strength and parameters of the dust or gas, such as reddening or atomic hydrogen column density. Some 130 DIBs are known in total, spanning the wavelength range 0.4–1.3 μm (see Herbig 1995 for a catalogue). Some of the most widely observed and discussed DIBs include those at 4428, 5780, 5797, 6177, 6203 and 6284 Å[5]. A portion of the DIB spectrum is illustrated in figure 3.17. Sharp features such as 5780, 5797 and 6203 (FWHM $\sim$ 2 Å) are juxtaposed with broad, shallow features such as 5778 and 6177 (FWHM $\sim$ 20 Å).

The identity of the DIB carrier(s) is a long-standing problem that has simultaneously fascinated and frustrated researchers for many decades. The various proposals are reviewed in detail by Herbig (1995). Although numerous, the DIBs are weak and the sum of their absorptions is very small, e.g. in comparison to the 2175 Å feature. Thus, the absorbers need not be very abundant. The sheer number of known DIBs, and their widespread distribution across the optical spectrum, strongly suggest that more than one carrier is involved: a single species of forbidding complexity would be needed to account for all of them (Herbig 1975). Further support for multiple carriers arises from intercorrelations of the features with each other and with reddening, suggesting the existence of several 'families' (e.g. Krelowski and Walker 1987, Moutou *et al* 1999). Origins in both dust grains and gaseous molecules have been proposed. Features produced by solid-state transitions in the large-grain population should exhibit changes in both profile shape and central wavelength with grain size, as previously discussed in the context of the 2175 Å feature (section 3.5), and emission wings would be expected for radii $>0.1\ \mu$m (Savage 1976); no such effects have been observed. There is also a lack of polarization in the features that might link them to the larger (aligned) grains (section 4.3.3). If the carriers are solid particles, they must be very small compared with the wavelength.

The possibility of a small-grain carrier for the DIBs may be examined further by searching for correlations between their strengths and parameters of the UV extinction curve. The ratio of equivalent width to reddening (W_λ/E_{B-V}) provides a convenient measure of DIB production efficiency per unit dust column in a

[5] By convention, each DIB is identified by its central wavelength in Å to four significant figures. One exception is that at 4428 Å, amongst the first to be studied, which traditionally has been rounded to 4430.

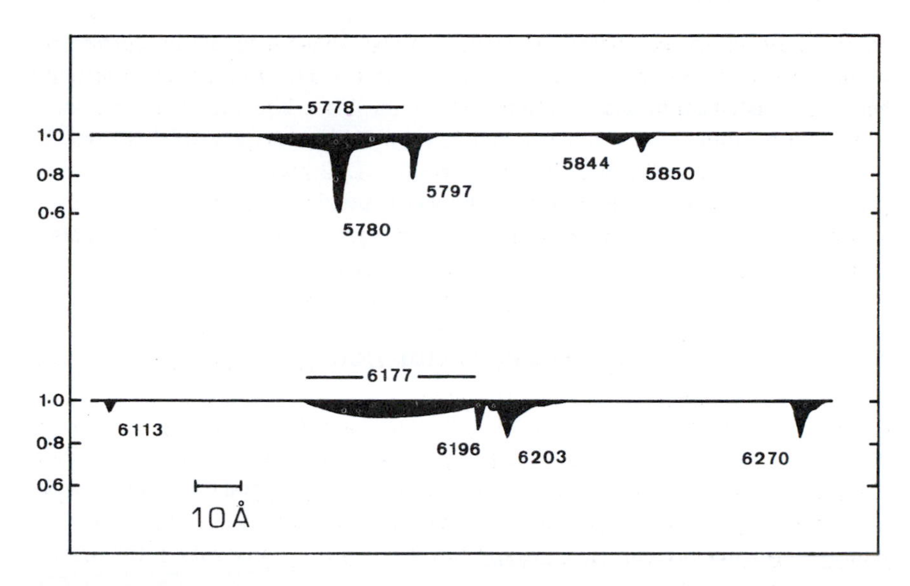

Figure 3.17. A schematic representation of diffuse bands in the yellow–red region of the spectrum, based on intensity traces for the reddened star HD 183143 (Herbig 1975). Interstellar absorptions are shown in the wavelength range 5730–5900 Å (top) and 6110–6280 Å (bottom). Photospheric and telluric features in the spectra are eliminated with reference to corresponding data for a comparison star (β Orioni) of similar spectral type and low reddening.

given line of sight[6]. Typically, this ratio displays a weak positive correlation with the corresponding relative strength of the 2175 Å bump and a weak negative correlation with the amplitude of the FUV extinction rise (Witt *et al* 1983, Désert *et al* 1995). These results clearly fail to establish any firm associations: on the contrary, it can be concluded that the DIB carriers and the bump carriers are *not* directly related, as the bump is less susceptible to variation than the DIBs and is still present in lines of sight where the DIBs are negligible (Benvenuti and Porceddu 1989, Désert *et al* 1995). The observations merely suggest that there is some correlated behaviour in their response to environment.

There has been a degree of consensus in the recent literature that the most plausible candidates for the DIBs are carbonaceous particles that might be classed (according to taste) as very small grains or large molecules – specifically, PAHs and fullerenes (e.g. Foing and Ehrenfreund 1997, Sonnentrucker *et al* 1997, Salama *et al* 1999, Galazutdinov *et al* 2000; see Herbig 1995 for a critical review of earlier literature). Ionized species are favoured over neutral species as they have stronger features in the visible. Observations show that the DIBs become relatively weak inside dark clouds (Snow and Cohen 1974, Adamson

[6] Strictly speaking, W_λ / A_V is more appropriate.

et al 1991), consistent with a reduction in the abundance of the carriers in regions shielded from ionizing radiation. The weak negative correlation found between DIB strength and the amplitude of the FUV rise might be explained if they are produced by ionized and neutral PAHs, respectively (Désert *et al* 1995). However, identification of specific DIBs with specific species is problematic, as the techniques used to study their spectra in the laboratory introduce wavelength shifts and line broadening (Salama *et al* 1999) and this severely hinders comparison with interstellar spectra.

3.7 Modelling the interstellar extinction curve

To construct a model for interstellar extinction, one must assume a composition and a size distribution for the particles to be included. The composition is represented in the calculations by the complex refractive index $m(\lambda)$ (equation (3.10)), data for which must be available over the spectral range of interest. In practice, the size distribution function $n(a)$ is split into discrete intervals, with a treated as a constant within each interval. The extinction efficiency factor Q_{ext} is then calculated for each chosen value of a, λ and $m(\lambda)$ and the total extinction follows from equation (3.7). Calculation of Q_{sca} also yields the albedo (equation (3.14)). Two or more separate populations of particles (e.g. C rich and O rich) are generally included and results are summed at each wavelength. Computed extinction curves are compared with observations and fits are refined by adjusting free parameters such as the relative contributions of the different materials and their size distributions.

A good model for the extinction curve reproduces its form and variability, without violating constraints placed by other evidence, such as element abundances and depletions (sections 2.4–2.5) and scattering properties (albedo and asymmetry factor; section 3.3.2). The goal is to find a unique model that satisfies all known constraints but this ideal has yet to be accomplished (see section 1.6 for a review). Models that make different assumptions regarding the nature of the grains are capable of fitting the extinction curve equally well. In terms of composition, the most specific information we have is evidence for the presence of silicates and PAHs (from infrared observations) and of graphitic carbon (the most likely candidate for the 2175 Å bump). In terms of size, large classical grains are needed to account for the visual–infrared extinction and the underlying (linear) component of the UV extinction, whilst much smaller grains are needed to explain both the bump and the FUV rise.

The model formulated by Mathis *et al* (1977, MRN; see also Draine and Lee 1984, Weingartner and Draine 2001) postulates two distinct populations of uncoated refractory particles, composed of graphite and silicates. Each population follows a power-law size distribution between minimum and maximum particle radii:

$$n(a) \propto a^{-q} \qquad (a_{\text{min}} < a < a_{\text{max}}) \tag{3.47}$$

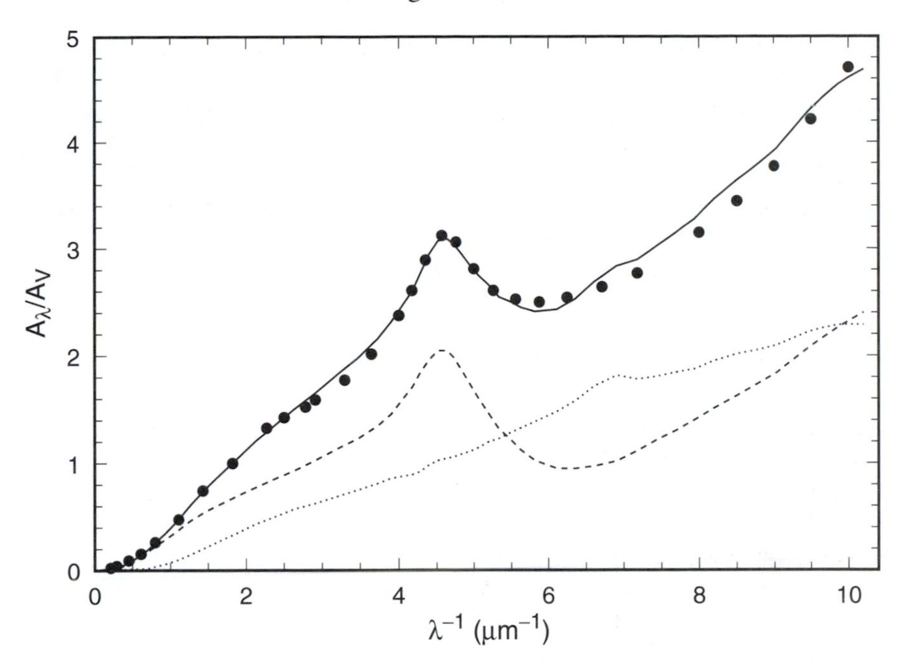

Figure 3.18. A fit to the extinction curve based on the 'MRN' two-component model (in the version of Draine and Lee 1984). The total extinction predicted by the model (continuous curve) is the sum of the contributions from graphite grains (broken curve) and silicate grains (dotted curve). The mean observational curve (table 3.1) is plotted as full circles.

with $n(a) = 0$ otherwise. An acceptable fit to the mean extinction curve (figure 3.18) is obtained with values of $a_{\rm min} \approx 0.005$ μm, $a_{\rm max} \approx 0.25$ μm and $q = 3.5$. Spatial variations may be accommodated by adjusting these parameters (Mathis and Wallenhorst 1981): larger values of $a_{\rm min}$ and/or $a_{\rm max}$ are typically needed to fit curves with $R_V > 3.1$. Graphite contributes to the extinction at all wavelengths in the MRN model and is largely responsible for the FUV rise as well as the bump. The power-law form of the size distribution is physically reasonable, as it is consistent with predictions for particles subject to collisional abrasion (Biermann and Harwit 1980). The assumption of a sharp cut-off size is unrealistic, however, and Kim *et al* (1994) propose a smooth transition to an exponential decay in $n(a)$ as a becomes large (see section 7.3.2 for further discussion).

In principle, the visible–infrared segment of the extinction curve may be explained by a single big-grain population rather than the summed contributions of two independent populations, as originally shown by Oort and van de Hulst (1946). However, the big grains must be compositionally heterogeneous, containing both silicates and some form of solid carbon: this is demanded by a

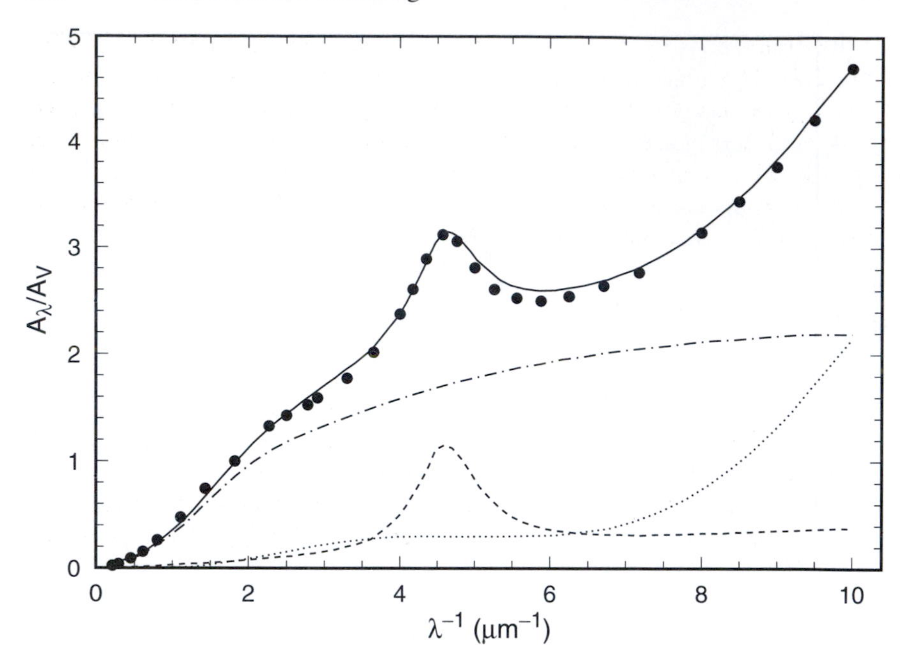

Figure 3.19. A fit to the observed extinction curve based on a three-component model (after Désert *et al* 1990). The total extinction predicted by the model (continuous curve) is the sum of the contributions from large silicate/carbon composite grains (dot–dash curve), small graphitic grains (broken curve) and PAHs (dotted curve). The mean observational curve (table 3.1) is plotted as full circles.

number of current observational constraints, including abundances, albedo data and the strength relative to A_V of the infrared silicate features. Composite silicate/carbon mixtures and silicate cores with carbonaceous mantles produce qualitatively similar results (Hong and Greenberg 1980, Duley *et al* 1989a, Mathis and Whiffen 1989, Désert *et al* 1990, Li and Greenberg 1997). For illustration, the model of Désert *et al* (1990) is shown in figure 3.19. This model adopts a power-law size distribution of big silicate/carbon grains, with $q = 2.9$ in equation (3.47), together with small graphite grains and PAHs to explain the bump and the FUV rise, respectively. Values of $a_{\rm min}$ and $a_{\rm max}$ are chosen to give a continuous size distribution, i.e. $a_{\rm max}$ for PAHs is set equal to $a_{\rm min}$ for graphite and $a_{\rm max}$ for graphite is set equal to $a_{\rm min}$ for big grains. Note that interpretation of the FUV rise in terms of PAHs is not unique; very small silicate grains (Mathis 1996a, Li and Draine 2001) may also contribute.

All of these proposals make heavy demands on the reservoir of available elements, as determined by abundance and depletion data. The MRN model, for example, requires ~250 ppm of C in graphite and ~32 ppm of Si in silicates (see Li and Greenberg 1997): it thus consumes essentially all of the available atoms for solar reference abundances and seriously exceeds availability for 63% solar (table 2.2). The model of Li and Greenberg (1997) makes less severe demands, apparently because of the cylindrical shape adopted for the big grains and the inclusion of oxygen in the organic refractory mantles. If the grains are composite, utilization of the elements is optimized in terms of opacity per unit mass if they are also *porous* (Mathis 1996a, Fogel and Leung 1998) and the extinction can then be accounted for with 80% solar reference abundances.

Recommended reading

- *Absorption and Scattering of Light by Small Particles*, by Craig F Bohren and Donald R Huffman (John Wiley and Sons, 1983).
- *Interstellar Dust and Extinction*, by John S Mathis, in Annual Reviews of Astronomy and Astrophysics, vol 28, pp 37–70 (1990).
- *Correcting for the Effects of Interstellar Extinction*, by Edward L Fitzpatrick, in Publications of the Astronomical Society of the Pacific, vol 111, pp 63–75 (1999).

Problems

1. The star o Scorpii has the following magnitudes in the standard Johnson notation: $V = 4.55$, $B - V = 0.82$, $K = 1.62$, and the following intrinsic colours apply to a star of its spectral type: $(B - V)_0 = 0.10$; $(V - K)_0 = 0.37$. Calculate the ratio of total-to-selective extinction (R_V) and visual extinction (A_V) in this line of sight. If the star lies in the Sco OB2 association (distance 160 pc) deduce its absolute visual magnitude. What would be the effect on your result of assuming the average value of $R_V = 3.05$ instead of your calculated value?
2. Show that the dust-to-gas ratio is related to the ratio of visual extinction to hydrogen column density by the approximation

$$Z_\mathrm{d} \simeq 1.6 \times 10^{19} s \left(\frac{m^2 + 2}{m^2 - 1}\right) \left(\frac{A_V}{N_\mathrm{H}}\right)$$

where s and m are the specific density and refractive index of the grain material and SI units are assumed.

3. (a) In the small particle approximation, the efficiency factor for absorption due to grains of refractive index $m = n - ik$ is given by

$$Q_\mathrm{abs} \simeq \frac{a}{\lambda} \left(\frac{48\pi nk}{(n^2 - k^2 + 2)^2 + 4n^2k^2}\right).$$

Suppose that the grains are composed of material with n constant and k a function of wavelength, such that $k(\lambda) \ll 1$ at all λ. What functional

Table 3.3. Q_{ext} values as a function of $X = 2\pi a/\lambda$ for spherical ice grains with constant refractive index $m = 1.33 + 0i$ (see problem 4).

X	Q_{ext}	X	Q_{ext}	X	Q_{ext}
0.5	0.022	5.5	3.824	10.5	1.890
1.0	0.127	6.0	3.916	11.0	1.675
1.5	0.400	6.5	3.934	11.5	1.626
2.0	0.835	7.0	3.736	12.0	1.671
2.5	1.341	7.5	3.626	12.5	1.824
3.0	1.723	8.0	3.275	13.0	1.934
3.5	2.330	8.5	2.989	13.5	2.088
4.0	2.848	9.0	2.725	14.0	2.330
4.5	3.100	9.5	2.374	14.5	2.442
5.0	3.516	10.0	2.198	15.0	2.637

form is implied for $k(\lambda)$ by the observation that the extinction A_λ follows a power law of index $\sim$1.8 in the infrared if we have absorption-dominated extinction in the small particle limit?

(b) By considering absorption and scattering produced by 'classical' ($a \sim 0.1\ \mu$m) sized grains with $n = 1.5$ and $k(2.2\mu\text{m}) = 0.1$, show that the assumption that extinction is dominated by absorption in the small particle limit is reasonable at $\lambda = 2.2\ \mu$m.

4. Table 3.3 gives values of the extinction efficiency factor Q_{ext} as a function of the dimensionless size parameter $X = 2\pi a/\lambda$, calculated using Mie theory for dielectric spheres of constant refractive index $m = 1.33 - 0i$ (appropriate to ice in the spectral range 0.16–2.5 μm). Use this table to plot a theoretical extinction curve for 'classical' ice grains of constant radius $a = 0.3\ \mu$m over a suitable range of wavenumber (λ^{-1}). Compare your result with the mean observed interstellar extinction curve (table 3.1). Note that it will be necessary to normalize the theoretical data in the same way as the observational data, i.e. to $E_{B-V} = 1$, interpolating where necessary. Deduce the ratio of total to selective extinction for your theoretical curve.

5. Comment on the suitability of classical dielectric particles as a component of models for interstellar extinction.

6. Show that the abundance of the carrier of the 2175 Å feature relative to hydrogen may be expressed in terms of the strength parameter A_{bump}, the oscillator strength f and the hydrogen gas to reddening ratio by

$$\frac{N_{2175}}{N_H} = 1.13 \times 10^{14} \frac{A_{bump}}{(N_H/E_{B-V})f}$$

assuming SI units.

7. (a) The star ρ Oph has a reddening of $E_{B-V} = 0.47$, a ratio of total-to-

selective extinction of $R_V = 4.3$ and a total hydrogen column density (measured from Lyman-α absorption) of $N_H = 7.2 \times 10^{25}$ atoms m^{-2}. Deduce the value of Z_d (equation (3.33)) for the line of sight to ρ Oph, assuming dielectric grains of refractive index $m = 1.50 - 0i$ and density $s = 2500$ kg m^{-3}. Compare your result with the average for the diffuse ISM. Summarize and critically assess the suggestion that the unusual value of Z_d toward ρ Oph is a consequence of grain growth in the dark cloud that obscures ρ Oph (see Jura 1980).

(b) Calculate the abundance of carbon required to be in small graphite grains toward ρ Oph if they are responsible for the 2175 Å feature in this line of sight, assuming that the oscillator strength per carbon atom in graphite is $f \sim 0.16$. The strength of the 2175 Å feature toward ρ Oph is listed in table 3.2. Compare your result with the Solar System abundance of carbon.

8. The polycyclic aromatic hydrocarbon coronene has an absorption feature at $\lambda \approx 3000$ Å with oscillator strength $f \sim 0.06$ per C atom. This feature is not observed in the interstellar extinction curve and an upper limit $W_\nu < 8 \times 10^{13} E_{B-V}$ (Hz) is set on its equivalent width toward reddened stars. Estimate an upper limit on the abundance of C in interstellar coronene. Express your answer as a fraction of the solar C-abundance. (*Note*: $(4\epsilon_0 m_e c)/e^2 = 3.8 \times 10^5$ in SI units.)

Chapter 4

Polarization and grain alignment

"...needle-like grains tend to spin end-over-end, like a well-kicked American football."

C Heiles (1987)

The interstellar medium is responsible for the partial plane polarization of starlight. The interstellar origin of this phenomenon is not in doubt as, in general, only reddened stars are affected and there is a positive correlation between polarization and reddening. The accepted model for interstellar polarization is linear dichroism (directional extinction) resulting from the presence of asymmetric grains that are aligned by the galactic magnetic field. If the direction of alignment changes along the line of sight, the interstellar medium also exhibits linear birefringence, producing a component that is circularly polarized. Studies of interstellar linear and circular polarization are important because they provide information both on grain properties (size, shape, refractive index) and on the galactic magnetic field. The identity of the alignment mechanism has proved to be an intriguing problem in grain dynamics that has teased theorists for many years. In this chapter, we begin by extending the discussion of extinction by small particles (chapter 3) to include the production of polarization in the transmitted beam. Observational results and their implications are discussed in detail in sections 4.2 and 4.3. Models for the spectral dependence of interstellar polarization are reviewed in section 4.4 and the problem of the alignment mechanism is considered in the final section. We are concerned here primarily with polarization of starlight over the same spectral range that was considered for extinction in chapter 3. Infrared spectral absorption features that display polarization enhancements are discussed in chapter 5 and polarization associated with infrared continuum emission is considered in chapter 6.

4.1 Extinction by anisotropic particles

A beam of initially unpolarized light transmitted through a dusty medium will become partially plane polarized if two conditions are met: (i) individual dust particles are optically anisotropic and (ii) there is net alignment of the axes of anisotropy. The most likely source of anisotropy is asymmetry in the shape of the particle. Another possibility is anisotropy of the grain material itself (as is the case for graphite, for example) but the optic axes of such particles will probably not become aligned unless they are also asymmetric in shape. In this section, it is assumed that asymmetric grains become aligned in the ISM. The mechanism that leads to alignment will be discussed later. As real interstellar grains may presumably assume an almost endless variety of shapes and structures, some generalization is inevitable when attempting to model their polarizing properties. It is convenient to assume simple, axially symmetric forms such as cylinders or spheroids, as their extinction cross sections may be calculated by a straightforward extension of the Mie theory for spheres (van de Hulst 1957, Greenberg 1968, Bohren and Huffman 1983). A more sophisticated approach is to use methods such as the discrete dipole approximation (section 3.1.4) to simulate the optical properties of grains of any desired shape.

To illustrate how starlight is polarized by dust in the ISM, consider an ensemble of elongated grains such as long cylinders. Suppose that each grain is orientated with its long axis perpendicular to the direction of propagation of incident radiation: we may define $Q_{\parallel}$ and $Q_{\perp}$ as the values of the extinction efficiency $Q_{\rm ext}$ (section 3.1.1) when the $\boldsymbol{E}$-vector is parallel and perpendicular to the long axis of the grain, respectively. The anisotropy in physical shape introduces a corresponding anisotropy in extinction: because the $\boldsymbol{E}$-vector 'sees' an apparently larger grain in the parallel direction, we have $Q_{\parallel} \geq Q_{\perp}$. Calculated values of $Q_{\parallel}$ and $Q_{\perp}$ for long dielectric cylinders are plotted against the dimensionless size parameter $X = 2\pi a/\lambda$ in figure 4.1. The quantity $\Delta Q = Q_{\parallel} - Q_{\perp}$ is a measure of the resulting polarization. Note that polarization is, in general, small compared with extinction ($Q_{\parallel} \sim Q_{\perp} \gg \Delta Q$; the ΔQ curve in figure 4.1 has been scaled up by a factor of eight for display). ΔQ would be reduced further for other angles of incidence and would become zero for propagation along the axis of symmetry of the grain.

Considering the parallel and perpendicular cases in turn, the extinction produced by a medium containing identical, perfectly aligned particles of column density $N_{\rm d}$ is

$$\begin{aligned} A_{\parallel} &= 1.086 N_{\rm d} \sigma Q_{\parallel} \\ A_{\perp} &= 1.086 N_{\rm d} \sigma Q_{\perp} \end{aligned} \tag{4.1}$$

by analogy with equation (3.6), where σ is the cross-sectional area of each particle

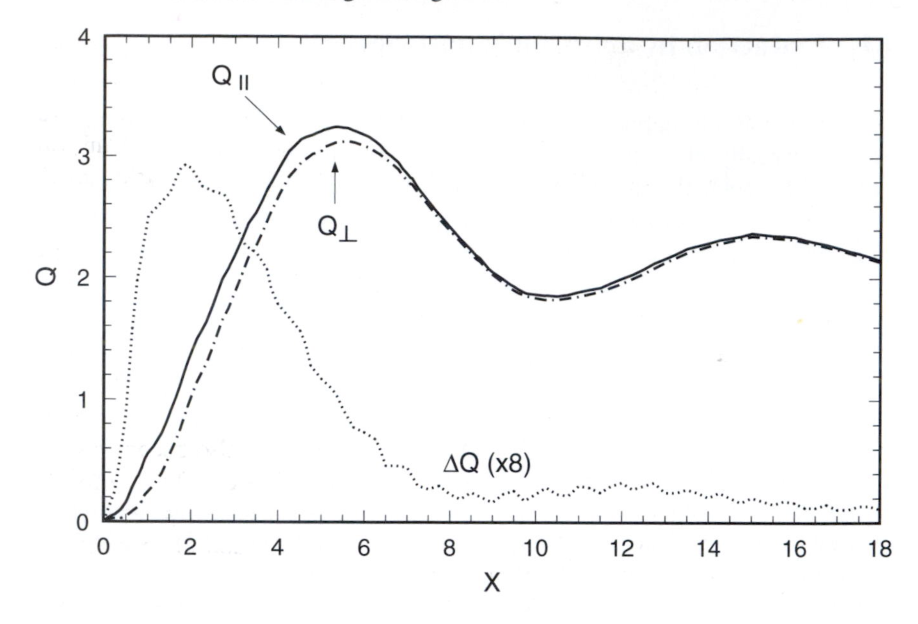

Figure 4.1. Extinction efficiency factors $Q_{\parallel}$ (continuous curve) and $Q_{\perp}$ (broken curve) plotted against $X = 2\pi a/\lambda$ for cylinders of radius a and refractive index 1.33 − 0.05i. The cylinders are assumed to be very long ('infinite') in comparison to their radii. Note that $Q_{\parallel} \geq Q_{\perp}$ for all values of X. The dotted curve is the difference, $\Delta Q = Q_{\parallel} - Q_{\perp}$, scaled up by a factor of eight. The equivalent calculation of Q_{ext} for spheres (section 3.1.1) would show a dependence on X qualitatively similar to the mean of $Q_{\parallel}$ and $Q_{\perp}$.

in the plane of the wavefront. The total extinction is

$$A = 1.086 N_{\mathrm{d}} \sigma \left\{ \frac{Q_{\parallel} + Q_{\perp}}{2} \right\} \tag{4.2}$$

and the amplitude of the resulting linear polarization is

$$\begin{aligned} p &= A_{\parallel} - A_{\perp} \\ &= 1.086 N_{\mathrm{d}} \sigma (Q_{\parallel} - Q_{\perp}). \end{aligned} \tag{4.3}$$

Thus, $p \propto \Delta Q$ for a given grain size. Both p and A may be evaluated as functions of wavelength from calculated values of $Q_{\parallel}$ and $Q_{\perp}$ for a chosen grain model.

The ratio of polarization to extinction

$$\frac{p}{A} = 2 \left\{ \frac{Q_{\parallel} - Q_{\perp}}{Q_{\parallel} + Q_{\perp}} \right\} \tag{4.4}$$

is a measure of the efficiency of polarization production. It depends on both the nature of the grains and the efficiency with which they are aligned in the line of sight. The most efficient polarizing medium conceivable is one that contains infinite cylinders with diameters comparable to the wavelength, perfectly aligned such that their long axes are parallel to one another and perpendicular to the line of sight. Calculations for this scenario place a theoretical upper limit on the polarization efficiency: at visual wavelengths, particles with dielectric optical properties give

$$\frac{p_V}{A_V} \leq 0.3. \tag{4.5}$$

The corresponding observational result is deduced in the following section.

It is informative to compare the behaviour of polarization and extinction in figure 4.1. Note that for constant grain size, X varies as λ^{-1}. Both polarization and extinction become vanishingly small at wavelengths long compared with a ($X \rightarrow 0$); however, in contrast to extinction, the polarization also becomes very small as the wavelength becomes short compared with a ($X > 8$ in this example). A peak in ΔQ appears at an intermediate value of X. Comparing results for dielectric cylinders of differing refractive index n, this is found to occur when $X(n-1) \sim 1$.

4.2 Polarimetry and the structure of the galactic magnetic field

4.2.1 Basics

In its simplest form, an astronomical polarimeter consists of a photoelectric photometer with a broadband filter and an analyser in the light path[1]. In this section, we consider observations obtained with a single colour filter such as the Johnson V (discussion of the spectral dependence of polarization is deferred to section 4.3). When partially plane-polarized light from a star is observed, intensity maxima and minima are recorded in orthogonal directions as the analyser is rotated. These measurements allow the amplitude and position angle of the polarization vector to be determined.

The amplitude, or degree, of polarization (P) is usually expressed as a percentage, defined by the equation

$$P = 100\left\{\frac{I_{\max} - I_{\min}}{I_{\max} + I_{\min}}\right\} \tag{4.6}$$

where I is the intensity. An alternative definition is the polarization in magnitude units, denoted by the lower-case symbol

$$p = 2.5 \log \frac{I_{\max}}{I_{\min}}. \tag{4.7}$$

[1] See Hough *et al* (1991) for a description of modern instrumentation.

This latter quantity is equivalent to the polarization defined in terms of model-dependent parameters in equation (4.3). We may easily show that P is proportional to p to a close approximation if the polarization is sufficiently small: from equation (4.7),

$$\frac{I_{\max}}{I_{\min}} \simeq 1 + \left\{\frac{\ln 10}{2.5}\right\} p$$

neglecting p^2 and higher powers; with $I_{\max} \simeq I_{\min}$ in equation (4.6),

$$\begin{aligned} P &\simeq 50 \left\{\frac{I_{\max}}{I_{\min}} - 1\right\} \\ &\simeq 46.05p. \end{aligned} \tag{4.8}$$

This approximation is generally valid for interstellar polarization.

The position angle of linear polarization is determined by the orientation of the analyser for maximum intensity, relative to some reference direction. This angle specifies the plane of vibration of the $\boldsymbol{E}$-vector in the transmitted beam projected on to the celestial sphere. For interstellar polarization, it is convenient to choose a reference direction with respect to the orientation of the Milky Way: the galactic position angle (θ_G) is defined as the angle between the $\boldsymbol{E}$-vector and the direction of the North Galactic Pole, measured counterclockwise (toward increasing galactic longitude) on the sky. Note that θ_G is bounded by the limits $0 \leq \theta_G \leq 180°$.

Polarization may be described more generally in terms of the Stokes parameters I, Q, U and V (e.g. Hall and Serkowski 1963, Spitzer 1978). For partially plane-polarized light, these are given by

$$I = I_{\max} + I_{\min} \tag{4.9}$$

$$Q = PI \cos 2(\theta_G - 90) \tag{4.10}$$

$$U = PI \sin 2(\theta_G - 90) \tag{4.11}$$

$$V = 0. \tag{4.12}$$

V/I denotes the circular polarization, which does, in reality, have a small but finite value for many lines of sight (section 4.3.5). The linear component is described by Q and U, which may also be expressed in magnitude units:

$$q = p \cos 2(\theta_G - 90) \tag{4.13}$$

$$u = p \sin 2(\theta_G - 90). \tag{4.14}$$

Spatial variations in these parameters provide a useful means of investigating alignment on the galactic scale.

4.2.2 Macroscopic structure

Our Galaxy is permeated by a magnetic field of mean flux density $B \approx 3\ \mu$G in the solar neighbourhood[2] and this field is responsible for grain alignment (section 4.5). Although some of the detailed physics involved is not yet fully understood, the general principles of magnetic alignment appear to be robust: grains align such that their longest axes are, on average, perpendicular to the mean field direction. The mean direction of the $\boldsymbol{E}$-vector in the transmitted beam is thus parallel to the mean field direction, i.e. the observed polarization traces the magnetic field on the sky and results for many stars yield two-dimensional maps of field structure. This technique has been used extensively to study the macroscopic structure of the magnetic field of our Galaxy (e.g. Mathewson and Ford 1970, Axon and Ellis 1976, Heiles 2000) and of others (Sofue *et al* 1986, Hough 1996, Alton *et al* 2000b).

A polarization map of the Milky Way is shown in figure 4.2. Consistent with the distribution of reddening material in our Galaxy (section 1.3), the most highly polarized stars generally lie within a few degrees of the galactic equator. Their polarization vectors show a tendency to align parallel to the equator in certain longitude zones (e.g. $\ell \approx 120°$) and to be randomly orientated in others (e.g. $\ell \approx 40°$), behaviour that hints at a correlation with galactic structure. At latitudes $|b| \sim 15°$, the most highly polarized stars tend to lie North of the equator toward the galactic centre ($\ell = 0°$) and South of the equator toward the anticentre ($\ell = \pm 180°$), consistent with an origin in dust associated with Gould's Belt. Stars polarized by dust in the Ophiuchus and Taurus dark clouds form prominent features at $(\ell,\ b) \approx (0,\ +15°)$ and $(175°,\ -15°)$, respectively. Considerable structure is also evident at high latitudes, such as a loop extending toward the North Galactic Pole from the Milky Way at $\ell \sim 30°$.

The behaviour of alignment with respect to galactic structure may be described in terms of the dependence of $\langle q \rangle$ on galactic longitude, ℓ, where $\langle q \rangle$ is the mean value of Stokes parameter q (equation (4.13)) for stars in a selected region of the Milky Way (Hall and Serkowski 1963, Spitzer 1978, Fosalba *et al* 2002). For alignment with the $\boldsymbol{E}$-vector predominantly parallel to the galactic plane ($\theta_G \simeq 90°$), we expect $\langle q \rangle$ to be positive and similar in magnitude to the mean value of p, whereas $\langle q \rangle$ is close to zero for random orientation and negative for net alignment of $\boldsymbol{E}$ perpendicular to the galactic plane. Figure 4.3 (upper frame) plots $\langle q \rangle$ against ℓ for various longitude zones along the Milky Way. A systematic variation is evident in $\langle q \rangle$ between near-zero and positive values, the sense of which is loosely represented by a sine wave with minima at $\ell \approx 45°$ and $-135°$. However, the intervening peaks have amplitudes that differ by a factor of about two. Also shown in figure 4.3 (lower frame) is the behaviour of the

[2] Note that magnetic flux densities are invariably expressed in cgs units in the astronomical literature: the equivalent SI unit is $1\ \mathrm{T} = 10^4\ \mathrm{G} = 10^{10}\ \mu\mathrm{G}$.

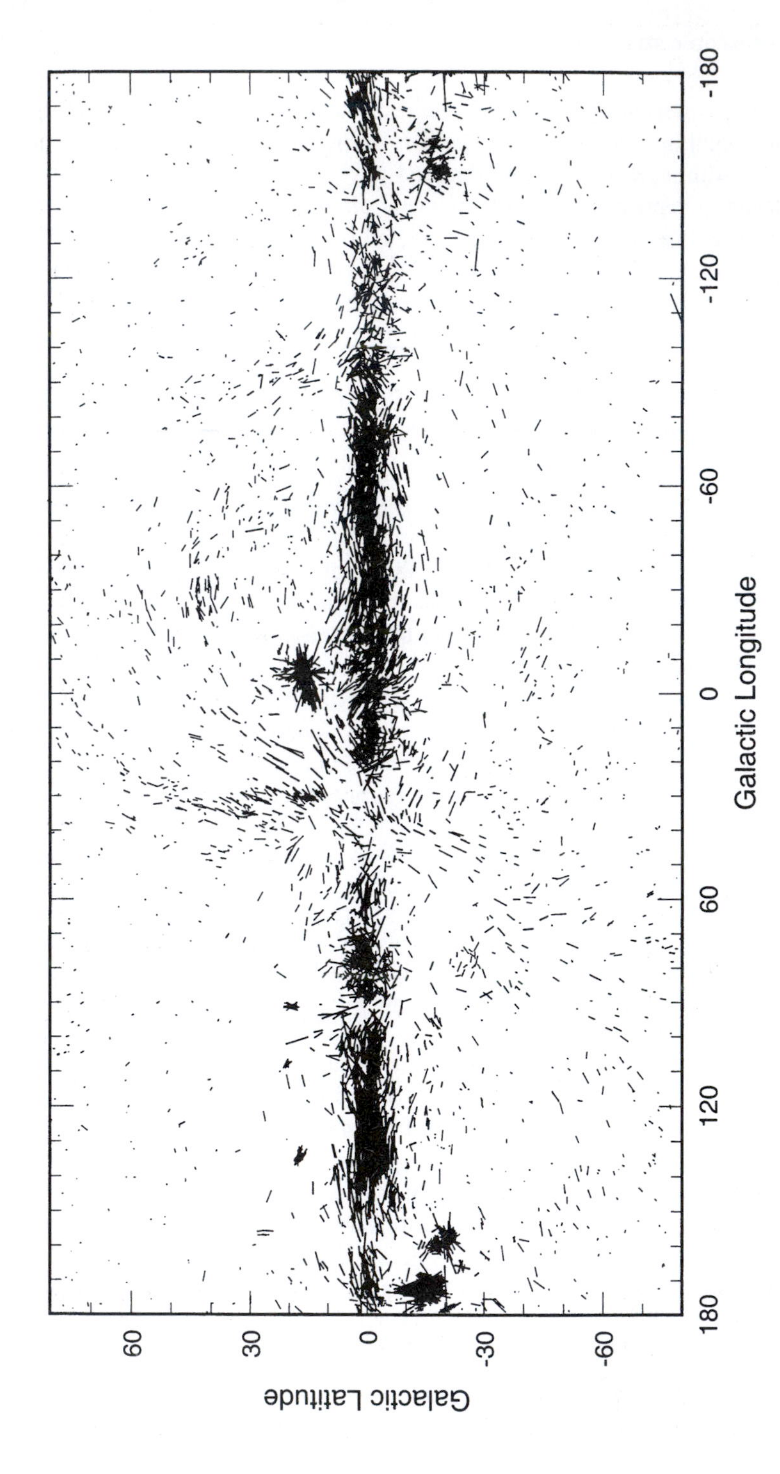

Figure 4.2. Distribution of linear polarization vectors in galactic coordinates. The length and orientation of each vector represents the polarization degree and position angle for a star at that locus. Data are from Heiles (2000) and references therein.

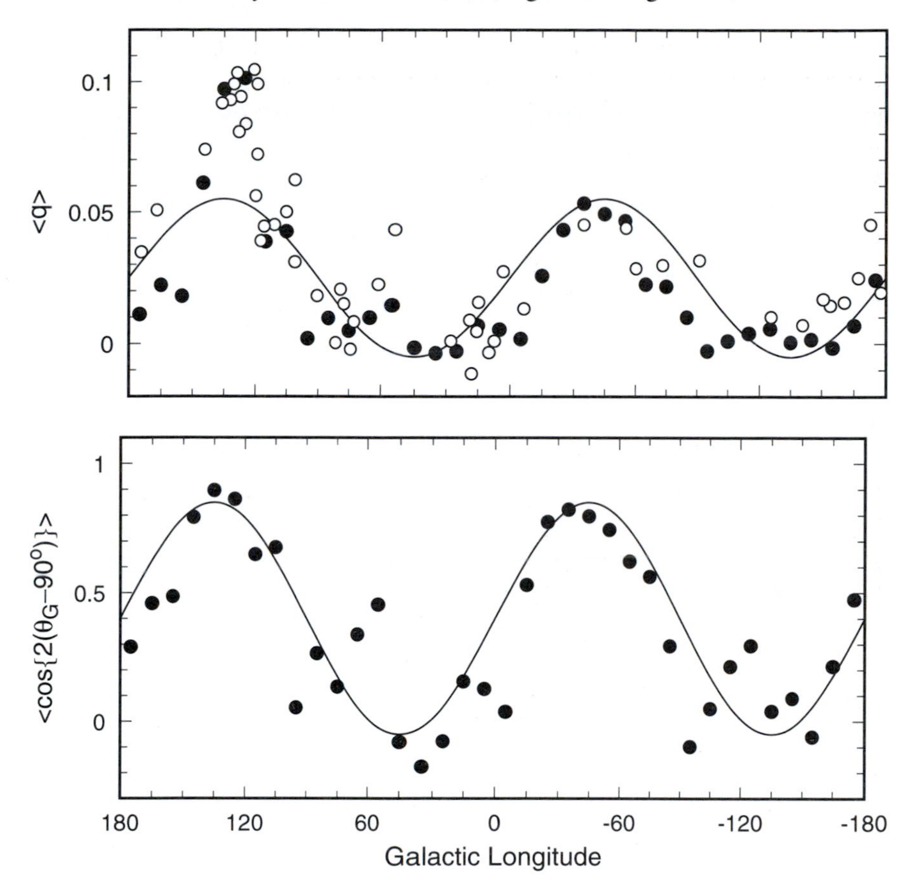

Figure 4.3. Plots of mean Stokes parameter q (upper frame) and its position-angle-dependent component (lower frame) against galactic longitude. Each point represents an average for many stars located within 10° of the galactic plane. Filled circles (both plots) are from the study of Fosalba *et al* (2002), which utilizes the database of Heiles (2000). Open circles (upper frame) are from Hall and Serkowski (1963). The sinusoidal fit in the lower frame (equation (4.16)) is also plotted (with appropriate scaling) in the plot for $\langle q \rangle$ above.

position-angle-dependent component of $\langle q \rangle$ alone, i.e.

$$f(\theta_G) = \langle \cos 2(\theta_G - 90) \rangle. \tag{4.15}$$

Variations in $f(\theta_G)$ with ℓ approximate much more closely to a sinusoidal form:

$$f(\theta_G) \approx 0.4 + 0.45 \sin 2(\ell + 90). \tag{4.16}$$

Note that the asymmetry in $\langle q \rangle$ must therefore arise from differences in polarization amplitude p. In any case, the sign of $\langle q \rangle$, which is determined by the sign of $f(\theta_G)$, confirms what may be surmised from figure 4.2: the polarization of starlight is consistent with alignment by a magnetic field directed predominantly parallel to the galactic plane. Peaks in $\langle q \rangle$ occur in directions that cross field lines, whereas $\langle q \rangle \simeq 0$ for directions along field lines. The minima in figure 4.3 occur in directions roughly parallel to the local (Cygnus–Orion) spiral arm, indicating that the magnetic field is directed along this spiral arm in the solar neighbourhood of the Milky Way. These results are consistent with data for other spiral galaxies, which show the same general correlation of field direction with spiral structure (e.g. Sofue *et al* 1986, Jones 1989a, Beck 1996).

4.2.3 Polarization efficiency

The degree of polarization in the visual waveband shows a distinct but highly imperfect correlation with reddening, illustrated in figure 4.4. The scatter is much greater than can be accounted for by observational errors alone and this plot thus demonstrates that the efficiency of the ISM as a polarizing medium is intrinsically non-uniform. A zone of avoidance is evident in the upper left-hand region of the diagram, the distribution of points being approximately bounded by the straight line

$$\frac{P_V}{E_{B-V}} = 9.0\%\ \mathrm{mag}^{-1}. \qquad (4.17)$$

This value of P_V/E_{B-V} represents an observational upper limit. With reference to equation (4.8), this may be expressed as an upper limit on the ratio of polarization to extinction:

$$\frac{p_V}{A_V} \leq 0.064 \qquad (4.18)$$

where $A_V = 3.05E_{B-V}$ (section 3.3.3) has been assumed. The result in equation (4.18) is consistent with the theoretical limit in equation (4.5): i.e. the maximum polarization efficiency p/A observed at visual wavelengths is a factor of roughly four less than the upper limit set by theory. If real interstellar grains resembled infinite cylinders, the efficiency of alignment would not need to be high to explain the observed polarization; more realistically, irregular or mildly anisotropic particles may suffice if alignment of their longest axes is fairly efficient.

Studies of interstellar polarization at visible wavelengths are limited by sensitivity considerations to lines of sight toward stars with modest degrees of extinction (typically $A_V < 5$ mag): the stars plotted in figure 4.4 thus sample mostly diffuse regions of the ISM in the solar neighbourhood. In order to investigate the behaviour of the magnetic field deep within molecular clouds and at large distances within the galactic plane, measurements at infrared wavelengths are needed. Extensive data are available in the K (2.2 μm) passband for a variety of galactic and extragalactic sources. At this wavelength, the extinction is a factor of about 11 less than in the visual (table 3.1). Figure 4.5 plots P_K against optical depth τ_K (where $\tau_K = A_K/1.086 \approx 0.084A_V$) for a variety

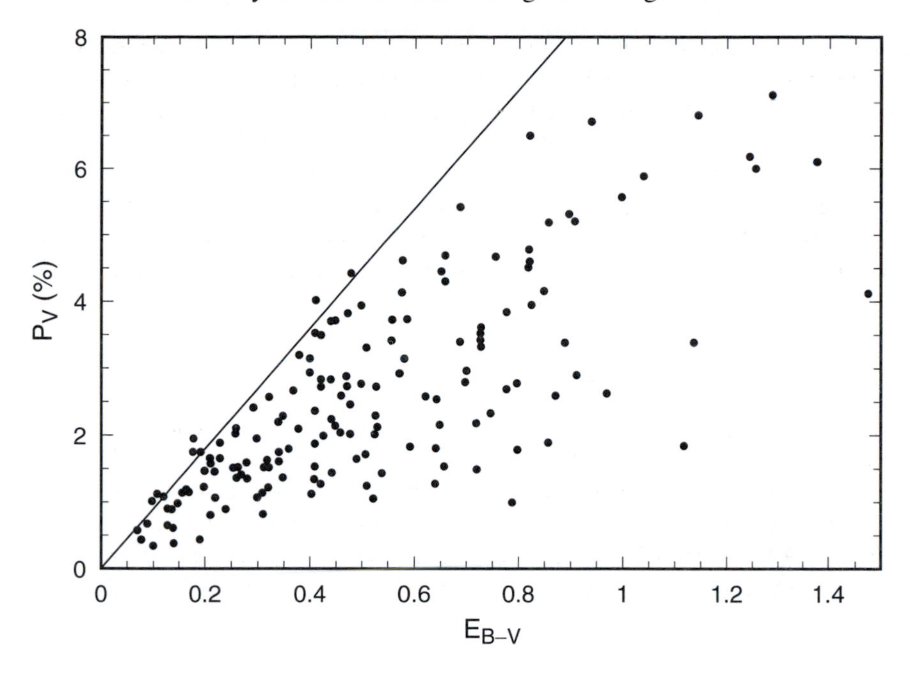

Figure 4.4. A plot of visual polarization (P_V) against reddening (E_{B-V}) for a typical sample of field stars in the Milky Way. The straight line represents maximum polarization efficiency (equation (4.17)). Data are from Serkowski *et al* (1975) and references therein.

of regions and environments. The total range in optical depth is equivalent to $0.6 < A_V < 36$. The average polarization efficiency is substantially below the maximum (represented by the upper curve in figure 4.5) in all regions. Nevertheless, the level of correlation is impressive, considering the diversity of environments sampled, suggesting that the alignment mechanism is generally effective in both dense and diffuse clouds.

Regional variations in polarization efficiency are potentially valuable as a diagnostic of the alignment mechanism and its relation to the magnetic field. Star-to-star variations in p/A might result from changes in physical conditions that affect alignment efficiency (section 4.5), such as temperature, density and magnetic field strength, or in grain properties such as their shape and size distribution and the presence or absence of surface coatings. However, apparent variations in p/A also occur due to purely geometrical effects. To illustrate this, consider the idealized situation where initially unpolarized radiation from a distant star is partially plane polarized by transmission through a single cloud with uniform grain alignment; the beam then encounters a second cloud of similar optical depth and uniform alignment in a different direction. The emergent light is, in general, elliptically polarized, the ISM behaving as an inefficient waveplate. This effectively introduces a weak component of circular polarization and causes

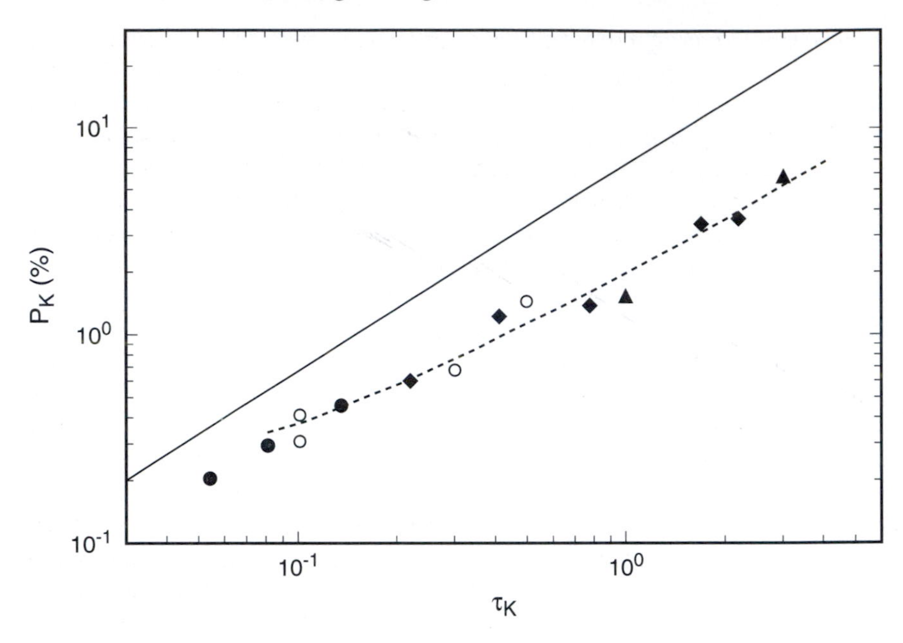

Figure 4.5. Plot of infrared polarization (P_K) against optical depth (τ_K) for diverse lines of sight (Jones 1989b, Jones *et al* 1992). Full symbols are averages for representative regions of the Milky Way: the diffuse ISM (full circles), dense clouds (diamonds) and the galactic centre (triangles). Open circles are observations of the dusty discs of other spiral galaxies. The full curve represents optimum polarization efficiency (i.e. equation (4.17), transformed to P_K/τ_K) and the broken curve is from the model discussed in section 4.2.4.

depolarization of the linear component. In the extreme case where the alignment axes of the two clouds are orthogonal, high extinction can be produced with no net polarization. Real interstellar clouds frequently exhibit complex magnetic field structure, thus changes in alignment geometry along a line of sight can arise because of twisted magnetic field lines within a single cloud, resulting in a reduction in the apparent efficiency of alignment (e.g. Vrba *et al* 1976, Messinger *et al* 1997). In view of these diverse effects, it is not unexpected that plots of linear polarization against reddening for heterogeneous groups of stars, as in figure 4.4, show scatter considerably in excess of observational errors. The straight line of equation (4.17) presumably represents optimum polarization efficiency for alignment by a uniform magnetic field perpendicular to the line of sight.

4.2.4 Small-scale structure

Although the macroscopic properties of interstellar polarization in the solar neighbourhood of the Milky Way (section 4.2.2) are described adequately in terms of a unidirectional magnetic field, this apparent uniformity breaks down on smaller size scales. The magnetic field may be described more realistically as the sum of a uniform component and a random component (Heiles 1987, 1996). The uniform component represents the general galactic magnetic field, upon which the random component is superposed. The latter is attributed to local structures in the magnetic field of scale size $\sim$ 100 pc in the diffuse ISM, possibly associated with old supernova remnants.

The general trend of polarization with optical depth (figure 4.5) is consistent with models based on this hypothesis in which the uniform and random components carry approximately equal energy (Jones 1989b, Jones *et al* 1992). The model illustrated was constructed by dividing the optical path into sequential segments, each of optical depth $\Delta\tau_K = 0.1$; in each segment, the magnetic field has a uniform component (constant position angle for all segments) and a random component (allowed to take any position angle). Vector addition of the two components results in a net field strength and direction. The grains are assumed to have alignment efficiency given by $Q_\perp/Q_\parallel = 0.9$. At low optical depth (few segments), the random component strongly affects the net polarization. As many segments are accumulated, the random component eventually averages out to a small net effect and the polarization is dominated by the uniform component. One implication of the model is that the typical optical depth interval over which the galactic magnetic field changes in geometry is $\tau_K \approx 0.1$ (equivalent to $A_V \approx 1$) in all environments. In the diffuse ISM, this extinction corresponds to pathlengths of a few hundred parsecs, in agreement with other estimates of the scale length for variations in the galactic magnetic field (Heiles 1987). In dense clouds, it corresponds to pathlengths of only a few parsecs or less.

4.2.5 Dense clouds and the skin-depth effect

The distribution of polarization vectors across the face of an interstellar cloud can, in principle, give a two-dimensional representation of the magnetic field within that cloud – see Moneti *et al* (1984) for a typical example. Probing field structure deep within star-forming clouds is particularly important, as the magnetic field appears to play a influential role in cloud collapse (chapter 9). However, investigations that attempt to do this by mapping the polarization of background starlight are hindered by a sampling problem: the observed polarization is often dominated by dust in the outer layers of the cloud. The problem is illustrated in figure 4.6, which plots polarization efficiency versus extinction in the K band for a single cloud. Clearly, P/A tends to be greatest for stars with low extinction and least for stars with high extinction. The data are consistent with a trend of very rapidly declining P/A with A in the outer layers of the cloud. Thus, stars observed through relatively opaque regions of the cloud often have degrees of polarization little or no greater than those of stars sampling much lower dust columns. If the polarization amplitudes are dominated by the outer layers, the position angles must likewise be biased and polarization maps thus contain little or no information on magnetic field structure within the cloud.

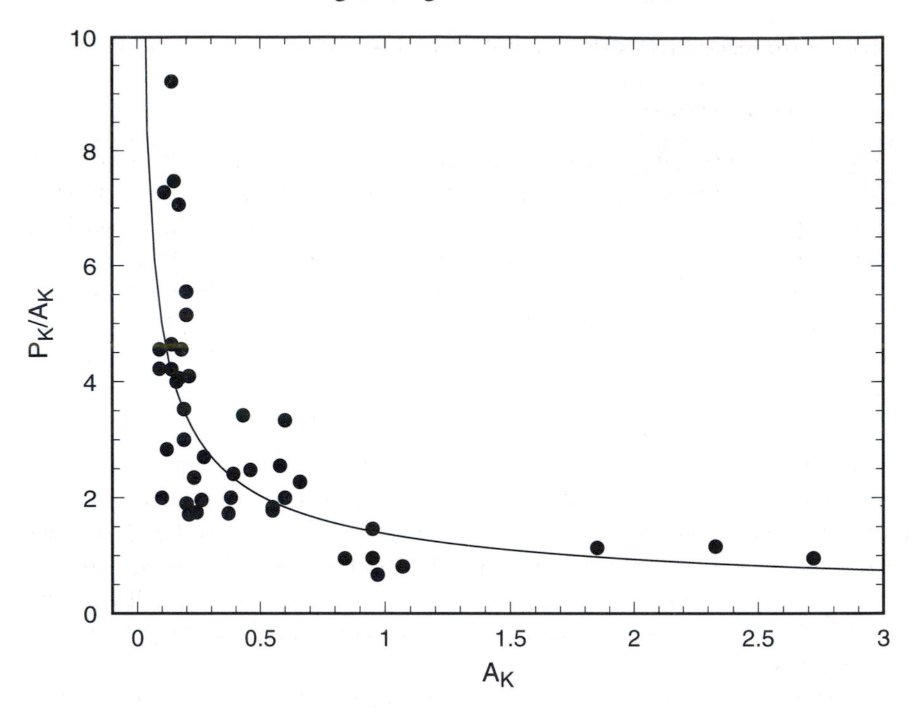

Figure 4.6. Plot of polarization efficiency P_K/A_K against extinction A_K for field stars observed through the Taurus dark cloud (Gerakines *et al* 1995b and references therein). The line represents an unweighted least-squares power-law fit to the data ($P_K/A_K = 1.38A_K^{-0.56}$).

Some authors have concluded that grains responsible for polarization of starlight are critically under-represented or even absent in the dense interiors of cold dark clouds (e.g. Goodman *et al* 1995, Creese *et al* 1995, Jones 1996). However, other observations clearly indicate that polarization is produced in regions of high density. Many objects embedded deep within molecular cloud cores show very high degrees of polarization (e.g. Jones 1989b) that cannot be explained purely in terms of grains in the outer layers of the clouds. The steady increase in polarization with optical depth (figure 4.5) is hard to explain if the inner regions of dense clouds contribute no polarization. Detection of far infrared polarized continuum emission from cold dust (section 6.2.5) also argues for significant alignment of dust within clouds. But perhaps the most direct evidence is provided by the detection of polarized absorption associated with spectral features of ices (section 5.3.8) that form as mantles on dust *only* inside molecular clouds. The most likely cause of the trend in figure 4.6 is a systematic reduction in the efficiency of grain alignment with density (Gerakines *et al* 1995b, Lazarian *et al* 1997, see section 4.5.7) but even in the densest regions some degree of alignment must be retained.

4.3 The spectral dependence of polarization

4.3.1 The Serkowski law

When the degree of polarization is measured through multiple passbands, systematic variations with wavelength are evident. The spectral dependence of linear polarization or polarization curve (usually plotted as P_λ versus λ^{-1}) displays a broad, asymmetric peak in the visible region for most stars. Two examples are shown in figure 4.7. The wavelength of maximum polarization, λ_{max}, varies from star to star and is typically in the range 0.3–0.8 μm with a mean value of 0.55 μm. The dependence of P_λ on λ is well described by the empirical formula

$$P_\lambda = P_{max} \exp\left\{-K \ln^2\left(\frac{\lambda_{max}}{\lambda}\right)\right\} \tag{4.19}$$

where P_{max} is the degree of polarization at the peak (Serkowski 1973, Coyne *et al* 1974, Serkowski *et al* 1975). Equation (4.19) has become known as the Serkowski law. The parameter K, which determines the width of the peak in the curve, was initially taken to be constant with a value of $K = 1.15$: an adequate description of polarization in the visible region can be achieved with K set to this value in equation (4.19).

Extension of the spectral coverage revealed discrepancies that led to a refinement of the empirical law. With K treated as a free parameter, least-squares fits of equation (4.19) to data for stars with a range of λ_{max} show that K and λ_{max} are linearly correlated:

$$K = c_1\lambda_{max} + c_2 \tag{4.20}$$

where c_1 and c_2 are constants (Wilking *et al* 1980, 1982, Whittet *et al* 1992, Clayton *et al* 1995, Martin *et al* 1999). This dependence of K on λ_{max} implies a systematic decrease in the width of the polarization curve with increasing λ_{max}. The optimum values of the slope and intercept are somewhat different depending on the spectral range under consideration: they were initially determined from fits to data in the visible to near infrared (VIR; $0.35 < \lambda < 2.2$ μm), for which the best current values are $c_1 = 1.66 \pm 0.09$ and $c_2 = 0.01 \pm 0.05$ (Whittet *et al* 1992). Equation (4.19) with this constraint on K, sometimes referred to as the Wilking law, yields excellent fits to data in the spectral range for which it was formulated. However, the Wilking law tends to underestimate the degree of polarization in the ultraviolet for stars with low λ_{max} (Clayton *et al* 1995): data in the visible to ultraviolet (VUV; $0.12 < \lambda < 0.55$ μm) are better matched with $c_1 = 2.56 \pm 0.38$ and $c_2 = -0.59 \pm 0.21$ (Martin *et al* 1999). This is illustrated in figure 4.7 for two stars with contrasted λ_{max} values. VIR-optimized, VUV-optimized and compromise fits are plotted for each star, the latter using the unweighted mean values of c_1 and c_2. For the longer-λ_{max} star, these fits are almost identical; for the shorter-λ_{max} star, none provides an ideal fit to the ultraviolet data. The overall agreement is nevertheless excellent.

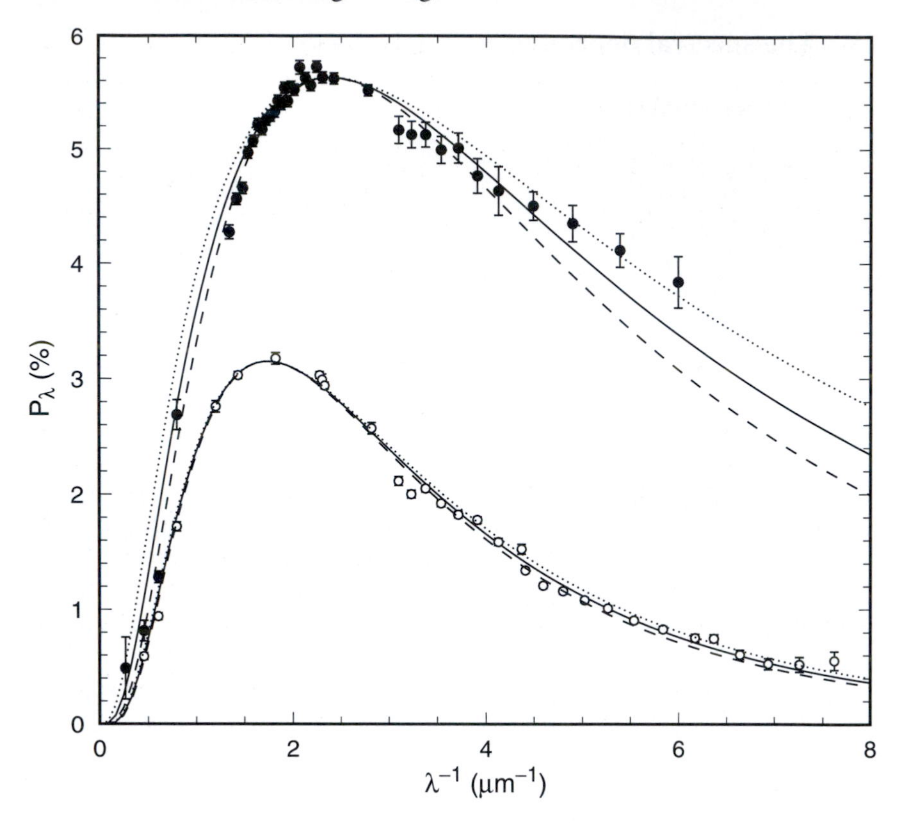

Figure 4.7. Interstellar linear polarization curves for two stars with different values of the wavelength of maximum polarization. Top: HD 204827 (full circles, $\lambda_{max} = 0.42$ μm); bottom: HD 99872 (open circles, $\lambda_{max} = 0.58$ μm). Observational data are from Martin *et al* (1999) and references therein. Also shown are empirical fits based on the Serkowski law: VIR-optimized fit (broken curve); VUV-optimized fit (dotted curve); compromise fit (full curve).

The mathematical representation of P_λ provided by the Serkowski law has practical applications allowing, for example, reliable interpolation to wavelengths other than standard passbands. However, it should be remembered that it is an *empirical* law and cannot be related directly to theory (although we may hope to reproduce it by appropriate choice of models). Its significance lies in the fact that the key parameter, λ_{max}, is a physically meaningful quantity, related to the average size of the polarizing grains: λ_{max} has status as a polarization parameter analogous to R_V for extinction (section 3.4.3). We noted in section 4.1 that for dielectric cylinders of radius a and refractive index n, polarization is produced most efficiently when the quantity $2\pi a(n-1)/\lambda$ is close to unity and hence

$$\lambda_{max} \approx 2\pi a(n-1). \tag{4.21}$$

Although strictly applicable to cylinders, a corresponding proportionality between λ_{max} and some characteristic particle dimension a may be assumed for polarizing grains of arbitrary shape. If $n = 1.6$ (appropriate to silicates) is used in equation (4.21), then the mean value of λ_{max} (0.55 μm) yields $a \approx 0.15$ μm, i.e. classical-sized grains. Star-to-star variations in λ_{max} (e.g. figure 4.7) thus suggest spatial fluctuations in the mean size of the polarizing grains and the trend of K with λ_{max} (equation (4.20)) may be interpreted as a narrowing of the size distribution in response to processes leading to growth.

4.3.2 Power-law behaviour in the infrared

The Wilking version of the Serkowski law was devised to improve the quality of fits to data in the near infrared (1.2–2.2 μm). However, subsequent observations at longer wavelengths demonstrated the presence of significant excess polarization compared with levels predicted by extrapolation of this formula (Nagata 1990, Martin and Whittet 1990, Martin *et al* 1992). This result holds for lines of sight that sample a range of environments with differing λ_{max} values: an example is shown in figure 4.8. An empirical law based on equation (4.19) cannot adequately describe the continuum polarization for $\lambda > 2.5$ μm ($\lambda^{-1} < 0.4$ μm^{-1}) for any reasonable choice of K. A better representation of the spectral dependence of polarization in the infrared is provided by an inverse power law

$$P_\lambda = P_1 \lambda^{-\beta} \tag{4.22}$$

where P_1 (the value of P_λ at unit wavelength) is a constant for a given line of sight. This form provides a good fit to data in the spectral range $1 < \lambda < 5$ μm, as shown in figure 4.8 in the case of Cyg OB2 no. 12. Comparison of results for different stars indicates that the index β is typically in the range 1.6–2.0 (similar to the value for extinction; section 3.3.3) and is uncorrelated with λ_{max} (Martin *et al* 1992). Changes in λ_{max} are evidently associated with variations in the optical properties of aligned grains active in the blue-visible region of the spectrum rather than in the infrared.

Polarization data for objects with very high visual extinctions are often obtainable *only* in the infrared and their λ_{max} values are thus unknown. The list includes dust-embedded young stars, field stars obscured by dense molecular clouds and sources associated with the galactic centre (e.g. Dyck and Lonsdale 1981, Hough *et al* 1988, 1989, 1996, Nagata *et al* 1994). In cases where sufficient spectral coverage is available, these sources exhibit continuum polarization consistent with the power law from of equation (4.22), with indices β similar in value to those determined for much less reddened stars (Martin and Whittet 1990). An example is shown in figure 4.9. A single mathematical form thus appears to be capable of describing the infrared continuum polarization in both diffuse and dense regions of the ISM.

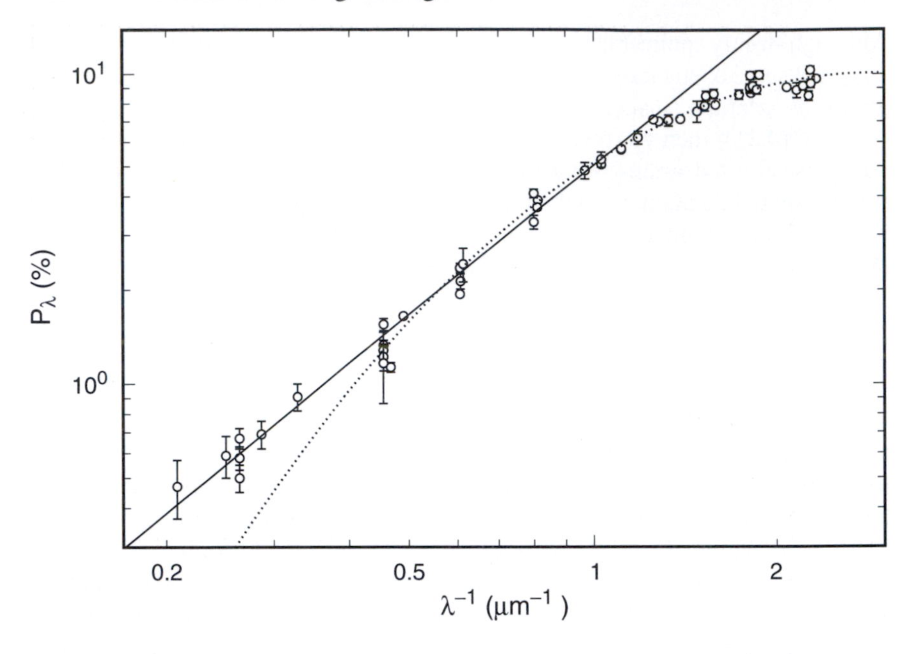

Figure 4.8. Polarization curve in the visual–infrared for the highly reddened hypergiant Cygnus OB2 no. 12, plotted in log–log format to illustrate power-law behaviour in the infrared. Observational data are from Martin *et al* (1992) and references therein. Two fits are shown: the dotted curve is the Serkowski-law fit to visible and near infrared data (0.45–2.3 μm^{-1}) with K treated as a free parameter. The straight line is a power-law fit to infrared data from 0.2 to 1 μm^{-1}. The parameters of the fits are: $P_{max} = 10.1\%$, $\lambda_{max} = 0.35\ \mu m$, $K = 0.61$ (Serkowski law) and $P_1 = 5.06\%$, $\beta = 1.6$ (power law).

4.3.3 Polarization and extinction

As polarization is differential extinction, these phenomena should exhibit analogous behaviour with respect to wavelength if the grains responsible for the extinction curve are aligned. Results discussed in section 4.3.2 clearly show that this is, indeed, true in the infrared: both polarization and extinction are well described by a power law of similar index, comparing lines of sight that sample extremes of environment over widely different pathlengths. The implication is that the largest grains have little variance in size distribution and are relatively easy to align.

This commonality is lost at shorter wavelengths, however. Figure 4.10 compares polarization and extinction data for the same line of sight, that toward HD 161056, a star with 'normal' extinction. Comparing upper and lower frames, an obvious difference is the lack of any feature near 4.6 μm^{-1} in the polarization curve corresponding to the 2175 Å extinction bump. More generally, the systematic decline in polarization in the ultraviolet ($\lambda^{-1} > 3\ \mu m^{-1}$) is in contrast

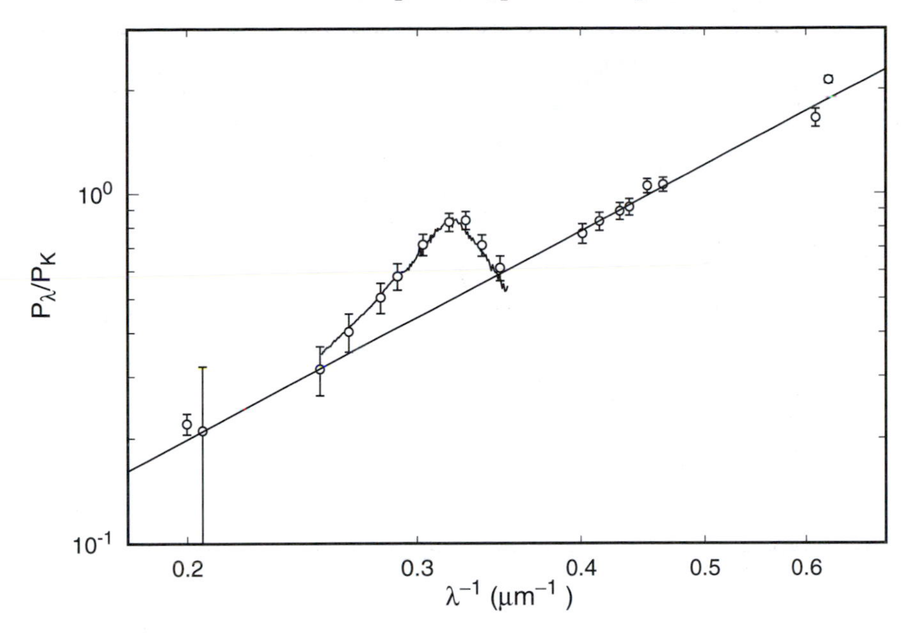

Figure 4.9. Near infrared polarization curve for the Becklin–Neugebauer (BN) object in the molecular cloud associated with the Orion nebula. Data are from Hough *et al* (1996) (curve) and the compilation of previous literature by Martin and Whittet (1990) (points with error bars). Values are normalized with respect to polarization in the K (2.2 μm) passband. The peak centred at 0.32 μm^{-1} is the polarization counterpart to the 3.0 μm H_2O-ice absorption feature (section 5.3) in the spectrum of BN. A power law of index $\beta = 1.97$ (diagonal line) is fitted to the continuum polarization.

with the relatively high levels of extinction, both within and beyond the bump. This behaviour is qualitatively consistent with the optical properties of small cylinders (section 4.1), i.e. $p_\lambda \rightarrow 0$ whilst $A_\lambda \rightarrow$ constant as λ becomes short compared with grain dimensions (see figure 4.1). However, this situation is only reached in the ultraviolet ($\lambda \sim 0.2$ μm) for *classical*-sized grains ($a \sim 0.1$ μm). Very small grains ($a \sim 0.01$ μm or less) should produce ultraviolet polarization if they are aligned: the observations indicate that, in general, they are either very poorly aligned or approximately spherical.

The spectral dependence of polarization across a dust-related absorption feature is a powerful diagnostic of the shape and alignment status of the carrier: excess polarization is expected if the carrier resides in polarizing grains (Aitken 1989). The presence of peaks in the infrared that correspond to ice and silicate features seen in molecular clouds (see figure 4.9 and section 5.3.8) attest to the fact that silicate grains and ice grains (or grains with ice mantles) are, indeed, being aligned. In contrast, as noted earlier, no counterpart to the 2175 Å bump feature is detected in polarization toward HD 161056 (figure 4.10). Ultraviolet polarimetric

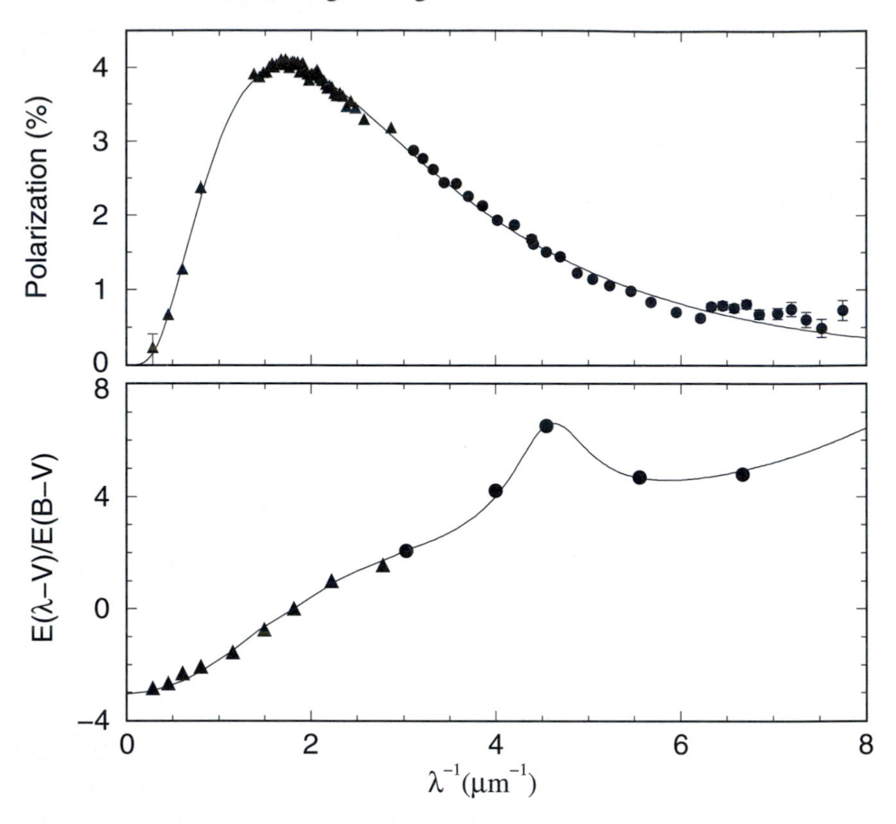

Figure 4.10. A comparison of interstellar polarization and extinction curves for the line of sight to HD 161056. Polarization data (upper frame) are from Somerville *et al* (1994) and references therein, obtained with the Hubble Space Telescope (circles) and ground-based telescopes (triangles). The fit to polarization is based on the Serkowski formula (equation (4.19)) with K as a free parameter ($P_{\rm max} = 4.03\%$, $\lambda_{\rm max} = 0.59\ \mu$m, $K = 1.09$). The extinction curve (lower frame) is based on data from the ANS satellite (circles) and ground-based telescopes (triangles). The extinction curve for this star has $R_V = 3.0$ and is closely similar to the interstellar average (also shown). Note the lack of any detectable excess polarization corresponding to the 4.6 μm^{-1} extinction bump.

data are available for some 30 stars, the large majority of which, like HD 161056, show no hint of a bump excess (two further examples appear in figure 4.7). Very low upper limits on the polarizing efficiency of the carrier grains are implied. However, two stars do show evidence for an excess (figure 4.11), suggesting that graphite particles are being partially aligned in these lines of sight (Wolff *et al* 1993, 1997). Yet, even toward these stars, the amplitude of excess polarization is very small compared with the extinction in the bump. It may be concluded quite generally that the bump grains are very inefficient polarizers. Similarly

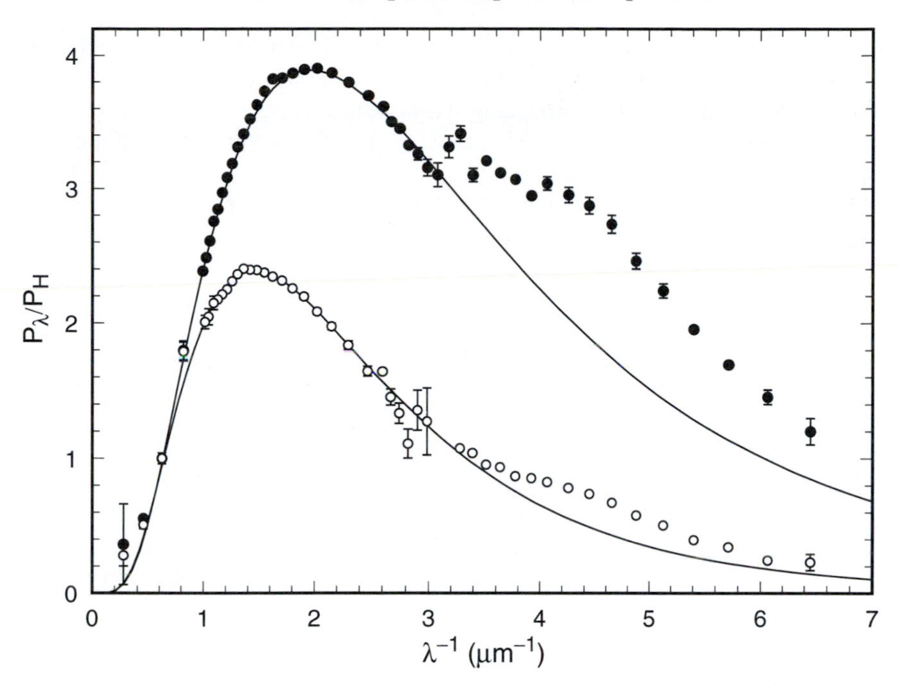

Figure 4.11. Polarization curves for two stars with polarized bumps: HD 197770 (top) and ρ Oph (bottom). Observational data are from the Wisconsin Ultraviolet Photopolarimeter Experiment and various ground-based telescopes (Wolff *et al* 1997) and are normalized with respect to polarization in the H (1.65 μm) passband. Empirical fits based on the Wilking form of the Serkowski law are shown in each case. The λ_{max} values are 0.51 μm (HD 197770) and 0.68 μm (ρ Oph).

negative results are found for the visible diffuse interstellar bands (section 3.6) and the 3.4 μm hydrocarbon infrared feature (section 5.2.4): all attempts to detect polarization excesses in these absorptions have been unsuccessful (Martin and Angel 1974, 1975, Fahlman and Walker 1975, Adamson and Whittet 1992, 1995, Adamson *et al* 1999). The carriers are inferred to be either gaseous molecules or very small grains that fail to align or lack optical anisotropy: they cannot be linked to silicate grains that produce polarization features in the mid-infrared, or indeed to any large classical grains that produce visible polarization.

It is evident from this discussion that the observed interstellar polarization is produced primarily by relatively large grains that also contribute to the visual extinction. The form of the extinction curve is characterized by the parameter R_V (section 3.4.3) and model calculations show that R_V is sensitive to grain size, specifically the number of smaller grains producing blue-visual extinction relative to larger ones producing visual–infrared extinction. Similarly, we showed in section 4.3.1 that λ_{max} is also a measure of grain size: variations in λ_{max} and its inverse correlation with the width parameter K can likewise be explained in terms of adjustments in the numbers of aligned grains producing blue-visual extinction relative to those producing extinction at longer wavelengths in the size distribution. A correlation between λ_{max} and R_V is thus expected and this has, indeed, been observed (Serkowski *et al* 1975, Whittet and van Breda 1978, Clayton and Mathis 1988, Vrba *et al* 1993, see figure 4.12). The data are broadly compatible with a linear correlation passing through the origin:

$$R_V = (5.6 \pm 0.3)\lambda_{max} \tag{4.23}$$

where λ_{max} is in μm (e.g. Whittet and van Breda 1978). This general trend is consistent with models for the growth of dielectric grains (McMillan 1978, Wurm and Schnaiter 2002). Care has been taken, in assembling the database plotted in figure 4.12, to exclude stars with shell characteristics[3]. Nevertheless, the degree of scatter in figure 4.12 is substantially greater than observational error, indicating real variations in the way these parameters respond to environment.

Is λ_{max} a more reliable grain-size parameter than R_V? It does have the advantage that no assumptions need be made as to the nature of the background star, save only that it is unpolarized, whereas R_V depends on evaluation of intrinsic colours from the star's spectral type, and this can be a major source of error. However, if such errors can be avoided, R_V gives a more direct measure of grain properties: whereas R_V is determined by the sum of all grains in the line of sight that contribute to extinction in the B, V and longer-wavelength passbands, λ_{max} is also dependent on alignment. In a dense cloud, P/A may decline systematically with extinction (e.g. figure 4.6) and this may lead to systematic changes in the ratio R_V/λ_{max}. This ratio will decrease if progressive failure of the alignment mechanism affects primarily the smaller grains, leading to an increase in λ_{max}, whilst the mean size (and hence R_V) stay the same. This may occur near the interface of diffuse and dense material in the outer layer of a cloud (Whittet *et al* 2001a). At higher density, R_V will tend to increase because of grain growth but increases in λ_{max} will be more modest if grains in regions of growth are poorly aligned (Whittet *et al* 1994, 2001a). Such situations can easily account for the scatter in figure 4.12.

4.3.4 Regional variations

Subject to the caveats discussed above, observed λ_{max} values may be used to study spatial variations in the size distribution of aligned grains. There are two reasons why this may be valuable. First, it provides an independent check on the validity of assuming a global average value for R_V in diffuse regions of the ISM; and second, it gives insight into grain growth processes inside dark clouds. In this

[3] As discussed in section 3.2, erroneously large values of R_V may be obtained for stars with fluxes contaminated by circumstellar infrared emission. Note, also, that an intrinsic component to the polarization may be produced by scattering in a circumstellar shell (Whittet and van Breda 1978, Clayton and Mathis 1988).

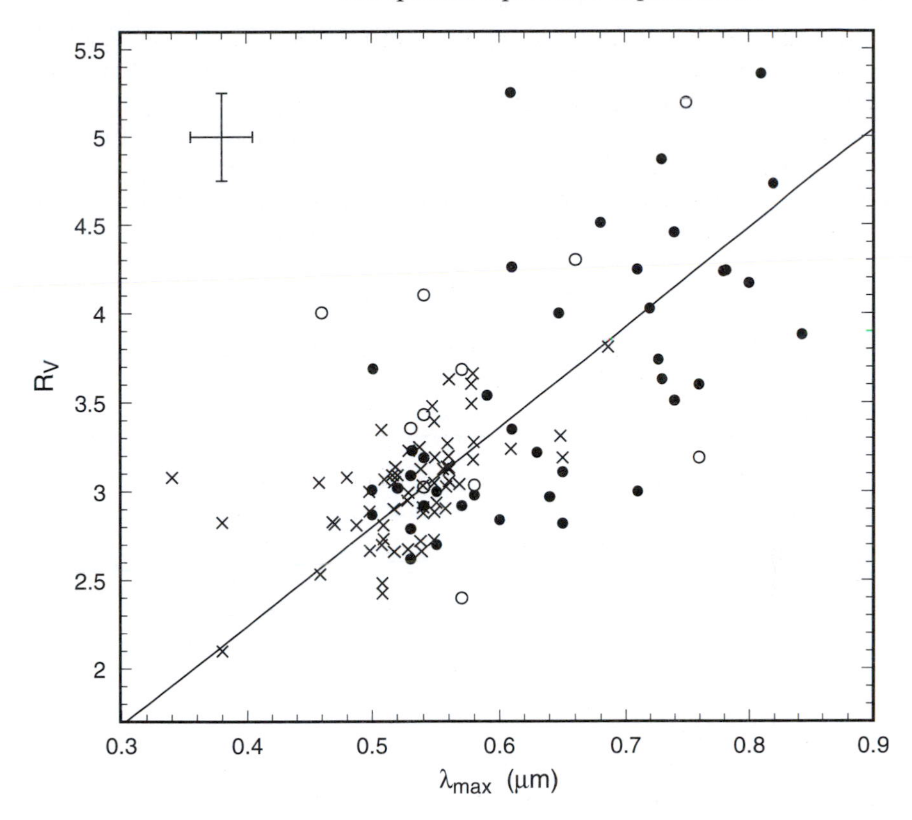

Figure 4.12. Correlation of λ_{max} with R_V. Filled circles, open circles and crosses represent lines of sight associated with dense clouds, H II regions and diffuse clouds, respectively. The large symbol at the upper left shows typical error bars. The data have been compiled from Whittet and van Breda (1978), Clayton and Mathis (1988), Larson *et al* (1996) and various papers cited in table 4.1. The continuous line is the relation $R_V = 5.6\lambda_{max}$ (Whittet and van Breda 1978).

section, we review evidence for both macroscopic and cloud-to-cloud variations in λ_{max}.

As noted in chapter 1 (section 1.5.4), systematic spatial variations in grain properties could be a serious source of error in photometric distance measurements, if not properly accounted for. Large, galaxy-wide variations in R_V, such as those claimed by Johnson (1968), have not been confirmed and appear to have arisen through systematic errors in the methods used to evaluate R_V. Measurement of λ_{max} provides an independent test. Average values of both R_V and λ_{max} are plotted against galactic longitude (ℓ) in figure 4.13. Clearly, the amplitude of any variation in these parameters is quite small (<10%). Nevertheless, a significant second-order effect is evident in the data for λ_{max}

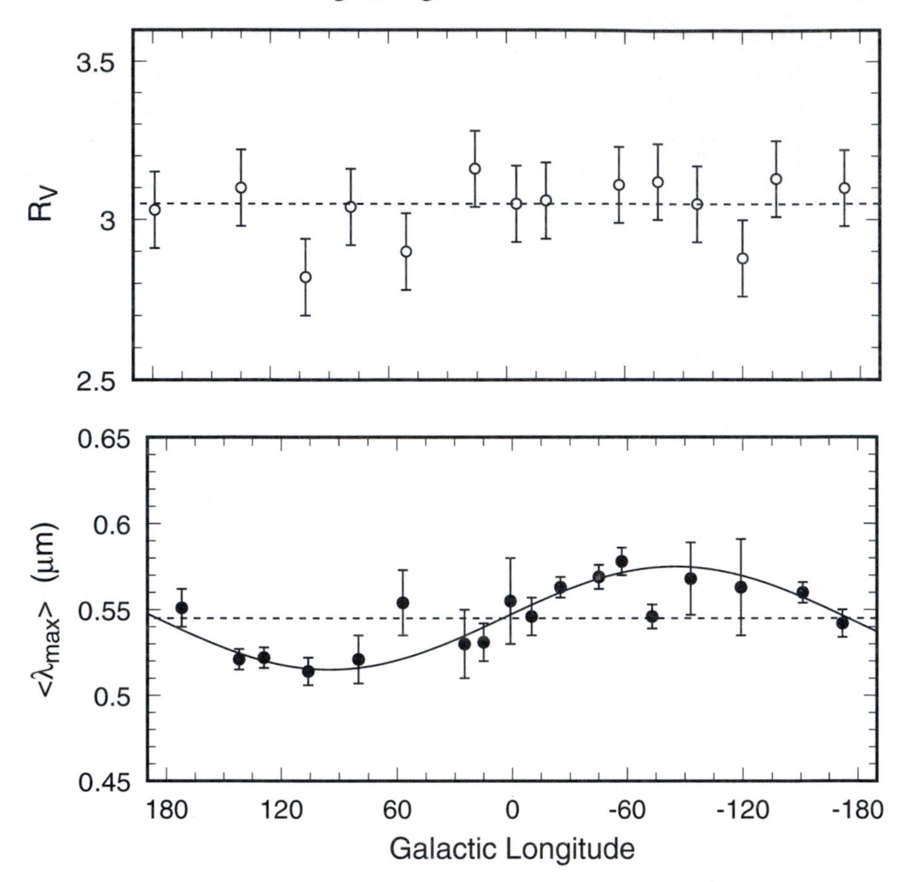

Figure 4.13. Plots of R_V and λ_{max} against galactic longitude (upper and lower frames, respectively). Data are from He *et al* (1995), Whittet (1977) and references therein. Each point represents the average for a region of the Milky Way. The horizontal broken lines represent average values ($R_V = 3.05$ and $\lambda_{max} = 0.545$ μm). The sinusoidal curve is a fit to the λ_{max} data (equation (4.24)).

(lower frame of figure 4.13). This takes the form of a systematic modulation with a 360° period, described by the empirical law

$$\lambda_{max} = 0.545 + 0.030 \sin(\ell + 175^\circ) \tag{4.24}$$

(Whittet 1977). There is no convincing evidence for an equivalent variation in R_V (upper frame) but the implied range ($2.9 < R_V < 3.2$; equations (4.23) and (4.24)) is comparable with the scatter in R_V.

Systematic variations in R_V have been proposed that are related to the alignment of the grains (Wilson 1960, Greenberg and Meltzer 1960, Rogers and Martin 1979): such variations are expected if we view a slightly different mean cross-sectional area per grain dependent on the viewing angle with respect to the net direction of alignment. However, the observed variation in λ_{max} (figure 4.13) is clearly different from that in q (figure 4.3), which has a 180° rather than a 360° period, ruling out the alignment mechanism as the source of the variations. Omission of stars more distant than 1 kpc increases the amplitude of the variation in λ_{max} (Whittet 1979), indicating that the phenomenon is peculiar to the solar neighbourhood. Its origin may be understood by considering the distribution of regions where λ_{max} is above average (Serkowski *et al* 1975, Whittet 1977, 1979, Clayton *et al* 1995). Concentrations of stars with high λ_{max} are to be found in clouds and associations such as Orion and Scorpius–Ophiuchus, located in Gould's Belt, the local sub-system of early-type stars. The direction of the centre of Gould's Belt coincides precisely with the direction of peak λ_{max} in the galactic plane (at $\ell = -85°$ in figure 4.13). Let us assume that the local ISM contains two populations of grains: 'normal' interstellar grains with $\lambda_{max} \approx 0.51\ \mu$m, distributed in the galactic plane; and 'large' grains with $\lambda_{max} > 0.6\ \mu$m, distributed in a disc tilted at 18°. The proportion of grains from each population in the column to a distant star in the galactic plane changes with direction. Because the Sun lies toward the edge of Gould's Belt, roughly along the line of intersection of the two planes, the resulting variation in λ_{max} with ℓ (figure 4.13) has a 360° period. As Gould's Belt contains regions of recent star formation, it is probable that the attendant grain population has been processed more recently through molecular clouds than the average interstellar population.

Table 4.1 presents a statistical summary of λ_{max} data for several individual regions of current or recent star formation in the Galaxy. The list includes dark clouds (R CrA, ρ Oph, Chamaeleon I, Taurus), H II regions (the Orion nebula and M17) and young clusters/associations (NGC 7380, Cyg OB2 and the α Persei cluster). Also included for comparison are results for the diffuse interstellar media of the Milky Way and the Large Magellanic Cloud. Corrections have been applied, where necessary, for foreground polarization. The regions are listed in order of descending mean λ_{max}. This appears to represent an evolutionary sequence: λ_{max} is highest in regions of active star formation and close to the general interstellar value in the more mature clusters and associations. Note that Vrba *et al* (1981) consider the α Per cluster (age $\sim$ 20 Myr) to be an older analogue of the R CrA cloud (age $<$ 1 Myr). The standard deviation of λ_{max} tends to correlate with the mean, reflecting a greater spread in λ_{max} (from normal values up to about 1 μm) in the dark clouds. There is also a tendency for λ_{max} to increase with A_V in dark clouds (Vrba *et al* 1981, 1993), supporting the view that grain growth is most efficient in the densest regions. Subsequent dispersal of the dense material following star formation eventually returns the mean grain size to its normal interstellar value.

Perhaps the most remarkable statistic in table 4.1 is the close agreement in the mean value of λ_{max} comparing the ISM in the Milky Way with that in the Large Magellanic Cloud. Clayton *et al* (1983, 1996) conclude from a detailed study of the polarization curves toward several LMC stars that aligned

Table 4.1. Statistical summary of λ_{max} data for nine regions of current or recent star formation, compared with the diffuse interstellar medium in our Galaxy (ISM) and in the Large Magellanic Cloud (LMC). Means and standard deviations are given in μm; n is the number of stars in each sample.

Region	$\langle\lambda_{max}\rangle$	σ	n	Reference
R CrA cloud	0.75	0.09	43	Vrba *et al* (1981), Whittet *et al* (1992)
ρ Oph cloud	0.66	0.08	60	Whittet *et al* (1992), Wilking *et al* (1980, 1982), Vrba *et al* (1993)
M17	0.63	0.13	11	Schulz *et al* (1981)
Orion nebula	0.61	0.08	19	Breger *et al* (1981)
Chamaeleon I cloud	0.59	0.07	50	Whittet *et al* (1994)
Taurus cloud	0.58	0.07	27	Whittet *et al* (2001a)
α Persei cluster	0.54	0.07	55	Coyne *et al* (1979)
NGC 7380	0.51	0.05	10	McMillan (1976)
Cygnus OB2	0.48	0.07	21	Whittet *et al* (1992), McMillan and Tapia (1977)
ISM	0.54	0.06	180	Vrba *et al* (1981)
LMC	0.55	0.10	19	Clayton *et al* (1983)

grains in the two galaxies have rather similar optical properties. Data for other galaxies are sparse; where available, they indicate that the general form of the polarization curve is not radically different from that in the Galaxy, albeit with a tendency toward lower λ_{max} values. A sample of five stars in the Small Magellanic Cloud yields $\langle\lambda_{max}\rangle = 0.45 \pm 0.08$ μm (Rodriguez *et al* 1997) and the integrated light from the globular cluster S78 in M31 (the Andromeda Galaxy) similarly yields $\langle\lambda_{max}\rangle = 0.45 \pm 0.05$ μm (Martin and Shawl 1982). The most precise evaluation of the polarization law in an external galaxy to date is that of Hough *et al* (1987) toward a supernova (SN 1986G) in NGC 5128 (Centaurus A). The supernova occurred within the equatorial dust lane of this galaxy, with an estimated reddening $E_{B-V} \simeq 1.6$; the observed polarization from 0.36 to 1.65 μm is consistent with the standard galactic Serkowski law with $\lambda_{max} = 0.43 \pm 0.01$ μm. This value of λ_{max} can be explained if the polarizing grains are ~20% smaller on average than those in the Milky Way if the refractive index is the same (equation (4.21)). The polarization efficiency ($P_V/E_{B-V} \simeq 3\%$ mag^{-1}) is lower than the maximum value (equation (4.17)) but consistent with the observed range for stars in the Milky Way (figure 4.4). Hence, within the limits of the available data, both the form of the P_λ curve and the P/A ratio are similar in other galaxies compared with our own.

4.3.5 Circular polarization

Small but measurable degrees of circular polarization (V/I) are predicted in cases where the direction of grain alignment changes along the line of sight. Observational data have been published for a number of stars (e.g. Avery *et al* 1975, Martin and Angel 1976 and references therein) but few have spectral coverage comparable with that available for linear polarization. For those in which the data are sufficiently extensive, V/I is found to vary strongly with wavelength, generally exhibiting opposite handedness in the blue and red spectral regions. A typical example is shown in figure 4.14, which compares observations of linear and circular polarization for o Sco. V/I changes sign at a distinct wavelength λ_c (the cross-over wavelength), which is close to the wavelength of peak linear polarization. The value of λ_c is determined reliably in only six lines of sight (Martin and Angel 1976; McMillan and Tapia 1977) and is found to be essentially identical to λ_{max} to within observational error:

$$\lambda_{max}/\lambda_c = 1.00 \pm 0.05. \tag{4.25}$$

However, a notable exception was found in the case of the reddened supergiant HD 183143, which has a λ_{max} value of 0.56 μm but no evidence for a cross-over in V/I within the wavelength range 0.35–0.8 μm (Michalsky and Schuster 1979).

Observations of the wavelength dependences of circular and linear polarization in the same lines of sight allow some discrimination between grain models (section 4.4). The linear birefringence that gives rise to circular polarization is uniquely related to the linear dichroism at all frequencies by a Kramers–Kronig integral relation (Martin 1974, Chlewicki and Greenberg 1990) and this leads to a prediction that the ratio λ_{max}/λ_c is sensitive to k, the imaginary component of the grain refractive index. A value of unity ($\lambda_{max} = \lambda_c$) occurs for $k = 0$, as illustrated by the model in figure 4.14. For increasing absorption ($k > 0$), λ_c is predicted to increase relative to λ_{max}; the precise relation depends on the real as well as the imaginary part of the refractive index but, typically, λ_{max}/λ_c is reduced to $\sim$0.9 for $k = 0.05$ and to $\sim$ 0.7 for $k = 0.3$ (see Aannestad and Greenberg 1983). Thus, the existing observations suggest $k < 0.03$, i.e. the aligned grains appear to be good dielectrics. The anomaly toward HD 183143 might indicate an unusual composition, although this seems implausible given that the star is a distant supergiant, presumably reddened by many discrete clouds along the line of sight. A component of polarization that is intrinsic to the star seems a more likely explanation.

No significant new observations of the spectral dependence of interstellar circular polarization have been published since the pioneering work of the 1970s. Although regrettable, this hiatus is not the severe impediment to progress that it might seem. The nature of the relationship between birefringence and dichroism is such that any diagnostic properties of circular polarization should be retrievable from the linear polarization alone, provided it is known with sufficient precision over a sufficiently wide range of wavelengths (Martin 1989). The dielectric nature

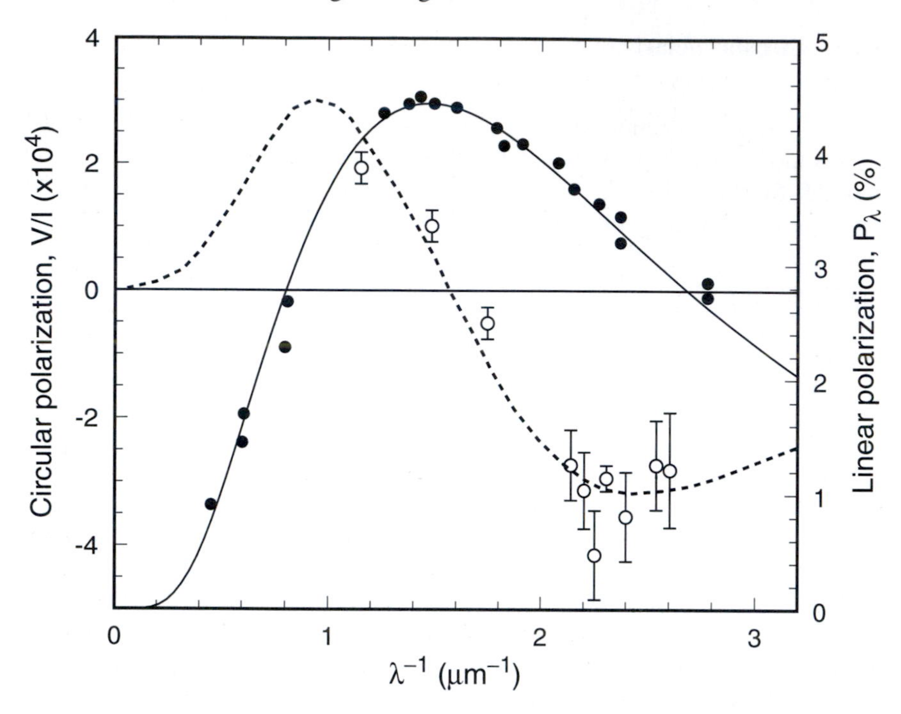

Figure 4.14. Linear and circular polarization curves for *o* Scorpii. Linear polarization data (filled circles, right-hand scale) are from Martin *et al* (1992) and references therein, circular polarization data (open circles, left-hand scale) are from Kemp and Wolstencroft (1972) and Martin (1974). The continuous curve is a Serkowski law fit to the linear data with K treated as a free parameter. The broken curve is a model for circular polarization from grains with $m = 1.50 - 0\mathrm{i}$ (Martin 1974). The horizontal line indicates zero V/I. Note that the wavelength at which V/I changes sign is similar to the wavelength of maximum linear polarization ($\lambda_c \approx \lambda_{max} \approx 0.65$ μm).

of the polarizing dust is, indeed, indicated by models for the spectral dependence of linear polarization, reviewed in the following section.

4.4 Polarization and grain models

Observations of interstellar polarization place useful constraints on models for interstellar grains. A successful model for extinction (section 3.7) should be capable of describing the spectral dependences of polarization as well, subject to assumptions concerning the shape and degree of alignment of the particles. We have already noted that the polarizing grains appear to be relatively large (section 4.3.3) and composed of materials with predominantly dielectric optical properties. It is difficult to reproduce the observed spectral dependence (e.g.

figure 4.7), with its smooth peak in the visible and monotonic decline in the infrared and ultraviolet, using a material in which the refractive index varies strongly with wavelength (see Martin 1978), a result that discriminates against metals and other strong absorbers such as graphite and magnetite. The observed albedo (section 3.3.2) leads us to conclude that at least one component of the dust is dielectric at visible wavelengths. The dielectric nature of the grains responsible for polarization is further supported by observations of circular polarization (section 4.3.5) and by the presence of polarization enhancement in the 9.7 μm feature (section 5.3.8) identified with silicates – a dielectric. On this basis, strongly absorbing materials are excluded from models for the general continuum polarization. Absorbers that contribute to the extinction are presumed to be in particles that produce no net polarization: either they approximate to spheres or they fail to align.

For reasons noted in section 4.1, it is usual to assume axially symmetric shapes such as spheroids or infinite cylinders in model calculations. Either form is capable of providing a reasonable simulation of interstellar polarization, although spheroids seem to produce the most realistic results (Kim and Martin 1994, 1995b). Attempts to match interstellar linear polarization with various extinction-based grain models are described by Wolff *et al* (1993). Bare silicate, core–mantle and composite models were considered. Bare silicates following the MRN size distribution (section 3.7) were generally found to be the most satisfactory. To obtain a good fit, only silicates with sizes above a certain threshold value are allowed to contribute to the polarization: both the smaller silicates and the entire size spectrum of graphite in the MRN model are assumed to be unaligned in a typical line of sight (Mathis 1979, 1986)[4]. Ability to account for the different levels of ultraviolet polarization found in stars with high and low λ_{max} (figure 4.7) as a function of size distribution appears to be an important discriminator between models: grains with carbonaceous (amorphous or organic refractory) mantles, in particular, are less successful in this respect than uncoated grains (Wolff *et al* 1993, Kim and Martin 1994).

Sample fits to polarization curves are illustrated in figure 4.15, taken from the results of Kim and Martin (1995b) for spheroidal silicate grains. Results for two contrasting values of λ_{max} are shown. The size distributions of the particles are derived rather than assumed. The fits are generally excellent over the entire spectral range from the infrared to the ultraviolet, although the larger grains needed in the long-λ_{max} case tend to produce wavelength-independent FUV polarization in excess of what is observed. Oblate spheroids were generally found to give somewhat better fits than prolate spheroids. Figure 4.16 plots distributions of grain mass with respect to size, corresponding to the polarization curves in figure 4.15. Note the general similarity of the two distributions for larger sizes and their divergence at smaller sizes. The latter is caused by the need for

[4] Lines of sight with polarized bumps (figure 4.11) require a small contribution from aligned graphite particles (Wolff *et al* 1993, 1997).

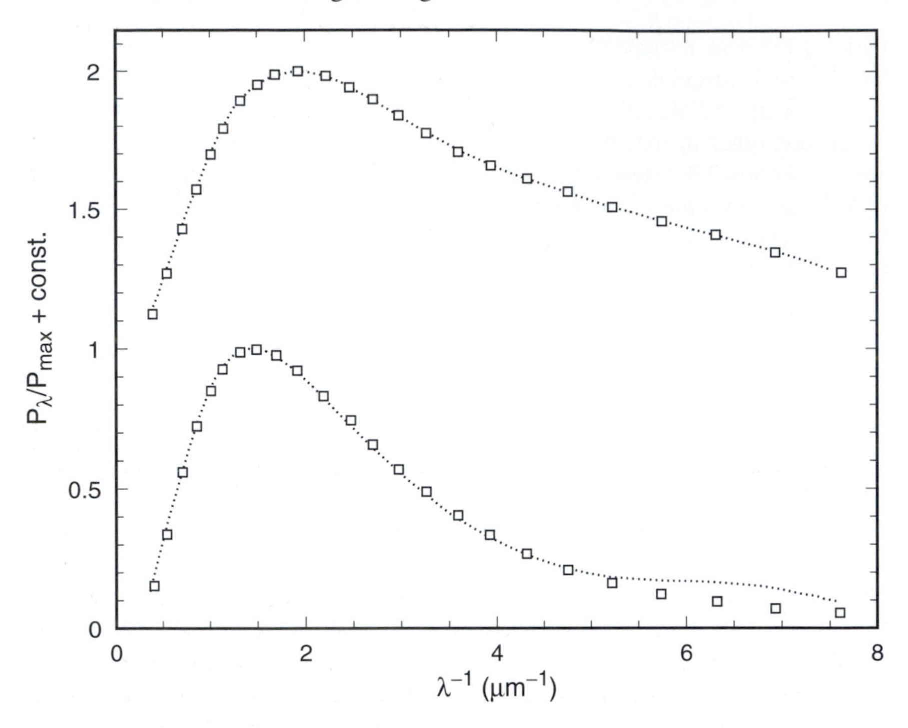

Figure 4.15. Models for interstellar linear polarization, based on calculations using the Maximum Entropy Method for aligned spheroidal grains composed of 'modified astronomical silicate' (Kim and Martin 1995a, b). Oblate spheroids of axial ratio 2:1 are assumed. The calculated polarization spectra (dotted curves) are fitted to representative observational data (squares) for two values of λ_{max}: 0.52 μm (top) and 0.68 μm (bottom). All data are normalized to $P_{max} = 1\%$; the upper curve is displaced upward by unity for display.

aligned grains in the 0.02–0.06 μm size range to produce relatively high levels of ultraviolet polarization in the short-λ_{max} case.

Finally, we compare in figure 4.17 the calculated mass distributions of silicates responsible for polarization and extinction in the Kim and Martin model. These calculations illustrate and confirm the dramatic deficiency in small particles amongst the polarizing grains, previously inferred from general arguments in section 4.3.3. Note that the vertical scale in figure 4.17 is absolute for extinction (as we may normalize the observed extinction to the hydrogen column density) but relative for polarization (as the degree of alignment is not uniquely determined). Nevertheless, we may estimate the degree of particle asymmetry by assuming perfect alignment (or conversely, the degree of alignment for a given degree of asymmetry). Kim and Martin (1995b) find that oblate spheroids with axial ratios as low as 1.4:1 are adequate if alignment is near perfect.

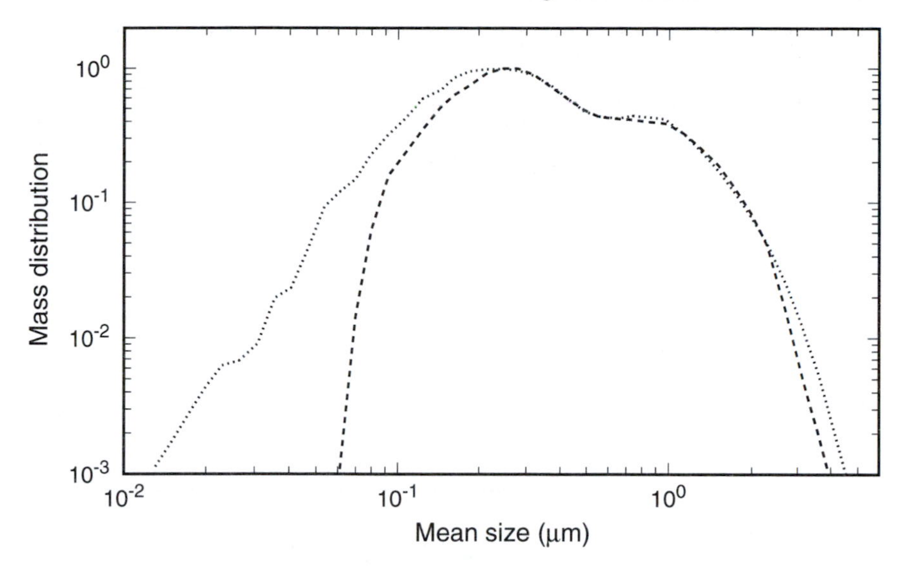

Figure 4.16. Calculated grain-mass distributions corresponding to the models shown in figure 4.15 (Kim and Martin 1995b). The mass distribution function is defined by analogy with the size distribution function (section 3.1.1) to be the mass of dust residing in particles with dimensions in the range a to $a + da$ (Kim and Martin 1994). Dotted and broken curves correspond to models for λ_{max} values of 0.52 μm and 0.68 μm, respectively.

4.5 Alignment mechanisms

Results reviewed in the preceding sections of this chapter provide a firm basis for concluding that polarization of starlight is produced by aligned, asymmetric grains that constitute a subset of all interstellar grains responsible for extinction. The physical processes responsible for alignment are constrained in several ways by the observations. The distribution of polarization vectors on the sky (figure 4.2) is highly consistent with a magnetic origin for alignment. Other important constraints include size selectivity (large grains appear to be much more efficiently aligned than small grains), compositional selectivity (silicate grains appear to be much more efficiently aligned than carbon grains) and sensitivity to environment (alignment efficiency is typically much higher in diffuse regions of the ISM than in dense clouds).

In this section, we review general principles of alignment theory and discuss the principal mechanisms. Table 4.2 provides a summary of the various proposals, together with a guide to the relevant literature. In addition to references cited in table 4.2, general reviews of grain alignment theory may be found in Aannestad and Purcell (1973), Spitzer (1978), Johnson (1982), Hildebrand (1988a, b), Roberge (1996) and Lazarian *et al* (1997). The magnetic properties of interstellar dust are reviewed by Draine (1996) and the properties of the galactic magnetic

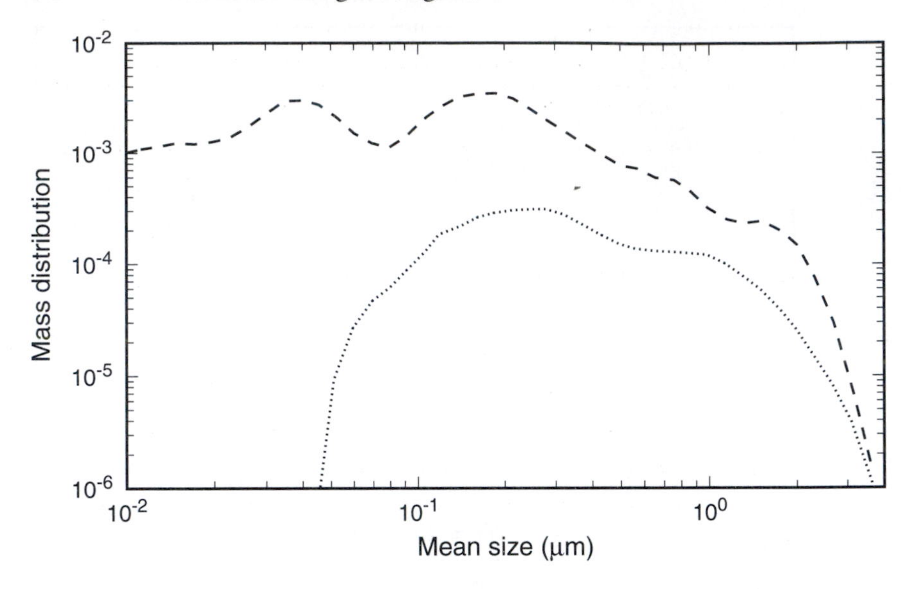

Figure 4.17. Comparison of grain-mass distributions calculated from fits to extinction and polarization curves with $R_V = 3.1$ and $\lambda_{max} = 0.55$ μm, respectively, assuming spheroidal silicate grains (Kim and Martin 1995b). The vertical scale is normalized to mass per unit H-atom for extinction (top curve). Note that only the silicate component of extinction is included. The vertical placement of the polarization curve (bottom) is arbitrary (dependent on the degree of alignment).

field are reviewed by Heiles (1987, 1996).

Magnetic alignment mechanisms (four out of the five listed in table 4.2) are based upon interactions between the spin of a grain and the ambient magnetic field ($\boldsymbol{B}$). The alignment process may be regarded as two distinct steps: coupling between the orientation of the principal axis of the grain and its angular momentum vector $\boldsymbol{J}$, by rotational dissipation, and alignment of $\boldsymbol{J}$ with respect to $\boldsymbol{B}$ by magnetic relaxation. We consider these effects in turn.

4.5.1 Grain spin and rotational dissipation

There are several processes that can contribute to a grain's spin but we shall begin by limiting the discussion to thermal collisions, as in classical alignment theory. Consider an initially stationary grain immersed in a gas containing atoms with a Maxwellian distribution of velocities at some temperature T_g. Random collisions with gas atoms impart impulsive torques that give the grain rotational as well as translational energy. Its angular speed (ω) increases until it becomes limited by rotational friction with the gas itself. If the collisions are elastic, the rotational

Table 4.2. Summary of grain alignment mechanisms.

Mechanism	Description	Reference
Davis–Greenstein (DG)	Alignment of thermally spinning grains by paramagnetic relaxation	Davis and Greenstein (1951), Jones and Spitzer (1967), Purcell and Spitzer (1971), Roberge and Lazarian (1999)
Superparamagnetic (SPM)	Alignment of thermally spinning grains by superparamagnetic relaxation	Jones and Spitzer (1967), Purcell and Spitzer (1971), Mathis (1986)
Purcell	Alignment of suprathermally spinning grains by paramagnetic relaxation (spin-up principally by H_2 formation)	Purcell (1975, 1979), Spitzer and McGlynn (1979), Lazarian (1995a, b)
Radiative torques	Alignment of suprathermally spinning grains by paramagnetic relaxation (spin-up by radiative torques)	Draine and Weingartner (1996, 1997)
Mechanical (Gold mechanism)	Mechanical alignment of thermally or suprathermally spinning grains in supersonic flows	Gold (1952), Roberge *et al* (1995), Lazarian (1995c, 1997)

kinetic energy of the grain about a principal axis with moment of inertia I is

$$E_{\rm rot} = \tfrac{1}{2} I \langle \omega^2 \rangle = \tfrac{3}{2} k T_{\rm g}. \tag{4.26}$$

A typical value of the mean angular speed may be estimated by taking an average moment of inertia $I = \frac{2}{5} m_{\rm d} a^2$ (appropriate to a sphere of radius a and mass $m_{\rm d}$); equation (4.26) then gives

$$\omega_{\rm rms} \approx \left(\frac{1.8 k T_{\rm g}}{a^5 s} \right)^{\frac{1}{2}} \tag{4.27}$$

where s is the density of the grain material. Assuming $a \sim 0.15\ \mu$m, $s \sim 2000$ kg m^{-3} and $T_{\rm g} \sim 80$ K (for a diffuse cloud; table 1.1), equation (4.27) yields $\omega_{\rm rms} \sim 10^5$ rad s^{-1}.

For a grain of arbitrary shape in collisional equilibrium with the gas, the rotational energies associated with spin about each of the three principal axes of inertia are equal. As $E_{\rm rot} = J^2/(2I)$, the angular momentum must then be greatest along the axis of maximum inertia. However, this state of energy equipartition will be disturbed if rotational energy is dissipated by internal processes (Purcell 1979). Frictional stresses within a rapidly spinning grain may dissipate rotational energy as heat. In the presence of a magnetic field, dissipation arises principally from the Barnett effect (the spontaneous alignment of atomic

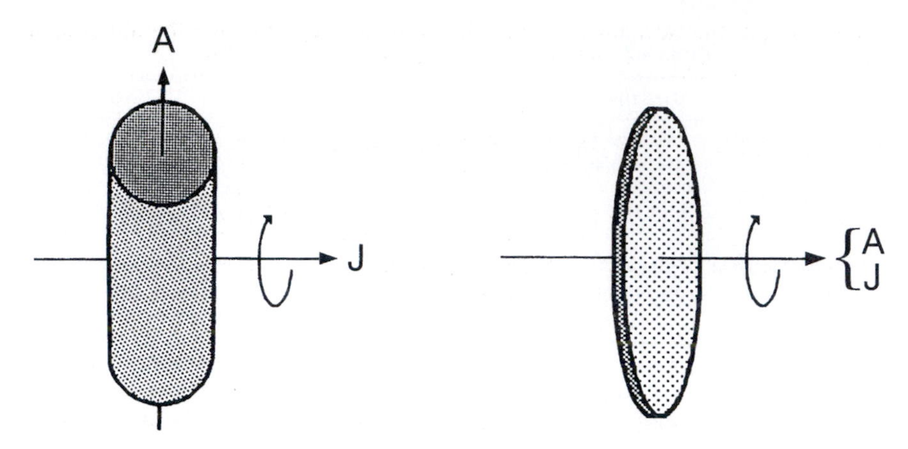

Figure 4.18. Schematic representation of the orientation of angular momentum $\boldsymbol{J}$ with respect to symmetry axis $\boldsymbol{A}$ for spinning grains subject to internal dissipation of rotational energy. The cylinder (left) spins 'end-over-end', with $\boldsymbol{J}$ perpendicular to $\boldsymbol{A}$, whilst the disc (right) spins like a wheel, with $\boldsymbol{J}$ parallel to $\boldsymbol{A}$. Other symmetric shapes with equivalent axial ratios, such as prolate and oblate spheroids, will behave in the same way. In either case, polarization consistent with observations will result if $\boldsymbol{J}$ becomes aligned with the magnetic field.

dipoles: Dolginov and Mytrophanov 1976, Purcell 1979, Lazarian and Roberge 1997a). Because these processes are internal, angular momentum must be conserved and the grain is driven toward a state of minimum energy with rotation about the principal axis of inertia. Elongated grains such as cylinders or prolate spheroids then spin 'end-over-end', with their rotation axes perpendicular to their axes of symmetry, whilst flattened grains such as discs or oblate spheroids spin like wheels (or planets), with their rotation axes parallel to their axes of symmetry (see figure 4.18). The timescale for this situation to be reached is $\sim (10^5/\omega)^2$ yr (Hildebrand 1988b), i.e. only about one year for the thermally spinning grain discussed earlier (and much less if spin is suprathermal; section 4.5.4). For either flattened or elongated grains (figure 4.18), alignment consistent with the observed polarization will be achieved if their angular momenta become orientated parallel to the magnetic field.

In reality, the angular momenta of spinning grains are expected to *precess* about the magnetic field (Roberge 1996), independent of the alignment mechanism. If a spinning grain bears electrical charge, it is endowed with a magnetic moment that leads to precession (Martin 1971). It was realized subsequently that the magnetic moment induced by the Barnett effect will impose precession even for an uncharged grain (Dolginov and Mytrophanov 1976, Purcell 1979). The precession period is expected to be short compared with timescales for other dynamical effects acting on the grain, including randomizing collisions and the alignment mechanism itself. On a time-average, the distribution of spin axes will be symmetric about the magnetic field.

4.5.2 Paramagnetic relaxation: the DG mechanism

The classical theory of alignment by paramagnetic relaxation was formulated by Davis and Greenstein (1951) soon after the discovery of interstellar polarization by Hall and Hiltner in 1949. It was quickly realized that grains do not behave ferromagnetically (like compass needles). Even pure iron grains cannot maintain ferromagnetic alignment when subjected to collisional torques (Spitzer and Tukey 1951) and, in any case, ferromagnetism would orientate the grains with their long axes parallel to the field direction, i.e. orthogonal to the direction implied by observations. The DG mechanism predicts alignment with the correct orientation by paramagnetic dissipation of rotational kinetic energy in thermally spinning grains.

The presence of an external magnetic field of flux density $\boldsymbol{B}$ causes the induction of an internal field (i.e. within the grain), the strength of which depends on the magnetic susceptibility of the material. In a static situation, the internal and external fields would be perfectly aligned. However, for a spinning grain, it is impossible for the internal field to adjust itself instantaneously to the direction of the external field, and so there is always a slight misalignment. This results in a dissipative torque that slowly removes components of rotation perpendicular to $\boldsymbol{B}$, tending to bring the angular momentum into alignment with $\boldsymbol{B}$. Alignment is opposed by gas–grain collisions, which tend to restore random orientation. A measure of the efficiency with which grains can be aligned is thus given by the quantity

$$\delta = \frac{t_{\rm c}}{t_{\rm r}} \tag{4.28}$$

where $t_{\rm c}$ is the collisional damping time, defined as the time taken by a grain to collide with a mass of gas equal to its own mass, and $t_{\rm r}$ is the timescale for paramagnetic relaxation. To achieve significant net alignment, we require δ to be greater than or of order unity.

A second condition that must be satisfied for DG alignment to operate was first demonstrated by Jones and Spitzer (1967). The mechanism is analogous to a heat engine operating between gas and dust: if the temperatures of these reservoirs are the same, no work can be done. However, we expect the gas to be warmer than the dust in diffuse regions of the ISM. This situation exists because gas–grain collisions are unimportant as a source of heat exchange at such low densities and so gas and dust are each in independent thermal equilibrium with the interstellar radiation field. Equilibrium temperatures $T_{\rm d} \sim 15$ K are predicted for the dust (section 6.1.1), compared with $T_{\rm g} \sim 80$ K (table 1.1) for the gas. The DG mechanism and, indeed, any mechanism that operates in an analogous way *requires this temperature difference to exist to achieve alignment.*

If the gas and dust temperatures equilibriate, the mechanism becomes inoperative (section 4.5.7).

The paramagnetic relaxation time, t_r, is determined by the magnitude of the dissipative torque relative to that of the angular momentum of the grain about an axis perpendicular to $\boldsymbol{B}$ (e.g., Spitzer 1978: pp 187–9):

$$t_r \propto \frac{I\omega}{V\chi'' B^2 \sin\theta} \tag{4.29}$$

where V is the volume of the grain, θ is the angle between $\boldsymbol{B}$ and $\boldsymbol{\omega}$, and χ'' is the imaginary part of the magnetic susceptibility ($\chi = \chi' + \mathrm{i}\chi''$). The latter is related to the grain's angular speed:

$$\chi'' = K_m \omega \tag{4.30}$$

where K_m is a constant (the magnetic dissipation constant) for a given grain material at a given temperature: for typical paramagnetic materials, $K_m \approx 2.5 \times 10^{-12}/T_d$. If spherical grains of radius a and density s are again assumed for simplicity[5] and if $\sin\theta$ is set to unity, then

$$t_r \approx 1.6 \times 10^4 \left(\frac{a^2 s T_d}{B^2}\right) \tag{4.31}$$

in SI units. The collisional damping time is given by

$$t_c = \frac{4as}{3(1.2 n_H m_H \langle v_H \rangle)} \tag{4.32}$$

where

$$\langle v_H \rangle = \left(\frac{8kT_g}{\pi m_H}\right)^{\frac{1}{2}}. \tag{4.33}$$

The factor 1.2 in equation (4.32) allows for the fact that ~10% of the impinging atoms are in fact He, with mass $4m_H$ and mean speed $0.5\langle v_H \rangle$. Combining equations (4.28) and (4.31)–(4.33), the requirement $\delta \geq 1$ leads to a lower limit on the magnetic flux density (in SI units) for efficient DG alignment:

$$B \geq \{2.3 \times 10^4 n_H a T_d (kT_g m_H)^{\frac{1}{2}}\}^{\frac{1}{2}}. \tag{4.34}$$

For grains of dimensions $a \sim 0.15 \times 10^{-6}$ m and temperature $T_d \sim 15$ K immersed in an H I cloud of density $n_H \sim 3 \times 10^7$ m^{-3} and temperature $T_g \sim 80$ K, we obtain $B \geq 1.5 \times 10^{-9}$ T, or about 15 μG. The average value of the interstellar magnetic field in the solar neighbourhood of the Galaxy, determined from observations of Faraday rotation, synchrotron radiation and

[5] The quantity a may be regarded here as the radius of an equivalent sphere, i.e. with the same volume as the aspherical grain that produces polarization.

Zeeman splitting, is $\sim$3 μG (Heiles 1987, 1996). Thus, the minimum required flux density exceeds the measured flux density by a factor of about five.

We must conclude that the DG mechanism fails quantitatively to predict significant alignment of classical-sized grains in the ambient magnetic field. Moreover, small grains are predicted to be better aligned than large ones (since $\delta \propto a^{-1}$), contrary to observational evidence. However, a positive feature of the DG mechanism is that it correctly predicts the *geometric* properties of the observed polarization. This strongly suggests that some analogous process is operating in the ISM. We now discuss developments of DG alignment, in which either the magnetic susceptibility or the rotation speed of the grain is enhanced.

4.5.3 Superparamagnetic alignment

If a paramagnetic grain contains clusters of ferromagnetic atoms or molecules (e.g. metallic Fe, Fe_3O_4 or other oxides or sulphides of iron, with $\sim$100 Fe atoms per cluster), the value of χ'' may be enhanced by factors up to $\sim 10^6$ over that typical of paramagnetic materials (Jones and Spitzer 1967). This effect is termed superparamagnetism (SPM) and the clusters are referred to as SPM inclusions. Alignment then proceeds in exactly the same way as for the DG mechanism described in section 4.5.2, but the relaxation time t_r (equation (4.29)) is reduced in proportion to the increase in χ''. Hence, alignment is efficient ($t_r \ll t_c$) in $B \sim 3$ μG magnetic fields for any reasonable choice of the other relevant parameters, provided that the temperature difference between gas and dust is maintained. SPM alignment is thus 'robust' in the sense that the alignment process is not marginal over a range of physical conditions. Only when $T_g \to T_d$ will it fail.

If the polarizing grains are composite particles, formed by coagulation of smaller units (Wurm and Schnaiter 2002), the number of SPM inclusions in each grain will be proportional to its volume: a large grain may contain many, a small grain may contain none at all. This provides a physical basis for understanding why only the larger grains tend to be aligned. Mathis (1986) postulates that a grain is aligned if, and only if, it contains at least one SPM inclusion. However, in the context of the MRN extinction model (sections 3.7 and 4.4), it is necessary to assume that the graphite grains lack SPM inclusions, irrespective of their size (unless they are isotropic in shape). The wavelength of maximum polarization is then determined by the average size of a silicate grain containing SPM inclusions.

As Fe is one of the major elements that can contribute to interstellar dust (sections 2.4–2.5), its presence in polarizing grains is not unexpected. The apparently arbitrary assumption that only silicate grains acquire Fe-rich inclusions might be justified on the basis of their different origins. Whereas carbonaceous dust forms in C-rich stellar atmospheres, silicate dust forms in O-rich environments that may promote the simultaneous growth of iron oxides or other Fe-rich solids (section 7.1). Indirect support is provided by studies of 'GEMS' (glasses with embedded metal and sulphides) within interplanetary dust grains. GEMS do, indeed, contain SPM units within a silicaceous matrix (Bradley 1994), the spatial frequency of which is consistent with the requirements of the Mathis model (Goodman and Whittet 1995). Although GEMS are not necessarily unmodified interstellar grains, they may have formed under similar conditions.

4.5.4 Suprathermal spin

The Purcell 'pinwheel' mechanism postulates alignment of suprathermally spinning grains by paramagnetic relaxation. Spin is said to be suprathermal if the rotational kinetic energy is much greater than would result from random thermal collisions. Real interstellar grains are unlikely to have smooth, uniform surfaces and collisions between gas and grains are unlikely to be elastic. In particular, a hydrogen atom colliding with a grain may stick and subsequently migrate across the surface until it combines with another to form H_2 (section 8.1), with the release of its binding energy (4.5 eV). Ejection of the molecule from the surface simultaneously imparts angular momentum to the grain. As discussed by Hollenbach and Salpeter (1971), molecule formation is likely to occur preferentially at active sites (defects or impurity centres) on the grain surface. A migrating H atom, which would be held only by van der Waals forces elsewhere on the grain, becomes trapped at an active centre until recombination with another migrating atom occurs. The distribution of active centres over the surface will determine the spin properties of the particle (Purcell 1975, 1979). The systematic contributions to angular momentum arising from a series of recombination events at a limited number of active sites will lead to angular speeds well in excess of those predicted by random elastic collisions; with a favourable geometry, they could be as high as $\omega \sim 10^9$ rad s^{-1}.

There are two important respects in which the Purcell mechanism differs from classical DG alignment. First, because the rotational energy of a suprathermally spinning grain greatly exceeds kT, where T is any kinetic temperature in the system, the mechanism does not depend on the existence of a temperature difference between gas and dust. The heat engine obtains its 'fuel' from the binding energy released when H_2 is formed. Second, a suprathermally spinning grain is far less vulnerable to disruption of its orientation by random collisions, because the energy imparted by those collisions is small compared with its rotational energy. Note that increasing the angular speed does not reduce the timescale for paramagnetic relaxation: the angular momentum $I\omega$ and the magnetic damping torque $V\chi''B^2 \sin\theta$ (equations (4.29) and (4.30)) both increase linearly with ω and so t_r is independent of ω (equation (4.31))[6]. The time available for alignment is limited not by gas damping but by the time the grain continues to be driven in the same direction. This depends on the survival time of the surface features that lead to suprathermal spin.

[6] Indeed, at rotational speeds as high as 10^9 rad s^{-1}, χ'' approaches the static susceptibility, i.e. the value of χ' at $\omega = 0$ (Spitzer 1978); equation (4.30) is then no longer applicable; χ'' becomes constant; and t_r will *increase* with ω.

Active sites that promote H_2 formation may be 'poisoned' by attachment of atoms other than H, especially oxygen (Lazarian 1995a, b). The timescale for this to occur is uncertain. One approach is to consider the growth time of a thin surface coating or mantle:

$$t_m = \frac{2.5s\Delta a}{\xi n(kT_g m)^{\frac{1}{2}}} \tag{4.35}$$

(see section 8.3.2), where Δa is the thickness of the surface layer formed by accretion of atoms of mass m and number density n, and ξ is the sticking coefficient (the probability that impinging atoms stick to the surface of the grain). If Δa is set to typical molecular dimensions ($\approx$3.7 Å) in equation (4.35), t_m becomes the time to accrete a *monolayer* (Aannestad and Greenberg 1983). We may then show, by repeating the calculation in section 4.5.2 but for $t_m/t_r \geq 1$ instead of $t_c/t_r \geq 1$, that alignment of suprathermally spinning paramagnetic grains is possible in magnetic flux densities consistent with observations (see problem 7 at the end of this chapter for a representative calculation). If the suprathermally spinning grains are also superparamagnetic, their alignment efficiency is further enhanced.

Can Purcell alignment account for the preferential alignment of large grains? The formulation presented here would suggest size selectivity of the wrong sense, as t_r depends on a^2 and t_m is independent of a (equations (4.31) and (4.35)). However, the number of active sites on a grain is obviously related to its surface area and Lazarian (1995a, b) argues that a critical number is needed to maintain suprathermal spin. The accreting O atoms are somewhat mobile and will tend to seek out and poison active sites. Whereas a large grain with many active sites can maintain spin-up for times that approach the monolayer accretion time, poisoning dominates for smaller grains and spin-up is short lived.

4.5.5 Radiative torques

We assumed in section 4.5.4 that H_2 recombination is the dominant process by which a grain acquires suprathermal spin. Purcell (1975, 1979) considered two other processes – surface variations in the sticking coefficient and photoelectric emission – but found both to be unimportant relative to H_2 recombination. A further possibility is that spin may be enhanced by the interaction of a grain with the ambient radiation field.

Harwit (1970) first drew attention to the fact that absorption of starlight might transfer angular momentum to a grain anisotropically. A grain may be subject to a highly anisotropic irradiation field in a number of situations: examples include the outer layers of a dark cloud, where the grain is shielded from the general interstellar radiation field in the direction of the cloud, and the envelope of a dust-embedded star, where radiation from the star itself is dominant. Harwit considered the intrinsic angular momentum of the photons themselves, concluding that it would be transferred such that prolate grains tend to align with their long axes transverse to the direction of propagation of the light, an effect that would be most efficient for the smallest grains. However, under typical interstellar conditions, such alignment is overwhelmed, not only by collisions with gas atoms but also by isotropic emission of low-energy photons from the grains (Purcell and Spitzer 1971, Martin 1972).

Another way in which photons may impart angular momentum was proposed by Dolginov and Mytrophanov (1976). If a grain were helical in shape, it would absorb and scatter left-handed and right-handed circular components of polarization in a transmitted beam differently and anisotropic irradiation by unpolarized light could then impart spin. A helix is not, of course, a very plausible shape for a real interstellar grain but irregular grains of arbitrary shape will, in general, have some non-zero average 'helicity'. In a detailed analysis of this mechanism, Draine and Weingartner (1996, 1997) show that the spin-up of classical-sized grains from radiative torques might equal or exceed that arising from H_2 formation.

4.5.6 Mechanical alignment

Gold (1952) pointed out that grains in relative motion through a gaseous medium will tend to align with respect to their direction of motion. Gas atoms colliding with a grain contribute angular momentum preferentially perpendicular to the drift velocity and perpendicular to the long axis of the grain. Thus, spinning grains will tend to align as depicted in figure 4.18 for streaming perpendicular to the page. The process is most efficient when the drift speed exceeds the thermal speed of the gas atoms (given by equation (4.33)). Examples of processes that may cause streaming include cloud–cloud collisions and differential acceleration of gas and dust by radiation pressure.

The Gold mechanism does not depend on magnetism to produce alignment but the dynamics of spinning grains are nevertheless highly constrained by magnetic fields (section 4.5.1). As interstellar grains tend to acquire charge by the photoelectric effect, significant drift speeds are generally reached only in directions parallel to $\boldsymbol{B}$. Streaming would then tend to produce polarization with the position angle orthogonal to that predicted by magnetic alignment, which is incompatible with the observed distribution of polarization vectors (figure 4.2). To account for this distribution, it would be necessary to invoke systematic, galaxy-wide streaming motions in a net direction perpendicular to the galactic disc. This seems highly implausible.

Although the Gold mechanism cannot account for the macroscopic pattern of alignment, streaming is undoubtedly important in some situations. Lazarian (1995c) has argued that mechanical alignment becomes important during 'spin-down' or 'cross-over' periods, when suprathermal spin is inoperative. In weakly ionized clouds, charged grains tend to drift through the gas in a direction *normal* to $\boldsymbol{B}$ due to ambipolar diffusion, at speeds that may be sufficient to produce significant mechanical alignment (Roberge *et al* 1995).

4.5.7 Alignment in dense clouds

So far in this section, we have been concerned primarily with demonstrating a viable mechanism for grain alignment in diffuse regions of the ISM, where the observed polarization efficiency P/A is generally highest. Magnetic alignment appears to satisfy major observational constraints (e.g. geometric pattern, size selectivity) at realistic field flux densities, provided that the grains are superparamagnetic and/or suprathermally spinning. We conclude this section with a discussion of alignment in denser regions.

Observations of polarization in dark clouds indicate a systematic decline in polarization efficiency with increasing gas density (section 4.2.5 and figure 4.6). Grains are thus, in general, poorly aligned inside dense clouds, yet high degrees of polarization are observed in some protostellar cores (sections 5.3.8 and 6.2.5). The decline in P/A with density is easily understood, as it is predicted by all the major alignment mechanisms discussed earlier (Hildebrand 1988a, b, Lazarian *et al* 1997). Grains will cease to spin suprathermally if H_2 formation sites are inactivated by mantle growth, if $H \rightarrow H_2$ conversion is complete, and (in the case of radiative torques) if the ambient radiation field becomes weak and/or isotropic. The magnetic field density increases with gas density as

$$B = B_0 \left(\frac{n}{n_0} \right)^{\kappa} \tag{4.36}$$

where B_0 and n_0 refer to the external values of B and n and the index κ is typically $\sim$0.4 (Mouschovias 1987). For DG alignment of paramagnetic or superparamagnetic grains, the polarization efficiency depends on δ (equation (4.28)) and on the relative temperatures of dust and gas:

$$\frac{P}{A} \propto \left(1 - \frac{T_d}{T_g} \right) \delta \tag{4.37}$$

(see Vrba *et al* 1981). As δ varies as B^2/n (equations (4.31) and (4.32)), we thus predict $P/A \propto n^{-0.2}$ for grains of a given size if T_d and T_g stay the same. However, the trend toward lower gas temperatures inside dense clouds (section 1.4.2) will impose a more rapid decline in P/A with n. If $T_g \rightarrow T_d$, then $P/A \rightarrow 0$ for all values of the other relevant parameters.

The real challenge is to understand how alignment is *possible* within dense molecular clouds. That alignment does occur is demonstrated most clearly by the occurrence of polarization excesses that correspond to ice absorptions (section 5.3.8) and by polarized far-infrared emission from cloud cores (section 6.2.5). Although scattering might contribute to the observed 3.0 μm excess in some lines of sight (Kobayashi *et al* 1999), the dominant cause appears to be dichroic absorption by aligned, H_2O-ice-mantled grains (Aitken 1989). A polarization excess has also been observed in the solid CO feature at 4.67 μm in the protostar W33A (Chrysostomou *et al* 1996). This is significant because CO is much more volatile than H_2O: the absorption should, therefore, occur exclusively in cold, dense regions along the line of sight, where grains are fully mantled and the gas and dust temperatures are closely coupled. Neither superparamagnetic nor Purcell alignment seem viable in such regions and we must seek alternatives. One possibility is that cosmic rays might enhance the rotational energies of the grains. Purcell and Spitzer (1971) showed that cosmic rays have little or no effect on grain alignment under typical interstellar conditions but in molecular clouds they might be a significant source of energy. Sorrell (1995) suggested that ejection of H_2 from hot-spots formed on a mantled grain after passage of a cosmic ray might lead to spin-up, but a quantitative evaluation by Lazarian and Roberge (1997b) showed that the resulting torques are insufficient to cause suprathermal spin.

Observations of polarized emission in the far infrared suggest that grains are being aligned selectively in warm, dense cores associated with luminous young stars (Hildebrand *et al* 1999). The implication is that an embedded star can impose alignment on dust in the surrounding medium. This might arise in several ways. Supersonic flows associated with winds from the star may induce mechanical alignment, or the radiation field may force mechanical alignment by streaming driven by radiation pressure on the dust. Alignment resulting from spin-up of the grains by radiative torques is also a good possibility. However, in the case of W33A, it is difficult to understand how a molecule as volatile as CO can remain in the solid phase sufficiently close to the source to become aligned by such processes. Further research is needed to fully explore the possibilities.

Recommended reading

- *The Polarization of Starlight by Aligned Dust Grains*, by L Davis and J L Greenstein, in Astrophysical Journal, vol 114, pp 206–40 (1951).
- *Scattering and Absorption of Light by Nonspherical Dielectric Grains*, by E M Purcell and C R Pennypacker, in Astrophysical Journal, vol. 186, pp 705–14 (1973).
- *Magnetic Fields and Stardust*, by R H Hildebrand, in Quarterly Journal of the Royal Astronomical Society, vol 29, pp 327–51 (1988).
- *Polarimetry of the Interstellar Medium*, ed W G Roberge and D C B Whittet (Astronomical Society of the Pacific Conference Series, vol 97, 1996).

Problems

1. Figure 4.4 plots visual polarization versus E_{B-V}, the diagonal line representing optimum polarization efficiency. Explain carefully *all* possible factors that can result in the locus of an individual star in this diagram falling below and to the right of this line.
2. Stokes parameters $q = 0.0243$ and $u = 0.0140$ (magnitudes) are measured in the V passband for the star HD 203532 ($E_{B-V} = 0.30$). Calculate the

degree and position angle of its linear interstellar polarization and deduce the polarization efficiency, expressing your answer in terms of both P_V/E_{B-V} and p_V/A_V (assuming $R_V = 3.1$). Comment on your result with reference to figure 4.4.

3. The stars κ Cas ($\ell \approx 121$, $b \approx 0$) and HD 154445 ($\ell \approx 19$, $b \approx 23$) have position angles for visual polarization $\theta_G \approx 88°$ and $151°$, respectively. With reference to figure 4.2, comment on whether the observed polarization in these lines of sight is consistent with the general galactic trend in $\boldsymbol{E}$-vectors.
4. Observations in the visible and near infrared have shown that the star HD 210121 has an exceptionally short wavelength of maximum polarization ($\lambda_{max} = 0.38 \pm 0.03$ μm, with $P_{max} = 1.32 \pm 0.04\%$). What percentage polarization would you predict this star to have at a wavelength of 0.15 μm in the ultraviolet?
5. Explain why a 'flattened' grain such as a disc tends to spin with its angular momentum vector parallel to its axis of symmetry, whereas an 'elongated' grain such as a long cylinder tends to spin with its angular momentum vector orthogonal to its axis of symmetry.
6. Derive expressions equivalent to equation (4.27) for the angular speed of (i) a disc of radius r and thickness $r/2$ and (ii) a cylindrical rod of radius r and length $20r$, in each case assuming the particle to be spinning about its axis of principal inertia in thermal equilibrium with the gas. Estimate ω_{rms} in each case, assuming $r = 0.2$ μm and 0.06 μm for the disc and the rod, respectively, that each are composed of material of density 2000 kg m^{-3} and that the gas temperature is 80 K. Compare your results with the estimate for a spherical grain given in the text (section 4.5.1). (*Note*: the dimensions of the disc, rod and sphere have been chosen such that the volume is approximately the same in each case.)
7. Estimate the minimum magnetic flux density required to align suprathermally spinning grains of mean radius $a = 0.15$ μm, density s = 2000 kg m^{-3} and temperature $T_d = 15$ K by paramagnetic relaxation. Assume that the grains are immersed in a gas of number density $n_H = 3 \times 10^7$ m^{-3} and temperature $T_g = 80$ K and that alignment is limited by the time (equation (4.35)) to accumulate a monolayer of surface oxygen atoms that become hydrogenated to H_2O-ice (density 1000 kg m^{-3}). Take the thickness of the monolayer to be 3.7×10^{-10} m and the sticking coefficient to be 0.5. The gas-phase number density of atomic oxygen may be determined with reference to information in table 2.2.
8. What would happen to the pattern of polarization $\boldsymbol{E}$-vectors on the sky (figure 4.2) if the galactic magnetic field were to undergo a sudden reversal in direction?

Chapter 5

Infrared absorption features

"It seems to me that in order to determine the composition of the dust, we must turn to the infrared part of the spectrum..."

J E Gaustad (1971)

Spectroscopy of solid-state absorption features provides a powerful and direct technique for investigating the composition and evolution of dust in the galactic environment. One prominent absorption feature attributed to interstellar dust – the ultraviolet extinction bump centred at 2175 Å– is discussed in some detail in chapter 3. In this chapter, we turn our attention to the infrared region of the spectrum, where the continuum extinction is much lower and more readily separable from discrete features. The vibrational resonances of virtually all molecules of astrophysical interest occur at frequencies corresponding to wavelengths in the spectral region from 2.5 to 25 μm. Astronomical infrared spectroscopy was an area of tremendous growth in the later part of the 20th century, from exploratory observations with low-resolution spectrophotometers in the 1970s to the development of grating spectrometers with detector arrays in the 1980s, culminating with the launch of the Infrared Space Observatory (ISO) in 1995. Observations with ground-based telescopes are hindered by strong telluric absorption at certain wavelengths: indeed, several of the species we may wish to study in the ISM (e.g. H_2O, CO_2, CH_4) are also 'greenhouse' gases responsible for infrared opacity in the Earth's atmosphere. This problem may be alleviated by placing telescopes at high altitude: the mean scale height of water, $\sim$2 km, is sufficiently low (thanks to precipitation!) that considerable advantage is gained by observing from a mountain-top site such as Mauna Kea (altitude 4.2 km) or an airborne observatory (cruising altitude $\sim$12 km). The infrared spectrum of interstellar dust has been studied extensively with such facilities through the available 'windows' in the atmospheric opacity. However, some telluric features are so strong that the spectral regions they block are accessible only to observation from space: good examples are the CO_2 features near 4.3 and

15 μm. Observations with ISO have allowed these gaps to be filled, enabling us to assemble a complete inventory of major interstellar condensates available to study by infrared techniques.

This chapter begins (section 5.1) with a brief discussion of the principles of infrared spectroscopy and the methods used to study interstellar dust. Subsequent sections review the results and their implications for dust in diffuse and dense phases of the ISM (sections 5.2 and 5.3, respectively). Infrared spectroscopy proves to be a sensitive diagnostic of the thermal history as well as the composition of the dust, a topic we shall return to in chapters 8 and 9.

5.1 Basics of infrared spectroscopy

5.1.1 Vibrational modes in solids

Absorption features at infrared wavelengths result from molecular vibrations within the grain material. The frequency of vibration for a given molecule is determined by the masses of the vibrating atoms, the molecular geometry and the forces holding the atoms in their equilibrium positions. Consider for simplicity a diatomic molecule containing atoms of masses m_1 and m_2. To a good approximation, the vibrations of the molecule may be represented by those of a harmonic oscillator in which the masses are joined by a spring obeying Hooke's law. The fundamental frequency of vibration is then given by

$$\nu_{\rm F} = \frac{1}{2\pi}\left\{\frac{k}{\mu}\right\}^{\frac{1}{2}} \tag{5.1}$$

where $\mu = m_1 m_2/(m_1 + m_2)$ is the reduced mass and k is the force constant of the chemical bond[1]. The vibrations of a typical molecule lead to spectral activity centred at frequency $\nu_{\rm F}$ or wavelength $\lambda_{\rm F} = c/\nu_{\rm F}$ in the electromagnetic spectrum. For gas-phase molecules, rotational splitting of vibrational energy levels results in molecular 'bands' composed of many closely spaced lines. This is illustrated in figure 5.1(*a*) for the case of CO: separate bands, termed P and R branches, arise from application of the rule $\Delta J = \pm 1$, where J is the rotational quantum number (Banwell and McCash 1994). All rotation is suppressed in the solid phase, however, and the P and R branches are replaced by a single, continuous spectral feature centred near $\lambda_{\rm F}$, as shown in figure 5.1(*b*). The solid-state feature is broader than the individual gas-phase lines, due to interactions between neighbouring molecules in the solid, but narrower than the entire band. Solid- and gas-phase absorptions are thus easily distinguishable: this is generally true even when the spectral resolution is insufficient to resolve the individual

[1] As real molecules are not perfect harmonic oscillators, the actual vibrational frequency differs from $\nu_{\rm F}$ by a factor that depends on the properties of the molecule and the vibrational quantum number of the energy level considered (Banwell and McCash 1994). At the low energy states considered here, this factor is close to unity.

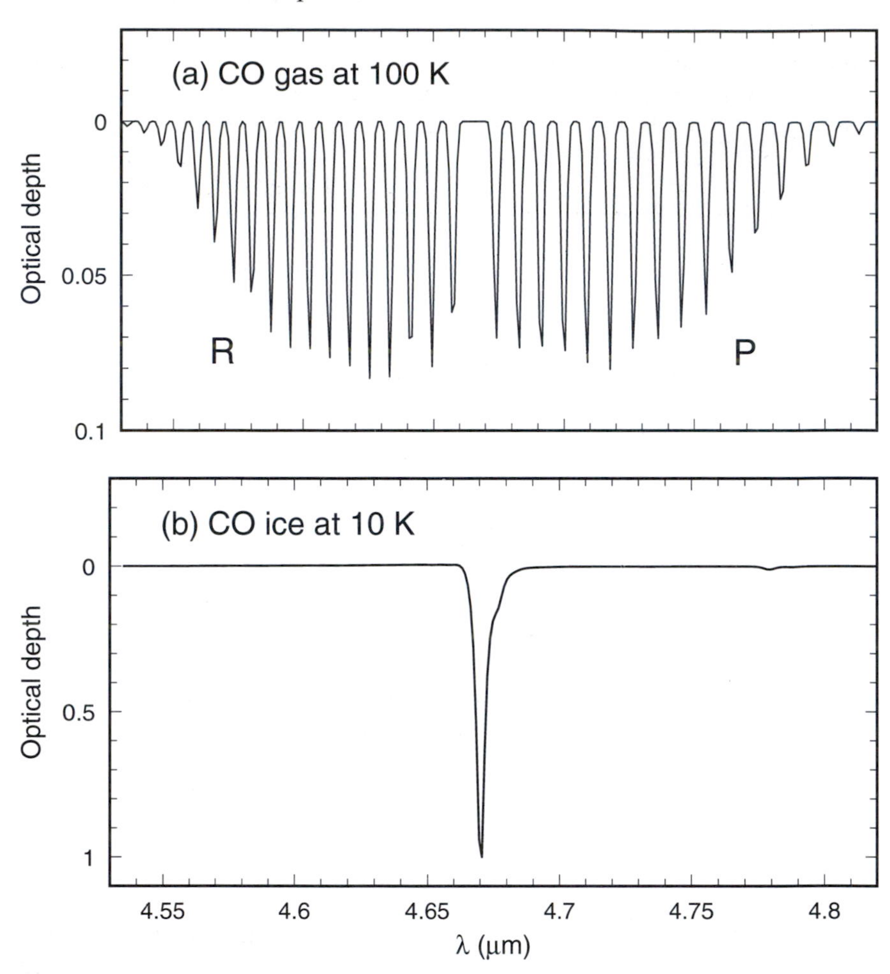

Figure 5.1. Infrared spectra of the fundamental vibrational mode of carbon monoxide: (*a*) gaseous CO at $T = 100$ K, showing the P and R branches caused by rotational splitting of the vibrational transition (Helmich 1996); and (*b*) solid CO at $T = 10$ K (Ehrenfreund *et al* 1996). Figure courtesy of Jean Chiar.

lines in the molecular bands. Note that vibrations are also possible at the harmonic (overtone) frequencies $2\nu_F$, $3\nu_F$, ..., but these do not generally produce observable features in interstellar dust.

Polyatomic molecules clearly have more possibilities for vibrational motion. A simple linear species such as CO_2 has vibrations associated with bending of the O=C=O structure in addition to longitudinal stretching of each C=O bond. At a slightly higher level of complexity, methanol (CH_3OH) has a correspondingly richer infrared spectrum, including distinct features arising from stretching of

Table 5.1. Molecular vibrational modes giving rise to absorptions in some refractory solids of astrophysical interest. Values of the mass absorption coefficient (κ) are from the following sources: hydrogenated amorphous carbon (HAC), Furton *et al* (1999); organic residue, unpublished data (see Whittet 1988); amorphous silicates, Day (1979, 1981), Dorschner *et al* (1988); silicon carbide, Whittet *et al* (1990) and references therein.

Material	Mode	λ (μm)	κ ($m^2\ kg^{-1}$)
HAC	C–H stretch	3.4	30–690
Organic residue	C–H stretch	3.4	40–80
$MgSiO_3$ (enstatite)	Si–O stretch	9.7	315
	O–Si–O bend	19.0	88
$(Mg, Fe)SiO_3$ (bronzite)	Si–O stretch	9.5	300
	O–Si–O bend	18.5	165
$FeSiO_3$ (ferrosilite)	Si–O stretch	9.5	210
	O–Si–O bend	20.0	82
Mg_2SiO_4 (fosterite)	Si–O stretch	10.0	240
	O–Si–O bend	19.5	86
Silicon carbide	Si–C stretch	11.2	660

the C–H, C–O and O–H bonds and structural deformation of the CH_3 unit. Vibrational modes that give rise to absorption in these and other species regarded as candidates for interstellar solids are listed in tables 5.1 and 5.2.

A general prerequisite for the production of a spectral feature is that the dipole moment of the molecule oscillates during the vibration. This is true for most molecules of astrophysical interest but there are two important exceptions, namely the homonuclear molecules O_2 and N_2. These species are infrared inactive, producing no features in their pure state. Weak features may be induced in a host matrix if interactions with neighbouring species perturb the symmetry of the molecule (Ehrenfreund *et al* 1992). However, searches for such features in interstellar ices have so far been unsuccessful (Vandenbussche *et al* 1999, Sandford *et al* 2001), yielding rather loose upper limits on abundances.

Comparison with laboratory data is the key to reliable interpretation of astronomical spectra. Assignments of solid-state features to specific molecules cannot always be made purely on the basis of wavelength coincidence. A given absorption is assigned initially to a chemical bond rather than to a specific

Table 5.2. Molecular vibrational modes giving rise to absorptions in some molecular ices of astrophysical interest. Values of the band strength $\mathcal{A}$ are from Schutte (1999) and references therein.

Molecule	Mode	λ (μm)	$\mathcal{A}$ (m/molecule)
H_2O	O–H stretch	3.05	2.0×10^{-18}
	H–O–H bend	6.0	8.4×10^{-20}
	libration	12	3.1×10^{-19}
NH_3	N–H stretch	2.96	2.2×10^{-19}
	deformation	6.16	4.7×10^{-20}
	inversion	9.35	1.7×10^{-19}
CH_4	C–H stretch	3.32	7.7×10^{-20}
	deformation	7.69	7.3×10^{-20}
CO	C=O stretch	4.67	1.1×10^{-19}
CO_2	C=O stretch	4.27	7.6×10^{-19}
	O=C=O bend	15.3	1.5×10^{-19}
CH_3OH	O–H stretch	3.08	1.3×10^{-18}
	C–H stretch	3.53	5.3×10^{-20}
	CH_3 deformation	6.85	1.2×10^{-19}
	CH_3 rock	8.85	1.8×10^{-20}
	C–O stretch	9.75	1.8×10^{-19}
H_2CO	C–H stretch (asym.)	3.47	2.7×10^{-20}
	C–H stretch (sym.)	3.54	3.7×10^{-20}
	C=O stretch	5.81	9.6×10^{-20}
	CH_2 scissor	6.69	3.9×10^{-20}
HCOOH	C=O stretch	5.85	6.7×10^{-19}
	CH deformation	7.25	2.6×10^{-20}
C_2H_6	C–H stretch	3.36	1.6×10^{-19}
	CH_3 deformation	6.85	6.0×10^{-20}
CH_3CN	C≡N stretch	4.41	3.0×10^{-20}
OCN^-	C≡N stretch	4.62	1.0×10^{-18}
H_2S	S–H stretch	3.93	2.9×10^{-19}
OCS	O=C=S stretch	4.93	1.5×10^{-18}
SO_2	S=O stretch	7.55	3.4×10^{-19}

molecule and ambiguities may occur when vibrational modes in different species arises at similar wavelengths: a prime example is the C–H stretch at $\lambda \sim 3.4$ μm, which will, of course, occur in any species that contains H bonded to C. The availability of laboratory spectra allows the possibility to distinguish between possible 'carrier' molecules for a given chemical bond. For example, the 3.53 μm C–H feature in methanol can generally be isolated from absorptions arising in other organic molecules at similar wavelengths (e.g. Grim *et al* 1991). The spectrum of a vibrating molecule is generally influenced by its molecular environment: both the composition and the structure of the host 'matrix' are important. For example, CO trapped in ice composed primarily of other species shows subtle differences compared with that of CO-ice in its pure state (Sandford *et al* 1988). Many classes of material, including both ices and refractory solids, show quite different absorption profiles in their ordered (crystalline) and disordered (amorphous) states. As interstellar grains are generally expected to form in an amorphous state and to become crystalline only if subjected to subsequent heating, infrared spectroscopy offers the possibility to explore their thermal evolution as well as chemical composition.

5.1.2 Intrinsic strengths

Laboratory experiments also provide information on the intrinsic strengths of the various vibrational features. Such data are needed to calculate the amount of a given absorber represented by the observed strength of its absorption. For a generic refractory material such as amorphous carbon or silicate (table 5.1), where many different molecular structures may be present, it is convenient to specify the intrinsic strength in terms of the mass absorption coefficient, defined as the absorption cross section per unit mass at the peak of the relevant absorption feature:

$$\kappa = \frac{C_{\rm abs}}{m} = \frac{3Q_{\rm abs}}{4as} \tag{5.2}$$

where $Q_{\rm abs} = C_{\rm abs}/\pi a^2$ (see section 3.1.1) and spherical grains of mass m, radius a and specific density s are assumed. Values of κ may thus be calculated from Mie theory if the optical constants of the material are known, or they may be measured directly for laboratory-generated smokes (e.g. Dorschner *et al* 1988 and references therein). The total mean density of an absorber along a column of length L that is required to account for the strength of a feature of peak optical depth $\tau_{\rm max}$ may then be estimated from the relation

$$\rho = \frac{\tau_{\rm max}}{\kappa L}. \tag{5.3}$$

Intrinsic strengths of absorption features in specific molecules may be expressed in terms of the integrated absorption cross section per molecule, or band strength,

$\mathcal{A}$. For an unsaturated absorption line, the column density N of the absorber is then given by

$$N = \frac{\int \tau(x)\,\mathrm{d}x}{\mathcal{A}} \tag{5.4}$$

where $x = \lambda^{-1}$. For features with Gaussian-like profiles, the approximation

$$N \approx \frac{\gamma \tau_{\max}}{\mathcal{A}} \tag{5.5}$$

may be used, where γ is the width (FWHM) of the profile in wavenumber units. Values of $\mathcal{A}$ for various laboratory ices are listed in table 5.2. For a description of measurement techniques, see Gerakines *et al* (1995a).

5.1.3 Observational approach

Apparatus for obtaining spectra in the laboratory typically consists of a radiation source, a sample chamber and an instrument to record the spectrum. These elements are also present in the astronomical context: the radiation source is a background star, the sample chamber is the interstellar medium and the recording instrument is a spectrometer attached to a telescope. The intrinsic spectrum of the radiation source may, of course, be measured directly in the laboratory situation, whereas it must be inferred or modelled in the case of the star. Two types of background star may be encountered, as illustrated schematically in figure 5.2. A field star is not directly associated with interstellar matter but its radiation is absorbed by foreground dust. The absorption may arise primarily in a single cloud (as depicted in figure 5.2) or in a number of clouds distributed along the line of sight. The alternative possibility is that the star observed is *embedded* in the cloud that produces the absorption. This is a common situation toward molecular clouds, as the brightest infrared sources are typically young stars still enclosed in opaque envelopes of protostellar material.

The observed spectra are usually displayed as plots of flux density in Janskys (1 Jy $= 10^{-26}$ W m^{-2} Hz^{-1}) against wavelength or wavenumber. As well as covering the entire spectral range of interest, it is important that the data should have sufficient resolution to detect any fine structure that may be present in the solid-state features. Resolving powers ($\lambda/\Delta\lambda \sim 1500$) needed to do this are now routinely available on ground-based telescopes and from the ISO archive. Examples are shown in figures 5.3–5.5. The flux spectra effectively show spectral energy distributions for each target star with foreground absorption features superposed. However, when making comparisons with laboratory analogues, it is important to display the data in a way that is independent of the characteristics of the background continuum source. This may be accomplished by fitting a mathematical representation of the continuum to regions of the spectrum that appear free of features. The adopted mathematical form may be a polynomial, a Planck function or indeed any function that seems to give a realistic description of the background source. If the observed spectrum and the adopted continuum

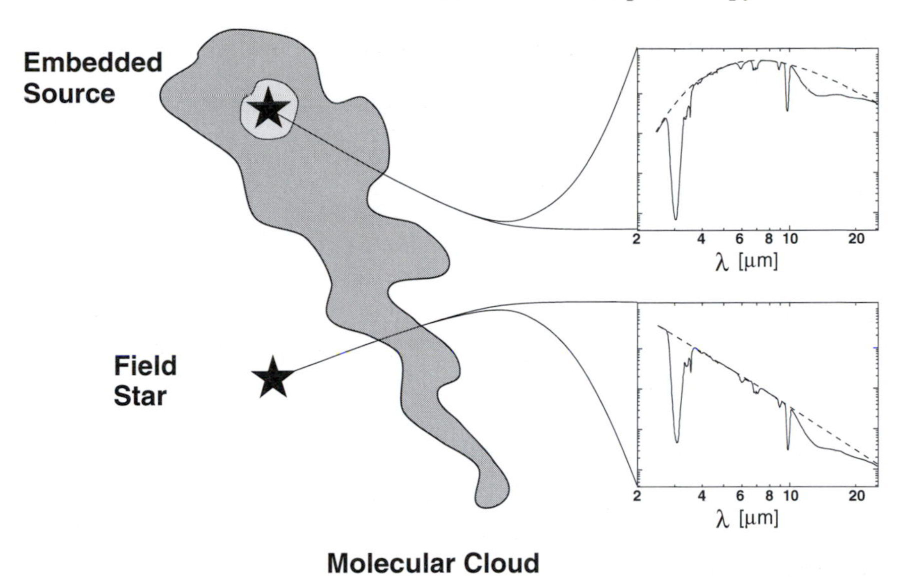

Figure 5.2. Schematic illustration of two types of infrared source and their spectra. Unlike the embedded source, the field star is not physically associated with the interstellar cloud that lies in the line of sight. Embedded stars tend to be brighter at mid- and far-infrared wavelengths because of circumstellar emission from warm dust, whereas the field star displays only a photosphere with foreground absorption and reddening. Figure courtesy of Perry Gerakines.

are represented by functions $I(\lambda)$ and $I_0(\lambda)$, respectively, then an optical-depth spectrum for the absorption features may be calculated (see equation (3.2)):

$$\tau(\lambda) = \ln\left\{\frac{I_0(\lambda)}{I(\lambda)}\right\}. \tag{5.6}$$

An example of the application of this procedure is shown in figure 5.5.

By careful characterization and selection of targets to observe and by applying the techniques described here, we can compare and contrast the properties of the dust in a range of environments. When this is done a general pattern is seen: whereas silicate absorption occurs in both diffuse and dense phases of the ISM (section 1.4.3), evidence for ices is found only in dense molecular clouds. This dichotomy is illustrated in figures 5.3 and 5.4: in each case, a diffuse-ISM-dominated spectrum is paired with a molecular-cloud-dominated spectrum. The features arising in these contrasting environments are discussed in the following sections.

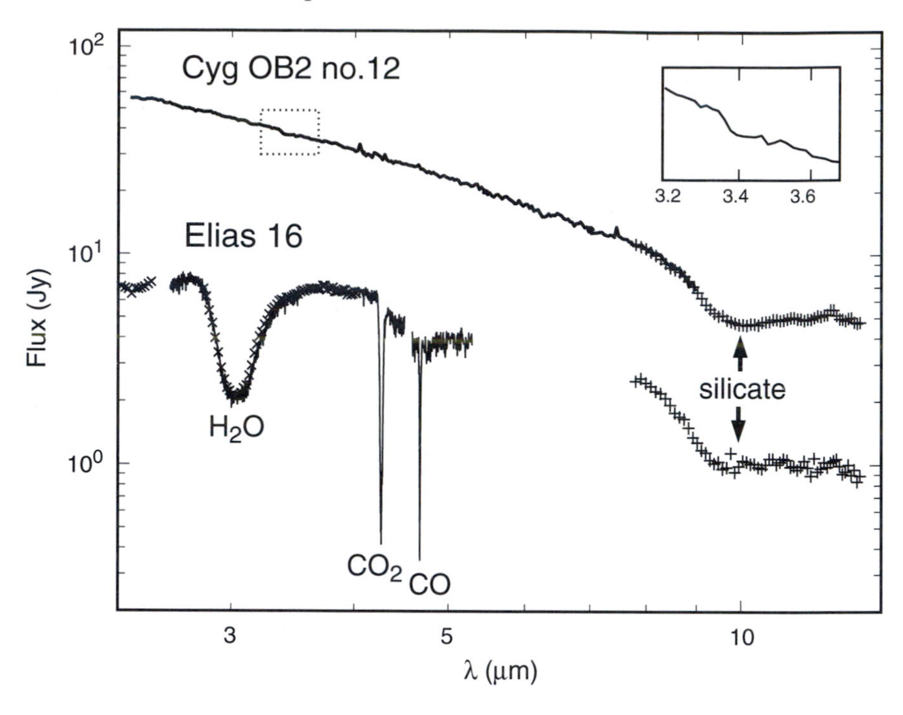

Figure 5.3. Infrared spectra from 2 to 15 μm of two highly reddened field stars that sample contrasting environments. Top: the blue hypergiant Cyg OB2 no. 12, reddened by predominantly diffuse-cloud material, shows silicate absorption near 10 μm but no ice absorption. The inset at the top right is an enlargement of the data in the dotted rectangle, containing the 3.4 μm hydrocarbon feature. Bottom: the red giant Elias 16 lies behind a clump of molecular material in the Taurus dark cloud and shows both silicate and various ice features. Data are from the ISO Short-Wavelength Spectrometer (SWS; Whittet *et al* 1997, 1998: continuous curves) and ground-based observations (Smith *et al* 1989: diagonal crosses; Bowey *et al* 1998: plus signs).

5.2 The diffuse ISM

5.2.1 The spectra

In attempting to characterize the spectrum of dust in the diffuse ISM, we are confronted with a sampling problem. The ideal continuum source in which to study absorption in diffuse clouds would lie close to the galactic plane at a distance great enough to ensure a large column density of dust, accumulated through the presence of many H I clouds (and no H_2 clouds) in the line of sight. There is, however, a paucity of known suitable candidates that are both sufficiently bright and sufficiently extinguished to give measurable optical depths in the features.

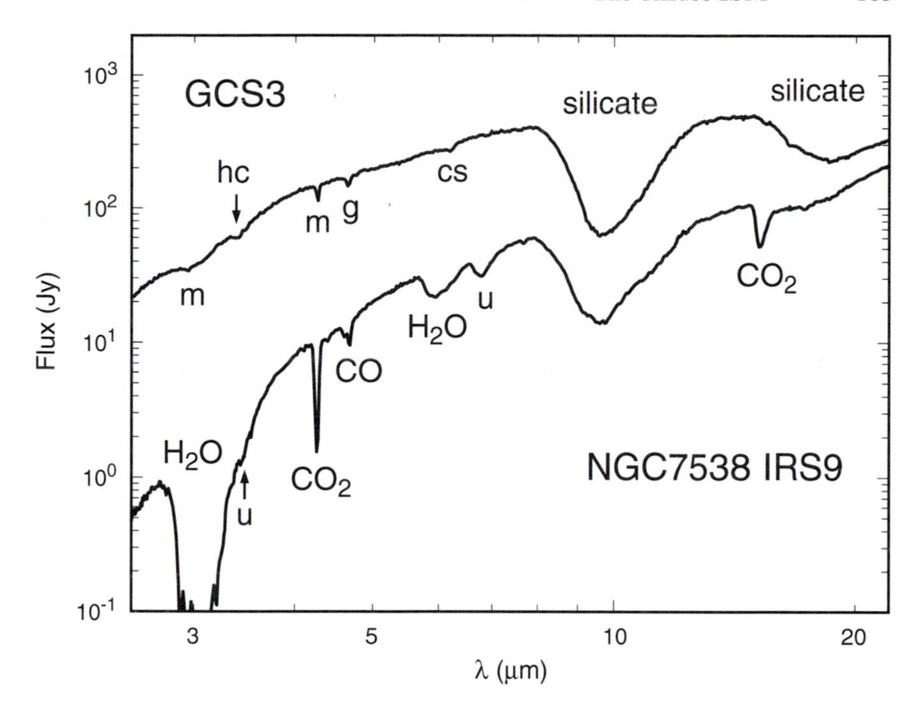

Figure 5.4. Infrared spectra from 2.5 to 20 μm of the galactic centre source GCS3 (top) and the dust-embedded protostar NGC 7538 IRS9 (bottom), obtained with the ISO SWS. The GCS3 spectrum has been scaled up by a factor of four for display. Several features discussed in the text are labelled. The 4.65 μm feature labelled 'g' in GCS3 is an unresolved molecular band of gas-phase CO; all other features arise in the solid phase. Features identified with ices in molecular clouds toward GCS3, labelled 'm', have direct counterparts in the NGC 7538 IRS9 spectrum. The feature labelled 'hc' at 3.4 μm is attributed to hydrocarbons in the diffuse ISM, whereas the feature labelled 'cs' at 6.2 μm is thought to arise in the circumstellar shell of GCS3. Unidentified features at 3.47 and 6.85 μm in NGC 7538 IRS9 are labelled 'u'.

The cluster of infrared sources associated with the centre of our Galaxy (GC) has long been regarded as a good approximation to this ideal (e.g. Becklin and Neugebauer 1975, Roche 1988). The sources in this cluster are believed to be luminous stars obscured by some 30 magnitudes of visual extinction accumulated along the 7–8 kpc pathlength. However, more recent studies indicate that a significant proportion of this extinction arises in molecular clouds. The spectrum of CGS3 (figure 5.4) is dominated by deep silicate absorption in both the stretching and bending modes, centred at 9.7 and 18.5 μm (table 5.1). Weaker absorptions at 3.0 and 4.3 μm, identified with H_2O and CO_2 ices, respectively, make the case for the presence of one or more molecular clouds along the line of sight (Whittet *et al* 1997). Their depths vary from source to source across the

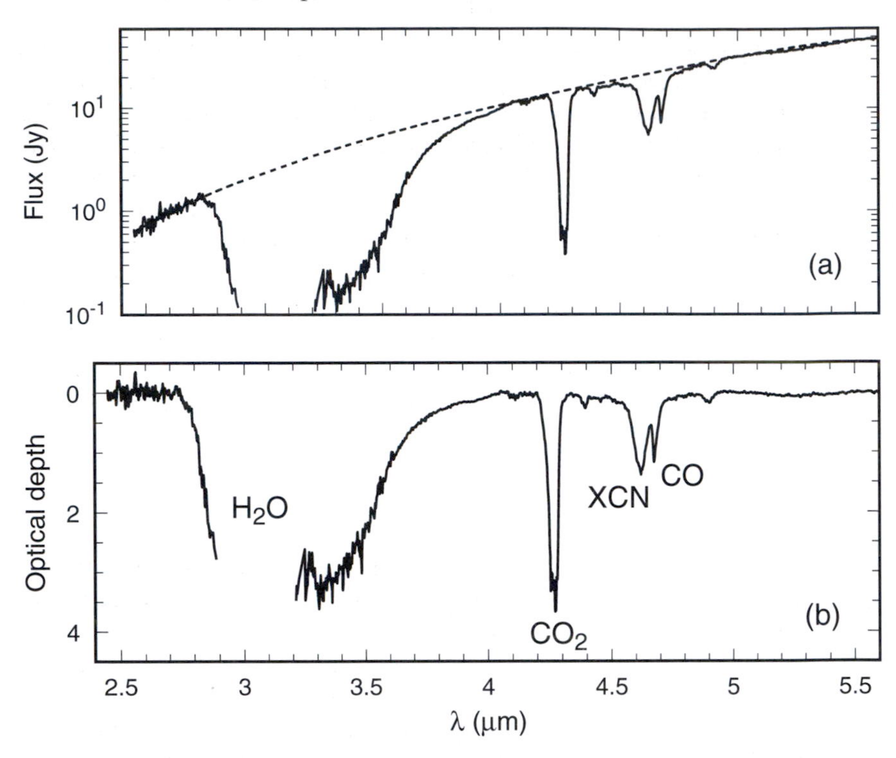

Figure 5.5. Infrared spectrum of the luminous protostar W33A obtained with the ISO SWS (Gibb *et al* 2000a): (*a*) flux spectrum, together with the adopted continuum obtained by fitting a fourth-order polynomial to the data in the wavelength ranges 2.5–2.7, 4.0–4.1, 4.95–5.1 and 5.5–5.6 μm; and (*b*) optical depth spectrum obtained from the ratio of the continuum flux to the observed flux (equation (5.6)). Note that the 3 μm H_2O-ice feature is saturated in this source (i.e. the flux falls below detectable levels in the trough of the feature).

GC cluster, indicating spatial variability in the opacity of the molecular material on a scale of a few arcminutes. Other features at 3.4 and 6.2 μm are attributed to C–H and C–C stretching in aliphatic and aromatic hydrocarbons, respectively (Sandford *et al* 1991, Schutte *et al* 1998). Whilst the 6.2 μm absorption may be at least partially circumstellar (Chiar and Tielens 2001), that at 3.4 μm appears to be a true signature of carbonaceous dust in the diffuse ISM (Adamson *et al* 1990, Sandford *et al* 1991, 1995, Pendleton *et al* 1994). A detailed comparison and deconvolution of the diffuse-ISM and molecular-cloud components of the GC spectrum may be found in Chiar *et al* (2000, 2002).

Other sources that have been used to investigate diffuse-ISM absorption features are the early-type stars, including both normal OB stars and Wolf–Rayet (WR) stars. Although the total extinction toward even the most reddened known examples are, in every case, considerably less than toward the GC sources and the absorptions correspondingly weaker, observations of these stars are very important: they allow us to distinguish between features peculiar to the GC line of sight and those that are genuine signatures of the widely distributed ISM. The best studied case is the B-type hypergiant Cyg OB2 no. 12 (e.g. Gillett *et al* 1975a, Adamson *et al* 1990, Sandford *et al* 1991): its infrared spectrum appears in figure 5.3. Broad 9.7 μm silicate absorption is clearly seen but otherwise the spectrum is remarkably free of substantial features (compare Elias 16, which lies behind a molecular cloud). The 3.4 μm C–H feature (inset) is weakly present but only upper limits can be set on absorptions arising in species containing O–H, C–O or C–N bonds (Whittet *et al* 1997, 2001b). We conclude from these results that spectral evidence exists for just two generic classes of grain material in the diffuse ISM: silicates and hydrocarbons.

5.2.2 Silicates

The 9.7 μm silicate absorption is easily the strongest and best studied infrared feature arising in the diffuse ISM. The profile of this feature helps to constrain the nature of interstellar silicates. Amorphous, disordered forms produce smooth, broad features, whereas crystalline silicates (common in terrestrial igneous rocks) produce profiles with sharp, narrow structure (Krätschmer and Huffman 1979, Day 1979, 1981, Dorschner and Henning 1986, Dorschner *et al* 1988, Hallenbeck *et al* 2000, Fabian *et al* 2000). The feature observed toward the galactic centre (figure 5.6) is generally smooth and lacking in structure: an excellent fit is obtained with laboratory data for amorphous olivine, whereas crystalline silicates produce structure that has no counterpart in the observations (a representative example is also shown in figure 5.6). Profiles for early-type stars are determined with lower precision but appear to be similarly devoid of structure to within observational limits (Roche and Aitken 1984a, Bowey *et al* 1998). Li and Draine (2001) estimate from these results that no more than 5% of the available Si can be in crystalline forms. Profile shapes observed in the diffuse ISM are closely similar to those seen in the dusty envelopes of red giants (chapter 7), consistent with a 'stardust' origin for interstellar silicates.

Although most naturally occurring terrestrial silicates are poor spectroscopic matches to the ISM because of their inherent crystallinity, there is one such group that can produce 9.7 μm profiles resembling the interstellar feature: the hydrated or layer-lattice silicates (Zaikowski *et al* 1975, Knacke and Krätschmer 1980). The degree of hydration present in interstellar silicates is thus called into question. This is open to observational test, by means of a search for absorption in the wavelength range 2.6–2.9 μm associated with the O–H groups they contain. As the galactic centre sources exhibit an ice feature that overlaps this spectral region (figure 5.4), the best constraints are provided by the early-type stars. Observations with the ISO SWS (Whittet *et al* 1997, 2001b) indicate that silicates contain no more than about 2% by mass of OH in the diffuse ISM toward Cyg OB2 no. 12,

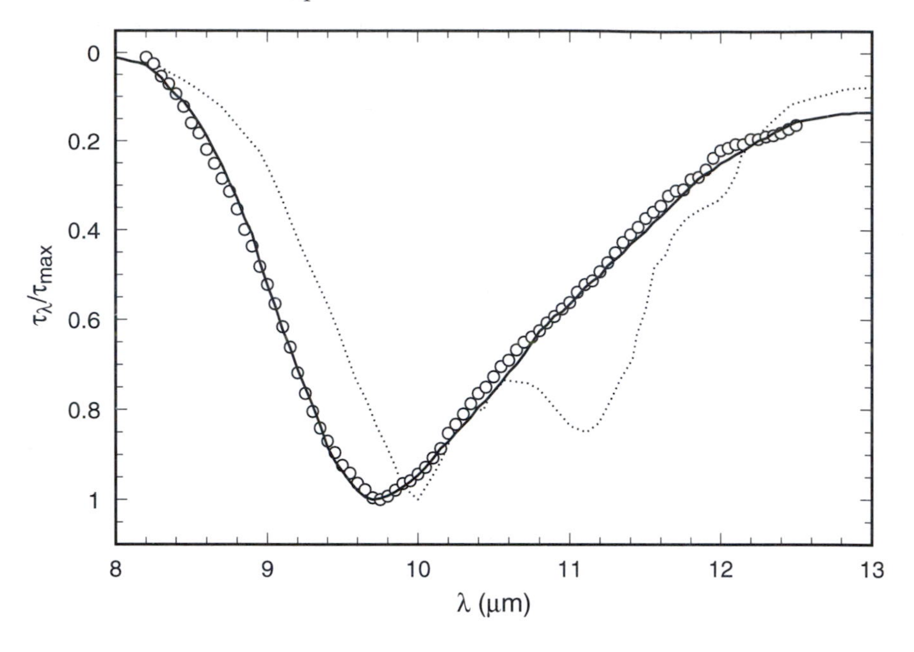

Figure 5.6. The 9.7 μm silicate profile. Observational data for the galactic centre source Sgr A obtained with the ISO SWS (circles) are closely matched by laboratory data for amorphous olivine ($MgFeSiO_4$, full curve; Vriend 2000). Also shown for comparison is a representative profile for a crystalline silicate, annealed fosterite (Mg_2SiO_4, dotted curve; Fabian *et al* 2000). All data are normalized to unit optical depth at peak absorption.

compared with 5–30% for terrestrial and meteoritic hydrated silicates. The latter are thought to form in aqueous environments that presumably do not exist in the ISM or in the circumstellar birth-sites of interstellar grains.

Spectropolarimetric observations give further insight into the nature of interstellar silicate dust. Figure 5.7 plots 8–13 μm polarization data averaged for two stars with high foreground extinction arising in the diffuse ISM. That silicate grains are being aligned (chapter 4) is confirmed by the existence of a polarization counterpart to the 9.7 μm absorption feature[2]. The theory of dichroic absorption by aligned grains (Aitken 1989) predicts that the position of peak polarization should be shifted to a somewhat longer wavelength compared with peak optical depth and this is indeed observed (the peaks occur at approximately 10.25 μm and 9.75 μm in P_λ and τ_λ, respectively, in these lines of sight). The polarization profile is especially sensitive to changes in optical properties associated with crystallinity. The lack of discernible structure in the P_λ profile (figure 5.7) is

[2] A counterpart to the 18.5 μm silicate feature is also expected and is indeed observed in some highly polarized protostars (section 5.3.8). However, this is difficult to observe in the diffuse ISM as early-type stars are intrinsically faint in the mid-infrared and interpretation of data for galactic centre sources is complicated by the superposition of emission and absorption components (see Roche 1996).

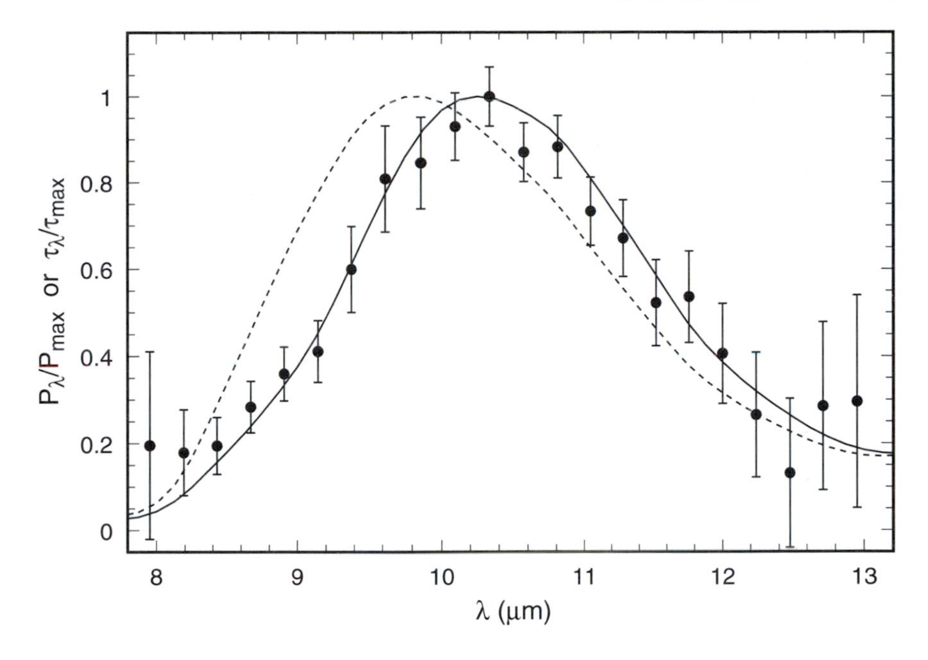

Figure 5.7. The polarization profile of the silicate Si–O stretch feature: points with error bars, polarization data averaged from observations of two early-type stars, WR 48A and GL 2104 (Smith *et al* 2000); full curve, calculated polarization for a model that assumes aligned oblate spheroids of size $a = 0.1$ μm and axial ratio $b/a = 2{:}1$, composed of amorphous olivine (Wright *et al* 2002). Also shown is the corresponding optical depth profile (broken curve). Data courtesy of Christopher Wright.

consistent with the amorphous nature of interstellar silicates implied by the optical depth profile. The observations are well matched by a model that assumes oblate spheroids composed of amorphous olivine.

On the basis of models for extinction (section 3.7), we expect silicates to be a subset of all grains responsible for visual extinction. A correlation between peak silicate optical depth ($\tau_{9.7}$) and A_V is thus expected if the various forms of dust are well mixed in the ISM. This plot is shown in figure 5.8. The early-type stars, which lie relatively close to the Sun (mostly within about 3 kpc, compared with ~7.7 kpc for the galactic centre), do, indeed, show a high degree of correlation, consistent with a straight line through the origin:

$$\frac{A_V}{\tau_{9.7}} = 18.0 \pm 1.0. \tag{5.7}$$

For comparison, the $A_V/\tau_{9.7}$ ratio for pure silicate dust is ~1 for very small ($a \sim 0.01$ μm) particles, rising to $A_V/\tau_{9.7} \sim 5$ for classical ($a \sim 0.15$ μm) grains (e.g. Stephens 1980, Gillett *et al* 1975a). Thus, silicates alone account for no more than about a third of the visual extinction in the solar neighbourhood, the

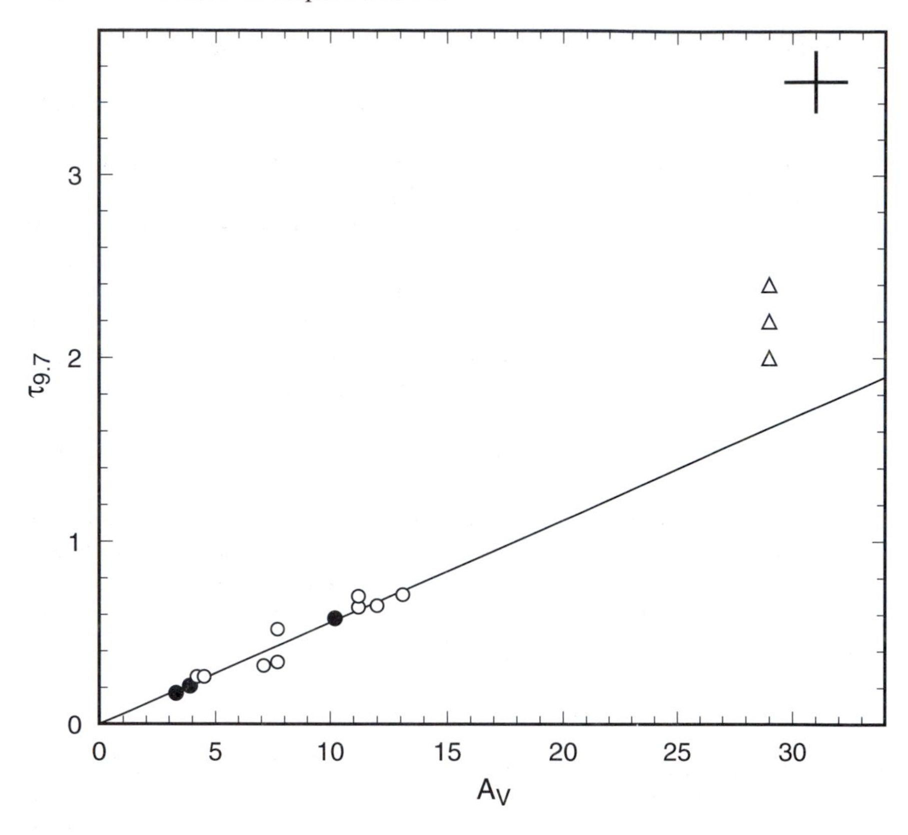

Figure 5.8. Plot of the peak optical depth in the 9.7 μm silicate feature against visual extinction. OB stars and WR stars are denoted by filled and open circles, respectively. Infrared sources in the Quintuplet and Sagittarius A regions of the galactic centre are denoted by triangles and a cross, respectively. The diagonal line is a fit through the origin to stars with $A_V < 15$ (equation (5.7)). Data are from Roche and Aitken (1984a, 1985), Rieke and Lebofsky (1985), Schutte *et al* (1998) and Chiar and Tielens (2001).

balance presumably arising in the C-rich component. Toward the galactic centre, the silicate feature is deeper than predicted by an extrapolation of equation (5.7) in figure 5.8, suggesting systematic variation in the relative abundances of silicate and C-rich dust with galactocentric radius (Roche and Aitken 1985, Thronson *et al* 1987; see section 2.3.2).

An estimate of the mean density of silicate dust required to account for the observed strength of the 9.7 μm feature may be obtained from equation (5.3). A mass absorption coefficient $\kappa_{9.7} \approx 290$ m^2 kg^{-1} is typical of amorphous magnesium silicates (table 5.1; note that crystalline forms have values typically a factor of 5–10 higher). Combining equation (5.7) with the average value of A_V/L in the galactic plane near the Sun (1.8 mag kpc^{-1}; equation (1.6)), we obtain $\tau_{9.7}/L \approx 0.1\,\text{kpc}^{-1} \approx 3.2 \times 10^{-21}\,\text{m}^{-1}$ and thus equation (5.3) gives

$$\rho_{\rm d}(\text{sil.}) \approx 11 \times 10^{-24}\,\text{kg m}^{-3}. \tag{5.8}$$

This is ~60% of the total dust density estimated from extinction (section 3.3.5). For the GC Sgr A region, values of $L \approx 7.7$ kpc and $\tau_{9.7} \approx 3.6$ (Roche and Aitken 1985) in equation (5.3) imply an enhancement in the line-of-sight average silicate density by a factor of about five compared with the solar neighbourhood.

An independent estimate of $\rho_{\rm d}$(sil.) may be obtained from the observed depletions discussed in chapter 2. Suppose that all of the cosmically available Si is tied up in silicates. With reference to equation (2.20), the silicate density allowed by the Si abundance $A_{\rm dust}^{\rm Si}$ in dust (table 2.2) is then

$$\rho_{\rm d}(\text{sil.}) = 10^{-6} A_{\rm dust}^{\rm Si} \rho_{\rm H} \sum n_{\rm X}\{m_{\rm X}/m_{\rm H}\} \tag{5.9}$$

where $n_{\rm X}$ is the relative number of element X per Si atom and the summation is carried out over elements X = Mg, Fe, Si, O. For a mixture of pyroxene and olivine structures that utilizes all the available Mg as well as Si (see section 2.5), $n_{\rm X}$ values are denoted by the generic formula $Mg_{1.07}Fe_{0.43}SiO_{3.5}$. For solar reference abundances, $A_{\rm dust}^{\rm Si} \approx 34$ ppm and hence

$$\rho_{\rm d}(\text{sil.}) \approx 8 \times 10^{-24}\,\text{kg m}^{-3}. \tag{5.10}$$

The two estimates of $\rho_{\rm d}$(sil.) in equations (5.8) and (5.10) are consistent to within the uncertainties (which could be as high as 20–30%, the most significant sources of error being the assumed values of $\kappa_{9.7}$ and the reference abundances, respectively). The implication is that essentially the full solar Si abundance must be in silicates to account for the strength of the 9.7 μm feature observed in the solar neighbourhood. More detailed calculations by Mathis (1998) show that the absorption per Si atom is optimized for porous, spheroidal grains containing >25% vacuum. In this case, it is possible to explain the observations with ~80% of the solar Si abundance in silicates.

5.2.3 Silicon carbide

One resonance listed in table 5.1 which is surprisingly absent in interstellar spectra is that of silicon carbide (SiC) at 11.2 μm. This feature is widely observed in emission in the spectra of C-rich red giants (chapter 7) and SiC is therefore presumably a component of the dust injected by such stars into the interstellar medium. Indeed, SiC particles of extrasolar origin have been isolated in meteorites (section 7.2.4), providing further circumstantial evidence for the existence of SiC at some level in interstellar dust. Its non-detection is thus remarkable, as the intrinsic strength of the feature associated with the Si–C bond is greater than that of the corresponding Si–O feature in silicates (table 5.1): if equal quantities of O-bonded and C-bonded silicon were present, one would expect the

11.2 μm feature to be *stronger* than the 9.7 μm feature. These absorptions are sufficiently broad that they overlap, but a detailed study of the observed profile (e.g. figure 5.6) allows any contribution from SiC to the observed silicate feature to be quantified. An upper limit $\tau_{11.2} < 0.1$ on the optical depth of SiC absorption toward the galactic centre has been determined (Whittet *et al* 1990), compared with $\tau_{9.7} \approx 3.6$ for silicates in the same line of sight.

The relative abundance of Si in silicate and silicon carbide dust in diffuse clouds may be estimated from this result. The column density of Si atoms contained within grain species X (where X represents SiC or silicates) is given by

$$N_{\mathrm{Si}}(\mathrm{X}) = \frac{f(\mathrm{X})\tau_\lambda}{28 m_{\mathrm{H}} \kappa_\lambda} \tag{5.11}$$

where $f(\mathrm{X})$ is the fraction by mass of Si in X. For silicon carbide, $f(\mathrm{SiC}) = 0.7$, whilst a mixture of magnesium and iron silicates yields $f(\mathrm{sil.}) \approx 0.2$. Substituting for $f(\mathrm{X})$ in equation (5.11) and using κ values listed in table 5.1, the ratio of Si atoms in SiC to Si atoms in silicates is given by

$$\frac{N_{\mathrm{Si}}(\mathrm{SiC})}{N_{\mathrm{Si}}(\mathrm{sil.})} \approx 1.5 \frac{\tau_{11.2}}{\tau_{9.7}}. \tag{5.12}$$

The observed limit on $\tau_{11.2}/\tau_{9.7}$ (Whittet *et al* 1990) then leads us to estimate that the number of Si atoms in interstellar SiC particles is $<5\%$ of those in silicates. This result is in accord with the requirement that essentially all the cosmically available Si should be in silicates to explain the strength of the interstellar 9.7 μm feature (section 5.2.2). It is also consistent with the dominance of silicates over SiC in meteorites (section 9.3). The scarcity of SiC in the ISM is nevertheless surprising if C-rich circumstellar shells (section 7.1) are a significant source of interstellar grains.

5.2.4 Hydrocarbons and organic residues

Models for interstellar dust require that C-rich material makes a substantial contribution to the grain mass in the diffuse ISM (see sections 2.5 and 3.7). It is evident from examination of the appropriate spectra (figures 5.3 and 5.4) that this material is relatively featureless in the infrared: the only feature confidently ascribed to it is the rather weak absorption centred near 3.4 μm, corresponding to C–H stretching vibrational modes (Duley and Williams 1981). Attempting to characterize a class of material on the basis of just one feature is obviously hazardous, yet considerable insight can be gained from its profile shape in comparison to those of laboratory analogues. Discrimination between candidate materials is also provided by the *lack* of observed features associated with other vibrational modes that commonly occur in carbonaceous solids.

Materials most widely discussed as carriers of the 3.4 μm feature are hydrogenated amorphous carbon (HAC) and organic refractory mantles (e.g. Moore and Donn 1982, Duley and Williams 1983, Sandford *et al* 1991, Pendleton

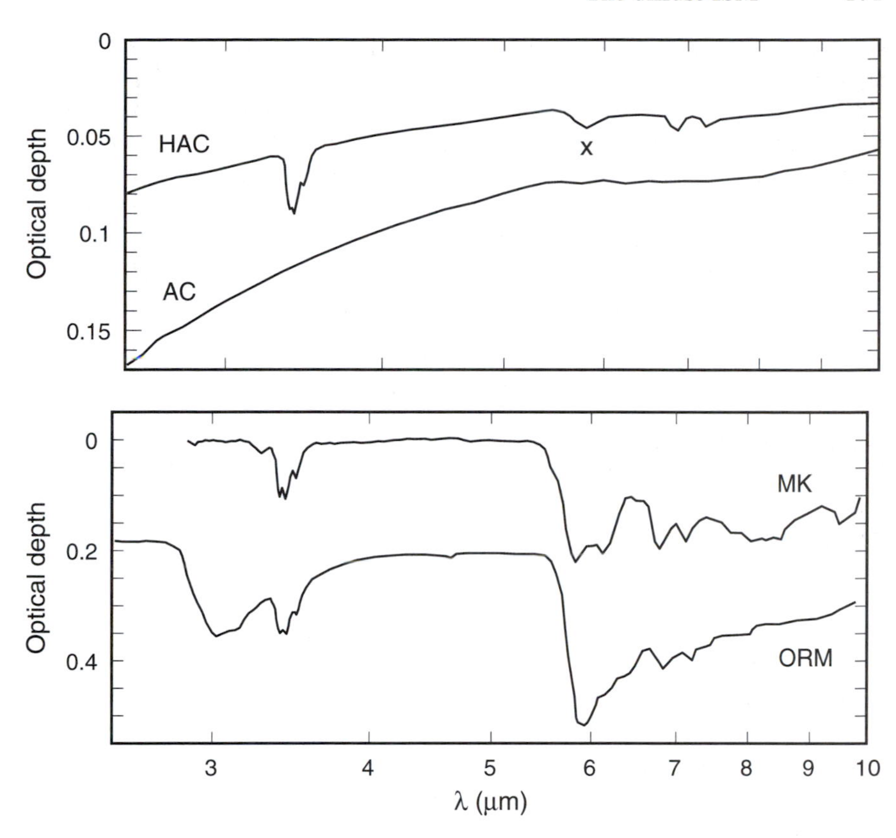

Figure 5.9. Infrared spectra of candidates for the C-rich component of interstellar dust in the diffuse ISM. Top frame: amorphous carbon nanoparticles before and after hydrogenation (labelled AC and HAC, respectively; Mennella *et al* 1999). The feature labelled x in the HAC spectrum is thought to be C=O absorption resulting from mild air contamination of the sample chamber. Bottom frame: organic refractory matter (ORM) produced by UV photolysis of a laboratory ice mixture (H_2O:CO:NH_3:C_2H_2 = 5:2:2:1; Greenberg *et al* 1995) and meteoritic kerogen (MK) from the Murchison meteorite (de Vries *et al* 1993). The ORM spectrum has been scaled to match the MK spectrum in the depth of the 3.4 μm feature.

et al 1994, Greenberg *et al* 1995). The wavelength of peak absorption clearly indicates that the vibrating atoms must be located in aliphatic (chainlike) rather than aromatic (ringlike) structures (the latter would produce features near 3.3 μm). If the host is an amorphous carbon, the carriers may take the form of aliphatic chains attached to the surface (Duley and Williams 1981). Infrared spectra of some candidate materials are shown in figure 5.9. Two organic refractory substances of different provenance are shown in the lower frame: an organic residue produced by radiative processing of a mixture of ices (ORM) and a kerogen extracted from the Murchison meteorite (MK). Both show 3.4 μm features that resemble the interstellar feature in position and shape (see Pendleton *et al* 1994). However, the spectra exhibit other features, as strong as or stronger than that at 3.4 μm, which have no counterpart in the ISM. In both cases, a sharp absorption edge appears between 5 and 6 μm, together with overlapping absorptions in the 6–10 μm range. These are attributed to various vibrational modes that naturally occur in complex organic matter, such as stretching of the C=O, C=C, C–OH and C–NH_2 bonds and deformations associated with CH and NH groups. The ORM spectrum also displays a strong feature at 3 μm: the O–H stretch wavelength. Non-detection of such features in the diffuse ISM indicates that the carrier must contain a predominance of C–H bonds over other bonds common in organic material; i.e. it is essentially a *hydrocarbon*.

The upper frame of figure 5.9 shows spectra of amorphous carbon nanoparticles produced by condensation of carbon vapour in the laboratory (Mennella *et al* 1999). Initially, the particles lack hydrogen and produce a continuous absorption spectrum devoid of sharp features. However, subsequent immersion in a hydrogen atmosphere leads to the development of a distinct absorption at 3.4 μm, together with structure at longer wavelength attributed to CH deformation modes at 6.85 and 7.25 μm. The latter are an order of magnitude weaker than the C–H stretch feature, consistent with their non-detection in interstellar spectra.

The observed profile of the 3.4 μm feature in the galactic centre is illustrated in figure 5.10. This was determined by deconvolving diffuse-ISM and molecular-cloud features in the spectra of several sources in the Sgr A region and averaging the results (see Chiar *et al* 2002 for details). The feature contains structure in the form of subpeaks centred at 3.38, 3.42 and 3.48 μm, together with a broad wing extending to shorter wavelength. The properties of the molecular groups responsible for the observed feature are constrained by the positions and relative strengths of these subpeaks. Saturated aliphatic hydrocarbons with the general formula CH_3–$(CH_2)_n$–CH_3 produce distinct features associated with symmetric and asymmetric C–H stretching vibrations in the CH_2 and CH_3 groups: Sandford *et al* (1991) assign the observed subpeaks at 3.38, 3.42 and 3.48 μm to the CH_3 asymmetric, CH_2 asymmetric and CH_3 symmetric modes, respectively. Their relative strengths suggest a CH_2:CH_3 ratio of about 2.5:1. However, the symmetric CH_2 mode (at 3.50 μm) is weak or absent in interstellar spectra. Laboratory studies show that this mode is suppressed if one of the CH_3 end groups is replaced by an electronegative group, such as OH or an aromatic ring. The CH_2 groups must be close to the electronegative group for suppression to be effective and so relatively short chains ($n \sim 2$–4) are implied. The surfaces of amorphous carbon particles provide an aromatic substrate to which the chains may become attached. The spectra of laboratory analogues (Mennella *et al* 1999, Schnaiter *et al* 1999) do, indeed, show subpeaks at 3.38, 3.42 and 3.48 μm that correspond to those in the observed spectrum. An example taken from the work of Mennella *et al* (1999) is illustrated in figure 5.10.

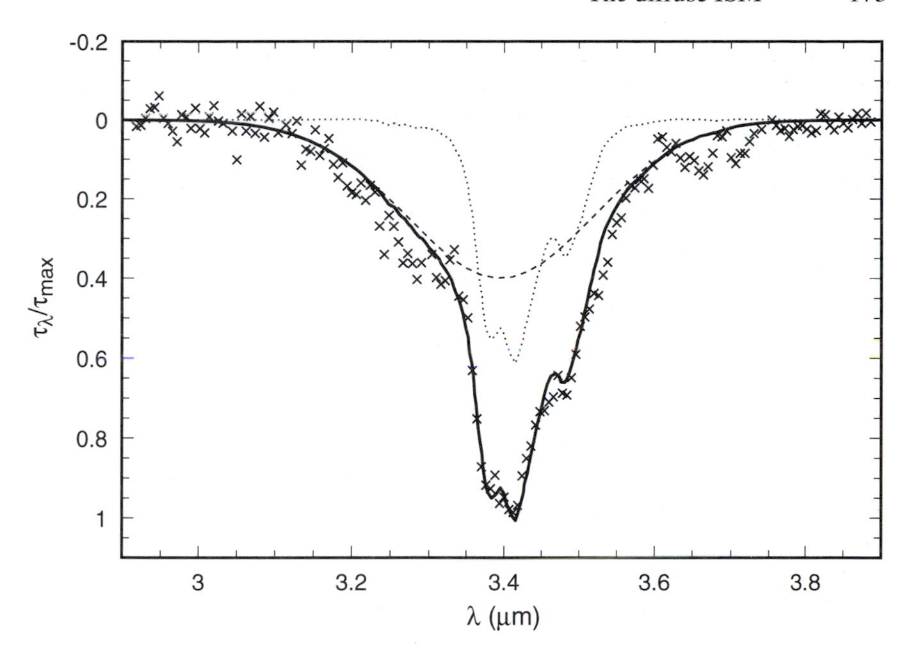

Figure 5.10. Spectrum of the 'C–H stretch' region in the galactic centre, illustrating the profile of the 3.4 μm feature. The observational data (points) represent a mean of several sources in the Sgr A region and the molecular cloud component has been extracted to leave only diffuse-ISM absorption (Chiar *et al* 2002). The continuous curve fit to the data combines absorption by hydrogenated amorphous carbon (dotted curve, Mennella *et al* 1999; see also figure 5.9) with a broader Gaussian component centred at 3.395 μm (broken curve).

It would be helpful to know what fraction of the available carbon in the ISM is tied up in the grains that produce the 3.4 μm feature. Unfortunately, this quantity cannot be readily estimated, as the intrinsic strength of the feature depends critically on the degree of hydrogenation, which is unknown for interstellar dust. Tielens *et al* (1996) adopted intrinsic strengths appropriate to CH_2 and CH_3 groups in pure aliphatic hydrocarbons. This leads to an estimate of ∼32 ppm of the total carbon abundance in aliphatics, which is about 15% of all C depleted into dust for solar reference abundances (table 2.2). In contrast, laboratory HAC samples that match the observed profile require ∼80 ppm (Duley *et al* 1998; Furton *et al* 1999), or about 35% of the depleted C. This discrepancy may be attributed to the different degrees of hydrogenation of the samples: the HAC considered by Furton *et al* has $N(\mathrm{H})/N(\mathrm{C}) \approx 0.5$, compared with 2–3 for pure aliphatic hydrocarbons. Some of the carbon in the HAC makes no contribution to the feature.

The degree of carbon hydrogenation in interstellar dust seems likely to depend on the environment and, indeed, there is observational evidence for spatial variation. The depth of the 3.4 μm feature is a factor of about two weaker per unit A_V toward early-type stars compared with the galactic centre (Adamson *et al* 1990, Sandford *et al* 1995). This difference may reflect changes in the optical properties of the absorber as a function of hydrogen content. Carbon solids that lack H are relatively opaque, producing strong continuous absorption in the visible and near infrared. As hydrogenation increases, deepening of the 3.4 μm feature is accompanied by a reduction in continuous absorption as the carbon becomes more polymeric (compare AC and HAC curves in figure 5.9). The observed behaviour of $\tau_{3.4}$ with A_V may thus be understood if the degree of hydrogenation is systematically higher toward the galactic centre compared with the solar neighbourhood (Adamson *et al* 1990).

The results discussed in this section favour hydrocarbons over organic residues as the carrier of the 3.4 μm feature. However, very high levels of energetic processing tend to 'carbonize' the organics, i.e. reduce them to amorphous carbon, so the distinction may be moot. One factor that may distinguish HAC models from ORM models is the mode of production: laboratory work on HAC has focused on vapour-condensed nanoparticles thought to resemble those forming in C-rich stellar atmospheres (e.g. Mennella *et al* 1999; Schnaiter *et al* 1999), whereas ORM production is by energetic processing of ices that simulate mantles on classical grains in molecular clouds (Jenniskens *et al* 1993). An observation capable of distinguishing between these scenarios is spectropolarimetry of the 3.4 μm feature. If the carrier resides in mantles on aligned silicate cores, the 3.4 μm feature should, like the 9.7 μm feature, display polarization enhancement. If, however, the carrier is associated with very small carbon grains, it would likely show little or no detectable polarization, consistent with the behaviour of the 2175 Å ultraviolet extinction bump and the optical diffuse bands. A study of the GC source Sgr A IRS7 by Adamson *et al* (1999) found no evidence for polarization enhancement at 3.4 μm, setting an upper limit well below that predicted for an absorber attached to silicates. This result thus favours carbon nanoparticles as the host. An origin in carbonized mantles is not excluded, however, if an efficient mechanism exists for dispersing them from the cores (see Greenberg *et al* 2000).

5.3 The dense ISM

5.3.1 An inventory of ices

Characteristic spectra of infrared sources that probe dense molecular clouds and protostellar envelopes are presented in figures 5.3 (Elias 16), 5.4 (NGC 7538 IRS9) and 5.5 (W33A). Elias 16 is a highly reddened background star observed through a dense cloud (see figure 5.2), whereas NGC 7538 IRS9 and W33A are young stellar objects (YSOs) or 'protostars' deeply embedded in the clouds from

Table 5.3. An inventory of interstellar ices based on infrared spectroscopy. Abundances are expressed as percentages of the H_2O abundance for three categories of sight-line. A range of values generally indicates real spatial variation; where followed by a colon, it may merely reflect observational uncertainty. Values for XCN, an unidentified molecule containing C≡N bonds, are based on an assumed band strength (Whittet *et al* 2001b). A dash indicates that no data are currently available.

Species	Dark cloud	Low-mass YSOs	High-mass YSOs
H_2O	100	100	100
NH_3	≤10	≤8	2–15
CH_4	—	<2	2
CO	25	0–60	0–25
CO_2	21	20–30	10–35
CH_3OH	<3	≤5	3–30
H_2CO	—	<2	2–6:
HCOOH	—	<1	2–6:
XCN	<1	0–2	0–6
OCS	<0.2	<0.5	0.2

which they formed[3]. It is obvious from a cursory comparison of figures 5.3–5.5 that dense clouds have much richer spectra than diffuse clouds. Silicate features at 9.7 and 18.5 μm are ubiquitous but features identified with ices (table 5.2) are prominent only in dense clouds. The observations are qualitatively consistent with a model in which refractory cores containing silicates (present in all types of cloud) act as substrates for the condensation of volatile molecular ice mantles in the more shielded environments of dense clouds. The mantles are composed predominantly of H_2O-ice, producing stretching and bending mode features at 3.0 and 6.0 μm. The strength of H_2O-ice absorption relative to silicate absorption is sensitive to environment, measuring the contribution of mantle material to the total grain mass. Taking the two spectra in figure 5.3 as representative examples, the ratio $\tau_{3.0}/\tau_{9.7}$ varies from less than 0.03 (Cyg OB2 no. 12, diffuse ISM) to $\sim$2 (Elias 16, dense ISM).

Table 5.3 presents an inventory of interstellar ices. Abundances calculated using equation (5.4) are expressed as a percentage relative to the value for H_2O. Mean results for three different classes of background star are included. Dark cloud values are based principally on Elias 16 in Taurus: examples of low-mass YSOs include Elias 29 (ρ Oph) and HH100–IR (R CrA) and examples of high-mass YSOs include NGC 7538 IRS9 and W33A. The abundances of

[3] The rigorous definition of a 'protostar' is a YSO still in the accretion phase and powered mainly by gravitational rather than thermonuclear energy (Beichman *et al* 1986; see section 9.1); however, the term is often used more generally to mean any young (pre-main-sequence) star that remains embedded in dust and emits most of its observable flux in the infrared.

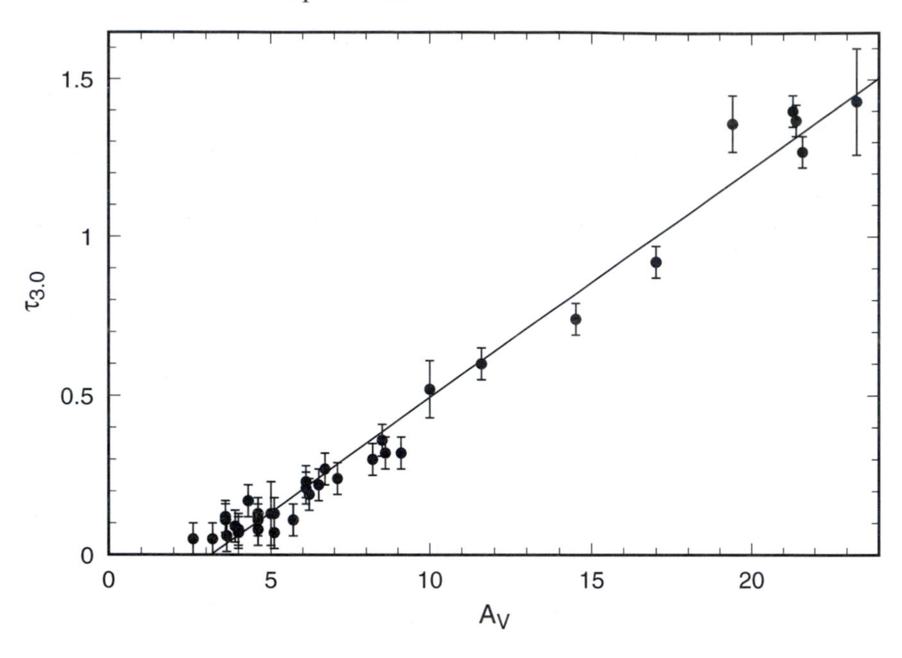

Figure 5.11. Plot of optical depth in the 3.0 μm ice feature against visual extinction for field stars located behind the Taurus Dark Cloud. The straight line is the least-squares fit (equation (5.13)). Data are from Whittet *et al* (2001a) and references therein.

most detected species lie in the 1–30% range relative to H_2O. Species with abundances $<1\%$ and typical band strengths are generally difficult or impossible to detect with current instrumentation. In some cases, blending of features greatly hinders detection and increases the uncertainty in the results: NH_3 and H_2CO are notable examples. Nevertheless, the results in table 5.3 suggest that a 'typical' interstellar ice might have a basic composition of H_2O:CO_2:NH_3:CH_4 roughly in the proportions 100:20:8:2, to which variable amounts of CO and CH_3OH may be added. The infrared-inactive species O_2 and N_2, absent from table 5.3, might also make substantial contributions.

5.3.2 The threshold effect

Quiescent dark clouds undisturbed by massive star formation represent ideal 'laboratories' in which to study grain mantles in a relatively pristine state. Insight into the development of ice mantles on the dust may be gained by observing ice features in the spectra of field stars lying behind such clouds across a range of extinctions. Figure 5.11 plots the optical depth ($\tau_{3.0}$) of the principal H_2O-ice feature against visual extinction (A_V) for stars toward the Taurus cloud. A close linear correlation is evident, described by the equation

$$\tau_{3.0} = q(A_V - A_0) \tag{5.13}$$

(Whittet *et al* 1988, 2001a) where $q = 0.072 \pm 0.002$ is the slope and the intercept $A_0 = 3.2 \pm 0.1$ represents the *threshold extinction* for detection of H_2O-ice in this cloud. Figure 5.11 may be contrasted with figure 5.8, which indicates a null intercept for silicates in the solar neighbourhood. This difference is easily understood if the presence or absence of ice mantles on the grains is governed by the prevailing physical conditions. Dark clouds typically contain clumps of molecular gas embedded in an envelope of more diffuse material (section 1.4.3). The external interstellar radiation field is pervasive in the envelope and the temperatures and densities encountered may be quite similar to those in a typical diffuse cloud. Accumulation of surface coatings in such regions will tend to be slow and opposed by photolysis and sublimation. A line of sight that does not intersect a clump should thus lack an ice feature and display low to modest extinction. However, the clumps are both denser (leading to more rapid growth) and effectively shielded from external radiation, providing an environment conducive to mantle growth and survival. The threshold extinction then represents some critical level of shielding. Once H_2O-ice mantles become established, the strength of the 3 μm feature should correlate with the column density of dust, for which A_V is a proxy, as is indeed the case in figure 5.11. If the scenario described here is broadly correct, a more volatile mantle constituent than H_2O should display a similar correlation but with a higher value of A_0. Observations of solid CO suggest that this is generally true (Whittet *et al* 1989, Chiar *et al* 1995, Shuping *et al* 2000) but with greater scatter, perhaps reflecting a greater sensitivity to local fluctuations in physical conditions. For an example, see problem 7 at the end of this chapter.

Threshold extinctions are less well determined for other clouds compared with Taurus but results nevertheless indicate significant cloud-to-cloud variations. The threshold for H_2O-ice in the R CrA cloud ($A_0 \approx 3.4$) is similar to that for Taurus, whereas that in Serpens ($A_0 \approx 5.5$) and ρ Oph ($10 < A_0 < 15$) are considerably higher (Whittet *et al* 1996, Eiroa and Hodapp 1989, Tanaka *et al* 1990). These differences are most likely governed by local physical conditions, especially the ambient radiation field (Williams *et al* 1992). Whereas Taurus and R CrA are regions of low-mass star formation, lacking early-type stars, the ρ Oph cloud is physically associated with an OB association.

The correlation in figure 5.11 allows us to estimate the abundance of H_2O molecules in grain mantles in the Taurus cloud. Using the appropriate value of the integrated absorption cross section per molecule from table 5.2, equation (5.5) becomes

$$N(H_2O) \approx 5 \times 10^{23} \gamma_{3.0} \tau_{3.0} \tag{5.14}$$

with $N(H_2O)$ in m^{-2} and $\gamma_{3.0}$ in μm^{-1}. Assuming the normal gas-to-reddening ratio (section 1.3.2), the column density of hydrogen for molecular cloud material above the ice threshold is

$$N_{\rm H} \approx 1.9 \times 10^{25}(A_V - A_0) \tag{5.15}$$

and combining equations (5.13), (5.14) and (5.15) gives

$$\begin{aligned} \frac{N(\rm H_2O)}{N_{\rm H}} &\approx 0.026 q \gamma_{3.0} \\ &\approx 7 \times 10^{-5} \end{aligned} \tag{5.16}$$

where $q \approx 0.072$ is the slope of the correlation line and $\gamma_{3.0} \approx 0.036\ \mu\text{m}^{-1}$ is the FWHM of the ice feature. For comparison, the solar abundance of oxygen is $N_{\rm O}/N_{\rm H} \approx 7 \times 10^{-4}$ (table 2.1), so only about 10% of the O appears to be tied up in H_2O on grains in this cloud.

5.3.3 H_2O-ice: the 3 μm profile

The profile of the 3 μm feature is illustrated in figure 5.12. The upper and lower frames show optical depth spectra for embedded and field stars, respectively, each compared with data for laboratory ices. The field-star spectrum is assumed to represent the quiescent dark-cloud medium (DCM), whereas the YSO spectra may include material in the circumstellar envelopes of the sources. The general form of the DCM profile is a broad, symmetric trough in the 2.8–3.2 μm region, with a long-wavelength wing extending to about 3.7 μm. The YSO profile is similar on the short wavelength side of the peak but the trough is broader and less symmetric and the long-wavelength wing is more pronounced.

Comparison of the observed profiles with those of laboratory analogues allows us to explore the properties of the ices. The full curve in each frame of figure 5.12 represents pure H_2O-ice at 10 K, i.e. representative of temperatures expected in quiescent molecular clouds. Ice deposited and maintained at such low temperatures is structurally amorphous, exhibiting a symmetric, Gaussian absorption profile that resembles the trough in the DCM spectrum. Also plotted are data that illustrate how the profile responds to a change in composition (the addition of ∼10% NH_3) and to an increase in temperature. N–H stretching vibrations contribute additional absorption in the 2.9–3.0 μm region and generally lead to a somewhat improved fit to the DCM profile at wavelengths below the peak. Ammonia molecules in a water-dominated ice matrix tend to form ammonium hydrate groups ($NH_3 \cdots OH_2$, where the dots represent the hydrogen bond) and the O–H stretch of H_2O molecules tied up in these groups is shifted to longer wavelength. The effect of this can be seen as additional absorption in the 3.2–3.7 μm region (comparing solid and dotted curves in figure 5.12). Finally, the effect of increasing the temperature of H_2O-ice from 10 to 80 K is to sharpen the 3.0 μm feature and shift its peak position to slightly longer wavelength. It is probable that temperature gradients exist in the lines of sight to YSOs and so the

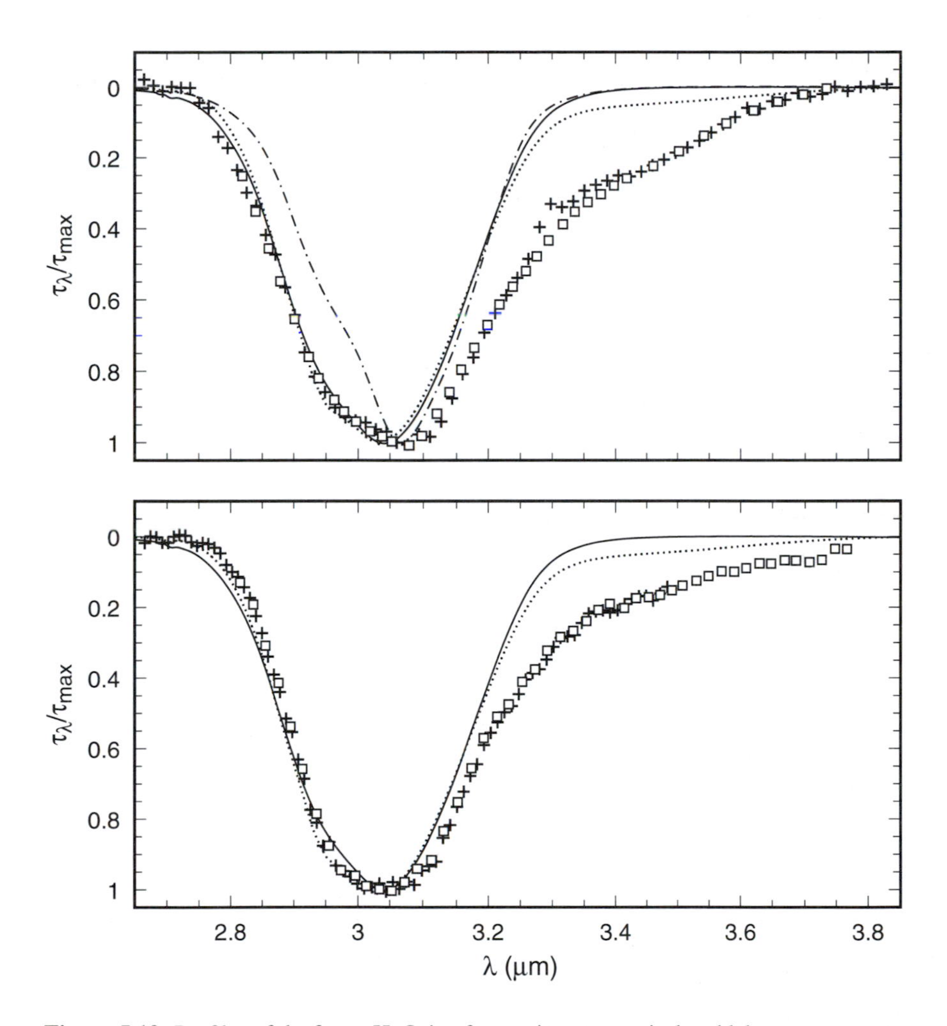

Figure 5.12. Profiles of the 3 μm H_2O-ice feature in astronomical and laboratory spectra. Top: YSOs; the crosses represent ISO SWS data (average for Elias 29, GL 490, NGC 7538 IRS1 and S140; Gibb *et al* 2000a) and the squares represent ground-based data (average for BN, GL 961, GL 989, GL 2591, Mon R2 IRS2, Mon R2 IRS3, S255 IRS1 and W3 IRS5; Smith *et al* 1989). Note that the profiles are in close agreement for these diverse groups of YSOs, except for a difference at 3.29 μm (attributed to a PAH emission feature in one source, NGC 7538 IRS1). The laboratory data are for pure H_2O-ice at 10 K (full curve), a 100:9 H_2O:NH_3 mixture at 10 K (dotted curve) and pure H_2O-ice at 80 K (dot–dash curve). Bottom: the field star Elias 16; again, crosses and squares indicate ISO SWS and ground-based data (Whittet *et al* 1998, Smith *et al* 1989), with 10 K ice spectra plotted as before. Data for ices are from the Astrophysics Laboratory at the University of Leiden.

observed profiles may include contributions from ices at different temperatures (Smith *et al* 1989).

The long-wavelength wing extending from 3.2 to 3.7 μm is not explained by the comparison spectra shown in figure 5.12. The O–H stretch in ammonium hydrate may make a contribution but cannot explain the entire wing. An NH_3 concentration of 20–30% would be required to match its average depth but this results in a deep 2.96 μm N–H feature that is inconsistent with the observed profile in most lines of sight (see van de Bult *et al* 1985, Smith *et al* 1989, Whittet *et al* 1996, Chiar *et al* 2000, Gibb *et al* 2001). Other possible explanations include absorptions arising in other molecules present in the ices and scattering effects in large ice-mantled grains. The wing overlaps the C–H stretch region of the spectrum (table 5.2), suggesting that it might arise from a blend of many discrete absorptions arising in various organic molecules. A specific case is CH_3OH, first detected by means of its C–H feature at 3.53 μm (Baas *et al* 1988, Grim *et al* 1991); other probable contributors in this spectral region include H_2CO and HCOOH. Further evidence for an absorptive component to the wing is provided by a broad, shallow feature centred near 3.47 μm, not securely identified but associated with carbonaceous material (Allamandola *et al* 1992; see section 5.3.7). However, these features make only minor contributions and cannot explain the wing as a whole. Absorptions in other unidentified species might be responsible but this seems unlikely, given the smoothness and continuity of the profile: no appreciable structure, beyond that already noted, is seen at resolving powers $\lambda/\Delta\lambda \sim 1000$.

The smoothness of the wing may be more readily understood if it is a scattering effect. Scattering becomes important if the grain dimensions are such that $2\pi a/\lambda > 1$, i.e. if their diameters are of order 1 μm or higher for $\lambda \approx 3$ μm. The absorption peak is then shifted to longer wavelength (see Smith *et al* 1989). Using a power-law (MRN) size distribution that provides good fits to the diffuse-ISM extinction curve (section 3.7), absorption profiles qualitatively similar to the observations can be obtained when the maximum radius a_{max} is increased to about 1 μm (Léger *et al* 1983, Pendleton *et al* 1990), compared with values of about 0.25 μm for fits to the diffuse ISM. Detailed modelling of both optical depth and polarization profiles (section 5.3.8) is needed to thoroughly test the scattering hypothesis for the wing.

We have focused in this section on the stretching mode of solid H_2O. The feature associated with the bending mode at 6.0 μm presents even greater challenges of interpretation. Like the 3.0 μm feature, it appears to be blended with absorptions of other species, including NH_3, H_2CO and HCOOH (Schutte *et al* 1996, Chiar *et al* 2000, Keane *et al* 2001). Column densities calculated from the 6.0 μm feature tend to be systematically higher than those obtained from other H_2O-ice features, even when blends with other ices are accounted for, suggesting that an additional absorption feature is present (Gibb *et al* 2000a, Gibb and Whittet 2002). We return to this topic in section 8.4.3.

5.3.4 Solid CO: polar and apolar mantles

Carbon monoxide is the most abundant gas-phase molecule in the ISM after H_2 and its presence as a condensate on interstellar dust was predicted (Duley 1974) some years prior to its first detection by Lacy *et al* (1984). The only available spectral signature of the dominant isotopic form is the C–O stretch at 4.67 μm. Observations of this feature allow us to calculate the abundance of CO in grain mantles. Results for different lines of sight show wide variation, with an extreme range of 0–60% relative to H_2O (Chiar *et al* 1998a). This is consistent with the volatility of solid CO compared with most other molecular ices (section 8.5). In pure form, CO sublimes at $\sim$17 K.

Studies of the 4.67 μm profile also provide an important clue as to the structure of grain mantles. Laboratory spectra of ice mixtures containing CO demonstrate that the precise position, width and shape of the feature are sensitive to the composition of the molecular mix containing the CO and to its thermal and radiative history (Sandford *et al* 1988, Palumbo and Strazzulla 1993, Ehrenfreund *et al* 1996). Pure solid CO produces a narrow (FWHM $\sim$ 0.005 μm) feature centred near 4.675 μm. Introduction of other species leads to broadening and shifts in position. The effect is slight whilst CO remains the dominant constituent. However, if CO becomes a minority constituent quite significant differences can occur, the scale of which depends on the dipole moment of the ice. CO itself is only very weakly polar and mixtures of CO with apolar molecules such as O_2 or N_2 have 4.67 μm profiles that differ very little from that of pure CO. However, mixtures of CO with polar molecules such as H_2O or NH_3 exhibit more substantial changes. All CO-bearing laboratory mixtures in which H_2O is the dominant constituent produce features shifted to significantly longer wavelength ($\sim$4.681 μm) and substantially broadened (FWHM $\sim$ 0.02 μm). Another interesting property of H_2O-dominated mixtures is that they can retain CO at temperatures well above its nominal sublimation temperature (up to about 100 K). The 4.67 μm profile shows substantial evolution with temperature in such ices (Schmitt *et al* 1989).

In most lines of sight, the observed CO profile displays a sharp minimum centred near 4.673 μm, accompanied by a wing extending to longer wavelength (see figure 5.13; the absorption on the short-wavelength side is the 'XCN' feature discussed later). This structure cannot be explained by any one CO-bearing ice mixture. However, it is readily explained by a model in which two distinct but overlapping absorption components are present that have different relative strengths in different lines of sight. One component of the model explains the sharp component of the profile, and this requires ices in which CO is either dominant or mixed with other molecules that have zero or near-zero dipole moments. The other component explains the wing and this is represented by CO in H_2O-dominated ices. We refer to these mixtures as 'apolar' and 'polar' ices, respectively (whilst noting that the 'apolar' component may, in fact, have a small dipole moment). Examples of fits are illustrated in figure 5.13. The apolar component usually dominates the spectrum but in some YSOs (e.g. W33A,

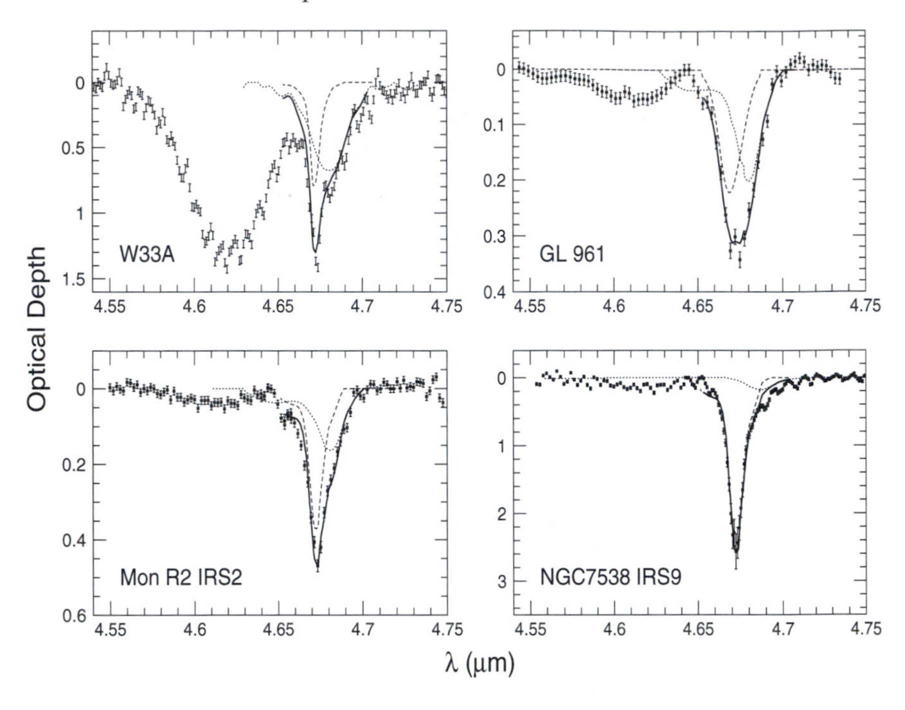

Figure 5.13. Spectra of four YSOs in the region of the 'XCN' and CO features at 4.62 and 4.67 μm. Differences in the strength of the XCN feature are thought to result from different degrees of energetic processing of the ices. Each CO profile is fitted with a two-component model based on laboratory data for polar and apolar ices (dotted and broken lines, respectively; the full line is the sum). The polar component is irradiated H_2O:CH_3OH = 2:1 (W33A), H_2O:CO = 20:1 (GL 961, Mon R2 IRS2) and H_2O:CO = 4:1 (NGC 7538 IRS9). The apolar component is pure CO (W33A), CO:CO_2 = 2:1 (GL 961), CO:H_2O = 10:1 (Mon R2 IRS2, NGC 7538 IRS9). All laboratory ices are at 10 K except for the polar components of the fits to GL 961 (20 K) and NGC 7538 IRS9 (100 K). Data from Chiar *et al* (1998a).

GL 961) the polar and apolar components are comparable. The observations are qualitatively consistent with hierarchical mantle growth, illustrated schematically in figure 5.14, in which CO-rich apolar mantles are deposited on top of H_2O-rich polar mantles. As they have different sublimation temperatures, spatial segregation may occur along the line of sight between grains bearing both polar and apolar mantles and those bearing only polar mantles.

5.3.5 Other carbon-bearing ices

C is present in the ices in a variety of forms, covering a range of oxidation/hydrogenation states from CO and CO_2 to HCOOH, H_2CO, CH_3OH

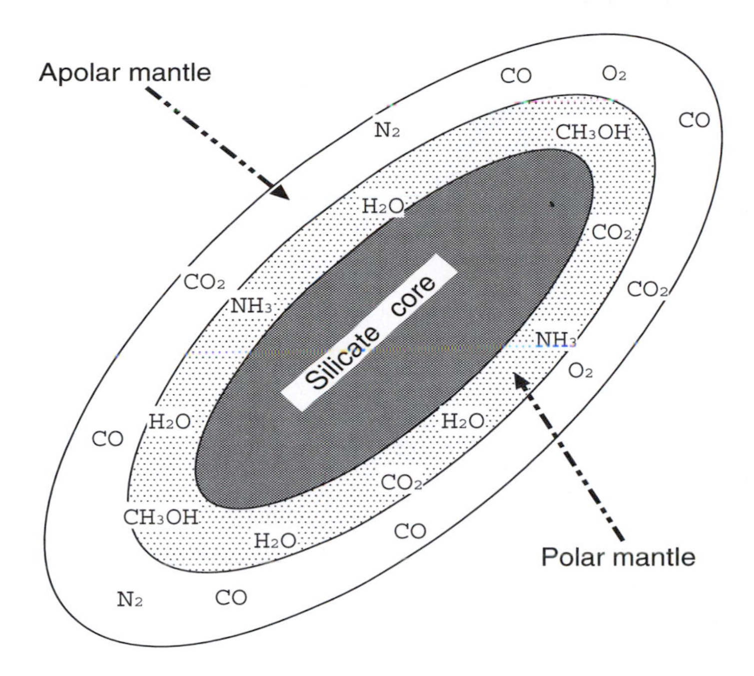

Figure 5.14. Schematic representation of an interstellar silicate grain with hierarchical mantle growth involving distinct polar and apolar layers. Figure courtesy of Erika Gibb.

and CH_4. The discussion in this section emphasizes CO_2 and CH_3OH, two well studied species that provide interesting constraints on mantle evolution. Of the remainder, HCOOH and H_2CO are difficult to study in detail because their principal spectral signatures are relatively weak and blended with other features such as the H_2O bending mode at 6 μm (Schutte *et al* 1996, 1999, Keane *et al* 2001): available data are consistent with abundances of a few per cent (table 5.3). CH_4 appears also to be a relatively minor constituent of the ices but its abundance is better constrained and its profile suggests residence in the polar component of the mantles (Boogert *et al* 1998). Hydrocarbons of higher molecular weight have been discussed, notably ethane (C_2H_6) following its detection in comets (Mumma *et al* 1996). Such species may contribute to the absorptions in the C–H stretch ($\sim$3.4 μm) and CH_3 deformation ($\sim$6.8 μm) regions but no specific identifications have yet been made.

The principal spectral features of CO_2 have diagnostic properties similar to that of CO (section 5.2.4). The stretching and bending modes of $^{12}CO_2$ (table 5.2) and also the stretching mode of $^{13}CO_2$ (at 4.38 μm) have been subject to thorough investigation (Gerakines *et al* 1999; Boogert *et al* 2000a). As an example, $^{12}CO_2$ data for GL 2136 are shown in figure 5.15. As with CO, a model with distinct polar and apolar components is needed to produce satisfactory fits (in this case, the 'apolar' component is an annealed H_2O:CO_2:CH_3OH = 1:1:1

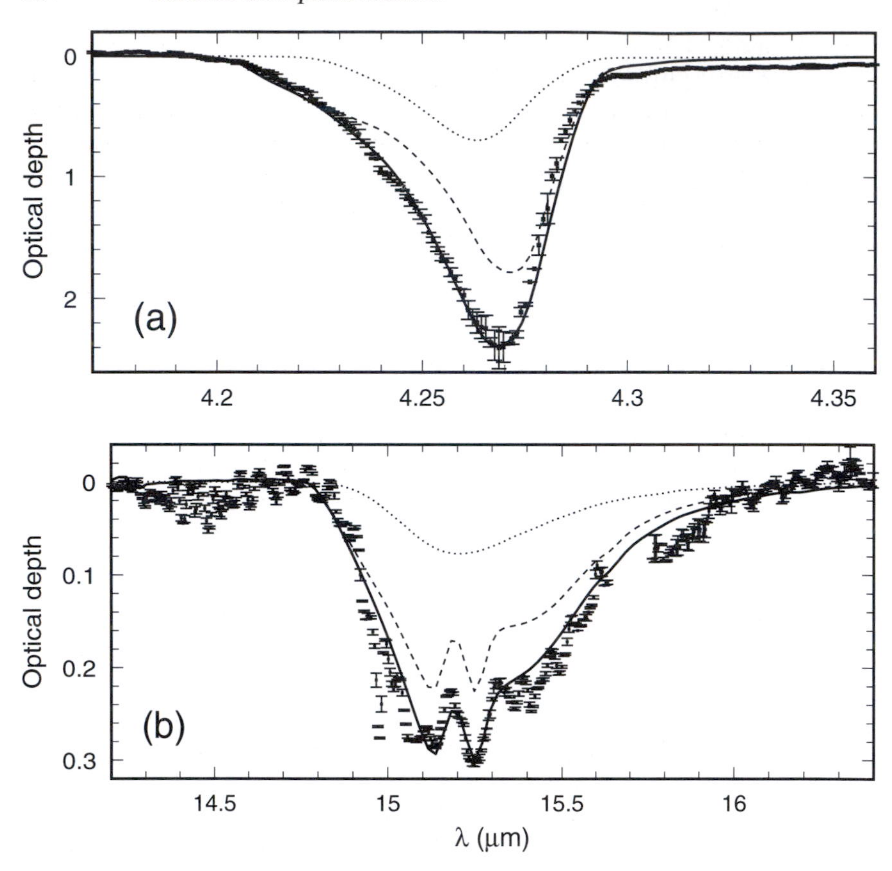

Figure 5.15. Spectra of the YSO GL 2136 in the region of the (*a*) stretching and (*b*) bending modes of CO_2. The profiles are fitted with a two-component model based on laboratory data for polar and annealed ices (dotted and dashed lines, respectively; the full line is the sum). The polar component is H_2O:CO_2:CO = 100:20:3 at 20 K. The annealed component is H_2O:CO_2:CH_3OH = 1:1:1 at 117 K. Data from Gerakines *et al* (1999).

mixture). Laboratory studies show that the structure evident in the profile of the CO_2 bending mode (figure 5.15(*b*)) is particularly sensitive to the composition and thermal history of the ices (Ehrenfreund *et al* 1999). CH_3OH is an important component of ices that give a good match to the data. Heating leads to the formation of CO_2:CH_3OH groups that tend to segregate from the polar (H_2O-dominated) ice layer. Thus, the annealed apolar ices may originate in quite a different way compared with the CO-rich apolar ices, which most likely condense directly from the gas. As expected, spectral evidence for thermal processing is seen preferentially in massive, luminous YSOs, whereas lines of sight that sample primarily the dark-cloud medium have profiles dominated by CO_2 in unannealed polar ice mixtures (Whittet *et al* 1998, Gerakines *et al* 1999, Nummelin *et al* 2001). The difference evident in the distribution of CO and CO_2 in the DCM (predominantly in apolar and polar ices, respectively) is somewhat surprising and presents a challenge to models for their formation (section 8.4.1).

CO_2 shows a degree of stability in its overall abundance in grain mantles: a concentration of about 20% relative to H_2O is typical of the DCM and many YSOs as well. In contrast, that of CH_3OH is quite variable, ranging from $<3\%$ in the DCM and 3–10% in most YSOs to $\sim$25–30% in the massive YSOs W33A and GL 7009S (Chiar *et al* 1996, Brooke *et al* 1999, Dartois *et al* 1999)[4]. These results suggest that CH_3OH is not a primary constituent of the ices that form in molecular clouds but that its formation is driven by subsequent processing. The position and width of the C–O stretch feature at 9.75 μm are most consistent with an origin in ices that contain CH_3OH as a dominant constituent (Skinner *et al* 1992), again suggesting segregation from the H_2O-dominant phase.

5.3.6 Nitrogen and sulphur-bearing ices

N and S are abundant elements (table 2.1) that remain predominantly in the gas phase in the diffuse ISM (sections 2.4.2 and 2.5). They are thus strong candidates for incorporation into icy mantles in dense clouds but there is a paucity of direct evidence to confirm this. In the case of N, some of the reasons have already been noted: detection of the dominant molecular forms predicted by chemical models is hindered by severe blending in the case of NH_3 and by lack of infrared activity in the case of N_2. However, scarcity of evidence is not evidence of scarcity. The effect of NH_3 on the 3 μm H_2O-ice profile (figure 5.12) is quite subtle for abundances $\sim$15% or less and so this feature in isolation does not provide reliable detections for most lines of sight (it can provide rather loose upper limits). Similar comments apply to the deformation mode near 6.2 μm, blended with the H_2O bending mode. The most dependable information on NH_3 abundances is obtained from the inversion mode at $\sim$9.3 μm (Lacy *et al* 1998, Gibb *et al* 2000a, 2001). This feature is superposed on the wing of the silicate profile, from which it can be extracted relatively easily, given data of adequate quality. The results are somewhat surprising. Although NH_3 is detected with an abundance of $\sim$15% in the spectra of a few massive YSOs, others yield upper limits well below this value. Data for the DCM are consistent with up to 10% but much lower values are not excluded. It is not yet clear whether NH_3 is a common constituent of interstellar ices at the $\sim$10% level or whether its abundance is generally much lower and becomes enhanced in certain environments.

[4] Methanol abundances of $\sim$50–100% relative to H_2O reported in early studies based on the deformation mode at 6.85 μm (e.g. review by Tielens and Allamandola 1987b) should be disregarded, as they are incompatible with lower values indicated by observations of other features such as the C–H and C–O stretching modes (e.g. Grim *et al* 1991, Schutte *et al* 1991, Gibb *et al* 2000a). CH_3OH evidently makes only a minor contribution to the observed absorption feature at 6.85 μm (Keane *et al* 2001).

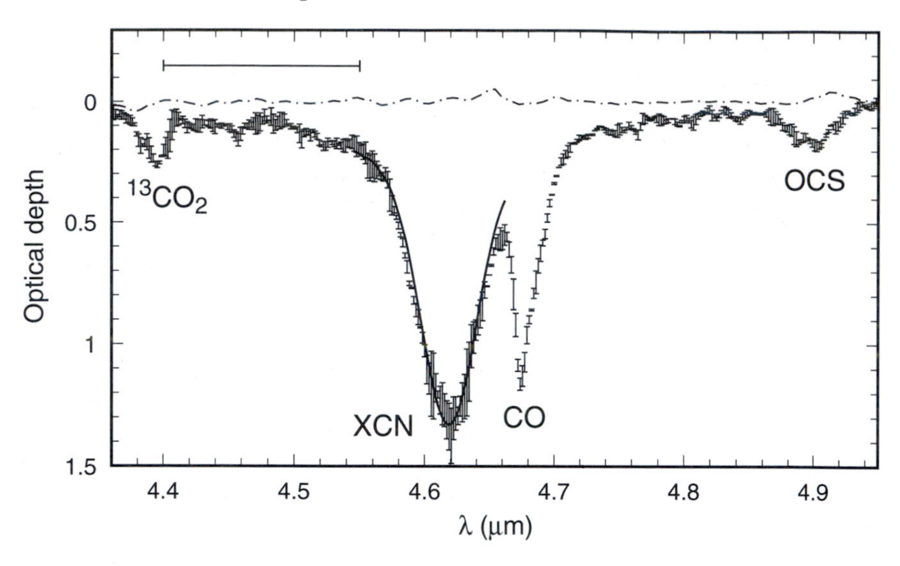

Figure 5.16. Spectrum of W33A in the 4.3–5.0 μm region (points; Gibb *et al* 2000a), where features associated with nitriles, cyanates and carbonyl sulphide are expected. Also shown for comparison are spectra of an object sampling the diffuse ISM (Cyg OB2 no. 12, dot–dash curve; Whittet *et al* 2001b) and of an ion-irradiated interstellar ice analogue (full curve overlaying the XCN profile; Palumbo *et al* 2000). The initial composition of the analogue was H_2O:N_2:CH_4 = 1:1:1. The horizontal bar indicates the spectral range of C≡N stretch absorptions occurring in nitriles.

Other N-bearing candidates include nitriles and cyanates. These have in common the presence of the C≡N triple bond. A nitrile has the generic formula X–C≡N, where X typically represents a hydrocarbon group such as CH_3 or CH_3CH_2, whereas a cyanate includes an O atom, i.e. X–O–C≡N. A spectral feature at 4.62 μm observed in several protostars has been associated with the C≡N stretch in such molecules; as the carrier lacks a specific identification, it is generally referred to as 'XCN' (Lacy *et al* 1984, Schutte and Greenberg 1997, Pendleton *et al* 1999, Whittet *et al* 2001b). Figure 5.16 illustrates this spectral region for W33A. The position of the observed feature discriminates against nitriles, as the latter typically absorb in the 4.40–4.55 μm spectral region (Bernstein *et al* 1997) where little absorption is seen in the interstellar spectrum. Isonitriles (X–N≡C) give better wavelength coincidence but they are less stable than nitriles and should be less abundant. Thus, identification with molecules containing OCN groups appears to be favoured (Whittet *et al* 2001b and references therein).

A good match to the observed 4.62 μm profile is obtained when interstellar ice analogues that initially contain N in the form of NH_3 or N_2 are subjected to energetic processing (UV photolysis or ion bombardment: Lacy *et al* 1984,

Bernstein *et al* 1995, Elsila *et al* 1997, Palumbo *et al* 2000). An example of an analogue subjected to ion bombardment is shown in figure 5.16. The energy imparted by irradiation breaks chemical bonds and generates radicals that subsequently recombine to form new species, some of which evidently include OCN groups. Formation of the carrier in this manner is highly consistent with the observed distribution of XCN features in lines of sight that sample different environments (Whittet *et al* 2001b): only upper limits are found in the DCM and toward most low-mass YSOs, whilst considerable variation is seen amongst higher-mass YSOs (see figure 5.13), suggesting different degrees of processing according to local physical conditions.

The absorptions of NH_3 and (where present) cyanates in interstellar ices represent no more than about 10% of the solar abundance of nitrogen. It is therefore reasonable to suppose that another important form of N remains undetected, of which N_2 is the obvious candidate. A similar problem arises for sulphur. Likely molecular forms include H_2S, OCS and SO_2 (table 5.2). Of these, OCS is observed in a few protostars (Palumbo *et al* 1997), including W33A (figure 5.16), and a tentative detection of SO_2 has also been reported (Boogert *et al* 1997). However, only upper limits have been set on H_2S (Smith 1991): a feature at 3.94 μm originally attributed to this species (Geballe *et al* 1985, Geballe 1991) has been shown to arise in a weak combination mode of CH_3OH (Allamandola *et al* 1992). Either S remains predominantly in the gas phase in dense clouds or it is present in the ices in some other form.

5.3.7 Refractory dust

The cores upon which ice mantles form in dense molecular clouds are presumed to be similar to dust grains existing in the diffuse ISM, i.e. composed of refractory materials such as amorphous carbon and silicates. In this section, we briefly review the observed spectroscopic properties of such materials in dense clouds and compare them with those of diffuse-cloud dust.

There are two distinct issues to consider when comparing the carbonaceous component of refractory dust in the two environments: the extent to which existing diffuse-ISM grains may be modified by incorporation into molecular clouds; and the possibility that new materials might be generated by energetic processing of the ices near newly born stars. The key signature of refractory carbon is the 3.4 μm C–H stretch (section 5.2.4). In dense clouds, this region is confused by the presence of the long-wavelength wing of the 3 μm H_2O-ice feature, but it is possible to extract the contributions of hydrocarbons and other organics from the wing by careful selection of an appropriate baseline (Allamandola *et al* 1992, Brooke *et al* 1996, 1999, Chiar *et al* 1996). A representative example is shown in figure 5.17. The profile is dominated by a broad, symmetric feature centred at $\sim$3.47 μm. This feature differs substantially in position, width and shape from the corresponding feature in the diffuse ISM (also shown), which has maximum optical depth at $\sim$3.41 μm. The identity of the

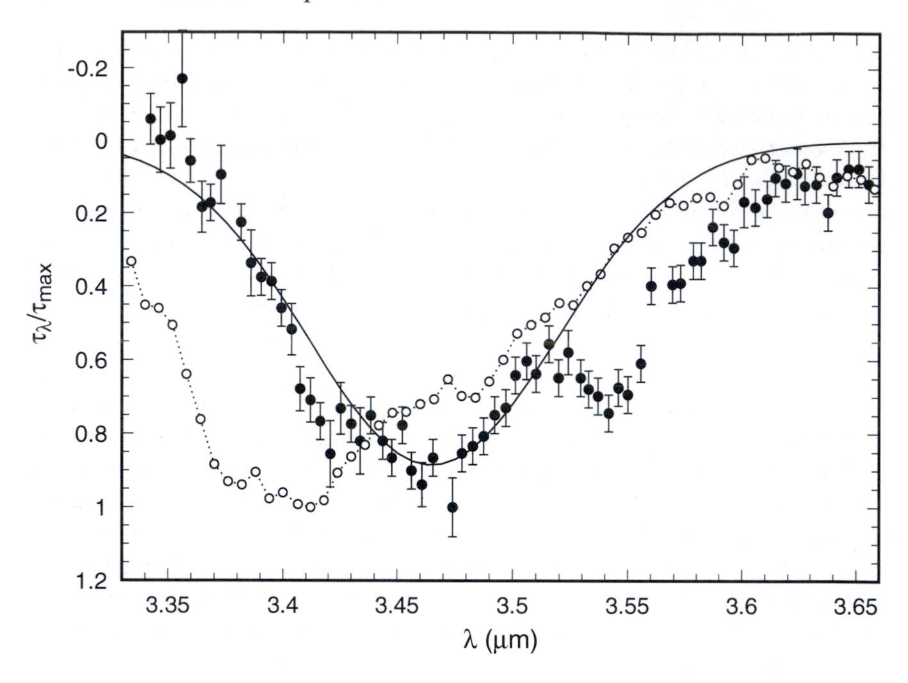

Figure 5.17. A comparison of absorption profiles in the C–H stretch region for dense and diffuse phases of the ISM. Points with error bars plot residual absorption after removal of the 3 μm ice feature (including the wing) from the spectrum of NGC 7538 IRS9 (Allamandola *et al* 1992). The absorption is dominated by the unidentified 3.47 μm feature, the profile of which is well represented by a Gaussian centred at 3.465 μm with FWHM 0.11 μm (curve). The trough centred near 3.54 μm is identified with CH_3OH. Open circles represent the corresponding data for the galactic centre (Chiar *et al* 2002; see also figure 5.10).

3.47 μm feature is not known with any certainty. Its position is most consistent with C–H bonds in diamond-like or tertiary carbon (i.e. with all other valence bonds of the C atom attached to other C atoms: Allamandola *et al* 1992, Grishko and Duley 2000). However, the carrier appears to be a component of the icy mantles rather than the refractory cores, as the absorption correlates in strength far more closely with the 3.0 μm ice feature than with the 9.7 μm silicate feature (Brooke *et al* 1996, Chiar *et al* 1996).

The lack of a counterpart to the diffuse-ISM 3.4 μm feature in dense clouds is puzzling and hard to reconcile with any model for its origin (section 5.2.4). Mennella *et al* (1999, 2001) attribute suppression of the feature in amorphous carbon nanoparticles to photolytic dehydrogenation; but this clearly cannot occur selectively within dark clouds, given the general attenuation of the UV radiation field, although it may be a factor in the outer layers. The carrier particles are likely to become coated with ices and/or accreted into the icy mantles of larger grains in dense clouds. The chemical processes that lead to ice formation on grain surfaces (section 8.3) may scavenge hydrogen (Grishko and Duley 2000) and thus selectively suppress those components of the C–H feature arising in hydrogen-rich CH_3 and CH_2 groups.

Regions of star formation are presumed to be the birthsites of organic refractory mantles, formed in the laboratory by photolysis and heating. Non-detection of the expected 3.4 μm feature argues against efficient production of such materials unless they are rapidly reduced to amorphous carbon (see section 5.2.4). Yet observations of the 4.62 μm XCN feature (section 5.3.6) provide strong circumstantial evidence for active processing of ices near some embedded stars. The residues that exhibit 4.62 μm absorption in laboratory analogues are, in some cases, quite refractory, suggesting that they might be capable of long-term survival. However, there is no trace of evidence for a C≡N-bearing absorber in the diffuse ISM, as illustrated in figure 5.16, which compares the spectrum of Cyg OB2 no. 12 with that of W33A. If organic refractory mantles are formed, any C≡N bonds they contain can account for no more than about 0.3% of the elemental nitrogen abundance in the diffuse ISM (Whittet *et al* 2001b). Again, the implication seems to be that any surviving residues must be quickly carbonized.

Silicates also exhibit a degree of spectroscopic dichotomy. The profile of the 9.7 μm absorption feature in dense clouds is systematically broader than that observed in the diffuse ISM (FWHM $\approx$ 3.3 and 2.5 μm, respectively: Roche and Aitken 1984a, Whittet *et al* 1988, Bowey *et al* 1998). Examination suggests that the difference can be attributed to an extension of the dense-ISM profile on the long-wavelength side of the peak (compare NGC 7538 IRS9 with GCS3 in figure 5.4). Whereas the diffuse-ISM profile (figure 5.6) is closely similar to that observed in emission in the outflows of dusty red giants (the so-called μ Cephei emissivity curve), the dense-ISM profile is consistent with that observed toward the Trapezium cluster in the Orion nebula (see section 6.3.1). Two explanations are possible: either the silicates themselves are different in structure and/or composition in molecular clouds; or additional dust components contribute to absorption at these wavelengths. The diffuse-ISM profile is well matched by a model based on amorphous olivine, as shown in figure 5.6, but no known silicate gives as good a fit to the profile seen in dense clouds (Wright *et al* 2002). Silicates may become partially annealed in the vicinity of an embedded YSO (chapter 9, see figure 9.9), such that the net profile is a combination of amorphous and crystalline components, but this does not readily account for the broadness of the feature in stars such as Elias 16 that sample exclusively DCM. Absorption in another dust constituent seems to be required.

Ices that have absorption features overlapping the silicate profile include NH_3, CH_3OH and H_2O (table 5.2). The NH_3 and CH_3OH features are relatively narrow (a few tenths of a micron) and have no effect on the apparent width of the silicate profile. The H_2O libration mode, however, is quite broad ($\sim$4 μm), with intrinsic strength ∼15% relative to the 3 μm O–H stretch. This feature is centred at 11.5 μm in pure crystalline ice and at 12.5 μm in pure amorphous ice (Kitta and Krätschmer 1983). When other species are included with H_2O, the peak of the feature shifts to yet longer wavelengths, typically 13–14 μm, and it becomes broader and shallower (Tielens and Allamandola 1987b). There is little doubt that the libration mode contributes to absorption in the 10–15 μm region in molecular clouds but the extent of its influence on the 9.7 μm profile is hard to quantify. In any case, the 'Trapezium' emissivity curve (section 6.3.1) that matches the dense-cloud absorption profile (Gillett *et al* 1975b, Whittet *et al* 1988) arises in dust recently heated by young OB stars: these grains are unlikely to have retained ice mantles, yet they have retained the property that causes broadening.

5.3.8 Spectropolarimetry and alignment of core–mantle grains

We have seen that silicates in the diffuse ISM exhibit a polarization counterpart to the 9.7 μm absorption feature, indicating a propensity for alignment (section 5.2.2; figure 5.7). Are silicates within dense clouds similarly aligned? Evidence for alignment in the cold dense-cloud medium is only circumstantial, as no data are currently available to test for 10 μm polarization excess in field stars, but there is ample evidence for alignment toward embedded protostars (e.g. Smith *et al* 2000). The most luminous objects are both sufficiently polarized and sufficiently bright in the mid-infrared to allow study of the weaker silicate bending mode feature in addition to the well studied stretching mode (Aitken *et al* 1988, 1989). Data for one such source are illustrated in figure 5.18.

Polarization represents the difference between two extinction efficiencies (equation (4.3)) that are independently related to the optical constants. Spectropolarimetry across an absorption feature is therefore highly sensitive to the optical properties of the relevant grain material and has the potential to tightly constrain them (Martin 1975, Aitken 1989, 1996). In the case of silicates, calculations show that the P_λ profiles of the principal resonances are sensitive to the composition, shape and porosity of the grains and to the presence or absence of mantles (Aitken *et al* 1989, Henning and Stognienko 1993, O'Donnell 1994b, Wright *et al* 2002)[5]. It is found that oblate spheroids generally give significantly better fits to the observations than prolate spheroids. Illustrative examples from the work of Wright *et al* (2002) are shown in figure 5.18: the 10 μm profile is quite well matched by porous olivine cores with carbon inclusions and ice mantles. However, no model published to date provides a satisfactory fit to the silicate bend mode at ∼19 μm. Polarimetric observations also place constraints on intrinsic strength: the amplitude of variation in the ratio P_λ/τ_λ with wavelength across a feature is determined primarily by its band strength. Observations of the silicate stretch mode in BN indicate $\kappa_{9.7} \approx 300$ m^2 kg^{-1} (see Aitken 1989), again

[5] Note that they are not sensitive to *size* as the grains are in the small particle limit at these wavelengths.

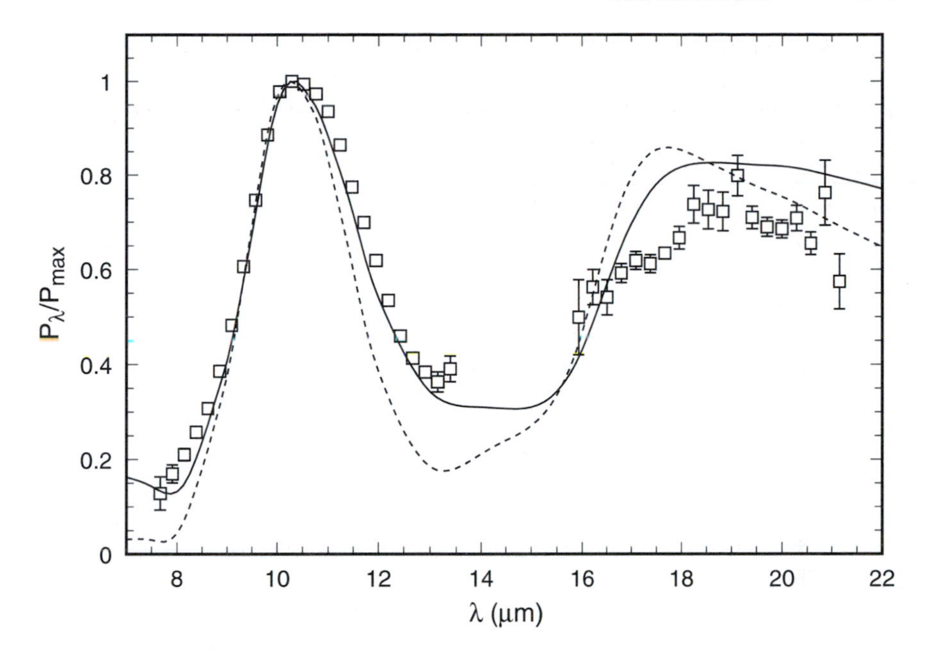

Figure 5.18. Silicate stretch and bend mode polarization profiles for the BN object compared with model calculations. The model assumes perfectly aligned oblate spheroids of size $a = 0.1$ μm and axial ratio $b/a = 2{:}1$ (Wright *et al* 2002): broken curve, pure amorphous olivine; full curve, composite cores with H_2O-ice mantles; the cores contain amorphous olivine (65%), carbon (10%) and vacuum (25%) by volume and the mantles have 25% of the volume of the cores. Observations and models are normalized to unity at the peak of the 10 μm feature.

consistent with laboratory data for amorphous silicates (table 5.1) and much less than expected for crystalline forms.

Ices that attach to aligned silicate grains should likewise exhibit polarization counterparts to their absorption features if mantle growth does not suppress alignment. In the case of the 3.0 μm H_2O-ice feature, this has been confirmed in several YSOs (Dyck and Lonsdale 1981, Hough *et al* 1989, 1996, Kobayashi *et al* 1999) and in the DCM toward the field star Elias 16 (Hough *et al* 1988). Polarization and optical depth profiles toward BN are compared in figure 5.19. These results are qualitatively consistent with dichroic absorption by aligned core–mantle grains. Detection toward Elias 16 confirms that the alignment mechanism is not entirely suppressed in the cold DCM (section 4.5.7). The polarization efficiency $\Delta P/\tau$ in the feature is dependent, in part, on the efficiency of grain alignment in the cloud and this ratio shows wide variation from star to star: from ~1% toward Elias 16 to ~4% toward BN and ~12% toward L1551 IRS5. In the latter case, the observed polarization may be dominated by scattered

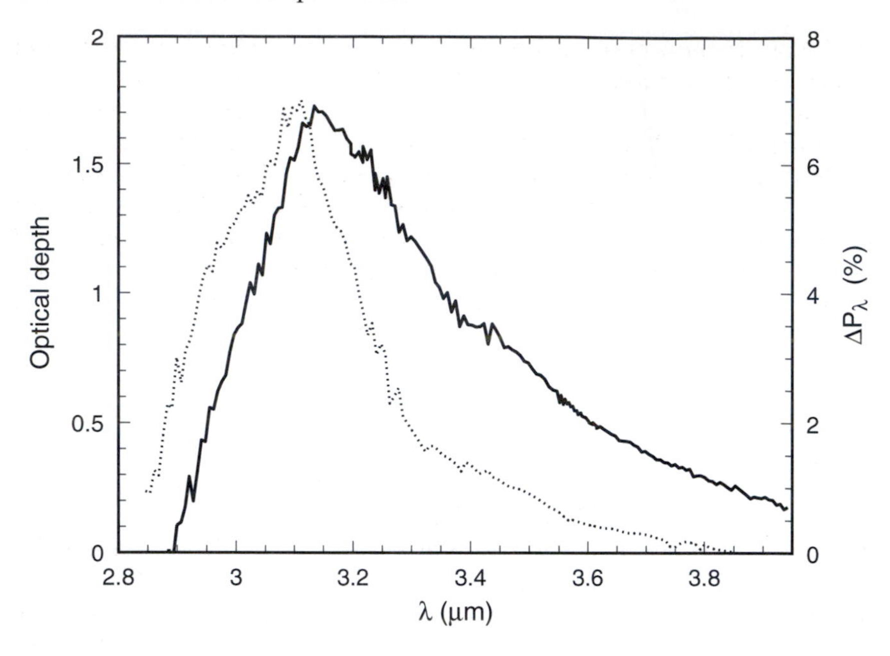

Figure 5.19. Polarization and optical depth profiles of the 3 μm H_2O-ice feature in the spectrum of the BN object (Hough *et al* 1996). Polarization in the feature (ΔP_λ, full curve, right-hand scale) is measured relative to a power-law fit to the continuum polarization (see figure 4.9). Note that optical depth (dotted curve, left-hand scale) peaks at significantly shorter wavelength.

radiation (Kobayashi *et al* 1999) but in the case of BN the simplest explanation is that icy grains are being aligned more efficiently in the dense molecular cloud (OMC-1) along this line of sight than in the dark cloud toward Elias 16.

The profiles plotted in figure 5.19 also provide information on the alignment status of the long-wavelength wing that accompanies the 3 μm ice feature. The presence of the wing in the P_λ curve, as indicated by the relatively high level of polarization in the 3.3–3.8 μm wavelength range, supports an origin by dichroic absorption in aligned grains. Scrutiny of the profile leads to the same conclusion for the weaker 3.47 μm feature, visible as substructure in the P_λ curve. This result supports the conclusion, based on spectroscopic arguments (section 5.3.7), that the carrier is intimately linked with the ices. Note that the polarimetric behaviour of the 3.47 μm feature is thus qualitatively different from that of the diffuse-ISM 3.4 μm hydrocarbon feature, thought to arise in small grains rather than coatings on classical grains (section 5.2.4).

Spectropolarimetric observations do not yet exist for most of the other ice absorption features discussed in this chapter. Given that 3 μm excesses appear to be common, polarization counterparts to any feature associated with the polar

(H_2O-rich) component of the ices are to be expected. However, detection of polarization counterparts to the solid CO and XCN features (Chrysostomou *et al* 1996) is more surprising. In the case of XCN, the grains are presumably situated relatively close to the star where the radiative environment needed for production of this mantle constituent prevails (section 5.3.6) and alignment may thus be strongly influenced by local conditions. In the case of CO, it is clear that grains with apolar mantles must be contributing to the observed polarization, and these should only exist in cold, quiescent regions of the molecular cloud. That polarization is produced in such environments is especially significant for grain alignment theory, as is discussed in section 4.5.7.

Recommended reading

- *Fundamentals of Molecular Spectroscopy* (fourth edition), by Colin Banwell and Elaine McCash (McGraw-Hill, London, 1994).
- *Infrared Spectroscopy in Astronomy*, ed B H Kaldeich (22nd ESLAB Symposium, ESA publication SP-290, 1989).
- *The ISO View of Interstellar Dust*, by D C B Whittet, in Astrophysics with Infrared Surveys, ed Michael D Bicay *et al*, Astronomical Society of the Pacific Conference Series, vol 177, pp 300–13 (1999).

Problems

1. (a) The C–O bond of carbon monoxide (in its most common isotopic form) has a force constant of 1870 N m^{-1}. Show that this is consistent with the presence of an absorption feature in solid CO at wavelength $\lambda \approx 4.67\ \mu m$, assuming that CO behaves as a simple harmonic oscillator.
   (b) If the rarer isotopic form, ^{13}CO, has the same force constant, estimate the wavelength at which you expect to see absorption due to ^{13}CO. What would the peak optical depth of the ^{13}CO feature be in a source with $\tau(^{12}CO) = 1.0$, if the band strength and linewidth are the same and the terrestrial $^{12}C/^{13}C$ ratio (89) applies?
2. Explain why hydrogenated amorphous carbon nanoparticles are thought to be probable carriers of the 3.4 μm feature in the diffuse ISM. What constraints can be placed on the nature of the absorbers from studies of the 3.4 μm profile shape?
3. Compare and contrast the spectroscopic properties of silicates in diffuse and dense regions of the ISM. How can we determine whether silicates are significantly hydrated in each environment?
4. Discuss briefly what is meant by the 'threshold effect' for molecular ices in dense clouds and explain its physical significance. Why is the threshold extinction for H_2O-ice different in different clouds? Why are the threshold extinctions for H_2O- and CO-ices different in the same cloud?

Table 5.4. Optical depth and visual extinction data for Taurus field stars with solid CO detections or upper limits (see problem 7).

Star	$\tau_{4.67}$	A_V
Elias 16	1.31 ± 0.03	23.7
Tamura 8	1.12 ± 0.08	21.7
Elias 15	0.53 ± 0.10	15.3
Elias 13	0.20 ± 0.03	11.8
Elias 3	0.38 ± 0.04	10.0
Elias 6	<0.04	6.8
Elias 14	<0.16	6.2
HD 29647	<0.06	3.6

5. Assume that NH_3 is present in ices in the Taurus dark cloud with a column density 10% of that of H_2O (see equation (5.16)). Estimate the fraction of the solar nitrogen abundance tied up in this form in the cloud.
6. (a) The stretch-mode of solid CO_2 in the spectrum of the highly reddened star Elias 16 has peak optical depth $\tau_{4.27} \sim 2.5$ and FWHM $\gamma \sim 0.002\ \mu\text{m}^{-1}$. Estimate the column density of CO_2 toward Elias 16, using the band-strength data listed in table 5.2.
   (b) The bend-mode of CO_2 has not yet been observed in Elias 16 because the star was too faint at 15.3 μm to be detectable with the spectrometer on board the Infrared Space Observatory. Make a prediction of the expected optical depth of this feature in Elias 16, based on the CO_2 column density calculated in (a) and assuming the FWHM to be the average of those detected in other objects ($\sim 0.002\ \mu\text{m}^{-1}$).
   (c) Discuss with reasoning whether you would expect the narrow subfeatures seen in the 15.3 μm CO_2 profile of GL 2136 (figure 5.15) to be also present in Elias 16.
7. (a) Table 5.4 lists peak solid CO optical depths ($\tau_{4.67}$) and visual extinctions (A_V) for several stars in the line of sight to the Taurus dark cloud. Plot a graph of $\tau_{4.67}$ versus A_V and deduce the most probable values of the threshold extinction A_0 and the slope q. Given that the mean observed FWHM of the solid CO feature is $7.5 \times 10^{-4}\ \mu\text{m}^{-1}$, deduce the abundance of solid CO relative to hydrogen in the Taurus cloud. Compare your result with the solar abundance of carbon.
   (b) Observations of mm-wave gas-phase CO emission suggest that

$$N(\text{CO})_{\text{gas}} \approx 1.2 \times 10^{21}(A_V - A_0)\ \text{m}^{-2}$$

   for $A_V > A_0$ in the Taurus cloud. Estimate the mean fraction of CO condensed onto dust in this cloud.

Chapter 6

Continuum and line emission

"...the dust that is merely opaque in the visible is self-luminous in the infrared and so in the midst of this optical darkness there has appeared a great infrared light."

N Woolf (1973)

Diffuse emission from interstellar dust was predicted by van de Hulst (1946) as a consequence of the presence of absorption: the balance of energy absorbed by the dust grains over the entire electromagnetic spectrum must re-emerge in the infrared. In a typical low-density interstellar environment, a dust particle gains energy mainly by absorption of ultraviolet photons from the ambient interstellar radiation field. A steady state is established such that the grain emits a power equal to that absorbed at some temperature $T_{\rm d}$ that depends on its size and composition; van de Hulst showed that for classical dielectric spheres of radii $a \sim 0.1$ μm, dust temperatures in the range 10–20 K are expected and emission should thus occur in the far infrared[1], a prediction confirmed some 25 years later (Pipher 1973). Dust emission constitutes a significant fraction ($\sim$10–30%) of the total radiative output of our Galaxy and therefore has a major effect on the detailed energy balance of the galactic disc.

Modern ground-based facilities for infrared and submillimetre astronomy, together with data from space missions such as the Infrared Astronomical Satellite (IRAS), the Cosmic-microwave Background Explorer (COBE) and the Infrared Space Observatory (ISO) provide unprecedented resources for studying diffuse emission from dust. As the interstellar medium is, in general, optically thin ($\tau \ll 1$) at FIR wavelengths, observed flux densities sample emission at all depths in a cloud with equal efficiency. Observations of the diffuse emission thus provide valuable information on the spatial distribution of dust in the interstellar medium

[1] For convenience, the term 'far infrared' (FIR) is taken here to mean the wavelength range 30–300 μm; similarly, 'near infrared' (NIR), 'mid-infrared' (MIR) and 'submillimetre' are 1–5 μm, 5–30 μm and 300–1000 μm, respectively.

as well as on grain properties (Hildebrand 1983). Whereas the detection of thermal FIR emission from cool dust had been anticipated, the discovery of strong excess emission at shorter infrared wavelengths was more surprising, indicating the occurrence of temperatures ($100 < T_d < 1000$ K) much greater than expected for classical particles in thermal equilibrium with their environment. This led to the development of models for dust emission that include contributions from very small grains (nanoparticles) subject to transient heating. At least some of these very small grains appear to be polycyclic aromatic hydrocarbons (PAHs), identified by means of spectral emission features that generally accompany the NIR and MIR continuum emission.

We begin in section 6.1 by discussing the theoretical basis for emission of radiation by dust. Observations of infrared continuum and line emission are reviewed in sections 6.2 and 6.3 and extended red emission in the visible is discussed in section 6.4. We are concerned here primarily with emission from interstellar dust heated by the ambient radiation field. The emissive properties of dust in circumstellar envelopes of red-giant stars are discussed further in chapter 7.

6.1 Theoretical considerations

6.1.1 Equilibrium dust temperatures

Interstellar grains exchange energy with their environment as a result of absorption and emission of radiation, collisions and exothermic surface reactions such as hydrogen recombination. In general, the equilibrium (steady state) dust temperature is determined primarily by radiative processes in diffuse regions of the ISM (collisions become more important in dense molecular gas and in hot plasmas; see Spitzer 1978, Dwek and Arendt 1992). We begin by considering a perfect spherical blackbody: its equilibrium temperature T_b is given by the Stefan–Boltzmann law

$$U = \frac{4\sigma}{c} {T_b}^4 \tag{6.1}$$

where U is the total energy density of photons in the interstellar radiation field (ISRF). The average value is $U \approx 5 \times 10^5$ eV m^{-3} $\approx 8 \times 10^{-14}$ J m^{-3} in diffuse H I clouds (e.g. Mathis 2000) and substitution into equation (6.1) gives $T_b \approx 3.2$ K, a result first deduced by Eddington (1926).

Now consider a spherical dust grain of radius $a \sim 0.1$ μm. The power absorbed from the ISRF is

$$W_{abs} = c(\pi a^2) \int_0^\infty Q_{abs}(\lambda) u_\lambda \, d\lambda \tag{6.2}$$

where Q_{abs} is the efficiency factor for absorption by the grain and u_λ is the energy density of the ISRF with respect to wavelength. Note that if the grain were composed of perfectly dielectric material, no energy would be absorbed

($Q_{abs} = 0$). However, all real solids absorb to some extent: even good dielectrics will have some residual absorption, induced by the presence of impurities. The power radiated by the grain is

$$W_{rad} = 4\pi(\pi a^2) \int_0^\infty Q_{em}(\lambda) B_\lambda(T_d)\, d\lambda \tag{6.3}$$

where $Q_{em}(\lambda)$ is the efficiency factor for emission from the grain (usually termed the emissivity) and

$$B_\lambda(T) = \frac{2hc^2}{\lambda^5} \frac{1}{\exp(hc/\lambda kT) - 1} \tag{6.4}$$

is the Planck function. It follows from Kirchhoff's law that $Q_{abs}(\lambda)$ and $Q_{em}(\lambda)$ are, in fact, identical at a given wavelength and we may replace them in equations (6.2) and (6.3) by a single function Q_λ. For equilibrium between the rates of gain and loss of internal energy, we have $W_{abs} = W_{rad}$ and thus

$$\int_0^\infty Q_\lambda u_\lambda\, d\lambda = \frac{4\pi}{c} \int_0^\infty Q_\lambda B_\lambda(T_d)\, d\lambda. \tag{6.5}$$

This equation may be used to deduce the temperature T_d of the grain if Q_λ can be determined. If we set $Q_\lambda = 1$ at all wavelengths, equation (6.5) leads to the Eddington result. Dust temperatures higher than T_b arise from the fact that most power is absorbed at wavelengths short compared with those at which most power is emitted, i.e. absorption occurs predominantly in the ultraviolet and emission occurs predominantly in the far infrared. Q_λ may be calculated from Mie theory (section 3.1) and is found to have widely different values in the wavelength domains of interest: for weakly absorbing materials, $Q_{UV} \sim 1$ and $Q_{FIR} \ll 1$. In the FIR, $a \ll \lambda$ and we may use the small-particle approximation (equation (3.13)) to specify Q_λ. In general, Q_λ follows a power law in the FIR, i.e.

$$Q_{FIR} \propto \lambda^{-\beta} \tag{6.6}$$

for some index β that depends on the nature of the material. Theoretically, we expect $\beta = 2$ for metals and crystalline dielectric substances and $\beta = 1$ for amorphous, layer-lattice materials (see Tielens and Allamandola 1987a and references therein). Considering a weak absorber with $\beta = 1$, van de Hulst (1946) showed that a dust temperature

$$T_d \sim T_s w^{\frac{1}{5}} \tag{6.7}$$

is predicted, where the ISRF is represented by a blackbody of temperature $T_s = 10\,000$ K (the equivalent stellar temperature), and $w = 10^{-14}$ is the dilution factor. Substitution of these values into equation (6.7) yields $T_d \sim 15$ K. Detailed calculations for specific materials (e.g. Greenberg 1971, Mathis *et al* 1983, Draine and Lee 1984) largely confirm the van de Hulst result for dielectric materials such as silicates in the low-density ISM. Strong absorbers such as graphite reach equilibrium at temperatures typically a factor of two higher than those of silicates (Mathis *et al* 1983). These results apply to the solar neighbourhood and are only weakly dependent on grain size for classical particles. It should be noted that the intensity of the ISRF is a function of galactocentric radius such that systematically higher grain temperatures are predicted in regions closer to the galactic centre (Mathis *et al* 1983).

Deep within a dense cloud heated only by the external ISRF, the ultraviolet and visible flux is severely attenuated. The grains are then heated primarily by absorption of infrared radiation and cosmic rays and by collisions with the gas, which will tend to equilibriate gas and dust temperatures. Their equilibrium temperatures may be influenced by changes in their optical properties, as may result from coagulation or mantle growth, as well as by changes in the radiative environment. Models for the transport of radiation within dense clouds allow dust temperatures to be calculated as a function of optical depth (Leung 1975, Mathis *et al* 1983). Consider, for example, an idealized spherical cloud with total visual extinction $A_V = 10$ through its centre. For a classical silicate grain, the predicted temperature falls from its external value of $\sim$15 K to $\sim$7 K at the cloud centre, a result that is not strongly influenced by the presence of absence of an ice mantle (Leung 1975). For an absorber such as graphite, however, the predicted temperature falls from $\sim$30 K externally to $\sim$17 K (unmantled) or 10 K (mantled) within the cloud. Mantle growth thus tends to reduce temperature differences between different grain materials. Note that grains as cold as 7–10 K will emit predominantly in the extreme FIR to submillimetre region of the spectrum (200–500 μm; see section 6.1.2) and are essentially undetectable at 100 μm, the longest wavelength available to IRAS.

The results discussed here apply to regions remote from individual stars. In the vicinity of an early-type star, dramatic enhancement of the ambient energy density at ultraviolet wavelengths naturally leads to much higher dust temperatures. Consider, for example, a particle situated in an H II region at a distance $r = 0.5$ pc from a star of radius $R_s = 1.1 \times 10^{12}$ m and surface temperature $T_s = 40\,000$ K (equivalent to spectral type O6). The dilution factor is $w = (R_s/2r)^2 = 1.27 \times 10^{-9}$ (Greenberg 1978) and substitution of these values into equation (6.7) gives $T_d \sim 650$ K. Ultimately, the temperature a grain can reach is limited by its sublimation temperature.

6.1.2 FIR continuum emission from an interstellar cloud

Consider an idealized cloud containing $\mathcal{N}$ spherical dust grains of uniform size, composition and temperature. As in section 6.1.1, we assume that the grains are classical spheres of radius $a \sim 0.1$ μm and that each grain is in thermal equilibrium with the ambient radiation field. A real interstellar cloud will, of course, contain grains with a range of sizes but the FIR emissivity should be described by the small-particle approximation over the entire size distribution for any reasonable grain model. Transient temperature fluctuations in very small grains, resulting in additional flux at shorter wavelengths, are considered in section 6.1.4. Assuming that the cloud is optically thin in the FIR, a flux density

$$F_\lambda = \mathcal{N}\left\{\frac{\pi a^2}{d^2}\right\} Q_\lambda B_\lambda(T_d) \tag{6.8}$$

is received, where d is the distance to the cloud and $(\pi a^2/d^2)$ is the solid angle subtended by an individual grain. The wavelength at which the flux spectrum peaks, deduced by analogy with the Wien displacement law, is related to the temperature by

$$\lambda_{peak} \approx 3000\left\{\frac{5}{\beta+5}\right\} T_d^{-1} \tag{6.9}$$

with λ_{peak} in μm. Setting $\beta = 0$ yields the normal form of Wien's law but for interstellar grains we expect $1 < \beta < 2$ and thus $\lambda_{peak} \sim 2300/T_d$ (μm).

First, consider the case of a quiescent dense cloud that contains no internal sources of luminosity. We noted in section 6.1.1 that internal dust temperatures of 10 K or less are predicted, resulting in peak emission at wavelengths >200 μm. Because of the inherent properties of the Planck function (equation (6.4)), such cold grains emit negligible flux at $\lambda \leq 100$ μm. However, a quiescent cloud will not be truly isothermal but will have an outer layer of warmer dust heated by the external ISRF to temperatures approaching those in unshielded regions; for $T_d \sim 23$ K (averaged over silicate and carbon grains), peak emission from the outer layer occurs near 100 μm. Thus, emission in (e.g.) the IRAS 100 μm passband depends only on the energy density of the ISRF and on the surface properties of the cloud; clouds of different mass may thus have similar surface brightness.

Few interstellar clouds are truly devoid of embedded sources. Molecular clouds are sites of star formation and a massive young star embedded in its parent cloud may generate intense local ultraviolet flux. Short-wavelength radiation cannot penetrate far into the cloud; absorption and re-emission by dust close to the source convert it into infrared radiation, to which the cloud is relatively transparent; and this tends to heat the dust elsewhere in the cloud to fairly uniform temperatures that depend on the total internal luminosity. The flux observed from an internally heated cloud may thus be used to deduce the mass of the cloud and to place constraints on grain properties, as discussed by Hildebrand (1983) and Thronson (1988). The total volume of dust in the cloud is given by

$$V = \mathcal{N}v \tag{6.10}$$

where $v = (4/3)\pi a^3$ is the volume of an individual grain. Substituting from equation (6.8) to eliminate $\mathcal{N}$, equation (6.10) becomes

$$V = \left\{ \frac{F_\lambda d^2}{\pi a^2 Q_\lambda B_\lambda(T_\mathrm{d})} \right\} v. \tag{6.11}$$

If the grains are composed of material of density s, equation (6.11) may be written in terms of the total dust mass

$$M_\mathrm{d} = Vs = \frac{4sF_\lambda d^2}{3B_\lambda(T_\mathrm{d})} \left\{ \frac{a}{Q_\lambda} \right\}. \tag{6.12}$$

In the small-particle approximation (equation (3.13)), the quantity Q_λ/a is independent of a and depends only on the refractive index m at wavelength λ. With reference to equation (6.8), the contribution of each grain to the observed flux spectrum is proportional to $a^2 Q_\lambda = a^3(Q_\lambda/a)$, i.e. to its *volume*. Adopting a suitably weighted average of Q_λ/a, equation (6.12) may thus be used to estimate the total mass of dust in a cloud from the observed flux density without detailed knowledge of the grain-size distribution.

Multiwaveband observations of an internally heated cloud in the FIR and submillimetre region allow the spectral index of the emissivity (equation 6.6) to be determined and, in principle, this may be used to place constraints on grain models. For example, if the emission at these wavelengths were predominantly from graphite, as in the Draine and Lee (1984) model, we would expect $\beta \approx 2$ (Tanabé *et al* 1983), whereas $\beta \approx 1$ for amorphous carbon. Silicates typically have intermediate values of β (Day 1979, 1981) which may vary with particle size (Koike *et al* 1987) as well as crystallinity. Ideally, in a cloud that is sufficiently warm, the Rayleigh–Jeans approximation

$$B_\lambda(T) \approx \frac{2ckT}{\lambda^4} \tag{6.13}$$

is valid in the FIR and the observed spectrum (equation (6.8)) will follow a simple power law independent of temperature, from which β may be evaluated (see problems 1 and 2 at the end of this chapter). However, temperatures $T_\mathrm{d} > 170$ K are required for the Rayleigh–Jeans approximation to be valid to an accuracy of ± 0.1 in β in the submillimetre region (Helou 1989). It is thus generally necessary to obtain an independent estimate of the temperature of a cloud in order to evaluate β, and the uncertainties are such that results do not provide a very sensitive discriminator between models for interstellar dust. This technique places more significant constraints on dust in circumstellar shells of evolved stars, in which dust temperatures are generally higher and the composition more homogeneous (section 7.2).

6.1.3 Effect of grain shape

Spherical grains are assumed in the preceding discussion; we briefly review here the implications of non-spherical shape (implied by observations of polarization) on the emissive properties of the dust. Greenberg and Shah (1971) investigated the effect of shape on calculations of dust temperatures and showed that, in general, the temperatures of non-spherical particles are somewhat lower than those of equivalent spheres. However, the difference is generally small and is always less than 10% for dielectric spheroids and cylinders; only highly flattened or elongated metallic particles show more significant departures. The total volume of dust determined from the infrared flux density emitted by a cloud (equation (6.11)) depends on the ratio of volume to projected cross-sectional area (v/σ) for an individual grain. The value of v/σ remains within a factor of two of its maximum value (equal to that of a sphere of the same cross section) for a wide range of grain shapes (Hildebrand 1983): serious discrepancies again arise only if the grains are highly flattened or elongated. Unless the bulk of the dust is made up of very thin needles or flakes, shape will have little influence on the determination of dust mass.

The most pertinent property of a non-spherical grain is that, for significant elongation, the emitted flux is linearly polarized. The $\boldsymbol{E}$-vector of the emitted radiation has its maximum value in the plane containing the long axis of the grain, which is perpendicular to the magnetic field direction for magnetic alignment (section 4.5). For an ensemble of aligned grains, the degree of polarization P_{em} (%) of the emitted radiation may be defined by analogy with equation (4.6):

$$P_{\mathrm{em}} = 100 \left\{ \frac{F_{\max} - F_{\min}}{F_{\max} + F_{\min}} \right\} \tag{6.14}$$

where $F_{\max}$ and $F_{\min}$ are the maximum and minimum flux densities measured with respect to the rotation of some analysing element. At a given wavelength, this quantity is directly related to the efficiency P_{λ} with which the grains introduce polarization to the transmitted radiation by differential extinction (section 4.1). Since the flux densities are proportional to the emissivities and, hence, by Kirchhoff's law, to the absorption efficiencies, equation (6.14) may be rewritten in terms of maximum and minimum values of Q_{λ} (Hildebrand 1988a). If the polarization is small (section 4.2.1), it can then be shown that

$$P_{\mathrm{em}}(\lambda) = -\frac{P_{\lambda}}{\tau_{\lambda}} \tag{6.15}$$

(see problem 4 at the end of this chapter). The negative sign in equation (6.15) arises because the position angles for emission and absorption are orthogonal (perpendicular and parallel to the magnetic field, respectively). It should be noted that P_{λ} and τ_{λ} become very small in the FIR where $P_{\mathrm{em}}(\lambda)$ is measured. However, in situations where polarized emission in the FIR and polarized absorption at shorter wavelengths can be observed for the same region, the planes of polarization are predicted to be orthogonal if the two effects originate in the same grain population.

6.1.4 Effect of grain size

So far, we have assumed that dust temperatures are determined by the time-averaged rates of absorption and emission of energy (equations (6.2) and (6.3)) and that individual quantum events are unimportant. Two conditions must hold for this assumption to be valid: (i) the total kinetic energy content of a grain must be large compared with that received when a single energetic photon is absorbed; and (ii) the energy of the absorbed photon must be distributed throughout the full volume of the grain (i.e. the grain must reach internal thermal equilibrium) on a timescale short compared with that for photon emission. However, the ISM contains a wide range of particle sizes from classical submicron-sized grains to those with molecular dimensions. Let us assume initially that condition (ii) holds for all sizes and examine the inevitable breakdown of condition (i) as progressively smaller particles are considered.

The thermal properties of solids are described by Debye's theory (1912). For a material with Debye temperature Θ, the heat capacity at constant volume varies as $C_V \propto T^3$ in the low-temperature approximation ($T \ll \Theta$). Values of $\Theta \sim 500$ K are expected for typical candidate grain materials (Duley 1973, Greenberg and Hong 1974, Purcell 1976). Purcell showed that the internal heat energy due to lattice vibrations for a grain of radius a (μm) and temperature T_d (K) is given in electron volts by

$$H \approx 17a^3 T_d{}^4 \tag{6.16}$$

assuming that the bulk heat capacity is applicable to small particles. (A more detailed treatment includes surface effects: see Aannestad and Kenyon 1979.) If the grain absorbs a single photon of energy ε_{ph}, the initial and final temperatures T_1 and T_2 are then related by

$$T_2 \approx \left\{ T_1{}^4 + \frac{\varepsilon_{ph}}{17a^3} \right\}^{\frac{1}{4}} \tag{6.17}$$

with ε_{ph} in eV and a in μm. For illustration, consider the amplitude ($\Delta T = T_2 - T_1$) of the increase in temperature induced by the absorption of individual 10 eV photons as a function of grain size, assuming an initial temperature $T_1 = 15$ K. The increase is minimal ($\Delta T < 0.5$ K) for classical grains ($a > 0.05\ \mu$m) but becomes progressively larger for smaller grains: e.g. $\Delta T \approx 13$ K for $a = 0.01\ \mu$m and $\Delta T \approx 32$ K for $a = 0.005\ \mu$m. The heated grain subsequently cools by emission of longer-wavelength radiation. Small grains thus undergo a sequence of temperature fluctuations or 'spikes' associated with the arrival of energetic photons, as shown schematically in figure 6.1 for $a = 0.005\ \mu$m.

At higher temperatures, equation (6.16) is no longer appropriate. The heat capacity becomes insensitive to temperature for $T > \Theta$ and is given by $C_V = 3Nk$ where $3N$ is the number of degrees of freedom in the grain. The amplitude of the temperature spike produced by a photon of energy ε_{ph} (eV) is

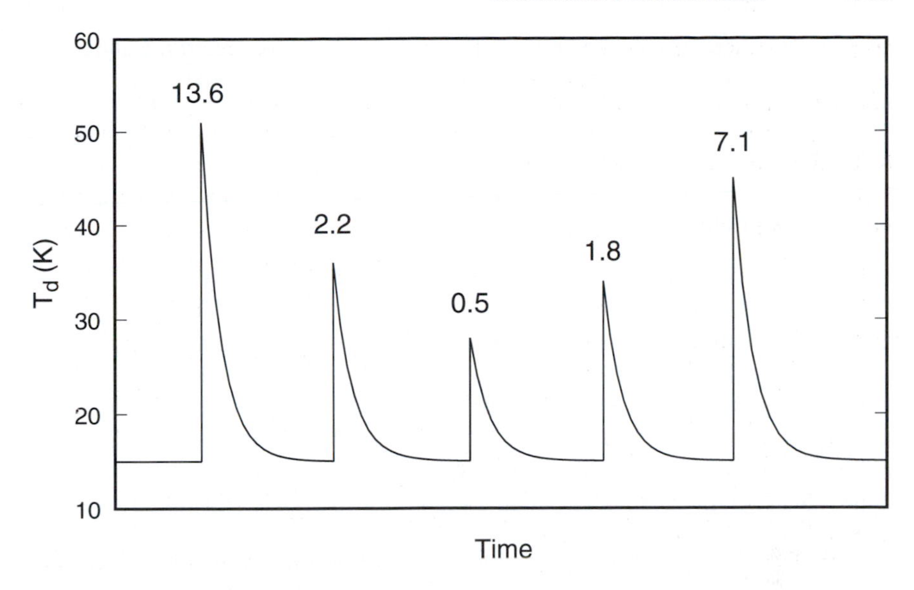

Figure 6.1. A schematic representation of temperature fluctuations in a small grain (radius $a = 0.005$ μm) induced by sequential absorption of individual photons. Each fluctuation is labelled by the photon energy in electron volts. (Adapted from Aannestad and Kenyon 1979.)

then

$$\Delta T \approx 4 \times 10^3 \varepsilon_{\rm ph}/N. \tag{6.18}$$

For a grain of radius 0.001 μm, $N \simeq 40$ and a spike of amplitude $\Delta T \simeq 1000$ K is predicted for a 10 eV photon (Sellgren 1984).

We showed earlier that the temperature spikes should have a negligible amplitude in classical ($a > 0.05$ μm) grains of homogeneous composition and structure. However, real interstellar grains may not behave as homogeneous solids in which vibrational energy is rapidly transmitted throughout the lattice. Grains formed by an accretion process will tend to be porous, containing numerous subunits that may be in poor thermal contact with one another. Amorphous carbon naturally exhibits such structure (Duley and Williams 1988), suggesting that PAH-like clusters within the framework of a larger amorphous carbon grain may undergo some degree of stochastic heating.

The principles and concepts discussed in this section provide a basis for understanding the observed emission spectrum of the ISM, reviewed in the following sections. The form and intensity of the spectrum will depend on a number of factors, including the nature of the radiation field (general ISRF or local embedded source) and the composition and size distribution of the particles. Whereas classical grains emit in the FIR, smaller grains may contribute continuum flux peaking at much shorter wavelengths, dependent on their time-averaged temperature. If specific vibrational resonances or electronic states are excited, discrete emission lines or luminescence bands will also be seen.

6.2 Galactic continuum emission

6.2.1 Morphology

Diffuse far infrared emission from our Galaxy was first detected in the early 1970s (Pipher 1973) and studied in detail following the launch of the IRAS mission in 1983 (e.g. Beichman 1987, Cox and Mezger 1989). A major discovery from IRAS was the detection of large-scale filamentary FIR emission, termed 'cirrus', arising not only from the plane of the Milky Way but also from regions of intermediate and high galactic latitude (Low *et al* 1984). Figure 6.2 illustrates the distribution of cirrus emission at 240 μm, as observed by the COBE satellite (Lagache *et al* 1998). In general, the intensity of emission is well correlated with other tracers of interstellar matter (section 1.3), including H I line emission and visual extinction (e.g. Terebey and Fich 1986, Laureijs *et al* 1987, Boulanger and Pérault 1988, Sodroski *et al* 1994, Boulanger *et al* 1996). The high-latitude cirrus originates in local clouds, typically as close as 100–200 pc. Although most cirrus clouds are diffuse, some have cores detected in CO line emission (e.g. Weiland *et al* 1986, Boulanger and Pérault 1988, Reach *et al* 1995a, Stark 1995), indicating the presence of a molecular phase. Contributions from individual clouds are superposed on a relatively smooth component of galactic FIR emission that follows a cosec b law (Beichman 1987). Toward the galactic plane, the cirrus naturally merges into a continuous sheet of diffuse emission (see figure 1.3) in which a typical line of sight intercepts many clouds.

6.2.2 Spectral energy distribution

The broadband spectral energy distribution of dust emission has been studied over virtually the entire infrared spectrum (1–1000 μm), in both the galactic disc and high-latitude cirrus clouds (e.g. Cox and Mezger 1987, Dwek *et al* 1997). The diffuse-ISM average spectrum, illustrated in figure 6.3, shows a broad maximum centred at 100–200 μm (the FIR peak) and a secondary maximum centred near 10 μm (the MIR excess). The FIR peak is attributed to classical-sized grains in thermal equilibrium with the ISRF (section 6.1.1) and its position implies temperatures $T_d \sim$ 15–20 K (equation (6.9)), in reasonable agreement with predictions of the graphite–silicate model (Draine and Lee 1984). The MIR excess accounts for ~40% of the total power radiated by dust at $\lambda <$ 120 μm (Boulanger and Pérault 1988) and its presence implies substantially higher temperatures for some component of the dust. It is now widely accepted that the excess is caused by continuum and line emission from very small grains that are stochastically heated (section 6.1.4) to mean temperatures typically in the range 50–500 K (Draine and Anderson 1985, Cox *et al* 1986, Weiland *et al* 1986,

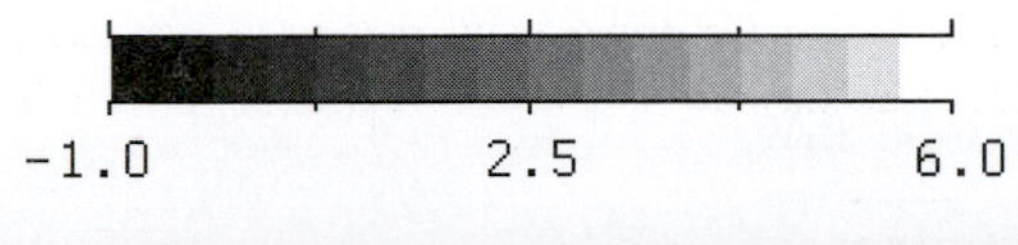

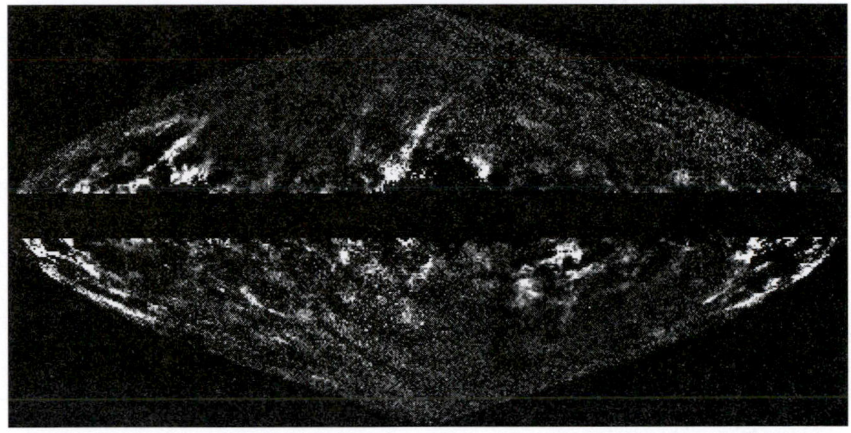

Figure 6.2. The distribution of cold dust emission at 240 μm at latitudes $|b| > 10°$ in the Galaxy, as determined from observations by the COBE satellite (Lagache *et al* 1998). The data are displayed in projection relative to galactic coordinates, with the galactic nucleus toward the centre of the diagram, north to the top and galactic longitude increasing to the left. The scale denotes intensity in units of MJy/sr.

Dwek *et al* 1997). In the model proposed by Dwek *et al* (1997), illustrated in figure 6.3, the MIR excess arises primarily in PAHs (section 6.3.2).

The earliest observational evidence for stochastic heating of small grains came not from the IRAS data but from ground-based studies of visible reflection nebulae surrounding early-type stars (Sellgren *et al* 1983, Sellgren 1984). Strong continuous emission with colour temperature $\sim$1000 K is typically seen in the NIR, as illustrated in figure 6.4. The intensity of the emission correlates closely with that of the visible nebula caused by scattered light from the illuminating star. However, the NIR flux cannot be explained by scattering: its surface brightness at 3 μm is two orders of magnitude higher than that predicted by extrapolation of scattering models fitted to the visible data for any reasonable value of the grain albedo (section 3.3.2). The colour temperature of the NIR emission does not fall off rapidly with distance from the star, as it should if it were emitted by grains in equilibrium with the ambient radiation field. These results prompted development of a model based on stochastic heating of very small grains (VSGs) in the nebular environment (Sellgren 1984).

Studies of other galaxies demonstrate the universality of the VSG phenomenon (Cox and Mezger 1987, Rice *et al* 1990, Rodriguez-Espinosa *et al* 1996, Dale *et al* 2001, Genzel and Cesarsky 2000). Distinct FIR and MIR

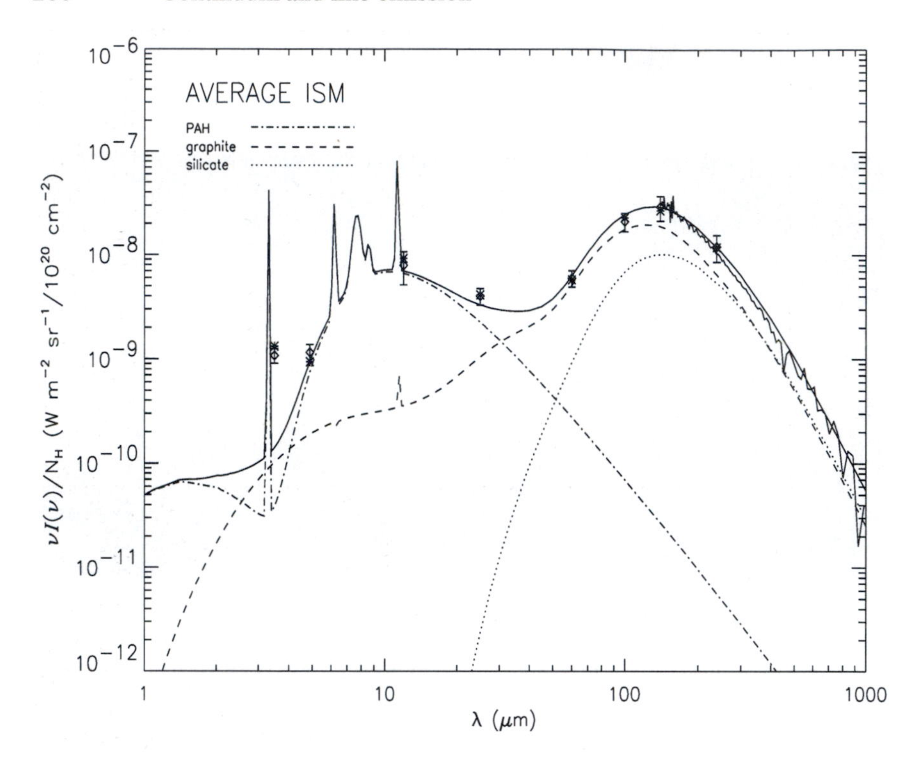

Figure 6.3. The average spectral energy distribution of diffuse emission from the diffuse ISM. The points and jagged curve represent COBE data obtained with the Diffuse Infrared Background Experiment (DIRBE) and the Far Infrared Absolute Spectrometer (FIRAS), respectively (Dwek *et al* 1997 and references therein). The continuous curve is a three-component model fit to the DIRBE data, combining emission from PAH molecules with that from uncoated graphite and silicate grains following MRN size distributions (see Dwek *et al* 1997 for details). The contribution from each individual component is also plotted. Figure courtesy of E Dwek, originally published in the *Astrophysical Journal*.

emission peaks associated with cold and warm dust are common features of both normal and active galaxies. As an example, figure 6.5 compares the broadband spectra of two Seyfert systems. Note the overall similarity in shape, comparing these spectra with the diffuse-ISM spectrum of our Galaxy (figure 6.3). The observations are well matched by a model that assumes emission from dust with a bimodal temperature distribution, allowing for variation in the relative strength of cold and warm components. This variation proves to be a useful tool for distinguishing between starburst galaxies and active galaxies (Genzel and Cesarsky 2000) as the MIR continuum is generally much stronger in the latter.

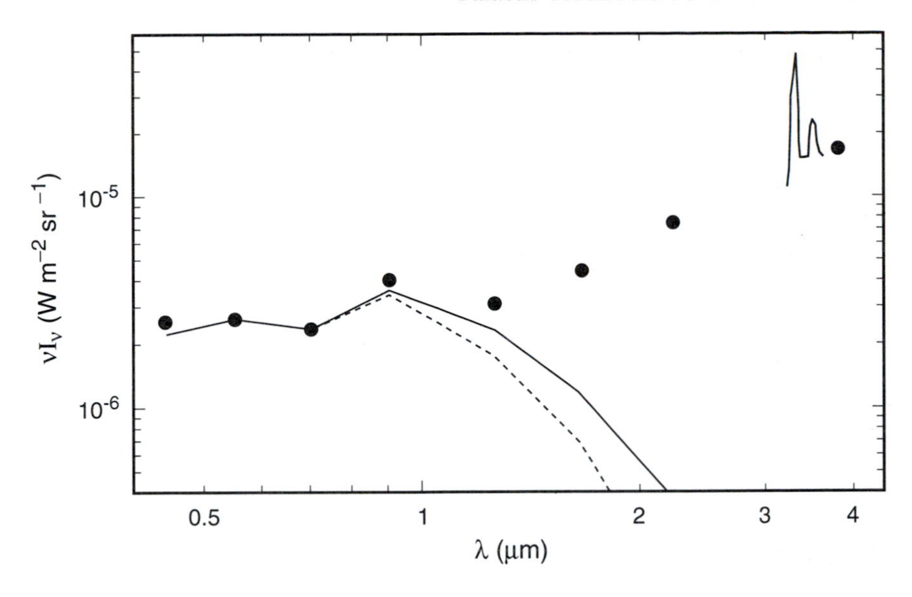

Figure 6.4. The visible-to-near-infrared spectrum of the reflection nebula NGC 2071 illuminated by the B-type star HD 290861a. The observations include broadband photometry (points) and low-resolution 3.2–3.5 μm spectrophotometry (curve). The vertical axis is νI_ν, where I_ν (W m^{-2} Hz^{-1} sr^{-1}) is the surface brightness measured at a locus 30″E, 30″S of the central star. Visual data are from Witt and Schild (1986), infrared data from Sellgren *et al* (1996). Note the presence of PAH emission features at 3.3 and 3.4 μm in addition to strong NIR continuum emission. Also shown are calculations of the brightness of scattered starlight (Sellgren *et al* 1996), assuming two different forms for the albedo: constant albedo (full curve) and albedo varying with wavelength according to the Draine and Lee (1984) model (broken curve).

6.2.3 Dust and gas

We showed in section 6.1.2 that infrared continuum emission is a useful quantitative measure of the distribution of dust mass. Although MIR excess emission from small grains appears to be ubiquitous throughout the Galaxy, it is the larger grains emitting in the FIR that contribute most of the mass. The observations may be compared with corresponding data for the principal tracers of interstellar gas, allowing the large-scale emission from dust associated with molecular, atomic and ionized phases of the ISM in our Galaxy to be distinguished and mapped (Sodroski *et al* 1987, 1989, 1994, 1997). Dust in the diffuse H I phase is found to contribute the dominant fraction ($\sim$70%) of the total FIR luminosity, whilst dust in cold molecular and warm ionized gas contributes $\sim$20% and $\sim$10%, respectively.

Comparison of FIR flux and total gas column in a given line of sight leads to

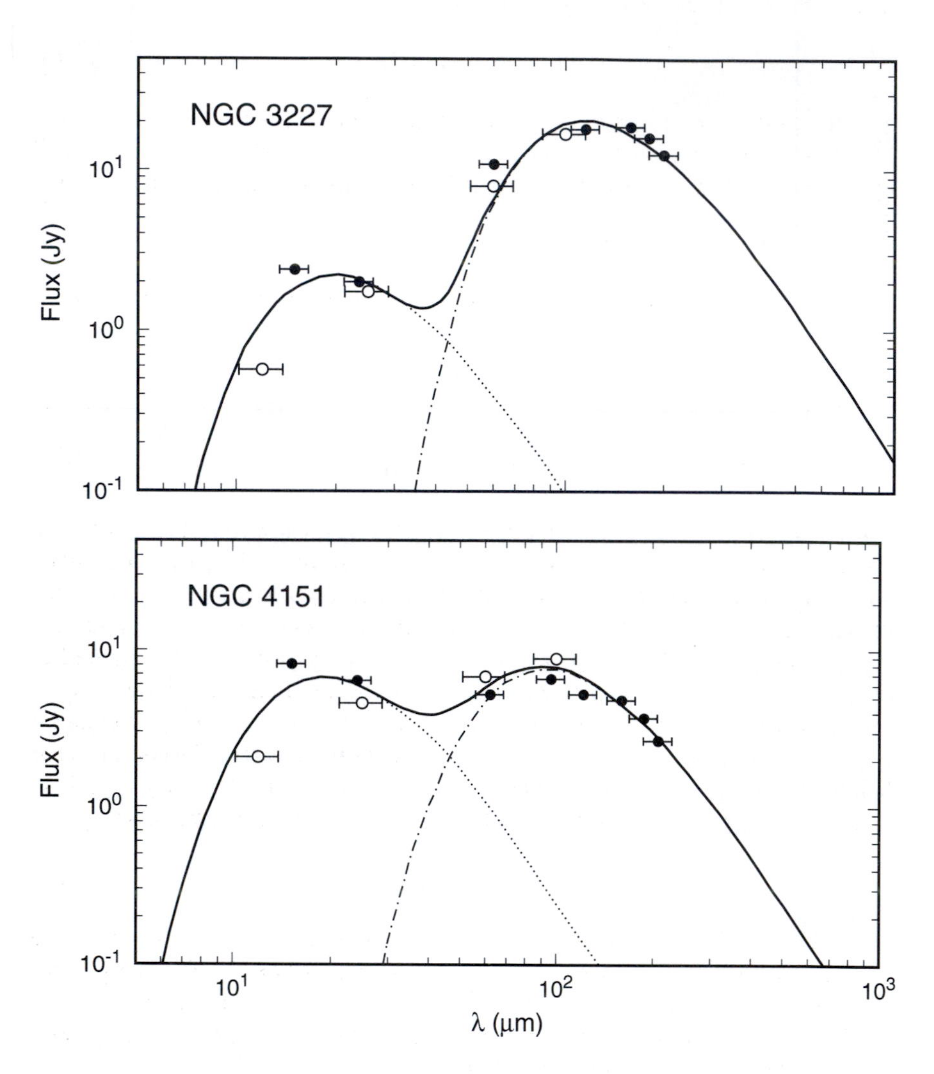

Figure 6.5. Spectral energy distributions of two Seyfert galaxies. Observations are from IRAS and ISO missions (open and filled circles, respectively). The full curves are models based on the sum of emission from warm dust (dotted curves) and cold dust (dot–dash curves). The temperatures of the warm and cold components are 161 K and 27 K (NGC 3227), 170 K and 36 K (NGC 4151). (Adapted from Rodriguez-Espinosa *et al* 1996.)

an evaluation of the dust-to-gas ratio[2]. The flux density at wavelength λ received in a beam of solid angle Ω may be expressed as

$$F_\lambda = \Omega \tau_\lambda B_\lambda(T_d) \tag{6.19}$$

assuming that $\tau_\lambda \ll 1$ in the FIR (Hildebrand 1983). The optical depth may thus be determined from the observed flux for dust at a given temperature (constrained by the spectral energy distribution). This is related to mean dust density ρ_d along pathlength L by

$$\tau_\lambda = \rho_d \kappa_\lambda L \tag{6.20}$$

where κ_λ is the FIR mass absorption coefficient of the grains; the dust-to-gas ratio is then

$$\begin{aligned} Z_d &= 0.71 \frac{\rho_d}{\rho_H} \\ &= 4.3 \times 10^{26} \frac{\tau_\lambda}{N_H \kappa_\lambda} \end{aligned} \tag{6.21}$$

where $N_H = \rho_H L / m_H$ is the total hydrogen column density in the line of sight and the factor 0.71 again allows for the contributions of noble gases (principally He) to the total mass of gas. The atomic and molecular contributions to N_H may be quantified by observations of 21 cm H I and 2.6 mm CO emission lines, respectively. The mass absorption coefficient at 100 μm is estimated to be $\kappa_{100} \approx 4.1$ m^2 kg^{-1} for the graphite–silicate grain model (see Sodroski *et al* 1987) and $\kappa_\lambda \propto \lambda^{-2}$ is generally assumed for extrapolation to longer wavelengths (e.g. $\kappa_{240} \approx 0.71$ m^2 kg^{-1}). Taking an average for the galactic plane in the solar neighbourhood, COBE observations of the dust emission at 240 μm (Sodroski *et al* 1994, 1997) indicate that $\langle \tau_{240}/N_H \rangle \approx 1.3 \times 10^{-29}$ m^2/nucleon, yielding

$$Z_d \approx 0.008. \tag{6.22}$$

There are several possible sources of error that might affect this result. The appropriate value of κ_λ is rather uncertain (Draine 1990), with discrepancies existing between values measured directly in the laboratory (Tanabé *et al* 1983) and those implied by Mie theory calculations (Draine and Lee 1984). Uncertainties are also associated with the evaluation of τ_λ from F_λ and its dependence on T_d (equation (6.19)) and with the CO $\rightarrow$ H_2 conversion. These factors may introduce random or systematic errors in Z_d that are hard to quantify but could be as high as 30–40% (Sodroski *et al* 1994). Any differences in spatial resolution between the infrared and radio observations of dust and gas will introduce systematic error. A more serious and less quantifiable source of systematic error will arise if the line of sight contains very cold ($T_d < 10$ K)

[2] Note also that the FIR emission may itself be used as a tracer of the gas for an assumed dust-to-gas ratio. This has the advantage that the results are independent of the CO $\rightarrow$ H_2 conversion factor and other uncertainties associated with evaluation of N_H (see Gordon 1995).

dust that contributes substantially to the mass but minimally to the FIR flux (see section 6.2.4). It is therefore reassuring to note that the result in equation (6.22) is quite consistent with those based on the extinction of starlight (equation (3.33)) and the elemental depletions (equation (2.22)).

Comparison of Z_d values determined for different locations in the galactic disc show a systematic decline with distance from the galactic centre (Sodroski *et al* 1997). To within uncertainties of measurement, the trend is consistent with the corresponding decline in metallicity (see figure 2.5 and the discussion in section 2.3.2). This result is not unexpected as it indicates a correlation between the abundance of dust and that of the dust-forming chemical elements.

6.2.4 The 'cold dust problem'

The contribution from the coldest regions of the ISM to the total mass is difficult to estimate from observations of dust emission in the FIR (see Bally *et al* 1991). As discussed in section 6.1.2, the flux from a molecular cloud devoid of internal sources of luminosity tends to be dominated by emission from the warm outer layers. Observations extending into the submillimetre are needed to sample the emission from cold internal dust with $T_d < 10$ K. In general, the 100–1000 μm spectrum of a molecular cloud will consist of emission from cold dust with λ_{peak} typically in the range 200–500 μm, superposed on the tail of the emission from warmer dust with $\lambda_{peak} \leq 100$ μm (e.g. Mathis *et al* 1983). In principle, the total mass of dust may be estimated by detailed modelling of the spectral energy distribution. However, because the amplitude of the emission at λ_{peak} declines with T_d, the observed submillimetre flux is rather insensitive to the total mass of cold dust (Helou 1989, Eales *et al* 1989, Draine 1990). A small error in flux can thus lead to a much larger error in the estimated dust mass.

An illustration of the problem is provided by comparison of results on the galactic dust-to-gas ratio from IRAS and COBE. Estimates of Z_d (section 6.2.3) based on IRAS data at 60 and 100 μm are typically a factor of two to three less than those based on COBE data at 140 and 240 μm (compare Sodroski *et al* 1987 and 1994). The IRAS passbands sample emission predominantly from a warm (20–30 K) component of the dust, whilst cooler (10–15 K) dust is missed because almost all of its flux is radiated at longer wavelengths. Does COBE fully sample the cold dust? The good agreement between Z_d values from independent methods (section 6.2.3) suggests that it does, at least for the diffuse ISM (see also Wall *et al* 1996). However, the existence of a much colder (4–8 K) component is difficult to exclude (Reach *et al* 1995b, Lagache *et al* 1998, Krügel *et al* 1998).

6.2.5 Polarization and grain alignment

Observations of polarization in the far infrared and submillimetre continuum emission from molecular clouds provide information on the shape and alignment of dust in the densest regions of the ISM. The polarization vectors observed at

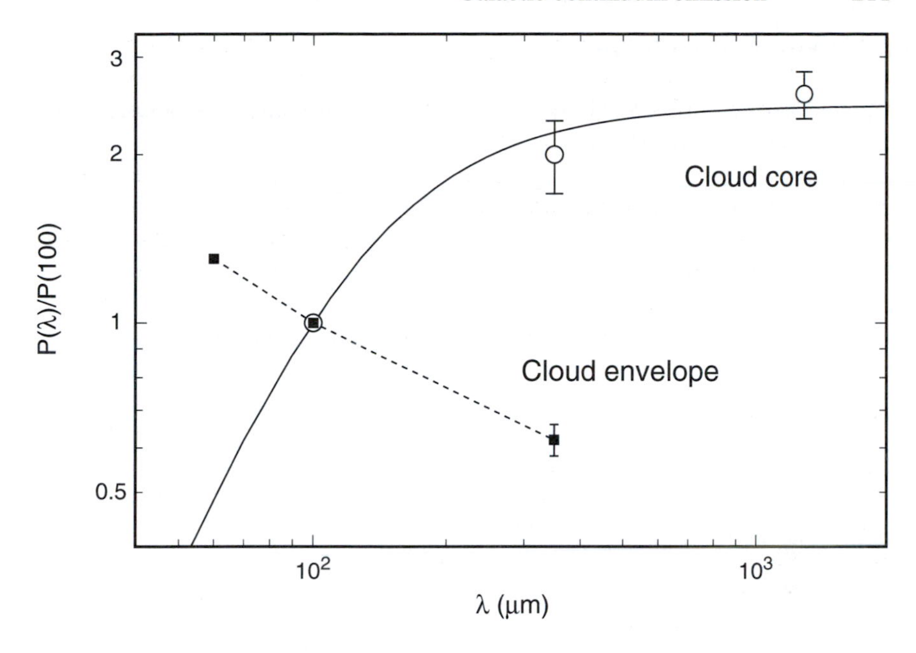

Figure 6.6. Spectral dependence of far-infrared–submillimetre polarization, based on observations of the Orion/KL and M17 molecular clouds (Hildebrand *et al* 1999 and references therein): full squares and dotted line, cold cloud envelopes; open circles, dense cloud core. The full curve is the predicted polarization for a uniform, self-absorbing cloud in which the optical depth varies with wavelength as an inverse power law (see Hildebrand *et al* 1999 for details). All data are normalized to unity at $\lambda = 100\ \mu$m.

these wavelengths are generally found to be orthogonal to those in the near- to mid-infrared (Hildebrand *et al* 1984, Dragovan 1986), as expected if the same population of aligned grains absorbs at the shorter wavelengths and emits at the longer (section 6.1.3). Calculations indicate that the amplitude and spectral dependence of $P_{\rm em}(\lambda)$ are consistent with emission by aligned oblate spheroids of axial ratio $\sim$1.5 (Hildebrand and Dragovan 1995). The distribution of polarized emission with respect to environment demonstrates that emitting grains are better aligned in the warm, compact cores of molecular clouds with active star formation than in colder regions remote from embedded stars (Hildebrand *et al* 1999). This is illustrated in figure 6.6, which compares the spectral dependence of polarization for these contrasting environments. The polarization at 350 μm is a factor of 3–4 higher in the cloud core compared with the envelope. Radiation or winds from embedded stars thus appear to amplify grain alignment in the surrounding medium – see section 4.5.7 for further discussion.

6.3 Spectral emission features

If characteristic vibrational modes within a grain lattice are excited, the corresponding infrared spectral features may appear in emission. This situation commonly arises in circumstellar shells and in ambient interstellar matter in close proximity to individual stars. However, few of the dust features seen in absorption in the ISM (chapter 5) have direct counterparts in emission. The primary reason for this is that volatile solids detected in absorption in molecular clouds (section 5.3) are generally sublimed in situations that would give rise to emission features. Also, populations of very small grains that are optically thin in absorption in the infrared may nevertheless be seen in emission if the relevant energy levels are efficiently 'pumped' by absorption at shorter wavelengths. The principal emission signatures attributed to cosmic dust arise in silicates and aromatic hydrocarbons.

6.3.1 Silicates

Silicates are ubiquitous and sufficiently robust to survive under a variety of physical conditions. Emission features corresponding to the stretching and bending vibrational modes at 9.7 and 18.5 μm (table 5.1) arise when silicate dust is heated to temperatures of a few hundred Kelvin or more. This commonly occurs in the envelopes of luminous young stars embedded in their parent molecular clouds and in the extended atmospheres of cool, evolved stars with O-rich circumstellar shells. By comparing the emission profiles in such lines of sight with the absorption profiles in colder regions of the ISM, one may hope to gain insight into the properties of silicates as a function of environment. It should be noted, however, that circumstellar emission by warm dust and foreground absorption by cold dust may be superposed in the same line of sight toward an embedded star. A general model for the spectral energy distribution must therefore include both emissive and absorptive terms (Gillett *et al* 1975b), such that the observed flux is

$$F_\lambda \propto Q_\lambda B_\lambda(T_d) \exp(-\tau_\lambda). \tag{6.23}$$

Two emissivity profiles that have been adopted as standards for the 9.7 μm silicate feature are compared in figure 6.7. These are determined from observations of the Trapezium cluster in the Orion nebula and the red supergiant μ Cephei. The emitting grains in the Trapezium region are ambient interstellar (molecular cloud) grains heated by O-type stars; those toward μ Cephei are presumed to have formed in the atmosphere of the star itself. In both cases, the profile is smooth and lacking in obvious structure, suggesting that the silicates are not highly annealed (section 5.2.2). However, there is an obvious difference in width, the Trapezium profile being substantially broader. These profiles have been used to construct models for the corresponding silicate absorption features in other lines of sight (Gillett *et al* 1975b, Roche and Aitken 1984a, Whittet *et*

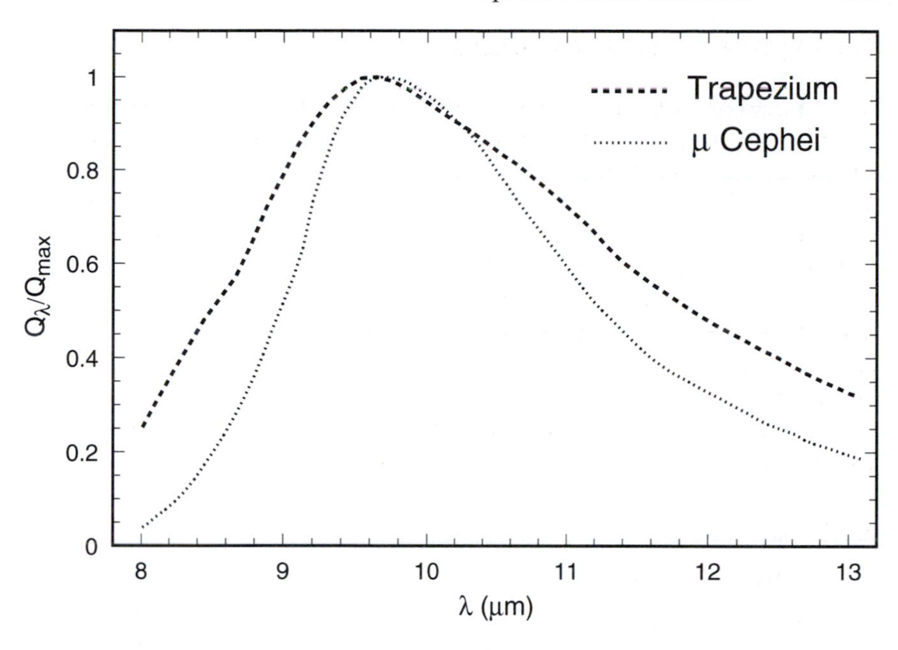

Figure 6.7. Comparison of the 8–13 μm silicate emissivity profile of the Trapezium cluster (broken curve, Gillett *et al* 1975b) with that of the red supergiant μ Cephei (Merrill and Stein 1976a, Roche and Aitken 1984a). In each case, $Q_\lambda \propto F_\lambda / B_\lambda(T_d)$ is calculated from the observed flux (equation (6.8)), assuming a temperature $T_d = 250$ K for the underlying continuum emission from dust.

al 1988, Bowey *et al* 1998). As a general rule, it is found that the μ Cephei profile gives the best fit to observations of the diffuse ISM, the Trapezium profile to observations of cold molecular clouds. It is widely concluded from this that silicates in the diffuse ISM are closely similar to those emanating from red-giant stars. However, the correspondence between the Trapezium profile and the absorption feature in molecular clouds may be fortuitous, as no allowance is made for the $\sim$12 μm ice libration feature that should contribute only in absorption. See section 5.3.7 for further discussion.

The silicate emission features observed in H II regions and circumstellar shells can be understood in terms of classical ($a \sim 0.1$ μm) sized grains that are nevertheless small compared with the wavelengths of the features. Do much smaller particles also contribute? Very small silicate grains are subject to transient heating: if they are widespread, their spectral features should be seen in emission not only in close proximity to stars but also in more widely dispersed regions of the ISM. We might expect, for example, to see 9.7 μm silicate emission superposed on the MIR continuum emission from the galactic disc (section 6.2.2). Whereas emission features attributed to hydrocarbons are, indeed, observed in such locations (see later), those of silicates are below current detection limits.

Calculations based on the graphite–silicate model for interstellar dust thus allow an upper limit to be estimated for the abundance of silicate grains small enough to undergo significant transient heating (Désert *et al* 1986, Li and Draine 2001). According to Li and Draine, no more than $\sim$10% of the interstellar Si can be in silicates with $a < 0.0015$ μm.

6.3.2 Polycyclic aromatic hydrocarbons

Emission features centred at wavelengths of 3.3, 6.2, 7.7, 8.6 and 11.3 μm were first noticed during the pioneering years of infrared spectroscopy, in the spectra of emission nebulae such as compact H II regions and planetary nebulae (e.g. Gillett *et al* 1973, Soifer *et al* 1976, Russell *et al* 1977). Lacking an obvious identification, they became known as the 'unidentified infrared' (UIR) features. A wealth of observational data has subsequently been acquired for them: some representative spectra are shown in figures 6.8 and 6.9. The features have widths in the range 0.03–0.5 μm (FWHM), intermediate between broad solid-state features (e.g. figure 6.7) and the gas-phase atomic and ionic lines characteristic of emission nebulae (e.g. NGC 7027 in figure 6.8). Statistical analyses suggest that they belong to a common generic spectrum, in the sense that when one feature is present, so generally are the others (e.g. Cohen *et al* 1986, 1989, Bregman 1989), but source-to-source variations do occur in their relative strengths. The total number of features and subfeatures in the family is at least 20 (Beintema *et al* 1996), although the weaker ones have been detected only in a few bright sources. The principal features (table 6.1) are observed in a wide variety of galactic environments, including not only emission nebulae but also reflection nebulae, high-latitude cirrus clouds and the galactic disc (e.g. Sellgren *et al* 1996, Lemke *et al* 1998, Mattila *et al* 1996). They are also conspicuous in the spectra of some external galaxies, especially those with vigorous star formation (Roche 1989b, Acosta-Pulido *et al* 1996, Genzel and Cesarsky 2000). The universality of the phenomenon is illustrated by the comparison in figure 6.9 between diffuse emission from the disc of the Milky Way and that from a starburst galaxy: after correction for relative Doppler motion, the spectra are almost identical.

The 'UIR' features remain unidentified, in the sense that specific lines lack specific molecular identifications. However, the UIR spectrum as a whole is confidently assigned to a *class* of molecules, i.e. to polycyclic aromatic hydrocarbons (PAHs) or to small grains containing PAHs (Duley and Williams 1981, Léger and Puget 1984)[3]. Because of this, the term 'aromatic infrared' is preferred as a description of the generic spectrum. PAHs are planar molecules composed of linked hexagonal rings, with H atoms or other radicals attached to the outer bonds of peripheral C atoms. The molecular structure of a single ring

[3] Since their first association with the UIR spectrum, a wealth of information has been published in the astronomical literature on the properties of PAHs. For extensive reviews, see Omont (1986), Allamandola *et al* (1987, 1989), Puget and Léger (1989), Duley (1993), Tielens (1993), d'Hendecourt (1997) and Salama (1999).

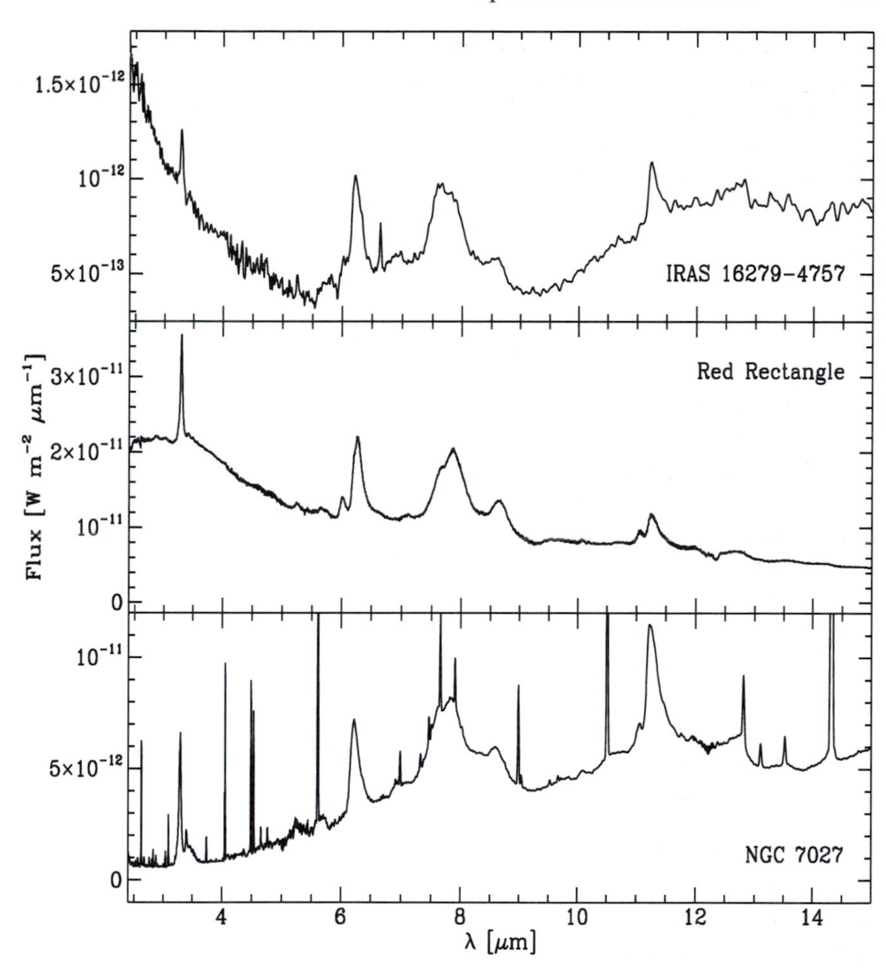

Figure 6.8. Infrared spectra of three evolved stars displaying aromatic infrared features: the post-AGB stars IRAS 16279–4757 and HD 44179 (the Red Rectangle) and the planetary nebula NGC 7027. The features listed in table 6.1 are visible in each spectrum. Additional sharp, narrow features prominent in NGC 7027 are gaseous nebular emission lines. The data were obtained with the SWS instrument on board the Infrared Space Observatory (Tielens *et al* 1999).

(benzene) and a simple four-ringed PAH (pyrene) are illustrated in figure 6.10. The infrared spectra of PAHs are characterized by a number of resonances, the strongest of which are listed in table 6.1. Identification of the interstellar features with PAHs is based on wavelength correspondence between observed and laboratory spectra (figures 6.11 and 6.12). It is of interest to note, however, that the observed emission features do not generally correspond to resonances

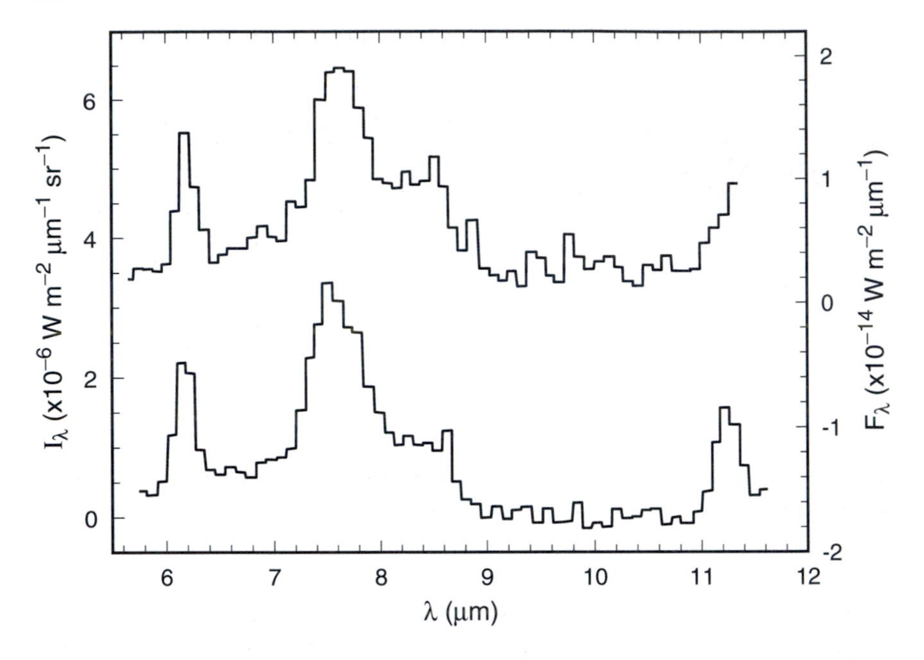

Figure 6.9. Aromatic emission features compared in two galaxies: lower curve (left-hand scale), surface-brightness spectrum of the disc of the Milky Way at $\ell = -30°$, $b = 0°$ (Mattila *et al* 1996); upper curve (right-hand scale), flux spectrum of the starburst galaxy NGC 6090 (Acosta-Pulido *et al* 1996). The wavelength-scale of the NGC 6090 spectrum has been corrected for the redshift of that system. PAH features centred at 6.2, 7.7, 8.6 and 11.3 μm are present. Both spectra were obtained with the ISOPHOT-S low-resolution spectrometer of the Infrared Space Observatory.

Table 6.1. Observed and laboratory wavelengths and assignments of the principal spectral features arising from vibrational modes in polycyclic aromatic hydrocarbons (Allamandola *et al* 1989).

λ_{obs} (μm)	λ_{lab} (μm)	Assignments
3.29	3.29	C–H stretch
6.2	6.2	C–C stretch
7.7	7.6–8.0	C–C stretch
8.6	8.6–8.8	C–H in-plane bend
11.3, 12.7	11.2–12.7	C–H out-of-plane bend

commonly seen in absorption in the ISM (chapter 5; see also Sellgren *et al* (1995) for an exception). This indicates that PAHs, although pervasive, are not major

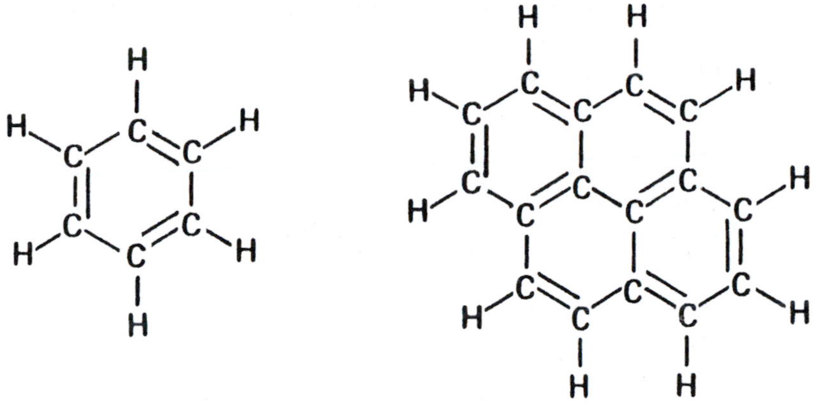

Figure 6.10. The molecular structure of benzene (left) and the four-ringed polycyclic aromatic hydrocarbon pyrene (right).

contributors to the mass density of interstellar grains. The emission spectrum can be understood if PAHs account for some 5–10% of the total carbon abundance of the ISM (e.g. Allamandola *et al* 1989, Giard *et al* 1994). A greater contribution from PAHs is probably excluded by the lack of detectable absorption features, not only in the infrared but also in the ultraviolet (see problem 8 at the end of chapter 3).

The common thread that links the various environments that give rise to the aromatic infrared features is the presence of carbonaceous dust exposed to ultraviolet radiation. Post-main-sequence stars generally lack an intrinsic source of UV, except in cases where a hot binary companion is present. However, as a red giant evolves from the asymptotic giant branch to become a planetary nebula (section 7.1), its expanding envelope is exposed to UV photons from the hot central core. Aromatic features are widely observed in PNe, with a strength that correlates strongly with the C/O abundance ratio of the envelope (Cohen *et al* 1986, Roche 1989a, Roche *et al* 1996). The features are invariably present in objects with $\mathrm{C} > \mathrm{O}$ but weak or absent when $\mathrm{C} < \mathrm{O}$, confirming the C-rich nature of the carrier. The widespread occurrence of the features in H II regions, reflection nebulae and cirrus clouds is an indicator that PAHs become well mixed with other forms of dust in the ISM (Giard *et al* 1994). The spectrum is produced most efficiently if the incident radiation is intense but relatively soft, i.e. if energetic photons beyond the Lyman limit are lacking. This occurs, for example, in the photodissociation regions surrounding H II regions as well as in reflection nebulae surrounding late B-type stars. Under such conditions, PAHs will be stochastically heated as described in section 6.1.4, reaching temperatures of up to $\sim$1000 K, and excitation of vibrational energy levels then leads to fluorescent emission of the aromatic spectrum. In the vicinity of ionization fronts, the strength of the features

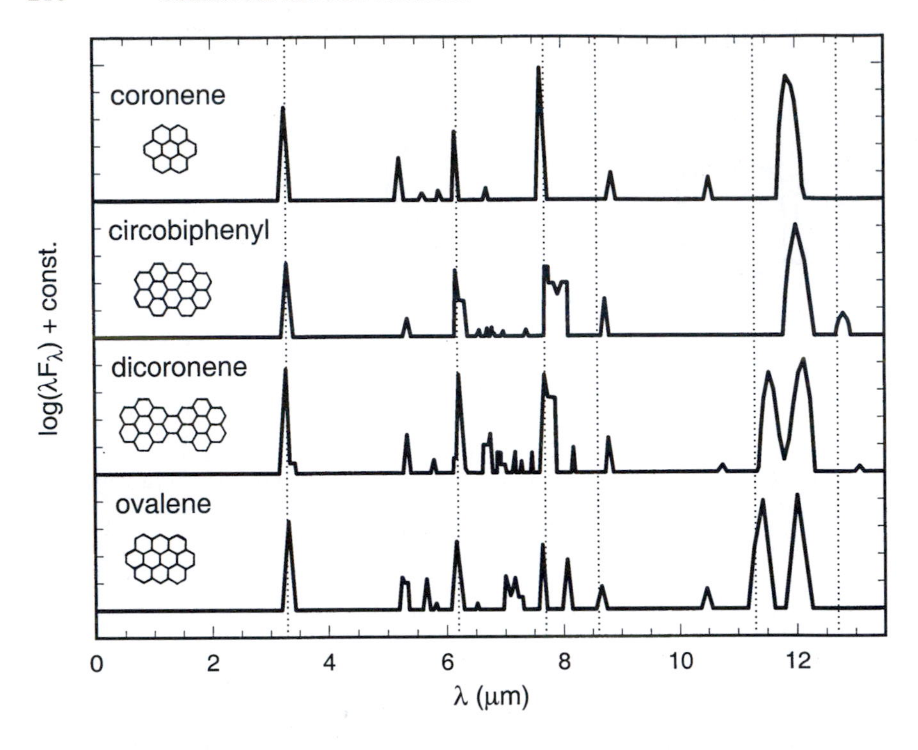

Figure 6.11. Predicted emission spectra of several PAHs at 850 K. The structure of each PAH is shown to the left of the spectrum. The vertical dotted lines indicate the positions of the principal observed features, at wavelengths 3.3, 6.2, 7.7, 8.6, 11.3 and 12.7 μm. (Adapted from Léger and d'Hendecourt 1988.)

is anticorrelated with that of nebular emission lines arising in the ionized gas (Roche *et al* 1989a), indicating that the carriers are depleted inside the boundaries of the H II region. The susceptibility of PAHs to destruction in such environments leads to an estimate of their typical size, which Geballe *et al* (1989b) find to be equivalent to 20–50 C atoms.

A wide variety of PAHs must presumably be present in astrophysical environments, although the most stable (symmetrical and condensed) forms will naturally tend to be favoured[4]. Individual PAHs in a given environment will be subject to different degrees of excitation depending on their absorption cross sections and molecular weights and on whether they are free or bound into larger structures; they will also be subject to differing degrees of ionization and hydrogenation (Allamandola *et al* 1989, Tielens 1993, Schutte *et al* 1993, Cook and Saykally 1998, Bakes *et al* 2001). All these factors may affect the observed

[4] An example of a PAH mixture composed of the most stable molecular forms is the soot produced by high-temperature combustion of hydrocarbons (Allamandola *et al* 1985), as in automobile exhaust.

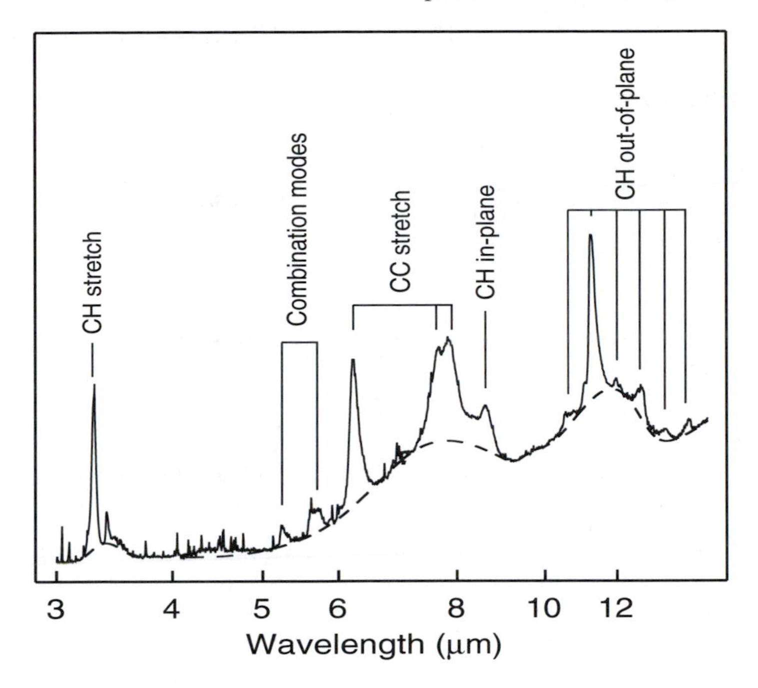

Figure 6.12. Assignment of aromatic features in the spectrum of the planetary nebula NGC 7027 to vibrational modes in PAHs. Figure courtesy of Emma Bakes, adapted from Bakes *et al* (2001).

spectrum. The spectra illustrated in figure 6.11 have a number of similarities but also some important differences. The features at 3.3 and 6.2 μm are relatively stable in position and should have a similar appearance in different mixtures of PAHs; variations from one PAH to another are more evident at the longer wavelengths. The 7.6–8.0 μm region contains several strong, overlapping C–C stretching modes that provide a natural explanation for the width and apparent duplicity of the observed 7.7 μm feature (see figure 6.12). Out-of-plane C–H bending modes in the 11–13 μm region of the spectrum are particularly sensitive to the geometry and degree of hydrogenation of the PAH molecule. Isolated peripheral H atoms tend to vibrate at higher frequency compared with those whose neighbouring sites are occupied (Allamandola *et al* 1989). The observed features at 11.3 and 12.7 μm may thus be associated with different degrees of hydrogenation, and the prominence of the 11.3 μm feature suggests that interstellar PAHs are not fully hydrogenated.

Detailed studies of the emission profiles reveal the presence of fine structure and underlying emission plateaux (Allen *et al* 1982, de Muizon *et al* 1986, Roche *et al* 1989a, 1991, Jourdain de Muizon *et al* 1990, Roelfsema *et al* 1996). However, for the most part, this complexity appears to arise from blending of

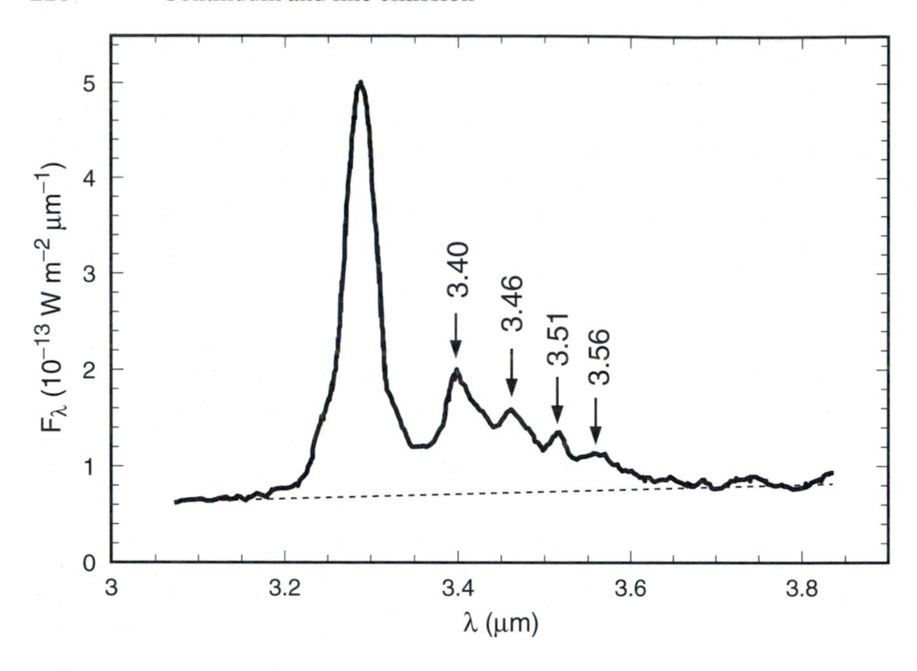

Figure 6.13. The 3.1–3.8 μm spectrum of the planetary nebula IRAS 21282+5050 at a resolution of 0.008 μm (Geballe *et al* 1994 and references therein). The dominant emission feature at 3.29 μm is accompanied by weaker features labelled by their wavelengths in microns. The broken line represents the assumed continuum underlying the emission features.

multiple features rather than intrinsic structure within individual features. The shape of each feature is well represented by the Lorentzian profile:

$$f(x) = \frac{1/(\pi\sigma)}{1 + \{(x - x_0)/\sigma\}^2} \tag{6.24}$$

where $x = \lambda^{-1}$ and x_0 and σ denote the position and width of the feature. Boulanger *et al* (1998) show that the observed spectrum in the 5–15 μm region can be explained entirely by superposition of multiple features with differing positions, amplitudes and widths but with identical profiles given by this functional form.

As an example of multiplicity, the 3.1–3.8 μm spectrum of the planetary nebula IRAS 21282+5050 is shown in figure 6.13. The strong 3.29 μm feature is accompanied by a weaker feature centred at 3.40 μm. Further structure is apparent in the 3.3–3.6 μm region, with weaker features at 3.46, 3.51 and 3.56 μm superposed on an underlying plateau. Comparison of these features in different sources suggests that a hierarchy governs their occurrence: considering the sequence (i) 3.29 μm, (ii) 3.40 μm, (iii) 3.3–3.6 μm plateau and (iv) the weaker features, each successive group requires the presence of all the previous ones to be seen. The principal feature at 3.29 μm is identified with the $v = 1 \rightarrow 0$ aromatic C–H stretch; excitation of higher vibrational levels should result in additional emissions in the 3.4–3.6 μm region, associated with transitions such as $v = 2 \rightarrow 1$ and $v = 3 \rightarrow 2$ (Barker *et al* 1987). It is thus tempting to associate the weaker observed features with these so-called 'hot bands', as their intensities relative to the principal feature should vary with environment, suggesting an explanation of the observed hierarchy (Geballe *et al* 1989b). However, detection of the $v = 2 \rightarrow 0$ overtone band at 1.68 μm leads to precise determination of the $v = 2 \rightarrow 1$ wavelength (3.43 μm; Geballe *et al* 1994), suggesting that it contributes no more than minor substructure to the long-wavelength wing of the 3.40 μm feature (figure 6.13). Emission from aliphatic subgroups attached to PAHs is a possible alternative (Duley and Williams 1981, Jourdain de Muizon *et al* 1990, Joblin *et al* 1996, Wagner *et al* 2000). Joblin *et al* propose that the 3.40 μm feature arises in PAHs with $-CH_3$ attachments.

A few sources display anomalous emission in the 3–4 μm spectral region. The primary characteristic of this group is the presence of a very strong feature centred at 3.53 μm, accompanied by weaker features at 3.43 and 3.29 μm. The positions and widths of the features are similar to those in sources with more typical 3–4 μm spectra (e.g. figure 6.13) but the relative strengths are reversed, with 3.53 μm very strong compared with 3.29 μm. This phenomenon was first seen in HD 97048, a pre-main-sequence Ae/Be star that illuminates a reflection nebula in the Chamaeleon I dark cloud (Blades and Whittet 1980). An extensive search revealed only a few additional cases, including another Ae/Be star and a protoplanetary nebula (Allen *et al* 1982, Whittet *et al* 1984, Geballe *et al* 1989a, Geballe 1997), suggesting that specialized physical conditions are necessary for it to occur. An important difference compared with normal PAH emission is that the 3.53 μm feature toward HD 97048 is not spatially extended in the nebula but appears to be formed very close to the star, presumably in its circumstellar shell (Roche *et al* 1986b). The profiles of the 3.43 and 3.53 μm features are well matched by laboratory data for formaldehyde (H_2CO) in a cold matrix (Baas *et al* 1983) but the presence of H_2CO on dust is highly implausible in an environment where temperatures $T_d > 1000$ K are expected (Roche *et al* 1986b). C–C overtone and combination bands in highly excited PAHs or amorphous carbon particles (Schutte *et al* 1990) may offer a more likely explanation.

We have seen that the aromatic infrared features provide interesting constraints on the physics and chemistry of dust in our Galaxy and we conclude this section with a brief assessment of how this knowledge may further our understanding in the broader context of extragalactic astronomy. The very occurrence of these features in the spectra of other galaxies (e.g. figure 6.9) proves the existence of PAHs or PAH-based dust grains in these systems that respond similarly to irradiation and are presumably created and destroyed in similar ways. Aromatic features appear to be universal spectral signatures of photodissociation regions driven by OB stars, the most luminous manifestations of star formation.

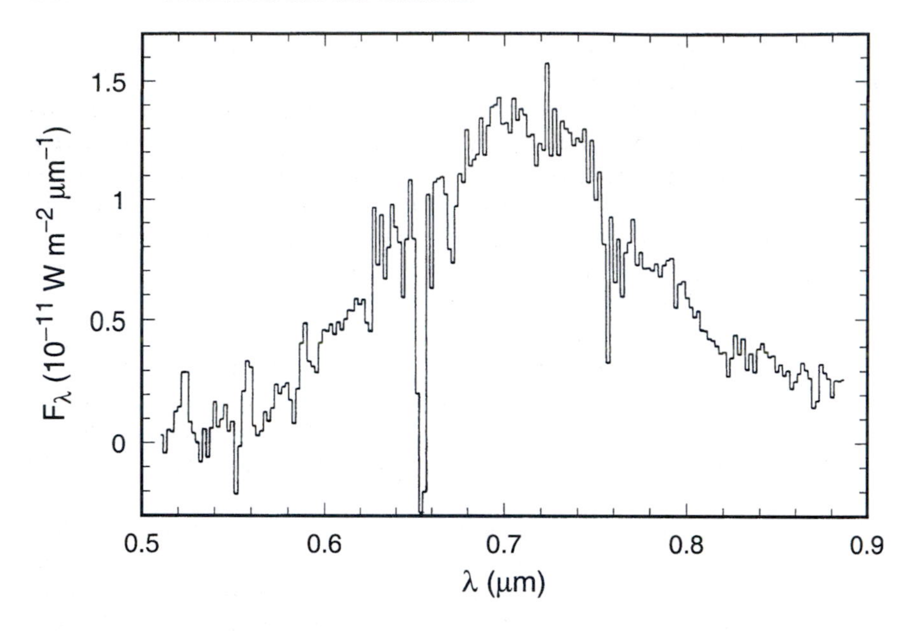

Figure 6.14. The profile of extended red emission observed in the reflection nebula NGC 2327, displayed after subtraction of the component due to scattering (Witt 1988).

Because of this, they are expected to occur in any system with significant star-formation activity and to provide a valuable tracer of such activity in galaxies generally (Mizutani *et al* 1989, 1994). They may be used, for example, as a discriminator between starburst galaxies (in which they are prominent) and active galaxies (in which they are weak or absent). Whereas the aromatic features are a defining characteristic of starburst systems (Genzel and Cesarsky 2000), the carriers are presumably destroyed by hard UV and x-ray radiation in the nuclear regions of active galaxies (Aitken and Roche 1985, Désert and Dennefeld 1988, Voit 1992). Finally, because of their visibility, not only in starburst galaxies but in normal galaxies with significant star formation, the PAH features are natural choices for evaluation of redshift in such systems at cosmological distances (Helou *et al* 2000).

6.4 Extended red emission

Many solids, including materials considered as probable constituents of interstellar grains, emit visible photoluminescence when exposed to energetic radiation (Witt 1988). A search for photoluminescence from interstellar dust revealed the presence of a broad emission band centred between 0.6 and 0.8 μm, now known as the extended red emission (ERE). An example of the observed profile is shown in figure 6.14. The wavelength of the peak shows spatial variations (Witt and Boroson 1990) and tends to increase with the hardness and intensity of the radiation field. An inventory of environments in which the ERE is observed is essentially the same as that for the aromatic infrared features discussed in the previous section: reflection nebulae, planetary nebulae, H II regions, cirrus clouds, the diffuse ISM and some external galaxies (see Witt *et al* 1998 for a review). The carriers of these two phenomena clearly coexist and share similar sources of excitation. However, they do not appear to be one and the same, as the intensities of their emissions are poorly correlated (Perrin and Sivan 1990, Furton and Witt 1992). Only neutral PAHs are expected to show photoluminescence (Allamandola *et al* 1987) and those occurring in nebular environments have a high probability of being ionized (Witt and Schild 1988). The ERE appears likely to be an important diagnostic of some component of interstellar dust. However, both C-rich and SiO-rich materials display photoluminescence and it is not yet clear which is the dominant source of the ERE.

The case for a carbonaceous carrier is strongly supported by the fact that, amongst a substantial sample of planetary nebulae, the ERE is seen preferentially in those that are C rich (i.e. $C > O$; Furton and Witt 1992). Although this result does not absolutely exclude an SiO-based carrier (Witt *et al* 1998), it is difficult to understand how such a material can be consistently favoured in a medium where the natural condensates are C rich and disfavoured in a medium where they are O rich (section 7.1). Laboratory investigations demonstrate that hydrogenated amorphous carbon (HAC) exhibits broad photoluminescence bands in the visible (Watanabe *et al* 1982) with a profile similar to that of the observed ERE. The degree of hydrogenation determines the position of peak emission by controlling the band-gap energy (ε_g) of the material (Duley and Williams 1988, 1990):

$$\lambda_{\text{peak}} \approx \frac{hc}{\varepsilon_g}. \tag{6.25}$$

Thus, energies in the 1.5–2.0 eV range are needed to explain the observations. The observed shift in position can be understood in terms of varying hydrogenation, requiring the grains to lose their H most effectively in the harshest radiation fields. However, this result leads to an objection to the HAC hypothesis. The efficiency as well as the spectral range of the photoluminescence is a function of hydrogenation and greatest efficiency is reached when the carbon is fully hydrogenated (Robertson 1996). But the band-gap energy is then $\sim$4 eV and so λ_{peak} is in the near ultraviolet, well beyond the range of the observed ERE. Dehydrogenation shifts the peak into the required range but this is accompanied by an exponential decline in efficiency, by a factor approaching 10^4. Thus, to explain both the spectral range and the amplitude of the ERE, it is necessary not only to place stringent constraints on the degree of hydrogenation but also to require that virtually all the carbon in interstellar dust contributes to the luminescence. For further discussion of issues related to the HAC hypothesis, see Seahra and Duley (1999), Gordon *et al* (2000) and Duley (2001).

A silicon-based carrier of the ERE was proposed by Witt *et al* (1998). Silicon nanoparticles produced in the laboratory have been found to be remarkably efficient emitters of photoluminescence in the 1.5–2.0 eV energy range (Wilson *et al* 1993). The particles consist of crystalline silicon cores with oxidized (SiO_2-coated) surfaces, a few nm in diameter. Witt *et al* hypothesize that they might resemble particles condensing in the outflows of cool, O-rich stars (section 7.1). Their size and structure appear to be crucial in governing the position and intensity of emission. Whereas bulk silicon is a rather inefficient source of luminescence peaking in the near infrared, these $Si{:}SiO_2$ nanoparticles have greatly enhanced efficiency, optimized in the red for particles $\sim$2 nm in diameter. Variation in λ_{peak} can be interpreted as evidence for variation in mean size and the particles are restricted to the size range 1.5–5 nm by the observed range in λ_{peak}. There are two important objections to the $Si{:}SiO_2$ hypothesis. The first has already been noted: selective occurrence of the ERE in C-rich planetary nebulae seems inconsistent with an O-rich carrier. The second issue concerns the size requirement, in relation not only to the ERE itself but also to emission features expected in the mid-infrared. As with silicates (section 6.3.1), observed limits on the strengths of emission features can be used to limit the abundance of small particles. To explain the ERE, Witt *et al* (1998) estimate that as much as $\sim$40% of the depleted Si must be in 1.5–5 nm nanoparticles and a substantial fraction of these atoms must be tied up in Si–O bonds that produce vibrational features near 10 and 20 μm. Li and Draine (2002) calculate the expected thermal emission spectrum of such particles and show that strong emission is, indeed, predicted (with a peak near 20 μm) that has no observed counterpart in regions where the ERE is detected. Free-flying $Si{:}SiO_2$ nanoparticles thus appear to be rare and to make a negligible contribution to the observed ERE. This objection might be alleviated for nanoparticles attached to the surface of a larger grain, as the temperature (and hence the MIR emission strength) would then be moderated by that of the host particle.

Recommended reading

- *The Determination of Cloud Masses and Dust Characteristics from Submillimetre Thermal Emission*, by R H Hildebrand, in Quarterly Journal of the Royal Astronomical Society, vol 24, pp 267–82 (1983).
- *A New Component of Interstellar Matter: Small Grains and Large Aromatic Molecules*, by J L Puget and A Léger, in Annual Reviews of Astronomy and Astrophysics, vol 27, pp 161–98 (1989).
- *The First Symposium on the Infrared Cirrus and Diffuse Interstellar Clouds*, ed R M Cutri and B Latter (Astronomical Society of the Pacific Conference Series, vol 58, 1994).
- *Detection and Characterization of Cold Interstellar Dust and Polycyclic Aromatic Hydrocarbon Emission from COBE Observations*, by E Dwek *et al*, in Astrophysical Journal, vol 475, pp 565–79 (1997).

- *A Primer on Far-Infrared Polarimetry*, by R H Hildebrand *et al*, in Publications of the Astronomical Society of the Pacific, vol 112, pp 1215–35 (2000).

Problems

1. Show that the flux density emitted by an interstellar cloud varies as $F_\lambda \propto \lambda^{-(\beta+4)}$ at wavelengths sufficiently long that the Rayleigh–Jeans approximation may be applied. Derive an equivalent expression for the frequency form of flux density (F_ν) as a function of ν.
2. The far infrared spectrum ($\log F_\lambda$ versus $\log \lambda$) of an interstellar clump is found to peak at $\lambda \approx 75$ μm. The flux declines steadily into the submillimetre region and a slope of -5.7 is measured at $\lambda = 500$ μm. Assuming that the Rayleigh–Jeans approximation is applicable at 500 μm, evaluate the spectral index of the dust emissivity and the mean temperature of the dust. Show or discuss with reasoning whether the Rayleigh–Jeans approximation is actually a reasonable assumption in this case.
3. Define the Debye temperature (Θ) of a solid and explain its significance with respect to the thermal properties of interstellar grains. Suppose that the ISM contains two types of (spherical) grain, silicates of radius 0.1 μm and carbon particles of radius 1 nm, each with $\Theta \sim 500$ K. Assuming that both types of dust initially have kinetic temperature $T_{\rm i} = 15$ K, calculate the temperature increase ΔT in each case when a 13.6 eV photon is absorbed. Hence estimate the wavelength of peak emission from each grain type, assuming that the effective mean temperature is $T_{\rm i} + \Delta T/2$ and that the dust emissivity is given by $Q_\lambda \propto \lambda^{-1.5}$.
4. Verify equation (6.15) by relating the degree of polarization for radiation emitted by an ensemble of aligned grains (equation (6.14)) to the ratio of polarization to extinction for transmitted radiation (equation (4.4)), assuming that the polarization is small.
5. Discuss briefly how far infrared emission from dust may be used to estimate the dust-to-gas ratio. What are the main sources of error in determinations of $Z_{\rm d}$ based on this method?
6. The COBE Diffuse Infrared Background Experiment measured a surface intensity $I_\nu = 160$ MJy/sr at wavelength $\lambda = 240$ μm in a beam centred on Galactic coordinates $\ell = 90°$, $b = 0°$. Calculate the optical depth of the dust at this wavelength in this line of sight, assuming a dust temperature of 18 K. Hence estimate the dust-to-gas ratio if the total hydrogen column density in the line of sight is measured from H I and CO data to be $N_{\rm H} = 1.2 \times 10^{26}$ atoms m^{-2}. Assume a mass absorption coefficient $\kappa_{240} = 0.71$ m^2 kg^{-1} for the dust. (See appendix A for the relation between surface intensity and flux density in their frequency and wavelength forms.)

Chapter 7

Dust in stellar ejecta

"...a complex interplay of gas dynamics, chemical kinetics and radiative transfer."

M Jura (1989)

The evolution of interstellar matter in galaxies is intimately linked with that of the stars they contain. Stars may be associated with dust at almost any phase of their lifecycle from protostellar youth to cataclysmic old age. They form in dense molecular clouds (chapter 9). Dusty circumstellar shells or discs are ubiquitous to the earliest stages of their evolution, emitting copious infrared radiation and providing raw materials for the accretion of planetary systems. Subsequently, much of this placental material is returned to the interstellar medium but, in a number of cases, unaccreted remnants of protostellar discs survive around mature main-sequence stars. After leaving the main sequence, stars may evolve to become significant sources of new grain material, formed in their winds during extended period of high mass-loss or in isolated cataclysmic events such as nova or supernova outbursts. The cool atmospheres of luminous red stars provide environments especially conducive to the rapid nucleation and growth of refractory dust grains. This 'stardust' may be C rich or O rich, dependent on the evolutionary state of the parent star. Once formed, the dust is driven outward by radiation pressure; ultimately, it reaches and merges with, the surrounding interstellar medium.

In this chapter, we begin (section 7.1) by examining the physical conditions and chemical reactions that control the formation of dust in stellar outflows. The observed properties of stardust are reviewed in section 7.2 and the final section assesses of the significance of dust in stellar ejecta as a source of interstellar dust.

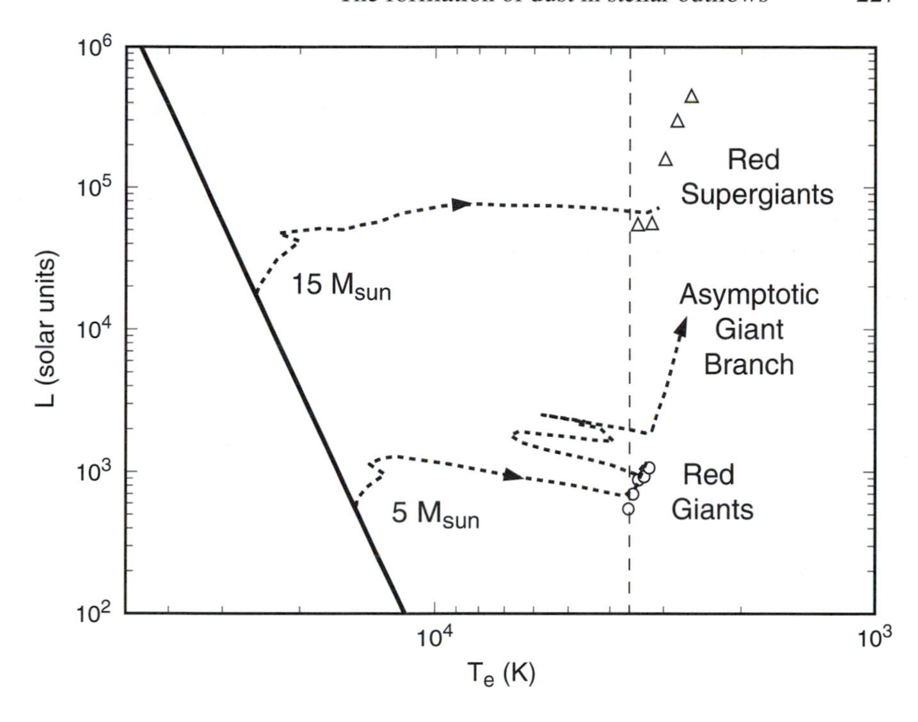

Figure 7.1. Hertzsprung–Russell diagram (effective temperature versus luminosity) showing the post-main-sequence evolution of typical dust-forming stars. The main sequence is represented by the full diagonal line. Evolutionary tracks are shown for stars of 5 and 15 solar masses (broken curves; Iben 1967). The mean loci of stars in the standard spectral-type sequence from M2 to M6 are plotted for red giants (luminosity class III, circles) and supergiants (luminosity class I, triangles). The vertical broken line indicates an effective temperature of 3600 K, below which stars typically begin to make dust. A 5 $M_\odot$ star may cross this threshold briefly during its initial (first ascent) red-giant phase but the principal episode of dust production occurs later when it reaches the asymptotic giant branch.

7.1 The formation of dust in stellar outflows

7.1.1 Theoretical considerations

The possibility of grain nucleation in stellar atmospheres was first discussed many years ago by Hoyle and Wickramasinghe (1962), Kamijo (1963) and Gilman (1969). Luminous post-main-sequence stars with photospheric temperatures between 2000 and 3500 K were identified as likely candidates, i.e. objects that lie in the upper right-hand region of the Hertzsprung–Russell (HR) diagram (figure 7.1). This group includes red giants, red supergiants and asymptotic giant branch (AGB) stars[1]. Close to their photospheres, the temperature is generally too high for solids to exist. Indeed, some red giants have chromospheres with temperatures as high as $\sim$10 000 K (Gail and Sedlmayr 1987). Beyond the chromosphere, the temperature declines monotonically with radial distance. The inner boundary of the dust shell is specified by the distance r_1 at which the gas temperature falls below the condensation temperature T_C of the dominant dust component. For a star of radius R_s forming grains with $T_C = 1000$ K, Bode (1988) estimates that $r_1 \approx 10R_s$. The outer boundary of the shell is specified by the distance r_2 at which the density and temperature of the circumstellar material become comparable with those of average interstellar matter, which may typically be 10^4–$10^5 R_s$. It is easily shown that the number density n of the gas in a stellar wind declines due to dilution with radial distance r from the star, such that $n \propto r^{-p}$, with $p = 2$ for uniform expansion at constant speed. Grain formation is thus expected to be most rapid in the zone immediately beyond the inner radius r_1: here, densities are typically $n \sim 10^{19}$ m^{-3}, many orders of magnitude greater than those characteristic of interstellar clouds. It is this combination of physical conditions – relatively high densities coupled with temperatures comparable with the condensation temperatures of many solids – that renders red-giant atmospheres particularly favourable sites for rapid grain formation. It should also be borne in mind that evolved stars often have pulsational instabilities, leading to periodic variations in the physical conditions at a given radial distance, and these may induce cyclic phases of grain nucleation, growth and ejection.

The growth of solid particles in an initially pure gas that undergoes cooling has been discussed in terms of classical nucleation theory (e.g. Salpeter 1974), originally formulated to describe the condensation of liquid droplets in the Earth's atmosphere. In this case, it must be assumed that the system maintains thermal equilibrium as the gas flows away from the star. Condensation of some species X then occurs when its partial pressure in the gas exceeds the vapour pressure of X in the condensed phase; and is most efficient at temperatures appreciably below the nominal condensation temperature. We refer to an individual unit of X in the gas phase (an atom or molecule) as a *monomer*. Random encounters between monomers lead to the formation of clusters; for equilibrium, the number density of clusters containing i monomers is

$$n_i = n_1 \exp(-\Delta E_i/kT) \tag{7.1}$$

(e.g. Dyson and Williams 1997: pp 54–5) where n_1 is the number density of individual monomers and ΔE_i is the thermodynamic free energy associated with the formation of the cluster. For small clusters, ΔE_i increases with i and thus n_i decreases. Above some critical size, however, the addition of further monomers is energetically favourable; the clusters become stable and grow rapidly to some

[1] For convenience, the term 'red giant' is sometimes used generically to refer to all types of cool post-main-sequence star. The term 'first ascent red giant' distinguishes those on the true red giant branch (luminosity class III) from red supergiants and AGB stars.

maximum size limited by the availability of monomers. Classical nucleation is thus a two-step process: (i) the formation of clusters of critical size and (ii) the growth of these clusters into macroscopic grains or droplets.

Although a useful starting point, classical nucleation theory is limited in its ability to realistically describe the formation of dust grains in stellar atmospheres. In a red-giant wind, where pressures are typically at least a factor of 10^6 less than in the terrestrial atmosphere, the timescale for nucleation may be longer than the timescale in which the physical conditions undergo appreciable change, and condensation is not then an equilibrium process (Donn and Nuth 1985, Nuth 1996). Moreover, solid particles are held together by strong valence bonds rather than by the weak polarization forces that bind liquid droplets, and the formation of grains thus involves chemical reactions as well as physical clustering (Salpeter 1974). Whereas classical theory describes the nucleation and growth of particles that are chemically identical to the original monomers, chemical reactions integral with the nucleation process may lead to the formation of solids (such as silicates) for which no equivalent monomer is available in the gas (Gail and Sedlmayr 1986). Because of the stochastic nature of the growth process, these solids will tend to be highly disordered and chemically unsaturated (Nuth 1996); crystalline or polycrystalline structure will appear only if the grain is heated, either during the growth process or at some later time.

7.1.2 The circumstellar environment

In general, the expanding envelopes of red giants contain a mix of ionic, atomic and molecular gas, with physical conditions controlling both the ionization state and the molecular composition as a function of radial distance from the star (Glassgold 1996). In the grain condensation zone, all relevant monomers are assumed to be neutral, i.e. there is no significant source of ionizing radiation. In cases where this is not so (e.g. where there is significant UV flux from the chromosphere or from an early-type binary companion), grain formation will tend to be inhibited because positively charged monomers repel each other. CO is the most abundant of all molecules containing heavy elements. It forms readily and is stable in the gas phase in stellar atmospheres at temperatures below about 3000 K. The binding energy of CO (11.1 eV) is sufficiently high that the bond cannot readily be broken in simple gas-phase reactions. Thus, despite its status as a radical, CO is chemically rather unreactive. It is also very volatile and cannot itself condense into solids in warm circumstellar environments. The abundance of CO in the gas is thus limited, in most situations, only by the abundances of the constituent atoms. This has dramatic consequences for the chemistry of dust condensation: whichever element out of C and O is less abundant will remain predominantly in the gas phase, within the CO molecules and will hence be unavailable to form solids. The abundance ratio $N(\mathrm{C})/N(\mathrm{O})$ (hereafter abbreviated C/O) is thus crucial in determining the composition of the dust.

Another molecule of some interest in the same context is N_2. Like CO, N_2

is volatile, stable and chemically unreactive, and is thus confined to the gas phase in conditions pertaining in circumstellar shells. Essentially all of the nitrogen present in the gas is, therefore, expected to be locked up in N_2 (irrespective of C/O ratio) and this abundant element is thus generally unavailable to form solids.

Red giants that show observational evidence for the presence of dust (see section 7.2) may be placed into two broad categories: those with approximately solar abundances (normal M-type stars with $C/O < 1$) and those enriched in carbon to the extent that $C/O > 1$. As discussed in section 2.1, carbon enrichment occurs when ^{12}C, the product of helium-burning, is dredged up to the surface. In the following discussion, it is assumed that a C-rich stellar atmosphere can be represented by a gas in which the abundance of ^{12}C is enhanced but that otherwise has approximately solar composition. In general, this is likely to be a reasonable approximation for non-binary carbon stars in the solar neighbourhood of the galactic disc, but it should be noted that some stars with carbon enrichment are also hydrogen deficient and some are metal poor. For solar composition, the C/O ratio is about 0.5 (table 2.1) and we thus expect all of the C and approximately half of the O to be tied up in CO in M stars, leaving the remaining O free to become bonded into molecules that may condense into solids. In a carbon star ($C/O > 1$), the roles are reversed. We discuss each of these cases in turn. As a red giant evolves, it may undergo successive phases of mass-loss, possibly involving distinct episodes of O-rich and C-rich grain formation. Note that if the degree of C-enhancement were such that C/O were precisely unity, both C and O would be tied up fully in gaseous CO and would effectively block each other from inclusion in solids.

7.1.3 O-rich stars

The temperature–pressure (T, P) phase diagram for an atmosphere of solar composition is illustrated in figure 7.2. CO is stable in the gas in all regions of the diagram above the thin dotted curve. Consider a pocket of gas which is steadily cooled, such that its locus in the T, P diagram follows the direction of the curved arrow in figure 7.2 as it expands outward from the star. The most abundant monomers that lead to the production of solids are expected to be Fe, Mg, SiO and H_2O. Initially, at $T > 1500$ K, these remain in the gas and only rare metals such as tungsten and refractory oxides such as corundum (Al_2O_3) and perovskite ($CaTiO_3$) are stable in the solid phase. Although contributing little in terms of mass, these high-temperature condensates might facilitate the deposition of more abundant solids by providing nucleation centres (Onaka *et al* 1989). The major condensation phase occurs at temperatures $1200 \rightarrow 800$ K with the nucleation and growth of amorphous SiO clusters. These clusters chemisorb other monomers and subsequent annealing may result in the growth of linked SiO_3 chains with attached Mg cations (enstatite, $MgSiO_3$) or individual SiO_4 tetrahedra joined by cations (fosterite, Mg_2SiO_4). Note that magnesium silicates appear to anneal at a faster rate than iron silicates. In an atmosphere of roughly solar composition,

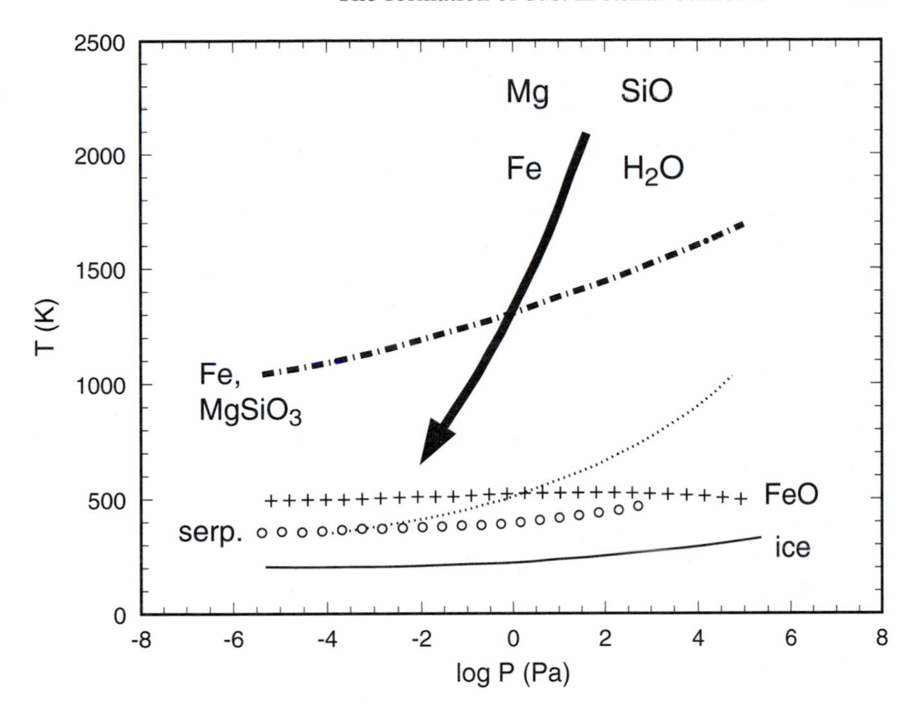

Figure 7.2. Temperature–pressure phase diagram illustrating stability zones of major solids in an atmosphere of solar composition (adapted from Salpeter 1974, 1977, Barshay and Lewis 1976). Above the thin dotted curve, gas-phase CO is stable and essentially all the carbon is locked up in this molecule. The most abundant gas-phase reactants that lead to the production of solids are Mg, SiO, Fe and H_2O. The curved arrow indicates the variation in physical conditions that may occur in the outflow of a typical red giant. Magnesium silicates and solid Fe condense below the bold dot–dash curve. At much lower temperatures, Fe is fully oxidized to FeO (below the curve marked +++) and may then become incorporated into silicates. Hydrous silicates such as serpentine are stable below the curve marked ∘ ∘ ∘. Finally, H_2O-ice condenses below the continuous curve.

one might expect olivine ($(Mg,Fe)_2SiO_4$) to form as the thermodynamically stable end-product, but a kinetically controlled formation process tends to favour SiO_2, $MgSiO_3$, Mg_2SiO_4 and metallic Fe because such species appear to form more rapidly, and there may be insufficient time to reach the most energetically favourable configuration (Nuth 1996, Rietmeijer *et al* 1999). At $T \sim 700$ K, essentially all the metallic elements are likely to have condensed into solids in some form or other. As the temperature falls further, iron is increasingly oxidized to FeO until little or no pure metallic phase remains below ∼400 K (figure 7.2). Finally, H_2O-ice may condense as the temperature drops below ∼200 K. A macroscopic grain emerging from such an atmosphere is thus likely to have a layered structure, dominated by Mg-rich silicate and Fe-rich oxide phases, perhaps deposited on a refractory 'seed' nucleus and coated with a thin surface layer of ice.

It is of interest to compare the condensation of solids in an O-rich red-giant wind with that in the early Solar System. As the prevailing physical conditions are analogous, the phase diagram in figure 7.2 is relevant to both situations and the solids predicted to condense are broadly similar (Barshay and Lewis 1976, Salpeter 1977). However, the dynamic evolution of the solar nebula was probably quite different for much of its existence and the condensation process may have been closer to thermodynamic equilibrium. This will lead to differences in the composition and structure of the predicted solids: under equilibrium conditions, for example, FeO and H_2O may react with magnesium silicates at $\sim$300–400 K to form minerals such as hydrated olivine and serpentine. The chemistry of the solar nebula is discussed further in chapter 9.

7.1.4 Carbon stars

The equivalent phase diagram for a C-rich atmosphere is shown in figure 7.3. Carbon is assumed to be enhanced such that its abundance exceeds that of oxygen by 10%. (For qualitative discussion, the actual degree of enhancement is not critical provided that $C/O > 1$.) As before, CO is stable in the gas above the dotted curve. Solid carbon is stable in the area enclosed by the bold dot–dash curve. The shape of this curve arises because different carbon-bearing monomers (C, C_2, C_3, C_2H_2, CH_4) predominate in different regions of the diagram (Salpeter 1974), as marked in figure 7.3. At pressures prevailing in red-giant atmospheres, acetylene (C_2H_2) is generally the dominant form. The kinetic processes that lead to the production of carbon dust in the winds of red giants appear to be closely analogous to soot production by combustion of hydrocarbons (Frenklach and Feigelson 1989, 1997). The basic unit of solid carbon is the hexagonal ring (see figure 6.10). However, the molecular structure of the available monomer (acetylene) is H–C≡C–H, i.e. it is the simplest example of a saturated *linear* molecule involving carbon bonding with alternate single and triple bonds. In order to produce aromatic hydrocarbons, it is necessary to replace the triple (sp) bond with a double (sp^2) bond. The same atoms present in acetylene can be rearranged to form the radical $C{=}CH_2$, which has two unpaired electrons. Such a metamorphosis may be brought about collisionally, involving, for example, the removal (abstraction) of an H atom:

$$C_2H_2 + H \rightarrow C_2H + H_2. \qquad (7.2)$$

The product (C=CH) constitutes a ring segment (figure 6.10) and the ring may be closed by chemical reactions that attach two further C_2H_2 molecules (Tielens 1990). Once formed, the ring is stable and can grow cyclically by abstraction of peripheral H atoms and attachment of further C_2H_2 units. This growth process

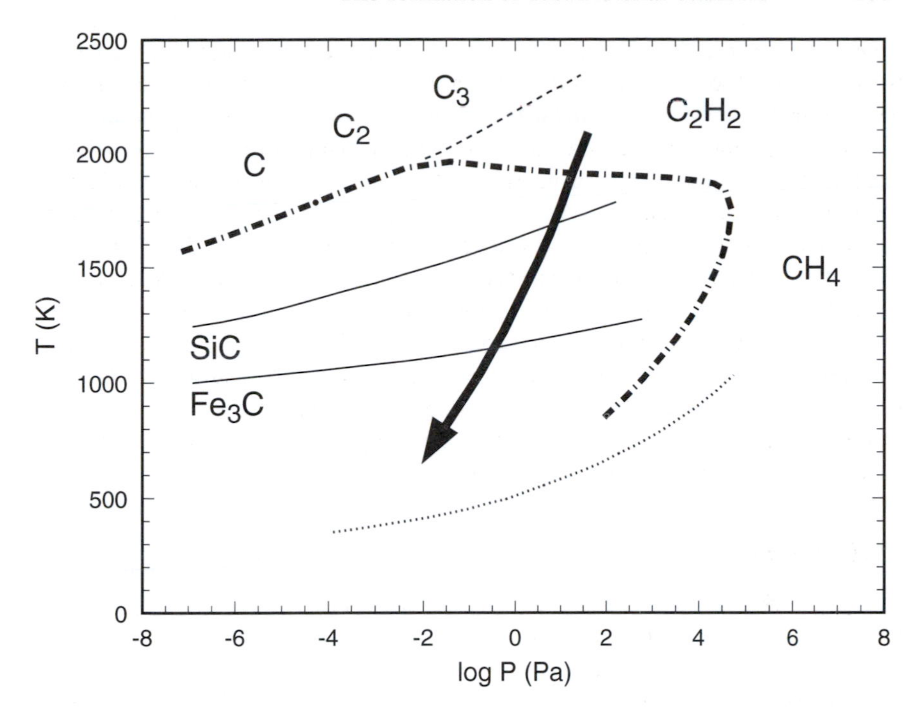

Figure 7.3. Temperature–pressure phase diagram illustrating stability zones of solids in a C-rich atmosphere (adapted from Salpeter 1974, 1977, Martin 1978). Solar abundances are assumed except that the abundance of carbon is enhanced to exceed that of oxygen by 10%. Above the dotted curve, gas-phase CO is stable and essentially all the oxygen is locked up in this molecule. Other gas-phase carriers of carbon (C, C_2, C_3, C_2H_2, CH_4) are most abundant in the regions labelled. The curved arrow indicates the change in physical conditions associated with a typical outflow from a red giant. Solid carbon is stable in the region enclosed by the bold dot–dash curve. Condensation curves for the carbides SiC and Fe_3C are also shown. The broken line above the centre represents the probable condensation curve for a hydrogen-deficient atmosphere in which C, C_2 and C_3 rather than C_2H_2 are the primary monomers.

may be represented symbolically by the pair of alternating reactions

$$C_nH_m + H \rightarrow C_nH_{m-1} + H_2 \tag{7.3}$$

$$C_nH_{m-1} + C_2H_2 \rightarrow C_{n+2}H_m + H \tag{7.4}$$

which may lead to the construction of planar PAH molecules containing several rings (Frenklach and Feigelson 1989). However, these reactions will be in competition with others that attach non-aromatic units to the rings. The likely outcome of the growth process is amorphous carbon, in which randomly grouped ring clusters are connected by bridging units with linear (sp) or tetrahedral (sp^3)

bonding (Tielens 1990, Duley 1993). This is, indeed, consistent with the known properties of soot particles. Note that amorphous carbon lacks the long-range order found in crystalline forms with sp^2 (graphite) or sp^3 (diamond) bonding.

In a C-rich atmosphere that is severely deficient in hydrogen, such as that of an R Coronae Borealis star, the nucleation process is likely to differ in detail from that described above. C, C_2 and C_3, rather than C_2H_2, will be the primary gas phase carriers of condensible carbon (see figure 7.3). In these circumstances, C and C=C monomers may assemble into ring clusters that accumulate into amorphous carbon grains with low hydrogen content.

The role of the metallic elements in the formation of dust in C-rich atmospheres appears to have received little attention. Unlike the situation in O-rich atmospheres, chemical reactions involving metals do not regulate the condensation of the primary solid phase. Nevertheless, a cooling gas containing a solar or near-solar endowment of metals must inevitably form metal-rich condensates. Abundant species such as Fe may condense as pure metals, as in the O-rich case, but silicates are prevented from forming because O remains trapped in CO. Some free C and S may be available to form carbides and sulphides, however. Phase transition curves for SiC and Fe_3C are shown in figure 7.3. Frenklach *et al* (1989) suggest that SiC grains may provide nucleation centres for condensation of carbon dust.

7.1.5 Late stages of stellar evolution

The ultimate fate of a single star in the upper right-hand region of the HR diagram (figure 7.1) depends on its mass. A 5 $M_\odot$ star will typically conclude its red-giant phase with the ejection of a planetary nebula, whilst its core evolves to become a white dwarf; in contrast, a 15 $M_\odot$ star will become a supernova (section 2.1.4). A further possibility for close binary systems is the occurrence of episodic nova eruptions as matter is transferred from one star to the other via an accretion disc.

Formation of a planetary nebula follows a period of intense mass-loss, the so-called 'superwind' phase of late-AGB evolution[2]. As the outer layers of the star dissipate, the mass-loss rate declines and the hot, compact core becomes visible. Once this occurs, energetic photons from the core begin to heat and ionize the envelope, which expands into a luminous, spheroidal nebula. As previously noted, grain nucleation tends to be inhibited in ionized gas and little or no new dust production is expected to occur in the PN itself. Expansion speeds typically reach 20–50 km s^{-1} in the nebular gas, compared with $\sim$10 km s^{-1} for red-giant winds. Material previously ejected during the AGB phase may thus be swept up and reprocessed, and some grain materials may be destroyed (Pottasch *et al*

[2] The term 'planetary nebula' (PN; plural PNe) is often confusing to those unfamiliar with astrophysical terminology. They are so-named by virtue of their disclike telescopic appearance and not through any physical association with planets. The problem is compounded by the fact that stars in transition between the AGB and PNe are termed 'protoplanetary nebulae' (PPNe), which likewise have no association with protoplanetary discs around young stars.

1984, Lenzuni *et al* 1989). However, Stasinska and Szczerba (1999) argue that the timescale for grain destruction in PNe is longer than their lifetimes.

A nova outburst is triggered when matter drawn from the surface of a main-sequence star undergoes thermonuclear ignition as it accretes onto the surface of a companion white dwarf, leading to ejection of a rapidly expanding shell of ionized gas. Grain formation commences when the shell cools to temperatures $\sim$1000 K at some radial distance from the star, at which an infrared 'pseudo-photosphere' develops as the dust becomes optically thick (Bode and Evans 1983, Bode 1988, 1989). The availability of neutral atomic and molecular monomers is limited by UV radiation from the white dwarf. Models suggest that the most feasible route to nucleation and growth is via a C-rich chemistry that follows loss of atomic O to CO (Rawlings and Williams 1989). Efficient dust production requires that the density is above some critical value at the condensation radius, and this depends on the total mass of ejected material (Gehrz and Ney 1987). Thus, some novae are rich sources of dust whilst others produce little or none.

As supernovae (SNe) manufacture many of the condensible elements, it is natural to presume that they are important sources of dust. Of the various classes, type II (involving core collapse a single massive star) is probably most important. A theoretical basis for dust condensation in their ejecta is described by Lattimer *et al* (1978). The expanding envelope may pass through an epoch of nucleation and growth analogous to that in novae (Gehrz and Ney 1987). The nature of the envelope will, of course, be highly dependent on the evolutionary state of the progenitor and may include compositionally distinct layers, but, on average, they are typically O rich overall (Trimble 1991). Clayton *et al* (1999) argue that C-rich dust may form, nevertheless, in SNe ejecta, as CO is dissociated by energetic electrons from ^{56}Co radioactivity.

7.2 Observational constraints on stardust

Astronomical techniques used to explore the nature and composition of dust in the shells of evolved stars include observations of infrared continuum emission, infrared spectroscopy of absorption and emission features and studies of ultraviolet extinction curves. These methods are fairly successful in identifying at least some of the grain materials: we review the principal results in section 7.2.1–7.2.3. Detection of pre-solar stardust in meteorites provides an important and complementary approach to the problem: these results are discussed in section 7.2.4.

7.2.1 Infrared continuum emission

Infrared continuum emission greatly in excess of that expected from the Rayleigh–Jeans 'tail' of a normal stellar photosphere is a defining characteristic of stars with dust shells. Ultraviolet, visible and near infrared radiation from the photosphere is absorbed by the grains and re-emitted at longer wavelengths. The spectral shape of the emission provides a useful diagnostic of grain composition on the basis of arguments presented in chapter 6 (section 6.1). The flux density emerging from an isothermal dust shell of temperature T_d is given by equation (6.8), i.e. $F_\lambda \propto Q_\lambda B_\lambda(T_d)$ where Q_λ is the grain emissivity and $B_\lambda(T_d)$ the Planck function. A real circumstellar shell will contain a range of grain temperatures resulting in a composite spectrum, the form of which depends on the wavelengths of peak emission (equation (6.9)) at the inner and outer boundaries of the shell and on the radial distribution of material in the shell (Bode and Evans 1983). However, at sufficiently long wavelengths, B_λ is described by the Rayleigh–Jeans approximation (equation (6.13)) independent of T_d and we have

$$\log F_\lambda = C - (\beta + 4) \log \lambda \tag{7.5}$$

where $Q_\lambda \propto \lambda^{-\beta}$ and C is a constant. The emissivity index β may thus be evaluated from the slope of the logarithmic FIR flux distribution. As discussed previously (section 6.1.2), this parameter constrains the degree of crystallinity of the grain material.

A wealth of observational data is now available on infrared emission from dusty post-main-sequence stars. In a few cases, the emission may originate primarily from ambient interstellar dust that happens to lie near the star: this appears to be the case toward certain first ascent red giants that might not otherwise be associated with dust (Jura 1999); but amongst more evolved objects with $T_{eff} < 3600$ K (figure 7.1), self-generated dust shells appear to be ubiquitous. The observed spectral energy distributions of both O-rich and C-rich stars can generally be explained by models that assume spherically symmetric expanding shells, in which dust characterized by β-values typically in the range 1.0–1.3 condenses at temperatures of order 1000 K (Campbell *et al* 1976, Sopka *et al* 1985, Jura 1986, Rowan-Robinson *et al* 1986, Martin and Rogers 1987, Le Bertre 1987, 1997, Groenewegen 1997, Wallerstein and Knapp 1998). An example is shown in figure 7.4. The implication is that the newly formed grains have essentially amorphous structure irrespective of C/O ratio, consistent with the predictions of a kinetic model for grain growth in stellar atmospheres (Donn and Nuth 1985).

The onset of dust condensation in cataclysmic objects such as classical novae is signalled by an abrupt increase in brightness at infrared wavelengths that typically occurs some 50–100 days after outburst. This increase in the infrared is accompanied by a corresponding decline in visual brightness as the grains absorb and scatter light from the progenitor binary star. An example is shown in figure 7.5. Such behaviour has been observed in the majority of classical novae for which contemporaneous visible and infrared data are available (Ney and Hatfield 1978, Bode *et al* 1984, Bode 1988, Gehrz 1988, Harrison and Stringfellow 1994), indicating that dust condensation is a common (but not inevitable) outcome of the eruption (section 7.1.5).

Do supernovae behave in an analogous way? Infrared imaging of the remnants of recent SNe in our Galaxy reveals evidence for intrinsic dust in

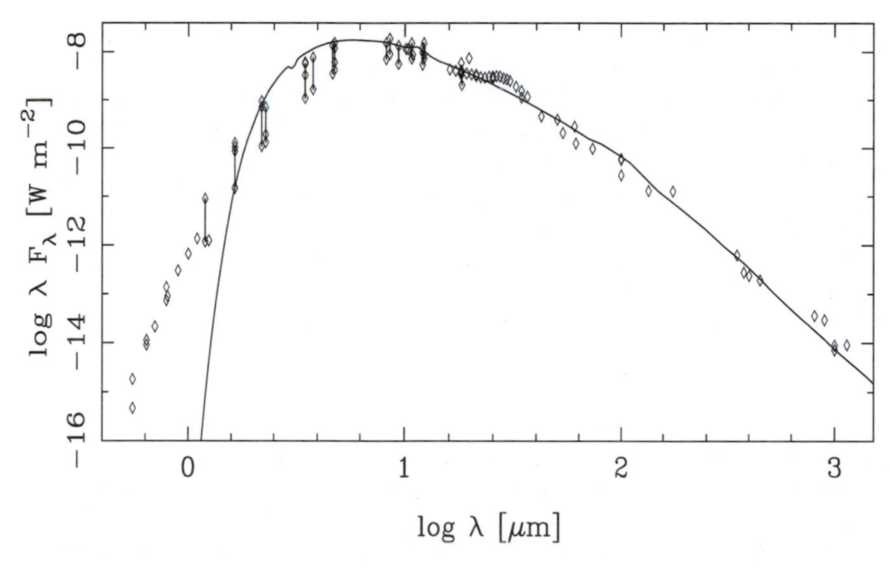

Figure 7.4. Spectral energy distribution of the carbon star IRC+10216. The observational data (points) are fitted with a model (curve) based on emission from dust composed of amorphous carbon and SiC in the ratio 100:3 (Groenewegen 1997). The vertical lines joining data points indicate variability in the star's flux between observations taken at different times. The model assumes a $\rho \propto r^{-2}$ density law and a grain size $a = 0.16$ μm (see Groenewegen 1997 for further details). Figure courtesy of Martin Groenewegen.

Cassiopeia A (Lagage *et al* 1996, Arendt *et al* 1999) but not in three other cases (Douvion *et al* 2001). Much attention has focused on the question of grain formation in the LMC supernova 1987A (see Dwek 1998). A systematic increase in infrared flux was observed ~400–600 days after outburst (Moseley *et al* 1989, Roche *et al* 1989b, 1993, Meikle *et al* 1993, Wooden *et al* 1993), but it has proven difficult to establish whether this was emitted primarily by dust created in the supernova itself or by pre-existing (ambient interstellar or circumstellar) dust that was merely heated by it. A steepening in the decline of the visual light curve has been attributed to extinction by newly formed dust (Gehrz and Ney 1990, Lucy *et al* 1991), but the effect is much less dramatic than in most novae and it seems possible to interpret the light curve without invoking a dust-forming event (Burki *et al* 1989). More compelling evidence for dust nucleation is provided by studies of Doppler components in the profiles of atomic and ionic emission lines in the optical spectrum (Danziger *et al* 1991): fading of the red-shifted component relative to the blue-shifted component approximately 530 days after outburst suggests creation of internal dust that naturally extinguishes the receding material behind the supernova more than the approaching material in front of it. It is thus probable that at least some of the emitting dust toward SN 1987A condensed within the ejecta. The subtlety of the effect on the visual light curve

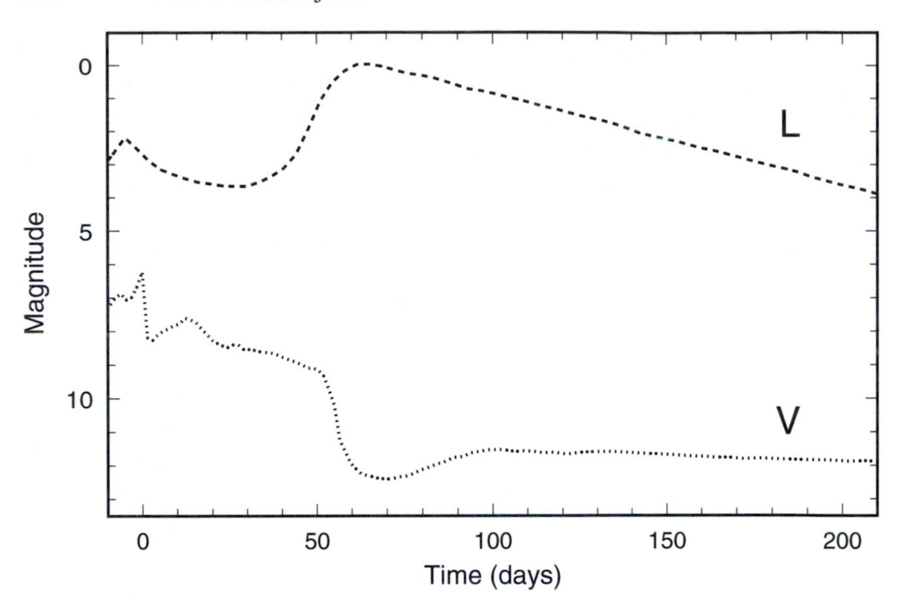

Figure 7.5. Visual and near infrared light curves for Nova Vulpeculae 1976. The V and L passbands are centred at 0.55 and 3.5 μm, respectively and time is measured from maximum brightness in V. The sharp drop in visual brightness 50–60 days past maximum is accompanied by a steep rise in infrared emission. (Adapted from Ney and Hatfield 1978.)

might be attributed to clumpiness in its spatial distribution. The dust temperature at this epoch was in the range ~400–800 K (dependent on the assumed emissivity index; Roche *et al* 1993), i.e. well below the expected condensation temperatures of refractory materials forming in stellar ejecta.

7.2.2 Infrared spectral features

Evolved stars have distinctive infrared spectra according to C/O ratio (e.g. Treffers and Cohen 1974, Merrill and Stein 1976a, b, Aitken *et al* 1979, Forrest *et al* 1979, Cohen 1984). The principal features observed in objects of each type are listed in table 7.1. Features at 9.7 μm and 18.5 μm, identified with the Si–O stretching and bending modes of amorphous silicates (see table 5.1 and section 6.3.1), generally dominate the MIR spectra of O-rich objects. These have interstellar counterparts in absorption but may be present in either net emission or net absorption in a circumstellar spectrum, depending on the optical and geometrical properties of the envelope. In C-rich objects, silicate features are generally lacking and replaced by an 11.2 μm emission feature identified with silicon carbide. The contrast between average MIR emission profiles of dust features in C-rich and O-rich red giants is illustrated in figure 7.6. Differences in composition are, of course, expected on the

Table 7.1. Dust-related circumstellar features observed in the infrared spectra of evolved stars and planetary nebulae. The columns indicate the wavelength, the proposed carrier, an indication of whether the feature is generally seen in absorption or emission (a/e), the presence or absence of a counterpart in the interstellar medium and the object type. Here 'PNe' includes both planetary nebulae and their immediate precursors, the hot post-AGB stars. Only the principal features of crystalline silicates (c-fosterite, c-enstatite) and PAHs are listed. Note that not all C-rich objects display exclusively C-rich dust features (see text).

λ (μm)	Carrier	a/e?	ISM?	Object type
O-rich objects:				
3.1, 6.0, 11.5	H_2O-ice	a	Yes	OH-IR stars
43, 62	H_2O-ice	e	No?	OH-IR stars
9.7, 18.5	silicates	a/e	Yes	M stars; PNe
19.6, 23.7, 27.5, 33.8	c-fosterite	e	No	M stars; PNe
19.4, 26.5–29.2, 43.2	c-enstatite	e	No	M stars; PNe
C-rich objects:				
3.3, 6.2, 7.7, 8.6, 11.3	PAHs	e	Yes	PNe
3.4	Aliphatic C	a	Yes	PPNe
6.2	Aromatic C	a	No?	WC stars
11.2	SiC	e	No	C stars; PNe
21	TiC?	e	No	PNe
30	MgS?	e	No	PNe

basis of arguments presented in section 7.1 and the observations clearly confirm this. Note, however, that of the principal condensates in the two environments, amorphous carbon and silicates, only the latter is detected directly. Amorphous carbon lacks strong infrared resonances, although it presumably contributes most of the continuum emission from carbon stars (section 7.2.1; e.g. figure 7.4). SiC emission at 11.2 μm is the only spectral imprint of dust commonly observed in normal C-type red giants (table 7.1), yet SiC appears to contribute only $\sim$10% or less of the dust mass in such objects (Lorenz-Martins and Lefèvre 1993).

The correspondence between the C/O abundance ratio and the form of the infrared spectrum (figure 7.6) is sufficiently strong that the dust features are sometimes used as a diagnostic of C/O ratio in stars too faint to be classified from visual spectra. The correlation is not perfect, however: in exceptional cases, of which at least 20 are currently known, silicate emission features are detected in stars classified as C rich (Willems and de Jong 1986, Little-Marenin 1986, LeVan *et al* 1992, Waters *et al* 1998). It seems unlikely that these objects violate the general principles described in section 7.1. Some might be unresolved binary systems containing both O-rich and C-rich components, but this possibility seems

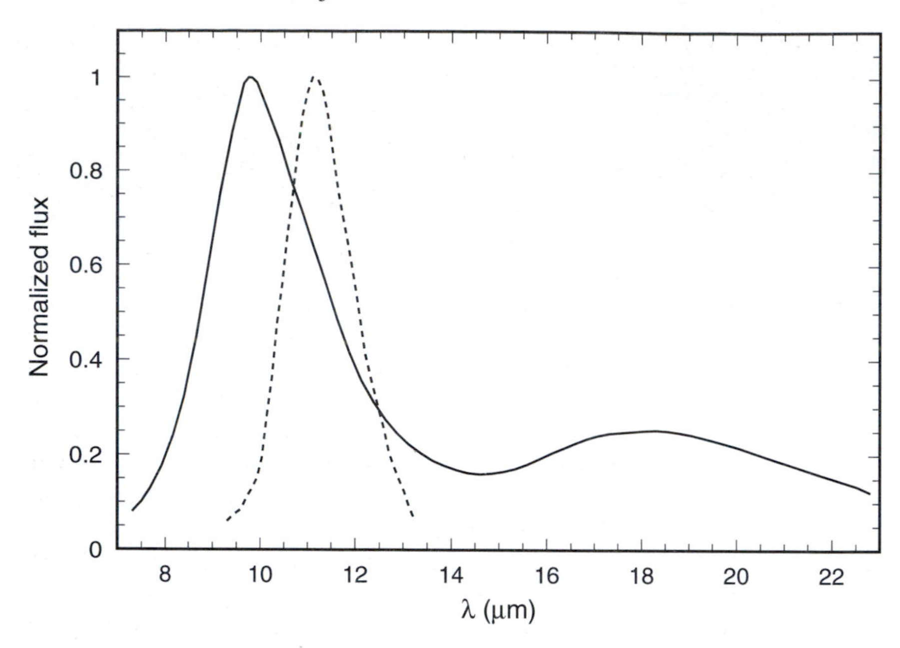

Figure 7.6. Smoothed mean profiles of 7–23 μm dust emission features observed in IRAS low-resolution spectra of red giants: full curve, silicate emission in M-type stars (C/O $<$ 1); broken curve, SiC emission in carbon stars (C/O $>$ 1). Each profile is normalized to unity at the peak.

inconsistent with detailed spectroscopic studies that fail to detect the expected signatures of M stars (Lambert *et al* 1990). Red giants are initially O rich and become carbon stars only if C-rich nucleosynthesis products are dredged to the surface from the interior (section 2.1.4). The composition of the atmosphere, and hence of the dust that forms, may thus evolve with time. C-enrichment is often associated with thermal pulses that induce temporary phases of high mass-loss (Zijlstra *et al* 1992): a 'recent' carbon star may thus be surrounded by expanding shells of O-rich material from earlier mass-loss episodes. However, the expanding shells will cool on timescales much shorter than the thermal pulse cycle (10^3–10^4 years; Iben and Renzini 1983) and they cannot therefore explain the presence of silicate *emission* features. If the mass-losing star is a member of a binary system, however, its ejecta may be captured into the circumstellar envelope of its companion, which could then act as a reservoir of warm silicate dust from earlier phases (Yamamura *et al* 2000).

Another class of object that displays ambivalence with respect to dust composition is the classical nova. Chemical models suggest that nova ejecta should be C rich (Clayton and Hoyle 1976, Bode 1989, Rawlings and Williams 1989) and this is supported by the detection of PAH-like emission features in some nova spectra (Evans and Rawlings 1994). However, some novae also display 9.7 μm and 18.5 μm emission that seems securely identified with silicate dust forming in their outflows (Bode *et al* 1984, Gehrz *et al* 1986, 1995, Evans *et al* 1997). The apparently simultaneous presence of both C-rich and O-rich condensates (Snijders *et al* 1987) might be understood if CO formation does not go to completion, such that only a minor fraction of both O and C is locked up in CO when the dust is formed (Evans *et al* 1997).

The profiles of the 9.7 μm and 18.5 μm silicate emission features in evolved stars are generally smooth and devoid of structure associated with crystallinity (figure 7.6). Indeed, the 9.7 μm profile seen in absorption in the diffuse ISM (section 5.2.2) closely resembles an inversion of the emission profile seen in red giants such as μ Cephei (Roche and Aitken 1984a; see section 6.3.1 and figure 6.7). The obvious inference is that silicate stardust is predominantly amorphous in structure. Results from the Infrared Space Observatory that provide convincing evidence for *crystalline* silicates therefore came as something of a surprise (Waters *et al* 1996, 1998, Jäger *et al* 1998, Molster *et al* 2001). In crystalline form, silicates produce several relatively narrow features in the 15–45 μm region and these have been observed in emission in the spectra of some highly evolved O-rich stars. An example is shown in figure 7.7. Such spectra provide important information on the mineralogy of the silicates as they are diagnostic of composition (the olivine/pyroxene and Fe/Mg ratios) as well as crystallinity. The Fe/Mg ratio appears to be very low, 0.05 or less, consistent with pure crystalline magnesium silicates $MgSiO_3$ and Mg_2SiO_4. In contrast, fits to the 9.7 μm profile in both circumstellar envelopes and the diffuse ISM are consistent with Fe/Mg $\sim$ 1 for the amorphous silicates (e.g. figure 5.6). The mass fraction of silicates in crystalline form appears to be $<10\%$ in most evolved stars but a few could have as much as 30%. Emission features in the 17–20 μm region are superposed on broader 18.5 μm absorption in some lines of sight, suggesting that the crystalline and amorphous silicates have different spatial distributions (Sylvester *et al* 1999).

Ice is expected to condense in an O-rich stellar atmosphere if sufficiently low-temperature environments exist (section 7.1.3). Such conditions naturally arise if the atmosphere becomes optically thick, such that its outer layers are protected from photospheric radiation and a steep temperature gradient is established. Thick envelopes are a common characteristic of intermediate-mass stars during the late phases of asymptotic-giant-branch evolution (figure 7.1); these are distinguished observationally by intense infrared continuum emission accompanied by OH maser emission (the OH-IR stars). Absorption features attributed to water frosts have been observed in the spectra of several objects of this type (Gillett and Soifer 1976, Soifer *et al* 1981, Roche and Aitken 1984b, Geballe *et al* 1988, Smith *et al* 1988, Omont *et al* 1990, Meyer *et al* 1998a). Features attributed to the stretching, bending and libration modes at $\sim$3.1, 6.0 and 12 μm (table 5.2) and to group resonances at longer wavelength ($\sim$43 and 62 μm) are observed. Spectra of OH 231.8+4.2 covering the 3.1 μm ice feature

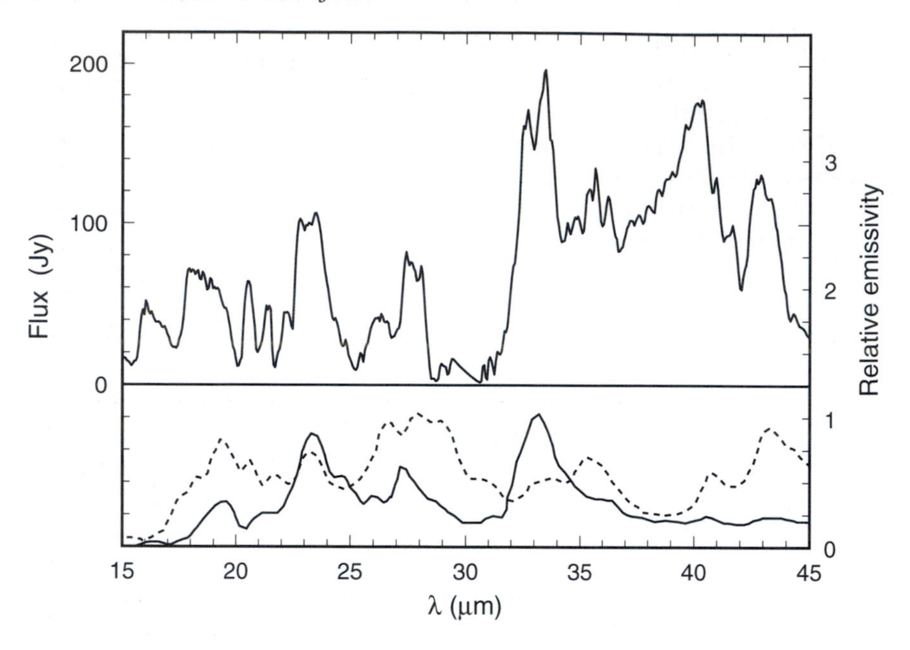

Figure 7.7. Infrared spectrum from 15 to 45 μm of the evolved binary system GL 4106, after subtraction of the continuum (upper frame, left-hand scale). The spectrum was obtained with the ISO short-wavelength spectrometer (Molster *et al* 1999). It is compared with spectra for crystalline silicates (lower frame, right-hand scale), calculated by multiplying laboratory-measured mass-absorption coefficients by a 100 K blackbody and normalizing to unity at the peak (full curve, fosterite; broken curve, enstatite; Jäger *et al* 1998). Prominent features at 23 and 33 μm in the observed spectrum are attributed primarily to crystalline fosterite.

and the 9.7 μm silicate feature are illustrated in figure 7.8. The silicate feature appears in absorption, as expected if it arises in an optically thick circumstellar shell (foreground interstellar dust will also contribute but this should make little difference to the profile shape). The profile of the 3.1 μm ice feature (upper frame) is well matched by a model that assumes pure H_2O-ice mantles with temperatures $\sim$80 K, deposited on silicate cores. This result is in contrast with fits to the corresponding interstellar feature seen in molecular clouds (see figure 5.12), which consistently fail to account for the long-wavelength wing seen in such spectra. Evidently, the properties of the ices that lead to the presence of the wing in interstellar clouds (section 5.3.3) are lacking in circumstellar envelopes. These differences may be compositional (condensation of the wing carrier(s) might be inhibited by the physical or chemical environment in the circumstellar envelope) or they might reflect a difference in the grain-size distribution. The profile of the silicate feature in OH 231.8+4.2 (figure 7.8, lower frame) is broadened to longer wavelength in a manner that suggests the presence of the expected water

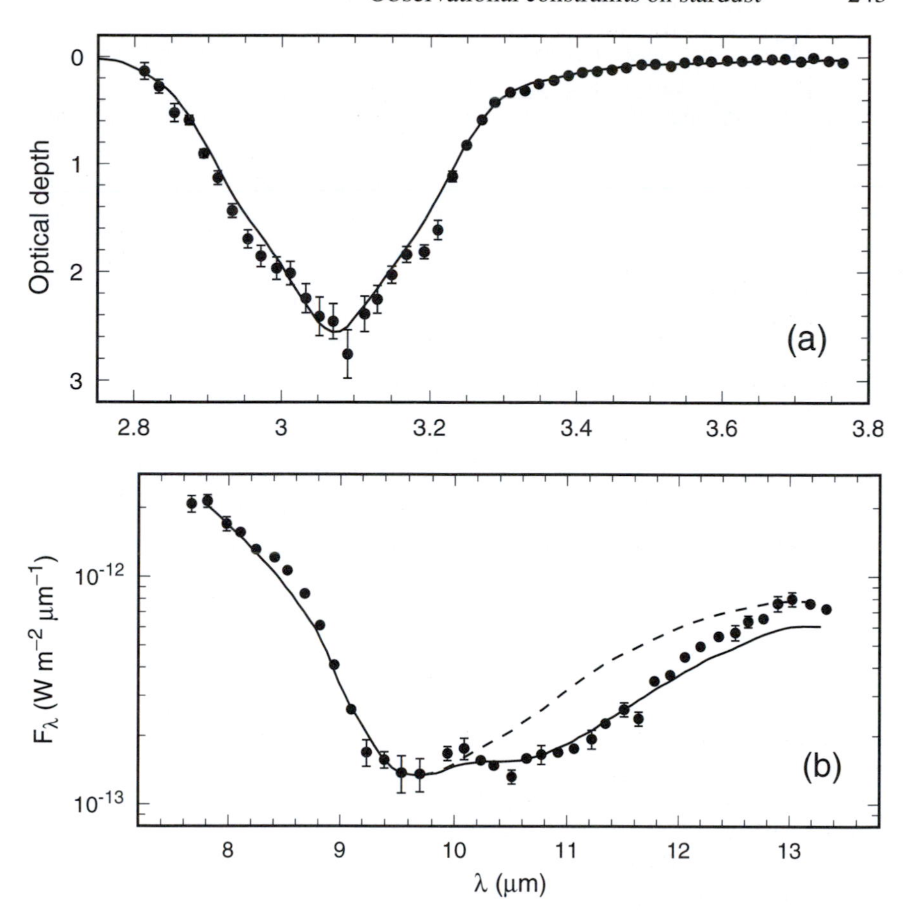

Figure 7.8. Infrared spectra of the OH-IR star OH 231.8+4.2, displaying ice and silicate absorption features. (*a*) Optical depth spectrum in the 3 μm region, showing the O–H stretching-mode feature of H_2O-ice. The observed profile (points) is fitted with a model based on spherical core–mantle grains: silicate cores of radius 0.1 μm are assumed to be coated with pure H_2O-ice mantles at 77 K, up to a maximum radius of 0.6 μm (Smith *et al* 1988). (*b*) Flux spectrum in the 10 μm region, showing the silicate Si–O stretching-mode feature. The observed profile is fitted with a model based on emission from hot dust combined with absorption by cold dust (full line, silicates with crystalline H_2O-ice; broken line, silicates alone; Soifer *et al* 1981).

libration-mode absorption, centred near 11.5 μm. The position and width of the libration feature are affected by both the crystallinity and the purity of the ices. The circumstellar profile is consistent with pure, crystalline H_2O-ice, whereas the corresponding interstellar feature, arising in impure, amorphous ices, is broader and shifted to longer wavelengths (see section 5.3.7).

Carbonaceous stardust is expected to be predominantly amorphous rather than graphitic for reasons discussed earlier (sections 7.1.4 and 7.2.1). A further constraint on the abundance of graphite may be deduced from a spectral feature expected to occur at 11.52 μm that arises in a lattice resonance. This feature should be seen in either absorption or emission, depending on the temperature and distribution of graphitic dust (Draine 1984). It is intrinsically weak and narrow (FWHM $\approx$ 0.01 μm), demanding high quality data with good spectral resolution to provide a significant constraint. Its absence in a few C-rich objects that have been sufficiently well studied (Glasse *et al* 1986, Martin and Rogers 1987) suggests that no more than about 5% of the solid carbon in stardust is graphitic.

The family of emission features attributed to polycyclic aromatic hydrocarbons (the aromatic infrared spectrum; table 6.1) was described in detail in chapter 6 (section 6.3.2). In the ISM, the aromatic features are emitted whenever interstellar PAH molecules or small carbonaceous grains containing PAHs are vibrationally excited by exposure to soft ultraviolet photons. In the circumstellar context, we expect PAHs to be abundant only in C-rich objects; and we expect a significant UV flux to be generally present only in objects that have evolved past the asymptotic giant branch. This is largely confirmed by observations: the aromatic spectrum is most prominent in C-rich planetary nebulae and their immediate precursors (see figure 6.8 for examples); it is generally weak or absent in physically similar objects that are O rich and in objects of all C/O ratio that lack a source of UV radiation.

Other dust-related spectral signatures observed in C-rich objects include absorption features at 3.4 and 6.2 μm and emission features at 21 and 30 μm (table 7.1). In terms of profile shape, the 3.4 μm feature seen in the C-rich protoplanetary nebula CRL 618 appears to be a direct circumstellar counterpart to the 3.4 μm interstellar feature attributed to aliphatic hydrocarbons (Chiar *et al* 1998b; section 5.2.4). Similarly, the 6.2 μm feature may be an absorption counterpart to the aromatic emission feature at the same wavelength; it has been attributed to interstellar dust (Schutte *et al* 1998) but Chiar and Tielens (2001) argue for an origin in the dusty envelopes of the WC stars in which it is observed.

Broad emission features at longer wavelengths, the most prominent of which arise at 21 and 30 μm, are commonly seen in the spectra of post-AGB stars (Forrest *et al* 1981, Kwok *et al* 1989, Hrivnak *et al* 2000; the 30 μm feature in IRC+10216 may be seen in figure 7.4). Remarkably, in some objects, as much as 25% of the total mid-infrared flux is emitted in these features. Their origins are not yet firmly established. Candidates for the 21 μm feature include diamonds (Hill *et al* 1998) and TiC nanocrystals (von Helden *et al* 2000), whilst the 30 μm feature has been attributed to MgS, possibly as a mantle on more refractory grain cores (Goebel and Moseley 1985, Nuth *et al* 1985, Szczerba *et al* 1999). Neither TiC nor MgS are likely to be major components of the stardust generally. The case for TiC as a trace component is supported by its presence in pre-solar meteoritic graphite grains (Bernatowicz *et al* 1996; section 7.2.4). The MgS concentration required in dust to explain the observed strength of the 30 μm feature appears to be at least a few per cent by mass (Martin and Rogers 1987), comparable with the SiC abundance implied by the 11.2 μm feature. However, the 30 μm band may not be a single feature (Hrivnak *et al* 2000) and might include contributions from other metal sulphides (e.g. CaS and FeS_2; Nuth *et al* 1985). Other likely condensates in C-rich atmospheres include FeS and Fe_3C but these are featureless in the infrared (Nuth *et al* 1985).

7.2.3 Circumstellar extinction

Circumstellar extinction curves can, in principle, provide information on both the composition and the size distribution of the dust. Traditionally, interstellar extinction curves are generated by the pair method (section 3.2), in which the spectral energy distributions of selected reddened stars of normal early type are compared with intrinsically similar objects suffering little or no reddening. This technique is generally much less reliable for circumstellar extinction in late type stars, as such objects often have highly complex photospheric spectra and may be intrinsically variable. Furthermore, practical considerations limit studies to objects that are relatively bright in the ultraviolet, whereas most evolved stars emit most flux in the near infrared. Finally, care must be taken to separate circumstellar and interstellar components of the measured extinction. To alleviate the problem of finding suitable comparisons, intrinsic spectra may be reconstructed with reference to model atmospheres (e.g. Buss *et al* 1989). However, in cases where stellar variability results from varying degrees of circumstellar extinction, the comparison between spectra at bright and faint phases of the light curve provides an extinction curve without reference to another object or model atmosphere (e.g. Hecht *et al* 1984). In serendipitous cases where a normal early-type star lies within or behind a late-type stellar envelope, the former may be used as a probe of the extinction in the latter using the standard technique (e.g. Snow *et al* 1987).

From an observational standpoint, the post-AGB stars are ideal choices for studies of circumstellar extinction as they are often intrinsically bright in the UV, unlike most red giants. Of these, the RCB class of variable stars is of particular interest as their variability is attributed to temporal changes in circumstellar dust opacity; circumstellar extinction curves may thus be derived using the variable extinction method subject to matching the stellar pulsational phase at high and low obscuration. Results show the presence of a broad peak centred near 2500 Å (4 μm^{-1}), in contrast to the well known 2175 Å (4.6 μm^{-1}) bump that characterizes interstellar extinction curves (Hecht *et al* 1984, Evans *et al* 1985, Drilling and Schönberner 1989, Drilling *et al* 1997; see section 3.5). Examples are shown in figure 7.9 (RY Sgr) and figure 7.10 (V348 Sgr). RCB stars are both C rich and hydrogen deficient and the observed 2500 Å feature seems likely to arise in glassy or amorphous carbon dust of low hydrogen content (Hecht *et al* 1984, Hecht 1991, Muci *et al* 1994). Figure 7.9 shows a sample fit to the circumstellar extinction curve of RY Sgr based on such material.

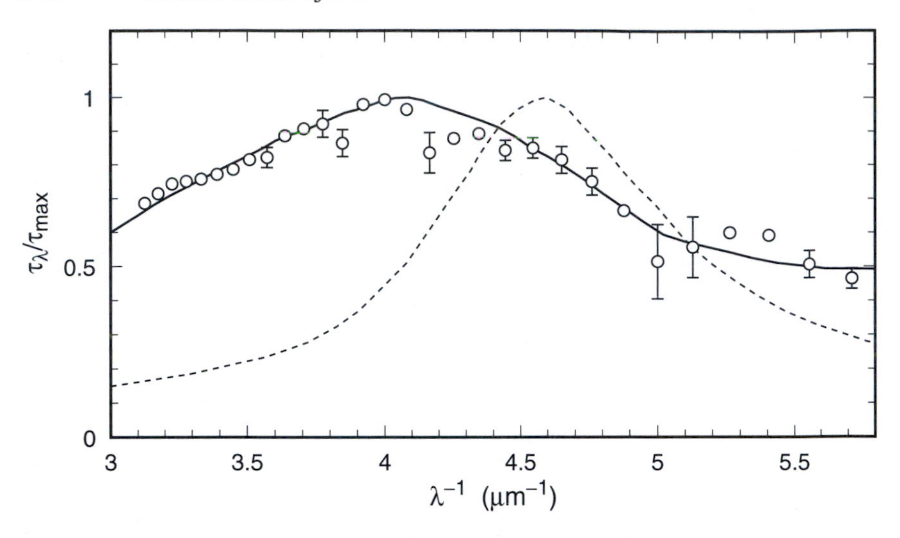

Figure 7.9. The profile of the mid-ultraviolet circumstellar absorption feature in the spectrum of the RCB star RY Sgr. Observational data (points) are fitted with a model (full curve) based on Mie calculations for glassy carbon spheres with radii that follow a power-law size distribution in the range 0.015–0.050 μm (Hecht *et al* 1984, Hecht 1991). The mean interstellar extinction curve is shown for comparison (broken curve).

The range of circumstellar extinction curves observed in C-rich post-AGB stars is illustrated in figure 7.10. In contrast to the RCB stars, HD 89353 is hydrogen rich and has strong PAH emission features in its infrared spectrum (Geballe *et al* 1989a, Buss *et al* 1993). Its ultraviolet extinction curve is featureless, devoid of peaks at either 2175 or 2500 Å, displaying instead an approximately linear rise with wavenumber into the FUV (Waters *et al* 1989; Buss *et al* 1989). This result is entirely consistent with the prediction that hydrogenation will suppresses the mid-UV absorption features in carbon dust (Hecht 1986). HD 213985 combines weak or absent PAH emission (Buss *et al* 1989) with a UV peak near 2300 Å, intermediate in wavelength between the normal interstellar feature and the circumstellar feature in RCB stars (figure 7.10). This star hints at a link between C-rich stardust and the carrier of the bump in the ISM, in which both the degree of hydrogenation and the internal structure of the particles are controlled by physical conditions (Muci *et al* 1994).

The majority of objects for which circumstellar extinction curves have been obtained to date are C rich but a few O-rich supergiants have been investigated by the 'nearby star' method (Snow *et al* 1987, Buss and Snow 1988, Seab and Snow 1989). To within the limits of the data, there is no evidence for either a mid-UV bump or a far-UV rise. The absence of mid-UV absorption is, of course, expected if dust forming in an O-rich environment lacks a carbonaceous component. In the best-studied case (α Scorpii), the extinction declines steadily

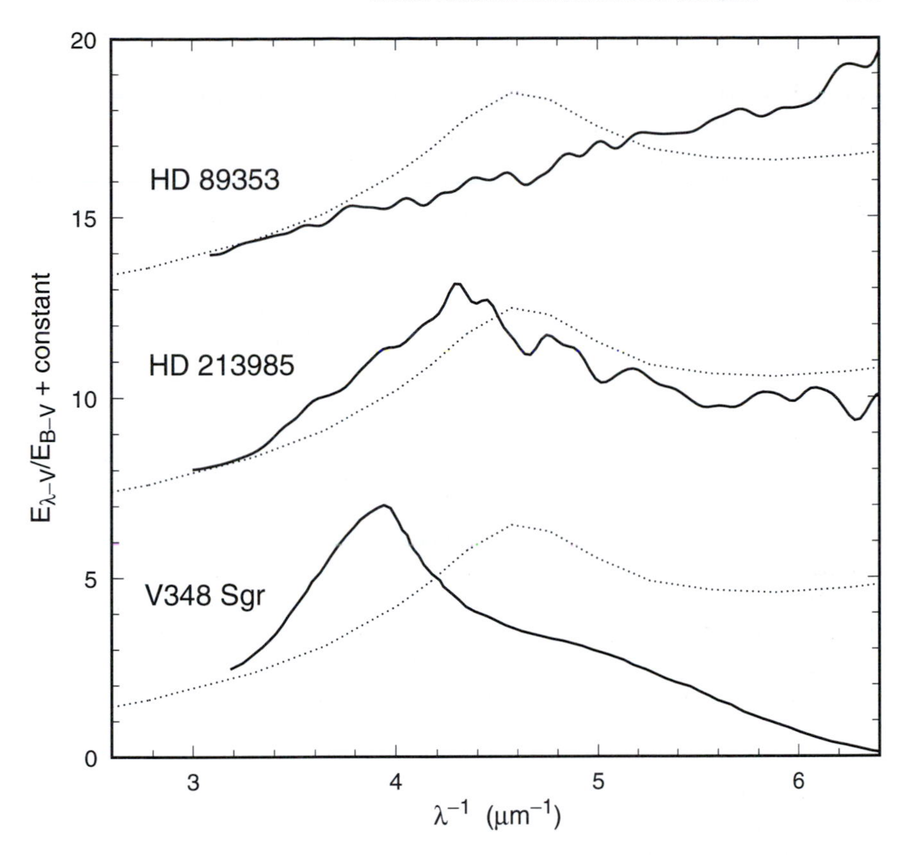

Figure 7.10. Contrasting circumstellar extinction curves for three post-AGB stars observed by the International Ultraviolet Explorer: the RCB star V348 Sgr (Drilling and Schönberner 1989) and the protoplanetary nebulae HD 213985 and HD 89353 (Buss *et al* 1989). The data for the latter two stars are displaced upward by 5 and 10 units, respectively, for display. In each case, the mean diffuse-ISM extinction curve is also shown for comparison (dotted curves; table 3.1).

with wavenumber between 3 and 8 μm^{-1}, suggesting a dearth of small grains. Calculations indicate that the data are consistent with the presence of relatively large ($a > 0.08$ μm) silicates (Seab and Snow 1989). If these results are typical, then O-rich stardust entering the ISM will tend to populate the upper end of the size distribution (section 3.7).

7.2.4 Stardust in meteorites

At one time, it was thought that all solid material in the Solar System condensed from a homogeneous, hot gas that erased the chemical and mineralogical

Table 7.2. Summary of stardust grains identified in primitive meteorites. Abundances are given by mass relative to the bulk meteorite. (Adapted from the review by Zinner 1998.)

Material	Size (μm)	Abundance (ppm)	Stellar source
Diamond	~0.002	500	SNe (+ others?)
SiC	0.3–20	5	AGB stars
'Graphite'	1–20	1	SNe, AGB stars, novae
SiC (type X)	1–5	0.06	SNe
Al_2O_3, $MgAl_2O_4$	0.5–3	0.03	M stars, AGB stars
Si_3N_4	~1	0.002	SNe

fingerprints of pre-solar material. The discovery of isotopic anomalies in primitive meteorites has completely altered this picture and opened up a new window on cosmic chemical evolution (e.g. Clayton 1975, 1982, 1988, Nuth 1990, Anders and Zinner 1993, Zinner 1998). Evidently, some fraction of the interstellar grains present in the solar nebula did not undergo vaporization and recondensation but accreted directly into planetesimals, some of which remain preserved in a relatively unaltered state in the asteroid belt. Meteorites derived from asteroidal parent bodies (Gaffey *et al* 1993) thus contain a 'fossil record' of pre-solar dust, much of which evidently formed in the ejecta of evolved stars. Some properties of this meteoritic stardust are summarized in table 7.2.

Identification depends on detailed isotopic analysis of grains extracted from carbonaceous chondrites. The bulk of the material in these meteorites has elemental and isotopic abundances indistinguishable from solar (section 2.2) but pre-solar grains display isotopic patterns that deviate from solar by amounts that cannot be reconciled with an origin in the solar nebula. Isotopic analysis is a powerful technique as it not only identifies exotic particles but also provides strong clues as to their origins. As an example, data on the $^{12}C/^{13}C$ ratio in three samples (two meteoritic and one stellar) are compared in figure 7.11. The distribution of values in SiC grains is highly consistent with that observed in C-rich stellar atmospheres, whereas that in graphite suggests a diversity of origins.

Nuclear reactions in stars (section 2.1.3) generate distinct isotopic abundance patterns according to physical conditions and the timescale on which they operate. Different patterns are predicted for certain elements dependent on whether the reactions proceed slowly (s-process) or rapidly/explosively (r-process). Insight is often gained from trace elements, including trapped noble gases, as well as from those that form the bulk of the solid. Only a few examples will be mentioned here (see Anders and Zinner 1993, Alexander 1997 and Zinner 1998 for reviews). Xenon (Xe) has no fewer than nine stable isotopes, some of which are s-process products whilst others are r-process products. A specific pattern of Xe isotopes, named Xe–S, is attributed to s-process nucleosynthesis in AGB

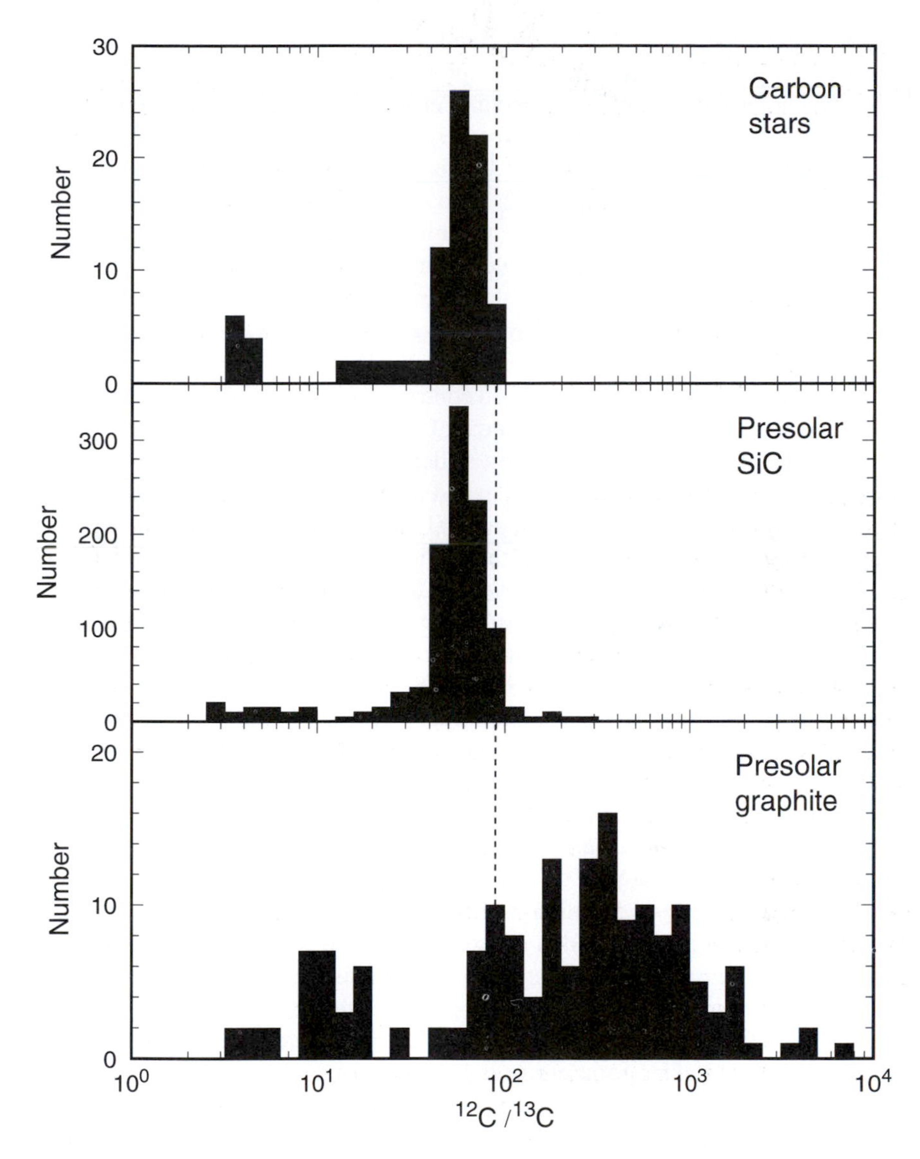

Figure 7.11. Bar charts comparing the distribution of the isotope ratio $^{12}C/^{13}C$ in the atmospheres of carbon stars and in two types of C-rich pre-solar dust grain extracted from primitive meteorites (SiC and graphite). The vertical broken line denotes the mean solar value (89). The $^{12}C/^{13}C$ ratio in evolved stars is influenced by mixing of products from the CNO cycle and the triple-α process (section 2.1.3), which favour ^{13}C and ^{12}C, respectively, but most carbon stars have values that are significantly subsolar (40–70). The SiC grains are consistent with an origin in carbon stars, whereas graphite appears to have many possible sites of formation.

stars; another, named Xe–HL, is identified with r-process nucleosynthesis. Pre-solar grains containing Xe–HL are thus presumed to have either originated in or been contaminated by supernova ejecta. Certain short-lived radioactive nuclides are also highly diagnostic. Large excesses of ^{44}Ca are attributed to ^{44}Ti decay with a half-life of about 60 years and ^{44}Ti is produced exclusively in supernovae. Similarly, ^{22}Ne excesses may arise from decay of ^{22}Na, with a half-life of 2.6 years. The distribution pattern of neon isotopes, like that of xenon, provides strong discrimination between possible origins.

By far the most abundant pre-solar grains identified to date are the diamond nanoparticles (table 7.2), first reported by Lewis *et al* (1987). Because of their small size, individual grains cannot be isotopically analysed and only bulk data are available. Their origins have proved controversial (Dai *et al* 2002), although a clear link to supernovae for at least some of them is established on the basis of a large Xe–HL excess (see the review by Anders and Zinner 1993). Note, however, that on average, only about one in 10^6 nanodiamonds will actually contain a xenon atom! So it is important not to draw general conclusions from data that might be heavily biased by a small fraction of the particles (Alexander 1997). The ^{12}C/^{13}C ratio is close to the solar one and this may imply that both red-giant winds and supernovae (which tend to produce subsolar and supersolar values, respectively) contribute to the mean. Formation by carbon vapour deposition might occur in either environment (Lewis *et al* 1987, 1989, Anders and Zinner 1993). As the free energy difference between diamond and graphite is quite small, chemical reactions yielding graphite as the thermodynamically stable product can also, in principle, yield diamond as a metastable product. Once formed, diamonds will be more stable than graphite in a hydrogen-rich environment. Shock processing of carbon dust in the ISM has also been discussed as a source of diamonds, a topic we will return to in chapter 8.

Most SiC grains identified in meteorites exhibit isotopic abundances consistent with an origin in C-rich red-giant winds (Bernatowicz *et al* 1987, Tang *et al* 1989; see figure 7.11). The formation of SiC dust in such outflows is, indeed, expected (section 7.1.4) and confirmed by observations (section 7.2.2), although the grains responsible for the observed 11.2 μm spectral feature appear to be in a different crystalline form compared with the meteoritic grains (Speck *et al* 1997). The reason for this difference is unknown. Note that SiC is a very minor constituent of carbonaceous chondrites overall: the fraction of all Si atoms tied up in SiC is only about 0.004% (Tang *et al* 1989), the vast majority being in silicates. This result is entirely consistent with the spectroscopic limit on SiC in interstellar dust (section 5.2.3; see Whittet *et al* 1990 for further discussion).

A minority of the meteoritic SiC is isotopically distinct from the 'mainstream' particles formed in C-rich red giants. These 'type X' particles (table 7.2) amount to about 1% by mass of all pre-solar SiC. Amongst other anomalies, they display a significant excess of pure ^{28}Si, thought to have formed deep within a supernova progenitor and to have subsequently mixed with a C-rich envelope. A supernova origin is thus indicated for this component of the

SiC. The type X SiC grains are isotopically similar to the even rarer Si_3N_4 grains (table 7.2), which seem likely also to originate in supernovae (Nittler *et al* 1995, 1996).

It will be seen from perusal of table 7.2 that, with the exception of diamond nanoparticles, the stardust grains identified in meteorites are generally quite large compared with typical interstellar grains. This is particularly notable in the case of graphite, as the meteoritic examples (Amari *et al* 1990, Bernatowicz *et al* 1996) are two orders of magnitude larger than those invoked to account for the 2175 Å bump in the interstellar extinction curve (section 3.5.2). This might seem disappointing, but there are several reasons to suppose that the size distribution of the meteoritic grains is unrepresentative of the ISM. Smaller grains may have been selectively destroyed prior to accretion into the meteorite parent bodies and further selection effects that tend to favour larger grains may arise during the extraction process. Meteoritic graphite is, in fact, rare in terms of fractional mass – even rarer, indeed, than SiC (table 7.2), which is below detectable levels in the ISM (section 5.2.3). The bulk of the C in carbonaceous chondrites is organic and this may have formed by surface reactions on graphite grains in the solar nebula (Barlow and Silk 1977b) that would naturally tend to consume the smallest particles most efficiently.

Graphitic particles in meteorites appear to have diverse stellar origins (Bernatowicz *et al* 1996, Nittler *et al* 1996). Isotopic signatures of both s-process and r-process products have been found, although elements other than carbon are generally too scarce to yield conclusive evidence for individual grains. About 70% have supersolar $^{12}C/^{13}C$ ratios (see figure 7.11) and it seems probable that many of these originate in supernovae. Most of the remainder may form in AGB winds, although there is no clear evidence for the expected peak in the distribution near $^{12}C/^{13}C \sim 50$ (in contrast to SiC; figure 7.11). Those with the lowest $^{12}C/^{13}C$ ratios may form in novae (Amari *et al* 2001). The grains display a distinctive internal structure, in which onion-like concentric layers of graphite encase a core composed of PAH clusters or amorphous carbon (Bernatowicz *et al* 1991, 1996). The cores appear to have condensed in isolation and then acquired graphitic mantles by vapour deposition in a C-rich atmosphere. About a third of the grains contain small (5–200 nm) refractory carbide crystals (TiC, ZrC, MoC). These carbides are embedded within the mantles and must have formed prior to mantle deposition: their presence (and the absence of embedded carbides with lower condensation temperatures, such as SiC) places constraints on the condensation sequence. Moreover, the densities required, both to grow $>1\ \mu$m sized grains within reasonable timescales and to explain the presence of the embedded carbide crystals, is a factor $\sim$100 higher than those expected in spherically symmetric AGB winds (Bernatowicz *et al* 1996). This apparent inconsistency does not exclude an origin in such stars if local density concentrations are present within the stellar envelope: millimetre-wave interferometric maps provide observational evidence that such concentrations can occur (see Glassgold 1996).

The vast majority of pre-solar grains identified in meteorites to date are C rich (table 7.2). The only known candidates for an origin in O-rich environments are oxides of aluminium, principally corundum (Al_2O_3) and in rare cases spinel ($MgAl_2O_4$). The absence of evidence for pre-solar silicates seems certain to be a selection effect, arising from the fact that silicates of *solar* composition are by far the most abundant minerals in the meteorites[3]. On the basis of their Al and O isotopic abundance patterns, pre-solar oxide grains are confidently assigned an origin in O-rich stellar winds (Nittler *et al* 1997). The presence of Al_2O_3 in significant quantities may affect emissivity in the 12–17 μm region between the silicate features (Begemann *et al* 1997). Careful searches have failed to detect any evidence for a component of supernova origin, contrary to an earlier proposal by Clayton (1981, 1982).

7.3 Evolved stars as sources of interstellar grains

That stardust contributes to the interstellar grain population is not in doubt. This is strongly implied by observational evidence that dust-forming stars undergo rapid mass-loss (e.g. Dupree 1986) and confirmed for some classes of particle by the identification of pre-solar stardust in the Solar System (section 7.2.4). We have seen that O-rich red giants produce amorphous silicates that are spectroscopically similar to interstellar silicates, whilst carbon stars produce a range of particles that may contribute to the observed extinction at various wavelengths and may explain the aromatic emission spectrum in the infrared. In this section, we examine the mass-loss process, review the properties of the stardust entering the ISM and attempt to assess its overall importance as an ingredient of interstellar dust.

7.3.1 Mass-loss

All stars lose mass to some degree in stellar winds driven by thermodynamic pressure. In the case of the Sun, this currently amounts to some 3×10^{-14} $M_\odot$ yr^{-1}, which is entirely negligible ($\sim$0.03% during its main-sequence lifetime). The mass-loss rate for a thermally driven wind is expected to increase with stellar luminosity. For first ascent red giants, Reimers (1975) proposes that

$$\dot{M} \propto \frac{L}{gR} \tag{7.6}$$

where $\dot{M} = \mathrm{d}M/\mathrm{d}t$ is the mass-loss rate for a star of mass M, luminosity L and radius R and $g = GM/R^2$ is the gravitational acceleration at the stellar surface.

[3] If searching for C-rich pre-solar grains in meteorites is likened to searching for needles in haystacks, then searching for pre-solar silicates is equivalent to searching for hay in haystacks, but of a rare and exotic strain. Techniques typically used in the former case that amount to "burning down the haystack to find the needle" (quote attributed to Edward Anders by Bernatowicz and Walker 1997) are clearly inappropriate in the latter.

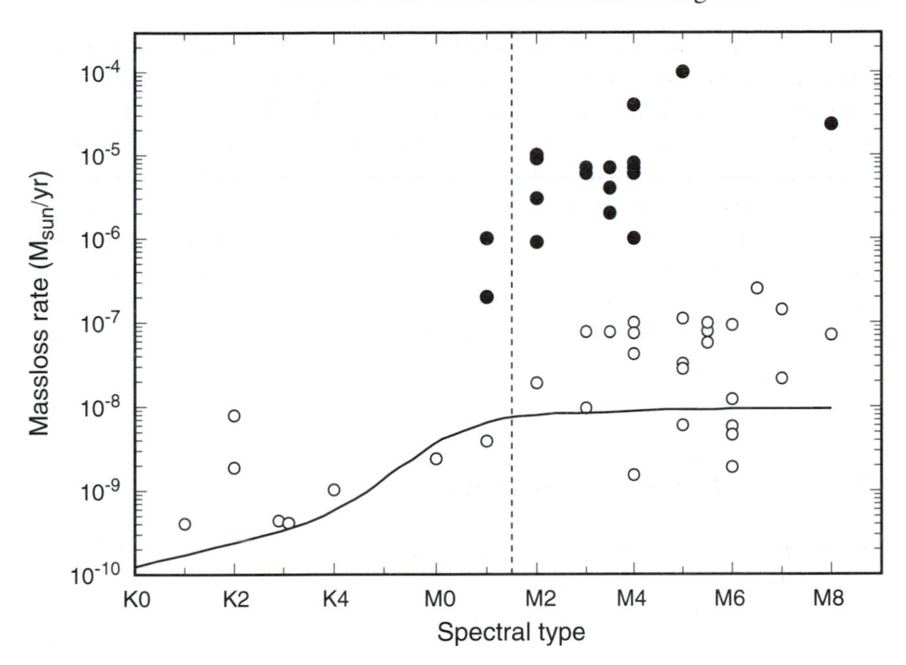

Figure 7.12. A plot of observed mass-loss rate against spectral type for O-rich red giants (luminosity classes II and III; open circles) and supergiants (luminosity class I; filled circles). Data are from Dupree (1986), Jura and Kleinmann (1990) and references therein. The vertical broken line indicates the spectral type equivalent to a photospheric temperature of 3600 K: dust formation typically occurs in stars lying to the right of this line. The full curve is the prediction of the Reimers model (equation (7.6)) for red giants of luminosity class III.

With all stellar quantities expressed in solar units, the constant of proportionality is estimated to be $\sim 4 \times 10^{-13}$ $M_\odot$ yr^{-1} (e.g. Dupree 1986). For a typical M-type star of luminosity class III, equation (7.6) thus predicts mass-loss rates in the range 10^{-8}–10^{-9} $M_\odot$ yr^{-1} (see problem 5 at the end of this chapter). However, the coolest and most luminous late-type stars are found to lose mass at rates that greatly exceed those predicted by this relation. Figure 7.12 plots observed mass-loss rate against spectral type for K- and M-type giants and supergiants. A distinct trend is seen: stars later than about M1–M2 (i.e. those with photospheric temperatures below about 3600 K) tend to have mass-loss rates ranging up to 10^{-4} $M_\odot$ yr^{-1}; and these cannot be explained in terms of gaseous winds driven entirely by thermodynamic pressure. However, the stars with the highest mass-loss rates often show independent evidence for the presence of dusty envelopes detected by their infrared emission, and this provides a strong hint that dust is the catalyst that induces high mass-loss rates in red giants (e.g. Wannier *et al* 1990).

Dust grains nucleating in stellar atmospheres are subject to outward acceleration due to radiation pressure. The rate at which a plane wave of intensity I carries linear momentum across a unit area normal to the direction of propagation is I/c. A grain that intercepts a portion of the wave experiences a net force of magnitude $(I/c)C_{\rm pr}$, where $C_{\rm pr}$, the cross section for radiation pressure, defines the effective area over which the pressure is exerted (Martin 1978: p 137). We may also define an efficiency factor $Q_{\rm pr}$ for radiation pressure as the ratio of radiation pressure cross section to geometrical cross section, by analogy with equation (3.5) for extinction, i.e. $Q_{\rm pr} = C_{\rm pr}/\pi a^2$ for a spherical grain of radius a. $Q_{\rm pr}$ depends on both the absorption and scattering characteristics of the grain and is related to the corresponding efficiency factors (section 3.1) by

$$Q_{\rm pr} = Q_{\rm abs} + \{1 - g(\theta)\}Q_{\rm sca} \tag{7.7}$$

where $g(\theta)$ is the scattering asymmetry parameter (equation (3.17)). Consider a spherical grain of mass $m_{\rm d}$ and radius a situated in an optically thin shell at a radial distance r from the centre of a star of luminosity L and mass M. The outward force due to radiation pressure is

$$F_{\rm pr} = \pi a^2 \langle Q_{\rm pr}\rangle \left(\frac{L}{4\pi r^2 c}\right) \tag{7.8}$$

where $\langle Q_{\rm pr}\rangle$ is the average value of $Q_{\rm pr}$ with respect to wavelength over the stellar spectrum. The opposing force due to gravity is

$$F_{\rm gr} = \frac{GMm_{\rm d}}{r^2}. \tag{7.9}$$

The ratio of these forces is of interest, as we require $F_{\rm pr}/F_{\rm gr} > 1$ for outward acceleration. Combining equations (7.8) and (7.9) and expressing $m_{\rm d}$ in terms of the specific density s of the grain material, we have

$$\frac{F_{\rm pr}}{F_{\rm gr}} = \frac{3L}{16\pi GMc}\left\{\frac{\langle Q_{\rm pr}\rangle}{as}\right\} \tag{7.10}$$

which is independent of r and varies with grain properties as the term in brackets.

$Q_{\rm pr}$ may be calculated as a function of wavelength from Mie theory for a given grain model, and the average value estimated with respect to the expected spectral energy distribution for a given stellar type. Martin (1978) estimates $\langle Q_{\rm pr}\rangle \approx 0.18$ for graphite and 0.003 for silicates, assuming spheres of constant radius $a = 0.05$ μm in the atmosphere of a red giant of luminosity 10^4 $L_\odot$ and mass 4 $M_\odot$. For discussion, we may take the graphite value as representative of solid (amorphous) carbon. The force imposed by radiation pressure is typically much greater for absorbing grains than for dielectric grains of the same size, because of the dominant contribution of $Q_{\rm abs}$ to $Q_{\rm pr}$ (equation (7.7)). It may be shown that dust is accelerated from red giants of luminosity $L \geq 10^3$ $L_\odot$ for a range of grain size and composition; in the previous example, $F_{\rm pr}$ exceeds $F_{\rm gr}$

by a factor of ~2000 for carbon grains and by a factor of ~40 for silicate grains (equation (7.10)).

The outward speeds of grains in a stellar atmosphere are limited by frictional drag exerted by the gas. Grains accelerated by radiation pressure thus impart momentum to the gas, driving it away from the star. This process appears to be an important, and often dominant, mechanism for mass-loss in luminous stars with optically thin dust shells (Knapp 1986, Dominik *et al* 1990).

7.3.2 Grain-size distribution

The terminal speed of a particle driven through a gas by radiation pressure depends on its size, such that large grains tend to overtake smaller ones. This results in grain–grain collisions at relative speeds typically a few kilometres per second, sufficient to cause fragmentation. Biermann and Harwit (1980) discuss the implications of such collisions for the size distribution of the particles in an expanding envelope. Multiple collisions lead to the imposition of a power-law size distribution, independent of the initial size distribution of the condensates provided that a range of sizes is present. On the basis of fragmentation theory (originally applied to asteroids), Biermann and Harwit argue that the emergent grains follow a size distribution of the form $n(a) \propto a^{-3.5}$.

The model of Biermann and Harwit (1980) provides a physical basis for understanding the nature of the size distribution for interstellar grains, as deduced by fitting the extinction curve (section 3.7). An important parameter that governs the goodness of fit to interstellar extinction is the upper bound of the particle radius, $a_{\max}$ (see equation (3.47)), which typically has a value ~0.25 μm for the diffuse ISM. However, we have direct empirical evidence from the meteorite studies (section 7.2.4) that at least some of the dust grains emerging from evolved stars are much bigger than this. A sharp cut-off to the size distribution is unphysical and it seems more reasonable to adopt a functional form that allows a smooth exponential decline in $n(a)$ as a becomes large:

$$n(a) \propto a^{-3.5} \exp(-a/a_0) \tag{7.11}$$

(Kim *et al* 1994, Jura 1994). The parameter a_0 in equation (7.11) then governs the typical size of a large grain (whilst allowing for the presence of some much larger grains; see figure 7.13). Jura (1994, 1996) finds that the properties of both C-rich and O-rich mass-losing red giants are consistent with a size distribution of this form with $a_0 \approx 0.10$–0.15 μm, similar to values obtained from fits to interstellar extinction. This agreement might be considered fortuitous, as grains are expected to be destroyed and reformed in the ISM on timescales shorter than their injection timescale (see section 7.3.5). Nevertheless, it hints that the physical processes governing the size distribution are similar from circumstellar to interstellar environments.

Observations of scattering around some late-type stars imply the presence of a more pronounced excess of large grains with radii >0.5 μm that may carry as

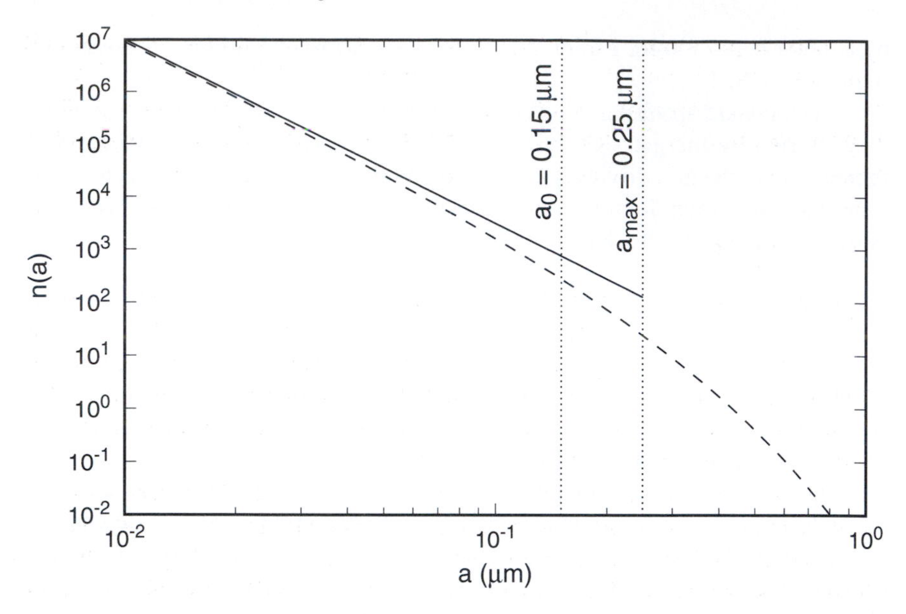

Figure 7.13. Comparison of the size distribution functions discussed in section 7.3.2. The full line represents the standard 'MRN' power law with a sharp upper bound at $a_{\max} = 0.25\ \mu$m. The broken curve represents the modified power law (equation (7.11)) with $a_0 = 0.15\ \mu$m.

much as ∼20% of the dust mass (Jura 1996). One possible explanation is that some dust-forming stars develop rotating equatorial discs, in which grains have more time to grow compared with those in an isotropically expanding shell.

7.3.3 Dust-to-gas ratio

The dust-to-gas ratio in an expanding circumstellar envelope may be determined, in principle, simply by comparing independent estimates of the mass-loss rate for dust and for gas. The former may be estimated from observations of far infrared continuum emission, the latter from millimetre-wave CO line emission (see Knapp 1985 and Olofsson *et al* 1993 for detailed discussion of techniques). Results may be subject to considerable error, arising principally from uncertainties in the drift speed of the dust relative to the gas, and possible systematic variations in this quantity with mass-loss rate. Knapp (1985) finds a mean value of $Z_d \approx 0.0063$ for O-rich stars. A marginally lower mean value of $Z_d \approx 0.005$ has been reported for carbon stars (Olofsson *et al* 1993), but given the degree of scatter and the possibility of systematic errors, this difference is probably not significant. We merely conclude that results for circumstellar envelopes are comparable with those for the diffuse ISM (see sections 2.4.2, 3.3.5 and 6.2.3).

Dust-to-gas ratios in planetary nebulae are typically an order of magnitude lower than those in the precursor stellar winds (Natta and Panagia 1981, Pottasch *et al* 1984). This is easily understood, as no new dust formation is expected in the nebula itself (section 7.1.5) and dust formed previously is being diluted by purely gaseous ejecta.

7.3.4 Composition

The primary stardust materials being injected into the interstellar medium are silicates and solid carbon (section 7.2); in general terms, these distinct components offer a natural basis for models that attribute interstellar extinction (section 3.7) and re-radiation (section 6.2) to a combination of such particles. Detailed comparisons yield differences, however, specifically in crystallinity, which we discuss briefly here.

As we have seen, the C-rich component of stardust seems to be predominantly amorphous. The observational evidence for this is quite strong and supported by theoretical predictions. The particles ejected by C-rich stars should thus range from PAHs and PAH clusters to sootlike amorphous carbon grains. The PAHs in stellar outflows are spectroscopically similar to interstellar PAHs in the infrared, whilst the soot produces mid-UV absorption resembling (but not identical to) the corresponding interstellar feature. In contrast, the meteoritic evidence points to the existence of ordered forms of carbon (diamond and graphite) in stardust at both extremes of the size distribution. Whilst amorphous carbon grains might become graphitized by interstellar processes, the graphitic mantles of the micron-sized meteoritic grains appear to have formed *in situ* in stellar ejecta. Probably these large grains are rare, whereas small, poorly graphitized carbon grains are ubiquitous. The more ordered forms of carbon stardust may originate primarily in supernovae rather than in red-giant winds.

Turning to the silicates, we face a similar problem. There is no evidence for crystalline silicates in the ISM, yet up to 30% of the silicates forming in stellar winds may be crystalline. It seems clear that this level of crystallinity should be seen, e.g by means of the structure it would introduce to the 9.7 μm profile, if it were routinely present in the ISM. However, the grains entering the ISM are not likely to remain crystalline indefinitely, as they are subject to long-term exposure by cosmic rays. This will result in destruction of long-range order within the particles and corresponding changes in their spectroscopic properties over time (see Nuth *et al* 2000 for further discussion).

7.3.5 Injection rate

An estimate of the rate at which stardust is injected into the ISM is desirable as a means of assessing quantitatively the contribution of stardust to interstellar dust. To accomplish this, it is necessary to evaluate number densities, mass-loss rates and dust-to-gas ratios for all types of star thought to contribute to the process.

Table 7.3. Estimates of integrated mass-loss rates in $M_\odot$ yr^{-1} for sources of stardust in the disc of the Galaxy. The second column indicates dust type (C rich or O rich). The dust injection rate (column 4) is calculated from the mass-loss rate (column 3) assuming a value of $Z_d = 0.006$ for the dust to gas ratio.

Stellar type	C or O	$[\dot{M}_G]$	$10^3\ [\dot{M}_G]_d$
O-rich AGB	O	0.5	3
C-rich AGB	C	0.5	3
Supernovae	both?	0.2	1 (?)
M giants	O	0.04	0.2
M supergiants	O	0.02	0.1
WC stars	C	0.01	0.06
Novae	both	0.003	0.02

Initially, we shall consider the entire mass-loss budget (gas and dust) and then deduce values for dust by factoring in the dust-to-gas ratio (see table 7.3). The integrated mass-loss rate for the galactic disc may be written

$$[\dot{M}_G] = A_G[\dot{M}_*]N_* \tag{7.12}$$

where $[\dot{M}_*]$ is the mean mass-loss rate in $M_\odot$yr^{-1} for stars of surface number density N_* (kpc^{-2}) and $A_G \approx 1000$ kpc^2 is the cross-sectional area of the Galaxy in the plane of the disc. The rate at which matter is returned to the ISM is the sum of $[\dot{M}_G]$ values for each type of star that makes a significant contribution.

We first consider intermediate-mass stars, i.e. those with main-sequence masses approximately in the range 1–8 $M_\odot$. Some dust may be produced on the first ascent red-giant branch (e.g. Omont *et al* 1999) but the mass-loss rates are modest. The entry for M giants in table 7.3 is estimated from their observed space density (e.g. Mihalas and Binney 1981) and an average mass-loss rate of 2×10^{-8} $M_\odot$ yr^{-1} (e.g. figure 7.12). The dominant mass-loss phase occurs on the asymptotic giant branch, culminating in the ejection of a planetary nebula (section 7.1.5). Estimates of $[\dot{M}_*]$ tend to be dominated by a relatively small number of stars with very high mass-loss rates (e.g. Thronson *et al* 1987). Completeness of sampling is a potential source of error, especially as the stars with the highest mass-loss rates naturally tend to have the thickest shells, rendering them invisible at shorter wavelengths. The possibility of sampling biases between O-rich and C-rich stars is also a concern: this might arise from differences in spatial distribution, or because carbon stars tend to have optically thicker shells for a given mass of dust compared with their O-rich counterparts (Thronson *et al* 1987, Epchtein *et al* 1990, Guglielmo *et al* 1998). To illustrate the problem, several authors have estimated that the mass-loss from AGB stars, summed over all C/O ratios, is in the range 0.3–0.6 $M_\odot$ yr^{-1} (Knapp and Morris

1985, Thronson *et al* 1987, Jura and Kleinmann 1989, Sedlmayr 1994, Wallerstein and Knapp 1998), yet Epchtein *et al* (1990) obtained $\sim$0.5 $M_\odot$ yr^{-1} for infrared carbon stars alone. I shall adopt the approach of assuming that the lowest estimates are, in fact, lower limits (because of incomplete sampling). Taking the Epchtein *et al* result for carbon stars and assuming that the contribution from O-rich AGB stars is at least comparable in the solar neighbourhood (see Jura and Kleinmann 1989), the total is $\sim$1 $M_\odot$ yr^{-1} for all AGB winds. Planetary nebula ejection adds a further $\sim$0.3 $M_\odot$ yr^{-1} (Maciel 1981), for a grand total of $\sim$1.3 $M_\odot$ yr^{-1} over the entire post-main-sequence lifetime of intermediate-mass stars.

An independent check on this result can be obtained from the observed formation rate for white dwarfs, assuming that all main sequence stars within the appropriate mass range ultimately become white dwarfs of mass $\sim$0.7 $M_\odot$ after passing through AGB and PN-ejection phases. This approach yields values in the range 0.8–1.5 $M_\odot$ yr^{-1} (Salpeter 1977, Jura and Kleinmann 1989), consistent with the previous estimate based on mass-loss rates and number densities.

Turning to more massive stars, we expect continuous mass-loss in radiatively driven winds, followed by explosive ejection of supernova remnants. Two types of star, red supergiants and Wolf–Rayet stars, are of interest during the wind phase. Although some red supergiants have high mass-loss rates ($\sim$10^5 $M_\odot$ yr^{-1} or more), their number density is low and consequently their overall contribution the the total mass-loss budget is quite small, $\sim$0.02 $M_\odot$ yr^{-1} (Jura and Kleinmann 1990), or only a few per cent of that from AGB stars. A similar situation arises for Wolf–Rayet stars. Only late WC stars, representing about 15% of the Wolf–Rayet population, appear to make dust, and data on their space density and mass-loss rates (Abbott and Conti 1987) lead to an estimate of $\sim$0.01 $M_\odot$ yr^{-1}. The contribution of supernovae may be estimated from the product of frequency and ejected mass. The average frequency for galaxies of similar Hubble type to our own is of order 1 per 50 years (section 1.3.2) and taking $\sim$10 $M_\odot$ as a typical value for the ejected mass, we obtain a rate of $\sim$0.2 $M_\odot$ yr^{-1}. These results, summarized in table 7.3, indicate that massive stars contribute considerably less than intermediate-mass stars to the overall mass-loss budget.

Novae occur in the Galaxy with a frequency $\sim$30 per year but the total mass ejected per event is typically no more than 10^{-4} $M_\odot$ (Bode 1988). These figures lead to a very small estimated contribution to the mass-loss budget (table 7.3).

The dust injection rate is simply calculated from estimates of $[\dot{M}_G]$:

$$[\dot{M}_G]_d = Z_d\,[\dot{M}_G] \tag{7.13}$$

where Z_d is the mean dust-to-gas ratio for the circumstellar material. Results appear in the right-hand column of table 7.3. A value of $Z_d \approx 0.006$ is adopted for stellar winds (section 7.3.3). Planetary nebulae are excluded from table 7.3 because, as previously discussed, they appear not to be important sources of new dust but merely recyclers of dust formed earlier, in the AGB phase; in any case, their dust-to-gas ratios are much lower than those in AGB winds (sections 7.1.5

and 7.3.3). The dust-to-gas ratio in SN ejecta is unknown but available evidence suggests that it is not particularly high. This is indicated both by studies of SN 1987A (reviewed in section 7.2.1) and searches for dust in young supernova remnants in the solar neighbourhood (Lagage *et al* 1996, Douvion *et al* 2001). A value of Z_d equal to that in stellar winds is assumed in table 7.3; this should probably be treated as an upper limit, although some investigators may disagree (see Dwek 1998).

Whilst bearing in mind that the data are subject to considerable uncertainty, we may draw a general conclusion from the results in table 7.3: AGB winds are the dominant source of new stardust in the ISM. This would still be true even if the lowest estimates of the AGB injection rate were to be adopted (see previous discussion). Only a substantial upward revision of the supernova injection rate would offer a serious challenge. The contributions of all other types of star are negligible compared with these.

The balance between C-rich and O-rich stardust is a question of some interest and uncertainty. Whilst the results in table 7.3 follow Jura and Kleinmann (1989) in assuming similar contributions from AGB stars of each type in the solar neighbourhood, others have argued that the contribution from O-rich stars is dominant overall (Thronson *et al* 1987, Bode 1988). The case for dominance by O-rich stardust is supported by the observation (section 5.2) that whilst silicates are ubiquitous in the ISM, SiC is evidently rare.

Is stardust a major component of interstellar dust? We can seek an answer to this question by comparing the injection timescale with that for destruction in the ISM. Summing the entries in column 4 of table 7.3 gives a total injection rate $\sim$0.007 $M_\odot$ yr^{-1}. Injection thus contributes to the mass density of interstellar dust at a rate

$$\dot{\rho}_d = \frac{\sum[\dot{M}_G]_d}{V_G} \sim 7 \times 10^{-33} \text{ kg m}^{-3} \text{ yr}^{-1} \tag{7.14}$$

with attention to units, where V_G is the volume of the galactic disc (taken to have radius 15 kpc and thickness 100 pc). The timescale for injection of stardust into the ISM is thus

$$t_{in} = \frac{\rho_d}{\dot{\rho}_d} \sim 2.5 \text{ Gyr} \tag{7.15}$$

where the diffuse-ISM value of $\rho_d \approx 1.8 \times 10^{-23}$ kg m^{-3} is assumed (section 3.3.5). Grain destruction in the ISM is dominated by shocks (see section 8.5.1 for a review). The timescale for destruction in shocks is estimated to be $t_{sh} \sim 0.5$ Gyr (Jones *et al* 1996), i.e. much shorter than that for injection: grains are apparently being destroyed more rapidly than stellar mass-loss can replenish them. The equilibrium mass fraction of interstellar dust originating in stars is

$$f_{sd} = (1 + t_{in}/t_{sh})^{-1} \tag{7.16}$$

and inserting these values gives $f_{sd} \sim 0.2$, i.e. stardust appears to account for only about 20% of the mass of refractory interstellar dust at any given time.

It is conceivable that the injection rates discussed in this section have been underestimated, perhaps by as much as a factor of two, if the census of high-mass-loss stars is incomplete. But halving our estimate of $t_{\rm in}$ in equation (7.16) increases $f_{\rm sd}$ to only 30%. Perhaps the destruction rate has been overestimated (section 8.5.1). If we take them at face value, these results strongly suggest that grain material is being replenished efficiently by *interstellar* processes.

Recommended reading

- *Formation and Destruction of Dust Grains*, by Edwin E Salpeter, Annual Reviews of Astronomy and Astrophysics, 15, 267–93 (1977).
- *Mass Loss from Cool Stars*, by A K Dupree, Annual Reviews of Astronomy and Astrophysics, 24, 377–420 (1986).
- *Grain Formation and Metamorphism*, by Joseph A Nuth, in The Cosmic Dust Connection, ed J M Greenberg (Kluwer, Dordrecht) pp 205–21 (1996).
- *Ancient Stardust in the Laboratory*, by Thomas J Bernatowicz and Robert M Walker, Physics Today, 50 (12), 26–32 (1997).
- *Carbon Stars*, by George Wallerstein and Gillian R Knapp, Annual Reviews of Astronomy and Astrophysics, 36, 369–433 (1998).

Problems

1. Explain the importance of the C/O ratio as a determinant of the composition of dust condensing in a stellar atmosphere. How is the C/O ratio likely to evolve with time in a typical star of intermediate mass?
2. Consider a red-giant atmosphere with $C/O = 1$ (carbon and oxygen of exactly equal abundance) and with other elements present at solar abundance levels. Discuss with reasoning which of the following statements are most likely to be true:
   (a) The star will produce only carbonaceous dust.
   (b) The star will produce only silicate dust.
   (c) The star will produce both carbonaceous and silicate dust.
   (d) The star will produce neither carbonaceous nor silicate dust but may produce some dust of purely metallic composition.
   (e) The star will produce no dust at all.
3. Show that the mass-loss rate of a star with a spherically symmetric expanding envelope is given by the equation

$$\dot{M} = 4\pi r^2 \rho(r) v$$

   where $\rho(r)$ is the density of circumstellar matter in the outflow at radial distance r from the centre of the star and v is the expansion speed.
4. The supergiant Betelgeuse (spectral type M2 Ib) is losing mass due to nucleation and growth of dust grains accelerated by radiation pressure to

a terminal speed of 15 km s^{-1} at a radial distance of 500 $R_{\odot}$ from the centre of the star. If the density of gas is 2×10^{-11} kg m^{-3} at this distance, calculate the mass-loss rate in $M_{\odot}$/yr (assuming spherical symmetry). How does your result compare with values for stars plotted in figure 7.12 of similar spectral type? How many stars like Betelgeuse would be needed in the galactic disc to explain the integrated galactic mass-loss rate from red supergiants?

5. Use equation (7.6) to estimate the mass-loss rate for a first ascent red giant of mass 1.3 $M_{\odot}$, radius 50 $R_{\odot}$ and luminosity 500 $L_{\odot}$.
6. Write a critical discussion of the arguments leading to the conclusion that sources other than stellar ejecta are needed to explain the abundance of dust in the interstellar medium.

Chapter 8

Evolution in the interstellar medium

"Once the newly formed grains are injected into the interstellar medium, they are subject to a variety of indignities..."

C G Seab (1988)

In this chapter, we examine the lifecycle of dust from its injection into the interstellar medium by evolved stars (chapter 7) to its incorporation into the envelopes of newly formed stars in molecular clouds. This lifecycle is illustrated schematically in figure 8.1. A vast range of physical conditions is encountered along the way (see section 1.4). The interaction of the dust with the ambient radiation field determines the grain temperature (section 6.1) in each phase and controls evaporation and annealing rates. Low-density intercloud regions of the ISM are permeated by shocks and energetic photons capable of destroying even the most hardy grain materials. Within a molecular cloud, however, the grains are temporarily shielded from both shocks and dissociative radiation. Here they may undergo rapid growth via coagulation and deposition of volatile surface coatings. Icy mantles provide both a repository for gas-phase atoms and molecules and a substrate for the production of new molecular species in the gas. The exchange of matter between gas and dust regulates the chemical evolution of the cloud as a whole. The mantles are composed, for the most part, of species formed by surface reactions and these reactions open up new pathways to molecule formation that are not possible in the gas phase. The mantles provide both sinks and sources for gaseous molecules as a function of time. The onset of starbirth exposes the grains once more to stellar radiation and winds, and this input of energy may drive chemical reactions toward greater molecular complexity. Understanding these processes is vital to the search for our origins because they govern the nature of the material available to form planets around newly born stars.

We begin this chapter by discussing the attachment of gas-phase atoms onto grain surfaces and the subsequent recombination of molecular hydrogen by surface catalysis (section 8.1). Important gas-phase reactions that influence the

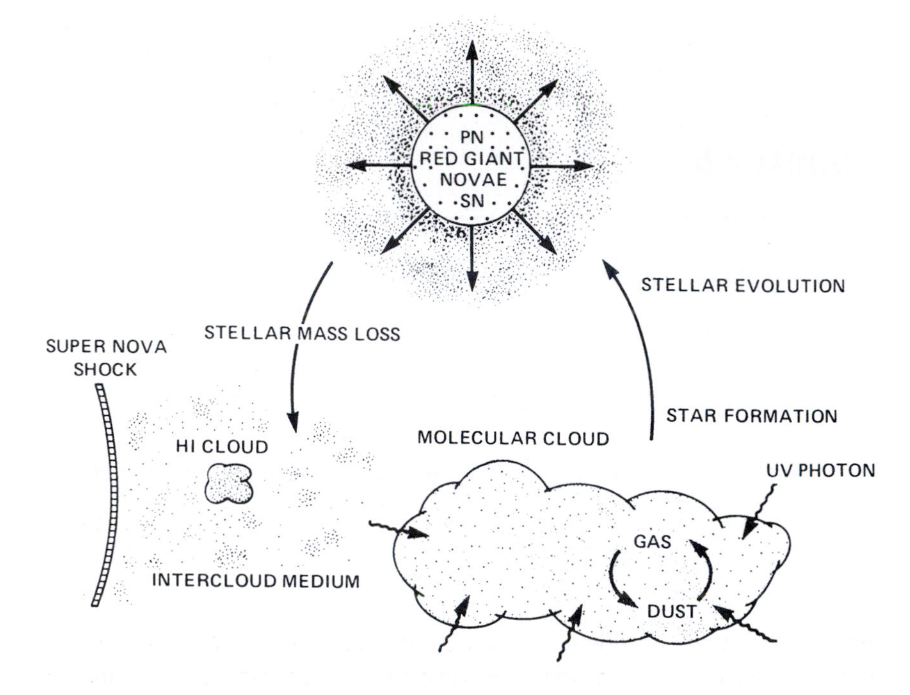

Figure 8.1. Schematic representation of the lifecycle of cosmic dust. Grains of 'stardust' originating in the atmospheres and outflows of evolved stars (red giants, planetary nebulae, novae and supernovae) are ejected into low-density phases of the interstellar medium, where they are exposed to ultraviolet irradiation and to destruction by shocks. Within molecular clouds, ambient conditions favour the growth of volatile mantles on the grains. Subsequent star formation leads to the dissipation of the molecular clouds. (From Tielens and Allamandola 1987b; reprinted by permission of Kluwer Academic Publishers.)

chemical evolution of the ISM are reviewed in section 8.2. We then consider mechanisms for the growth of dust grains in interstellar clouds (section 8.3) and discuss in detail the deposition and evolution of icy molecular mantles (section 8.4). Processes acting on refractory dust in diffuse phases of the ISM are reviewed in the final section.

8.1 Grain surface reactions and the origin of molecular hydrogen

The interaction of gas and dust is depicted schematically in figure 8.2. Atoms impinging upon the surface of a grain may become attached (adsorbed) and may subsequently migrate, interact with other atoms and desorb. Attachment may be physical or chemical and the binding energies involved are markedly different in

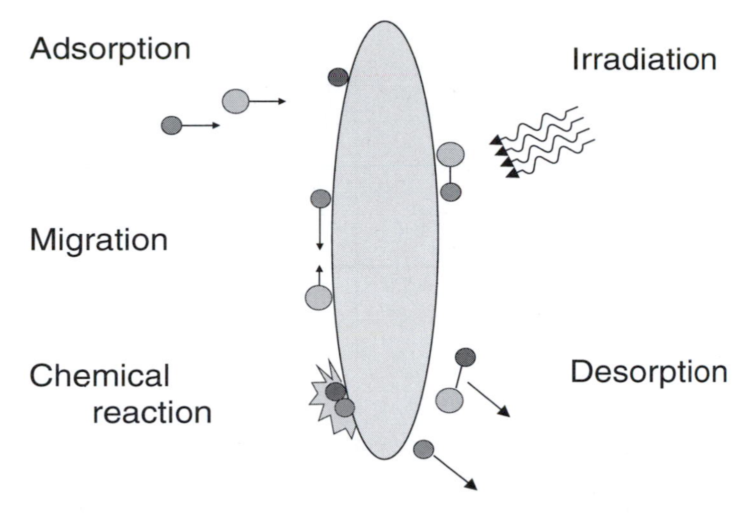

Figure 8.2. Schematic illustration of grain surface interactions. Figure courtesy of Perry Gerakines.

each case. Physical attachment is maintained by weak van der Waals forces with binding energies <0.1 eV, the surface acting merely as a passive substrate. The surfaces of real solids terminate in atoms or ions in states of low symmetry and low coordination and these tend to be chemically active. Moreover, interstellar grain surfaces are likely to be highly disordered, containing numerous lattice defects and impurities that enhance their chemical reactivity (see, for example, pp 92–5 of Duley and Williams 1984). Amorphous carbon dust may have a high concentration of active surface sites arising from the presence of carbon atoms with 'dangling' bonds and, in interstellar clouds, these will tend to become saturated with hydrogen. Similarly, silicate and oxide grains will have active surfaces arising from the presence of oxygen ions in low coordination sites and these will also tend to attach hydrogen. The binding energies for chemical adsorption are typically in the range 1–5 eV, i.e. at least an order of magnitude higher than those for physical adsorption.

Figure 8.3 illustrates potential energy diagrams for attachment of atoms and molecules onto a surface. Note that the forces for physical adsorption have longer range than those for chemical bonding and two distinct potential wells therefore occur: a particle approaching the surface first encounters a shallow well associated with physical adsorption forces, followed by a deeper well associated with chemical bonding. Chemisorption of accreting atoms is limited only by the availability of active sites and a surface in which all such sites are occupied may accrete further atoms into an outer layer by physical adsorption. The situation is different for accreting molecules, in that chemisorption is often inhibited by the presence of an activation energy barrier arising from the need to break or modify a molecular bond (Tielens and Allamandola 1987a). Chemisorption of molecules

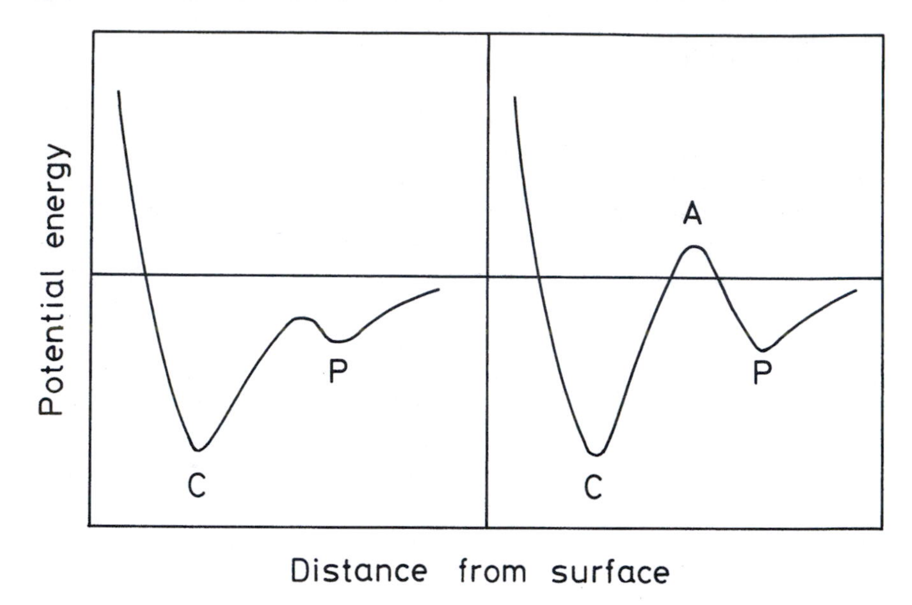

Figure 8.3. Schematic potential energy diagrams for the adsorption of an atom (left) and a molecule (right) onto the surface of a solid (adapted from Tielens and Allamandola 1987a). In each case, a particle approaching the surface (from right to left) encounters a relatively shallow negative potential well (P) associated with physical adsorption forces and a deeper well (C) associated with chemical bonding. In the molecular case, chemical bonding is inhibited by an activation energy barrier (A).

is thus generally unimportant in interstellar clouds.

The rate at which a dust grain adsorbs atoms or molecules from its environment may be expressed as a function of the kinetic temperature and density of the gas. Assuming a Maxwellian distribution of velocities, the mean speed of a particle of mass m is

$$\bar{v} = \left(\frac{8kT_{\mathrm{g}}}{\pi m}\right)^{\frac{1}{2}} \tag{8.1}$$

(Spitzer 1978), where T_{g} is the gas temperature. In diffuse interstellar clouds, we are concerned primarily with the adsorption of atomic hydrogen, and assuming $T_{\mathrm{g}} \approx 80$ K (table 1.1), equation (8.1) gives $\bar{v}_{\mathrm{H}} \approx 1.4$ km s^{-1}. Kinetic motion leads to collisions between the gas atoms and the grain surface. Assuming the grain to be spherical of radius a, immersed in a gas of number density n_{H}, the number of collisions per second is $\pi a^2 n_{\mathrm{H}} \bar{v}_{\mathrm{H}}$. An impinging atom may be adsorbed or returned to the gas; the probability of adsorption is denoted by ξ, the sticking coefficient, which has a value in the range $0 \leq \xi \leq 1$. The number of atoms adsorbed per second is therefore $\xi \pi a^2 n_{\mathrm{H}} \bar{v}_{\mathrm{H}}$ and the timescale (mean free time) for adsorption events is the reciprocal of this quantity,

$$t_{ad} = \frac{1}{\xi \pi a^2 n_H \bar{v}_H}. \tag{8.2}$$

The sticking coefficient depends on the nature of the surface and on the temperatures of both the gas and the dust. In a hot gas, in which the mean kinetic energy of the impinging atoms exceeds the adsorption energy, collisions tend to be elastic with low probability of attachment ($\xi \ll 1$). However, theoretical calculations and experimental investigations strongly suggest that $\xi \approx 1$ for chemical adsorption of atoms onto amorphous carbon or amorphous silicate grains in interstellar clouds (Tielens and Allamandola 1987a and references therein). With this value in equation (8.2), we estimate $t_{ad} \approx 800$ s for classical ($a \approx 0.1\ \mu$m) grains in diffuse clouds of density $n_H \approx 3 \times 10^7$ m^{-3}.

The lifetime of an adsorbed atom against evaporation depends on the dust temperature T_d and the binding energy ε and is given by

$$t_{ev} = t_0 \exp\left(\frac{\varepsilon}{kT_d}\right) \tag{8.3}$$

where t_0 is the oscillation time period of the adsorbed species in the direction perpendicular to the surface (typically 10^{-12} s). For chemical binding ($\varepsilon \sim 3$ eV), we have $\varepsilon \gg kT_d$ for any reasonable value of T_d, and so t_{ev} is generally long compared with cloud lifetimes: i.e. a chemisorbed atom is essentially stable with respect to thermal evaporation. For physical binding, t_{ev} is much shorter and critically dependent on temperature: if, for example, we set $\varepsilon = 0.03$ eV, then values of 1.3×10^3 s and 1.1×10^{-7} s are predicted by equation (8.3) for $T_d = 10$ and 30 K, respectively.

The accretion of atoms onto grains is of primary importance in interstellar chemistry because it provides the only plausible mechanism for the efficient production of molecular hydrogen (Hollenbach and Salpeter 1971, Duley and Williams 1984, Katz *et al* 1999). This process may be written symbolically as

$$\mathrm{H + H + grain \rightarrow H_2 + grain.} \tag{8.4}$$

The H atoms cannot recombine directly by two-body encounters in the gas phase because there is no feasible method of losing binding energy. The surface of a grain provides both a substrate on which the atoms become readily trapped and a heat reservoir that absorbs excess energy. Three distinct steps are involved (figure 8.2): (i) attachment of atoms to the surface, as described above; (ii) migration of an atom from one surface site to another until a second atom is encountered; and (iii) recombination, leading to the ejection of the resulting H_2 molecule from the grain surface.

The mobility of an atom attached to a dust grain is limited by the energy binding it to the surface, and this varies with position because of the discrete nature of the lattice (e.g. Tielens and Allamandola 1987a). The potential barrier against diffusion to an adjacent site is typically 30% of the binding energy for physical adsorption and 50–100% of the binding energy for chemical adsorption. The migration of H atoms involves quantum mechanical tunnelling, which occurs on a timescale of 3×10^{-9} s for physical adsorption. This is generally short compared with the timescale for evaporation at low temperatures ($T_d < 30$ K). Thus, migrating physically adsorbed atoms rapidly seek out and fill any vacant sites for chemisorption on the grain. H_2 formation is most likely to involve an encounter between a migrating physically adsorbed atom and a trapped chemisorbed atom. Binding energy of 4.48 eV is released on recombination, some fraction of which will be imparted to the newly formed molecule as rotational excitation and translational motion, ejecting it from the surface; any excess energy is absorbed by the grain as lattice vibrations.

8.2 Gas-phase chemistry

Interstellar clouds contain a wide variety of gaseous molecules: see van Dishoeck *et al* (1993) and Snyder (1997) for discussion and tabulations. Although grain surface reactions provide the only effective route to H_2, the abundances of many other observed species may be explained by chemical reactions taking place in the gas phase, given the prior availability of H_2. A detailed discussion of gas-phase chemical models is beyond the scope of this book, but it is appropriate to summarize the processes leading to the formation of important species that might subsequently condense onto the dust. For comprehensive accounts of gas-phase chemistry, see, for example, Herbst and Klemperer (1973), Mitchell *et al* (1978), Prasad and Huntress (1980), Duley and Williams (1984), Herbst and Leung (1986), Herbst (1987), Bergin *et al* (1995) and Lee *et al* (1996).

Under normal interstellar conditions, only binary (two-body) reactions need be considered, as the probability of three-body reactions in the gas phase is extremely low. In general, only exothermic reactions are likely to occur at a significant rate in quiescent clouds; certain weakly endothermic reactions are thought to be driven by shocks but these need not concern us here. Consider the formation of some molecule M by the schematic reaction

$$\mathrm{A} + \mathrm{B} \rightarrow \mathrm{M} + \mathrm{N}. \tag{8.5}$$

The reactants A and B may be atomic or molecular, neutral or ionic. The additional product N absorbs binding energy from the molecule M; N may be atomic, molecular or, in some instances, a photon. Reaction (8.5) will have some rate coefficient ζ associated with it that limits the abundance of M. In the diffuse ISM, molecules are destroyed predominantly by photodissociation and if the mean lifetime of molecule M against photodissociation, Δt, is short compared with cloud lifetimes, then the number density of M is given in terms of the number densities of A and B by the equilibrium equation

$$n_\mathrm{M} = \zeta n_\mathrm{A} n_\mathrm{B} \Delta t. \tag{8.6}$$

If the reactants bear no charge, the value of ζ is typically 10^{-17} m^3 s^{-1} (e.g. Duley and Williams 1984). However, if one of the reactants is ionized, much higher rates can occur, with ζ enhanced typically by a factor of 100. This arises because the electric field of the ion polarizes the neutral species, resulting in a net attractive Coulombic force. Exothermic ion–molecule reactions thus play an important and, indeed, dominant role in current models for gas-phase chemistry and such models are generally quite successful in reproducing the observed abundances of many species. A supply of ions is maintained by radiative and cosmic-ray ionization, with the latter dominating in dense clouds shielded from the external radiation field. It should be noted that in dense clouds, photodissociation timescales may become long compared with timescales for dynamical evolution, and molecular abundances may not reach a steady state.

As an illustration of ion–molecule chemistry, we will consider the formation of water, ammonia and methane. Water may be produced via a short sequence of reactions between O-bearing ions and molecular hydrogen. O^+ is assumed to be present, as a result of either direct photoionization or charge transfer reactions such as

$$O + H^+ \rightarrow O^+ + H. \tag{8.7}$$

O^+ is then processed to H_3O^+ by the ion–molecule sequence:

$$\begin{aligned} O^+ + H_2 &\rightarrow OH^+ + H \\ OH^+ + H_2 &\rightarrow H_2O^+ + H \\ H_2O^+ + H_2 &\rightarrow H_3O^+ + H \end{aligned} \tag{8.8}$$

and, finally, H_2O results from recombination of H_3O^+:

$$H_3O^+ + e^- \rightarrow H_2O + H. \tag{8.9}$$

An analogous set of reactions, substituting N for O in equations (8.7)–(8.9) and with one extra step in the sequence (8.8), leads to NH_3. However, carbon chemistry must be initiated in a different way, as formation of CH^+ by

$$C^+ + H_2 \rightarrow CH^+ + H \tag{8.10}$$

is endothermic. An alternative route to CH^+ involves the H_3^+ ion, formed via

$$H_2^+ + H_2 \rightarrow H_3^+ + H \tag{8.11}$$

which may then react with neutral carbon:

$$C + H_3^+ \rightarrow CH^+ + H_2. \tag{8.12}$$

CH^+ ions produced by equation (8.12) may subsequently react with H_2 to form CH_2^+, but CH_2^+ may also form directly by radiative association:

$$C^+ + H_2 \rightarrow CH_2^+ + h\nu. \tag{8.13}$$

In any case, CH_2^+ may undergo further processing via ion–molecule reactions analogous to sequence (8.8). Recombination may occur at any stage, leading to the production of CH, CH_2, CH_3 and, ultimately, CH_4. Equilibrium abundances of interstellar molecules depend on the rate coefficients of each step of the formation sequence and the dissociation and recombination cross sections of the products. Over a wide range of cloud densities and physical conditions, saturated molecules such as H_2O, NH_3 and CH_4 have relatively low abundances and account for only a minor fraction of the available O, N and C (e.g. Herbst and Leung 1986).

CO is by far the most abundant gas-phase molecule after H_2. CO is formed in diffuse clouds (van Dishoeck and Black 1987) by the ion–molecule reaction

$$C^+ + OH \rightarrow CO + H^+ \tag{8.14}$$

and by the alternative sequence

$$\begin{aligned} C^+ + OH &\rightarrow CO^+ + H \\ CO^+ + H_2 &\rightarrow HCO^+ + H \\ HCO^+ + e &\rightarrow CO + H. \end{aligned} \tag{8.15}$$

In dense clouds, CO is produced directly by the neutral exchange reaction

$$CH + O \rightarrow CO + H. \tag{8.16}$$

and by recombination of HCO^+ formed by the reactions

$$C^+ + H_2O \rightarrow HCO^+ + H \tag{8.17}$$

and

$$CH_3^+ + O \rightarrow HCO^+ + H_2. \tag{8.18}$$

Once created, CO is difficult to destroy because of its relatively large binding energy, as previously noted in our discussion of cool stellar atmospheres (section 7.1). In diffuse clouds, CO is destroyed principally by UV photodissociation; in dense clouds, the most important sink for gas-phase CO is probably adsorption onto grains.

Table 8.1 lists abundances for 24 atomic and molecular species, as predicted by the model of Herbst and Leung (1986) for a cloud of density $n = 2 \times 10^9$ m^{-3}, temperature $T_g = 50$ K and visual extinction $A_V = 2$ mag. These parameters are chosen to represent the intermediate, transitional state between a diffuse cloud and a dense cloud. Results may thus provide a reasonable guide to the abundances of atoms and molecules available to condense onto grains as the density reaches a critical value for the onset of rapid mantle growth. The abundances listed are generally in fair agreement with observations of lines of sight thought to have corresponding mean physical conditions. It is notable that, with the single exception of CO, all molecular species listed have very low abundances ($<5 \times 10^{-7}$) compared with those of atomic C, N and O.

Table 8.1. Predicted gas-phase abundances for atomic and molecular species containing the 'CHON' group of elements in a cloud of intermediate density, based on the model of Herbst and Leung (1986). Abundances are expressed as a fraction of $N_H = N(HI) + 2N(H_2)$.

Species	N_X/N_H	Species	N_X/N_H
C	2.1×10^{-5}	CH_2	1.9×10^{-7}
C^+	1.2×10^{-5}	NH_2	1.6×10^{-9}
N	2.1×10^{-5}	H_2O	1.0×10^{-8}
O	1.4×10^{-4}	HCN	1.1×10^{-9}
CH	3.7×10^{-7}	HCO	2.0×10^{-10}
NH	1.9×10^{-9}	HCO^+	1.3×10^{-10}
OH	3.0×10^{-8}	C_3	8.5×10^{-8}
C_2	4.9×10^{-7}	CO_2	3.2×10^{-11}
CO	3.9×10^{-5}	NH_3	4.7×10^{-12}
CN	1.5×10^{-7}	CH_3	4.2×10^{-10}
CS	3.2×10^{-8}	H_2CO	6.0×10^{-11}
H_3^+	4.7×10^{-9}	CH_4	7.0×10^{-10}

8.3 Mechanisms for growth

The size distribution of interstellar dust is determined by the balance of constructive and destructive processes acting in the prevailing environment. Grains may increase their average size and mass in two distinct ways: (i) by coagulation, as a result of low-impact grain–grain collisions; and (ii) by mantle growth, in which gas-phase atoms and molecules accumulate on their surfaces. In either case, the rate of growth increases with collision rate and, hence, with cloud density. There is ample observational evidence, reviewed in previous chapters, for selective growth of grains in dense clouds: this includes extinction and polarization curve variations (sections 3.4 and 4.3) and spectroscopic detection of mantle constituents (section 5.3). Enhanced adsorption rates in interstellar clouds are also implied by the correlation of depletion with mean density for certain elements (section 2.4.3; see figure 2.8). However, abundant refractory elements such as Mg, Si and Fe generally have high fractional depletions even in relatively low-density regions and there is little scope for significant growth by further adsorption of these elements. Grain mantles formed by adsorption from the gas are expected to consist of compounds predominantly involving C, N and O, and this is confirmed by spectroscopic studies of molecular ices. An important difference between coagulation and mantle growth is that whereas the former merely redistributes existing grain mass with respect to size, the latter involves the creation of new grain material. Complete deposition of all condensible elements into the mantles would increase the dust-to-gas ratio by a factor of about two compared with the diffuse ISM average (see section 2.4.2).

8.3.1 Coagulation

Both thermal Brownian motion and local turbulence will drive grain–grain collisions in interstellar clouds. These collisions may lead to coagulation or shattering of the incident particles, dependent on their structure, composition and relative speed. For typical interstellar grains, the critical impact speed above which grains will shatter is ~ 1 km s^{-1} (Jones *et al* 1996). For comparison, the average thermal speed of dust grains in an interstellar cloud of density $n_{\rm H} = 10^9$ m^{-3} is $\bar{v}_{\rm d} \sim 0.1$ km s^{-1} (Jura 1980). i.e. well below the critical value for shattering. In general, grains will thus tend to coagulate within clouds and shatter when exposed to shocks in lower-density regions. The timescale for collisions between grains of number density $n_{\rm d}$ and cross-sectional area $\sigma_{\rm d}$ is

$$t_{\rm col} = \frac{1}{n_{\rm d}\sigma_{\rm d}\bar{v}_{\rm d}} \approx \frac{10^{25}}{n_{\rm H}\bar{v}_{\rm d}} \tag{8.19}$$

where the approximation is based on the mean projected area of dust grains per H atom estimated by Spitzer (1978: pp 161–2). For the cloud considered above, equation (8.19) yields $t_{\rm col} \sim 3$ Myr, which is small compared with typical cloud lifetimes of 10–100 Myr. Coagulation may thus lead to substantial evolution of the grain-size distribution during the lifetime of a typical interstellar cloud.

Observations show that the grain-size parameter R_V, the ratio of total-to-selective extinction, is often enhanced in denser clouds (section 3.4). This effect is attributed to changes in the optical properties of the dust at blue-visible wavelengths, consistent with a reduction in the relative number of small grains. It is not necessarily accompanied by the appearance of spectral signatures of ice mantles (e.g. Whittet and Blades 1980). In the ρ Oph cloud, for example, stars with moderate extinction ($1 < A_V < 10$) show significant ($\sim$40%) enhancement in R_V compared with the diffuse ISM, whereas ice mantles are detected only toward sources with $A_V > 10$ (Tanaka *et al* 1990). High R_V values are thus associated with coagulation rather than with mantle growth: the relative number of small grains is naturally reduced as grains coagulate and this has a systematic effect on the shape of the extinction curve from the visible to the ultraviolet (Mathis and Whiffen 1989, Kim and Martin 1996, Mathis 1996b, O'Donnell and Mathis 1997, Weingartner and Draine 2001). The power-law form of the size distribution function implied by fits to extinction data (section 3.7) is consistent with this interpretation: it is predicted in any situation where grain size is regulated by coagulation and shattering (Biermann and Harwit 1980, Jones *et al* 1996), as previously discussed in the context of stellar outflows (section 7.3.2).

Independent evidence to support coagulation as the growth mechanism toward ρ Oph is provided by the observed ratio of hydrogen column density to extinction ($N_{\rm H}/A_V$), which exceeds the mean value for the diffuse ISM by a factor ~2. This enhancement is predicted by a coagulation model (Jura 1980, Kim and Martin 1996) in which grains tend to 'hide behind' one another, reducing the effective opacity in the visible. In contrast, a model based on mantle growth would predict a *reduction* in N_H/A_V.

The structure of grains produced by the coagulation process depends on the size and structure of the incident particles (Dorschner and Henning 1995; see Dominik and Tielens 1997 for detailed discussion of the relevant physics). Compact, quasi-spherical aggregates tend to be favoured, especially for impactors at the lower and upper limits of the size spectrum. As the smallest grains have the largest surface-to-volume energy ratios, they tend to form compact clusters that minimize their surface area. For the largest grains, energy transfer during a collision becomes comparable with internal binding energy, causing loose aggregates to shatter and favouring the most compact structures by 'natural selection'. In the size range between these extremes, loose 'fractal' aggregates are more likely to occur. The growth of clusters may proceed through sequential addition of individual subgrains to the cluster (particle–cluster aggregation) or through coagulation of clusters (cluster–cluster aggregation). Some sample computer simulations are shown in figure 8.4.

8.3.2 Mantle growth

Consider a spherical grain of radius a immersed in a gas of identical particles of mass m and number density n. The grain adsorbs particles at a rate $\xi\pi a^2 n\bar{v}$ (section 8.1), where ξ is the sticking coefficient and $\bar{v}$ the mean particle speed as defined in equation (8.1). The rate of increase in the mass of the dust grain is thus

$$\frac{\mathrm{d}m_\mathrm{d}}{\mathrm{d}t} = \xi\pi a^2 n(2.5kT_\mathrm{g}m)^{\frac{1}{2}} \tag{8.20}$$

where T_g is the kinetic temperature of the gas. If adsorption leads to the formation of a mantle of density s, then the rate of growth of the grain is given by

$$\begin{aligned}\frac{\mathrm{d}a}{\mathrm{d}t} &= \frac{1}{4\pi a^2 s}\frac{\mathrm{d}m_\mathrm{d}}{\mathrm{d}t}\\ &= 0.4\xi n s^{-1}(kT_\mathrm{g}m)^{\frac{1}{2}}.\end{aligned} \tag{8.21}$$

Assuming that the quantities on the right-hand side of equation (8.21) are constant with respect to time, integration gives the radius at time t simply as

$$a(t) = a_0 + \Delta a(t) \tag{8.22}$$

where a_0 is the radius at $t = 0$, i.e. the core radius, and

$$\Delta a(t) = 0.4\xi n s^{-1}(kT_\mathrm{g}m)^{\frac{1}{2}}t \tag{8.23}$$

is the thickness of the mantle. Note that $\Delta a(t)$ is independent of a: if this simple treatment is valid, all grains are expected to acquire mantles of equal thickness

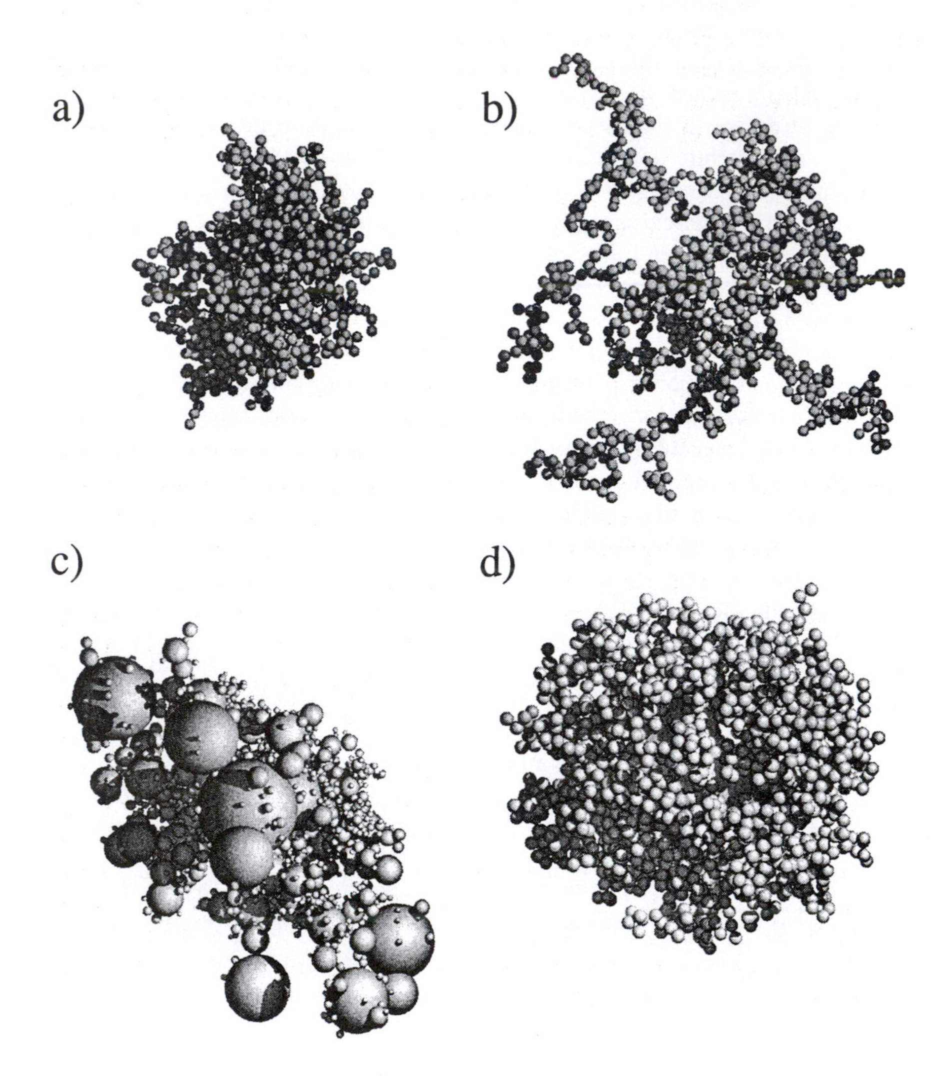

Figure 8.4. Computer simulations showing examples of particles formed by grain–grain coagulation (Dorschner and Henning 1995). (*a*) A compact particle produced by particle–cluster aggregation of 1024 identical small spheres. (*b*) A loose fractal particle produced by cluster–cluster aggregation of 1024 identical small spheres. (*c*) A composite particle produced by particle–cluster aggregation of 2001 spheres with sizes following a power-law distribution with exponent −3.15. (*d*) A compact particle produced by particle–cluster aggregation of 2000 identical small spheres onto one larger, spherical 'seed' particle.

irrespective of their initial size, leading to a shift of the distribution function to larger sizes whilst preserving its functional form. However, in practice, other factors may influence the outcome: for example, mantles will tend to desorb more rapidly from the smallest grains if a significant UV radiation field is present (see section 6.1.4); and, of course, coagulation may operate in concert with mantle growth.

If desorption rates are assumed to be small within dark clouds, then the timescale for growth of a mantle of a given thickness is deduced by rearranging equation (8.23):

$$t_{\mathrm{m}} = \frac{2.5 s \Delta a}{\xi n (k T_{\mathrm{g}} m)^{\frac{1}{2}}}. \tag{8.24}$$

We may use equation (8.24) to investigate whether the observed ice mantles result from 'frosting' (i.e. the direct condensation of pre-existing molecules from the gas) or from surface reactions involving adsorbed atoms and radicals. Following Jones and Williams (1984), we take the Taurus dark cloud as a basis for discussion, noting that the 3.0 μm spectral signature of H_2O-ice has a threshold visual extinction of about three magnitudes (section 5.3.2), interpreted in terms of the presence of thin ($\Delta a \approx 0.02$ μm) mantles of amorphous ice of density $s \approx 750$ kg m^{-3} (Whittet *et al* 1983, Jones and Williams 1984). Assuming cloud parameters and gas-phase abundances from the Herbst and Leung model (table 8.1), equation (8.24) predicts mantle accretion times of approximately 1 Myr, 5 Gyr and 10 Gyr if the process is limited by the abundance of O, OH and H_2O, respectively, assuming $\xi = 1$ for all species. As cloud life spans are less than 100 Myr, these calculations clearly show that the growth of ice mantles *cannot* result from frosting of gas-phase H_2O. Only surface reactions involving adsorbed atomic oxygen appear to be capable of producing such mantles on a realistic timescale. Even in much denser regions, where gas-phase H_2O abundances of 5×10^{-6} are expected (Herbst and Leung 1986), surface reactions may still be the dominant source of H_2O on grains. This conclusion seems likely to hold for most other known mantle constituents (table 5.3), including NH_3 and CH_4. Only in the case of CO is frosting likely to be important.

8.4 Ice mantles: deposition and evolution

8.4.1 Surface chemistry and hierarchical growth

The composition of the mantles forming in a given environment is critically dependent on the fraction of hydrogen that exists in molecular form. In the diffuse ISM, this fraction is very low; within a dense cloud it is close to unity (section 1.4.3). The formation of a dense cloud by contraction of diffuse gas is accompanied by a steady increase over time in the abundance of H_2. At early times, C, N and O atoms attaching to grain surfaces will encounter (and react with) H atoms on timescales much shorter than those in which they react with each other. Whereas newly formed H_2 molecules will generally desorb from the grain, the heavier products OH, CH and NH are more likely to remain attached to the surface and undergo further reaction. This initial phase of deposition will thus lead to the formation of hydrogenated ices, composed predominantly of H_2O together with NH_3 and CH_4 (d'Hendecourt *et al* 1985, Tielens and Hagen 1982). However, as the cloud evolves, a time will be reached at which $H \rightarrow H_2$ conversion is virtually complete in the densest regions, and surface reactions leading to the formation of hydrogenated molecules become much less frequent. A qualitatively different mantle material is then deposited: this secondary mantle is expected to be rich in molecules such as CO, O_2 and N_2. Most of the CO is probably accreted from the gas, whereas much of the O_2 and N_2 may form by surface catalysis of atomic O and N. In summary, a composite mantle structure is predicted for the densest regions of molecular clouds, in which an H-poor mantle is deposited on top of an H-rich mantle.

This model provides a clear basis for interpreting the observations reviewed in section 5.3.4. H-rich and H-poor ices are distinguished by very different dipole moments and this leads to observable differences in their vibrational spectra. The observations indicate the presence of distinct 'polar' and 'apolar' mantles, dominated by H_2O and CO, respectively, depicted schematically in figure 5.14. As expected, the polar mantles contain some NH_3 and CH_4. The apolar mantles are presumed to contain O_2 and N_2 in addition to CO but these homonuclear molecules are not available to direct spectroscopic observation in the infrared.

Although the partial segregation of H_2O and CO seems well understood, CO_2 has proven to be puzzling. Unlike CO, CO_2 is not predicted to have appreciable abundance in the gas phase (e.g. table 8.1) and this is confirmed by observations (van Dishoeck *et al* 1996). So frosting cannot lead to appreciable quantities of solid CO_2 in either polar or apolar phases. The substantial abundances of CO_2 detected in ice mantles in a variety of environments (up to $\sim$35% relative to H_2O; section 5.3.5) must therefore reflect processes occurring on the grains. CO_2 may form via surface reactions such as

$$CO + O \rightarrow CO_2 \tag{8.25}$$

although experiments suggest that this reaction possesses a small activation energy barrier that will tend to inhibit its progress at low temperature (Grim and d'Hendecourt 1986, Roser *et al* 2001). Under simulated interstellar conditions in the laboratory, CO_2 is easily produced in ices containing H_2O and CO by ultraviolet irradiation (d'Hendecourt *et al* 1986, Sandford *et al* 1988), via the sequence

$$\begin{aligned} H_2O + h\nu &\rightarrow OH^* + H \\ OH^* + CO &\rightarrow CO_2 + H \end{aligned} \tag{8.26}$$

where the asterisk denotes an excited state. The presence of CO_2 in ices surrounding embedded young stars might thus be explained by radiation-driven synthesis, but CO_2 is also abundant in the DCM remote from embedded sources of radiation (Whittet *et al* 1998, Gerakines *et al* 1999, Nummelin *et al* 2001). The external radiation field may induce some CO_2 formation in the outer layers of a cloud, and cosmic-ray excitation of H_2 can generate a weak UV field that might contribute at greater optical depth. However, calculations suggest that these mechanisms are insufficient to explain the observed CO_2 abundance (Whittet *et al* 1998): it seems most probable that CO_2 must, indeed, form by surface reactions such as (8.25) inside dark clouds. The existence of a small energy barrier is not the major impediment it might seem for diffusion-limited models, in which the production rate is limited by the arrival rate of reactants on the grain rather than by the reaction probability (Tielens and Hagen 1982, Tielens and Charnley 1997, Ruffle and Herbst 2001; see also Roser *et al* 2001 for experimental verification). Reaction (8.25) will compete for CO with the equivalent hydrogenation reaction

$$\mathrm{CO + H \rightarrow HCO} \tag{8.27}$$

which is the first step in the production of CH_3OH (Tielens and Charnley 1997) and also possesses an appreciable activation energy barrier. As CO_2 is more abundant than CH_3OH in the polar ices, reaction (8.25) must generally dominate. Models presented by van der Tak *et al* (2000) suggest that the abundance of CO_2 relative to CH_3OH is critically dependent on density.

8.4.2 Depletion timescales and limits to growth

The timescale upon which an atom or molecule with mean thermal speed $\bar{v}$ becomes depleted from the gas by adsorption onto grains of cross-sectional areas σ_d and number density n_d is (by analogy with equation (8.19))

$$t_{dep} = \frac{1}{\xi n_d \sigma_d \bar{v}} \approx \frac{10^{25}}{\xi n_H \bar{v}} \tag{8.28}$$

where again the approximation is based on the mean projected area of dust grains per H atom from Spitzer (1978). Consider the case of gaseous CO, extensively used as a tracer of molecular material in the ISM. For a cloud of density $n_H = 10^9$ m^{-3} and temperature 10 K, equation (8.28) gives an estimate of $t_{dep} \sim 1$ Myr if sticking is an efficient process ($\xi \approx 1$), as is generally assumed (e.g. Léger 1983, Leitch-Devlin and Williams 1985). Hence, attachment to grains should lead to almost total depletion of CO from the gas onto the dust in timescales short compared with cloud lifetimes, unless desorption is also efficient. The situation is essentially the same for any molecule with a comparable sticking coefficient and mean thermal speed, leading to speculation that clouds might exist that are undetectable by molecular rotational line emission (Williams 1985). In general, this seems not to be the case, as molecular emissions are usually quite well correlated with other tracers of dark clouds, such as dust extinction and far infrared emission; only in the densest cores ($n_H > 10^{10}$ m^{-3}) is there clear evidence for a short-fall in the gaseous CO abundance relative to dust (Willacy *et al* 1998, Caselli *et al* 1999, Kramer *et al* 1999). The most probable explanation is that molecules are being desorbed from grain mantles at rates roughly comparable with rates of adsorption in the DCM.

Accurate determinations of CO depletion along individual lines of sight can be made by observing gas-phase spectral absorption lines (section 5.1.1) in the same infrared sources as used to study the ices. In practise, however, this is quite difficult to do, requiring high spectral resolution and sensitivity (see Whittet and Duley 1991). Results available to date are thus limited to relatively bright protostellar sources where mantle desorption is being driven by energy from the source itself, a topic we defer to section 8.4.3. Estimates of CO depletion in quiescent dark clouds are currently available only from comparisons of millimetre-wavelength gas-phase emission lines with solid-state infrared absorption features (figure 8.5) and this method is intrinsically less reliable. As the emission lines of the commonest isotopic forms of CO are generally saturated, the rarer forms must be observed and the results scaled by standard isotopic abundance ratios. Another concern is that the volume of space being sampled may not be identical for gas-phase and solid CO: differences in beam size can introduce errors if the density changes across the line of sight; and, more seriously, material behind the source may contribute to the molecular emission (but not the solid-state absorption) and thus systematically reduce the apparent depletion. However, observations of background field stars are much less likely to be affected by sampling differences compared with those of embedded stars.

Figure 8.5 plots CO column density against visual extinction for field stars situated behind the Taurus dark cloud. A systematic trend of increasing $N(\mathrm{CO})$ with A_V above some threshold value is evident for both the solid and gaseous phases. The fractional depletion

$$\delta(\mathrm{CO}) = \frac{N_{\mathrm{dust}}(\mathrm{CO})}{N_{\mathrm{gas}}(CO) + N_{\mathrm{dust}}(\mathrm{CO})} \tag{8.29}$$

is deduced to be about 25–30% for $6 < A_V < 24$. It is notable that there is no obvious trend of increasing depletion with increasing optical depth, suggesting that desorption mechanisms operate fairly uniformly within the cloud over this range of extinction.

Processes that might remove molecules from grain mantles inside dark clouds include photodesorption and impulsive heating by x-rays and cosmic rays (Barlow 1978b, Léger *et al* 1985, Duley *et al* 1989b, Hartquist and Williams 1990, Willacy and Williams 1993, Bergin *et al* 1995). The existence of threshold extinctions for detection of H_2O and CO ices (section 5.3.2; figures 5.11 and 8.5) is easily understood if desorption is driven predominantly by the external radiation field. Deep within a dark cloud that lacks embedded stars, cosmic rays are likely to be the only significant source of energy contributing to desorption from grain surfaces. The ultraviolet field produced by cosmic-ray excitation of H_2

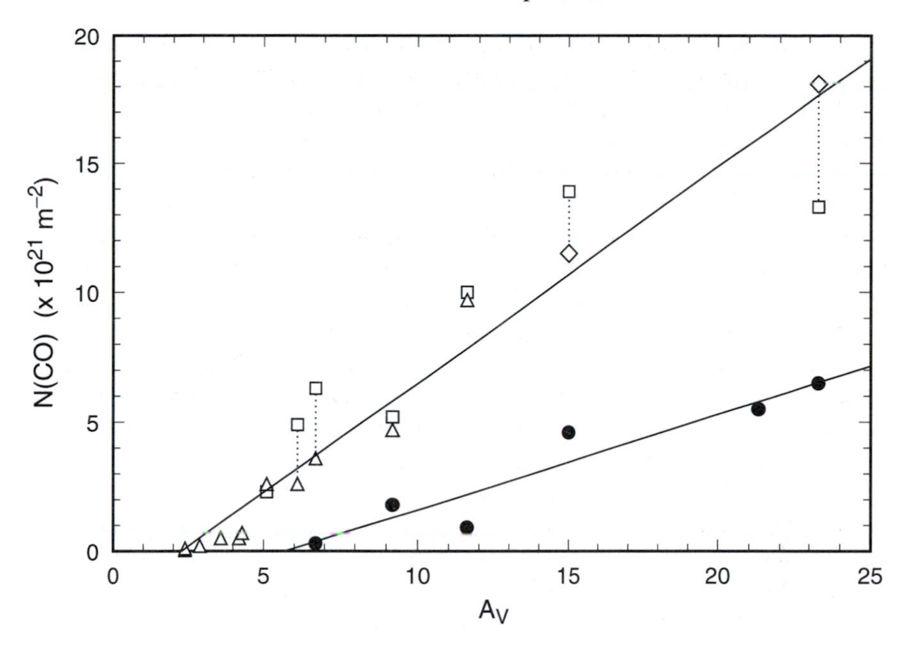

Figure 8.5. Plot of gas-phase and solid-state CO column densities against visual extinction for lines of sight toward field stars behind the Taurus dark cloud. Gas-phase data (open symbols) are from Frerking *et al* (1982): the results are based on observations of $^{12}C^{18}O$ (triangles), $^{12}C^{17}O$ (squares) and $^{13}C^{18}O$ (diamonds), using solar isotope ratios to convert to the usual form. Vertical dotted lines join data based on different isotopes observed in the same line of sight. Data for solid CO (black symbols) are from Chiar *et al* (1995) and include both polar and apolar ices. A_V values are estimated from infrared colour excesses using equation (3.37) with $r = 5.3$ (Whittet *et al* 2001a). The straight lines fitted to the data are merely to guide the eye; in reality the relationships are probably not linear.

(Prasad and Tarafdar 1983, Sternberg *et al* 1987) will promote photodesorption in regions where the external field is too weak to contribute. However, it is not well established that such mechanisms are efficient enough to maintain abundances in the gas phase at observed levels in dense cores (Hartquist and Williams 1990).

8.4.3 Thermal and radiative processing

When luminous stars form within a dense cloud, they inject energy into their local environment that may induce both physical and chemical changes in the surrounding material. Mantled dust grains may be warmed, leading to partial or complete sublimation or crystallization of the ices, and their composition may be altered by radiatively driven chemical reactions. Spectroscopy of the ices provides three methods for studying these changes: (i) evidence for volatility-dependent abundance variations, attributed to sublimation; (ii) detection of profile evolution as a function of temperature; and (iii) the occurrence of features identified with the products of energetic processing. We consider each of these in turn.

The sublimation zones surrounding a massive YSO are depicted schematically in figure 8.6. CO is the most volatile molecule commonly observed in the ice mantles and H_2O the least volatile with CO_2 also shown as an intermediate case. Heating should thus remove the CO-dominated apolar layer much more readily than the H_2O-dominated polar layer: whereas apolar ice is expected to vaporize above about 20 K, polar ice may survive up to at least 100 K. This expectation is confirmed by observations. Spectra of embedded stars at the wavelength of the CO fundamental band (figure 8.7) indicate considerable variation in the distribution of CO between solid and gaseous phases, implying source-to-source differences in levels of sublimation (Mitchell *et al* 1988, 1990, van Dishoeck and Blake 1998, Boogert *et al* 2000b, Shuping *et al* 2001). However, whereas the CO in the apolar ice is easily desorbed, CO may be retained as a minority constituent in the H_2O-dominated polar ices at temperatures up to $\sim$100 K (Schmitt *et al* 1989). Source-to-source variations in the profile shape of the solid CO feature (see figure 5.13) may thus be interpreted as differences in the relative strengths of the apolar and polar components as a function of dust temperature. In some sources, such as GL 4176 (figure 8.7), CO appears to be entirely in the gas phase. By comparing the strength of the 4.67 μm apolar CO component with that of the 3.0 μm H_2O-ice feature, Chiar *et al* (1998a) propose three general classes of object in an evolutionary sequence:

(i) those in which little or no sublimation of the apolar mantles has occurred, with ice-phase CO/H_2O abundance ratios $\sim$25–60%;
(ii) those in which moderate to high sublimation of the apolar mantles has occurred, with $CO/H_2O \sim$ 1–20%; and
(iii) those in which all CO has sublimed.

The spectra shown in figure 8.7 are representative of these three classes. It must be remembered, of course, that what we observe is a line-of-sight average: it would be possible to form a class (ii) spectrum by combining regions in which all and none of the CO has sublimed.

When they first form, the ices are amorphous. This is the case because they generally accumulate at temperatures well below the melting or sublimation points of the relevant molecules. Amorphous solids contain a distribution of molecular environments that produce broad, Gaussian line profiles. If the material is warmed, the molecules arrange themselves into more energetically favourable orientations, resulting in evolution of line profiles toward the sharper features generally seen in crystalline solids. Observational evidence for crystallization of the polar mantles is provided most readily by profile studies of the 3.0 μm H_2O-ice feature (section 5.3.3; see figure 5.12). Whereas quiescent dark-cloud environments such as that toward Elias 16 produce broad features consistent with ices maintained at 10–20 K, many YSOs show sharper features indicating that the ices have been heated to 70–100 K (Smith *et al* 1989). These sources generally

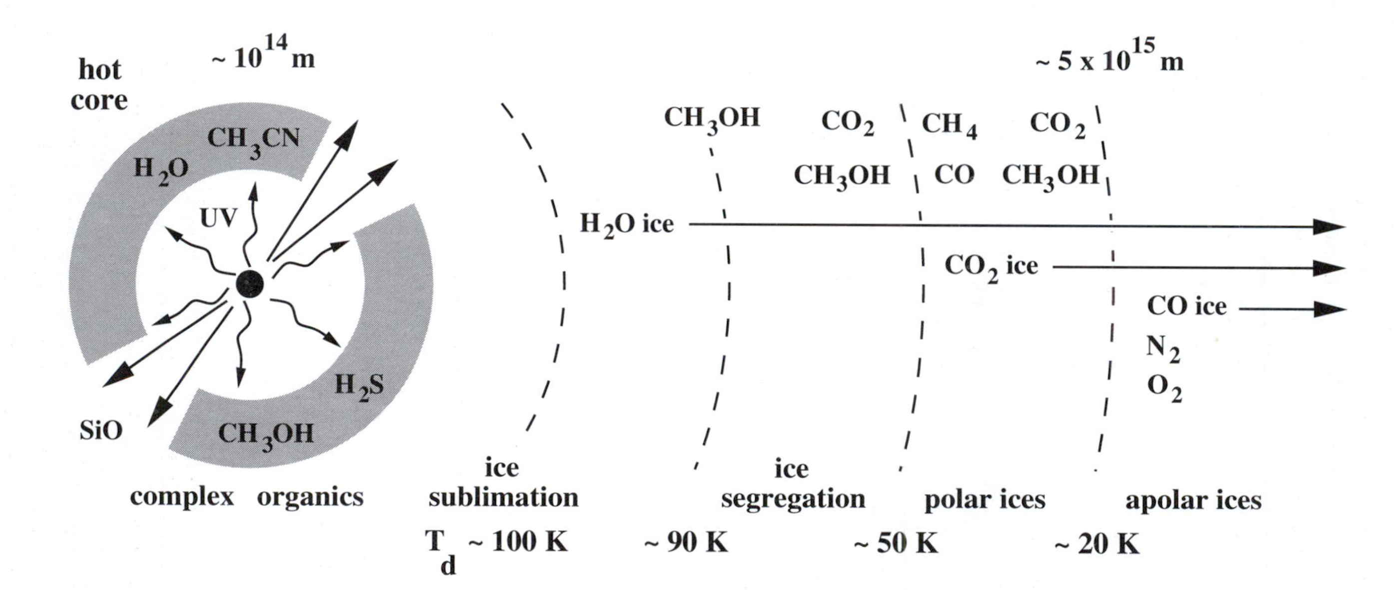

Figure 8.6. Schematic representation of the chemical environment of a massive young star embedded in a dense molecular cloud. The occurrence of various molecules in the icy mantles is indicated with respect to dust temperature (T_d) as a function of radial distance from the star. Approximate size scales are indicated for the gaseous hot core surrounding the young star and for the zone within which the apolar ices are sublimed. Figure courtesy of Ewine van Dishoeck, adapted from van Dishoeck and Blake (1998).

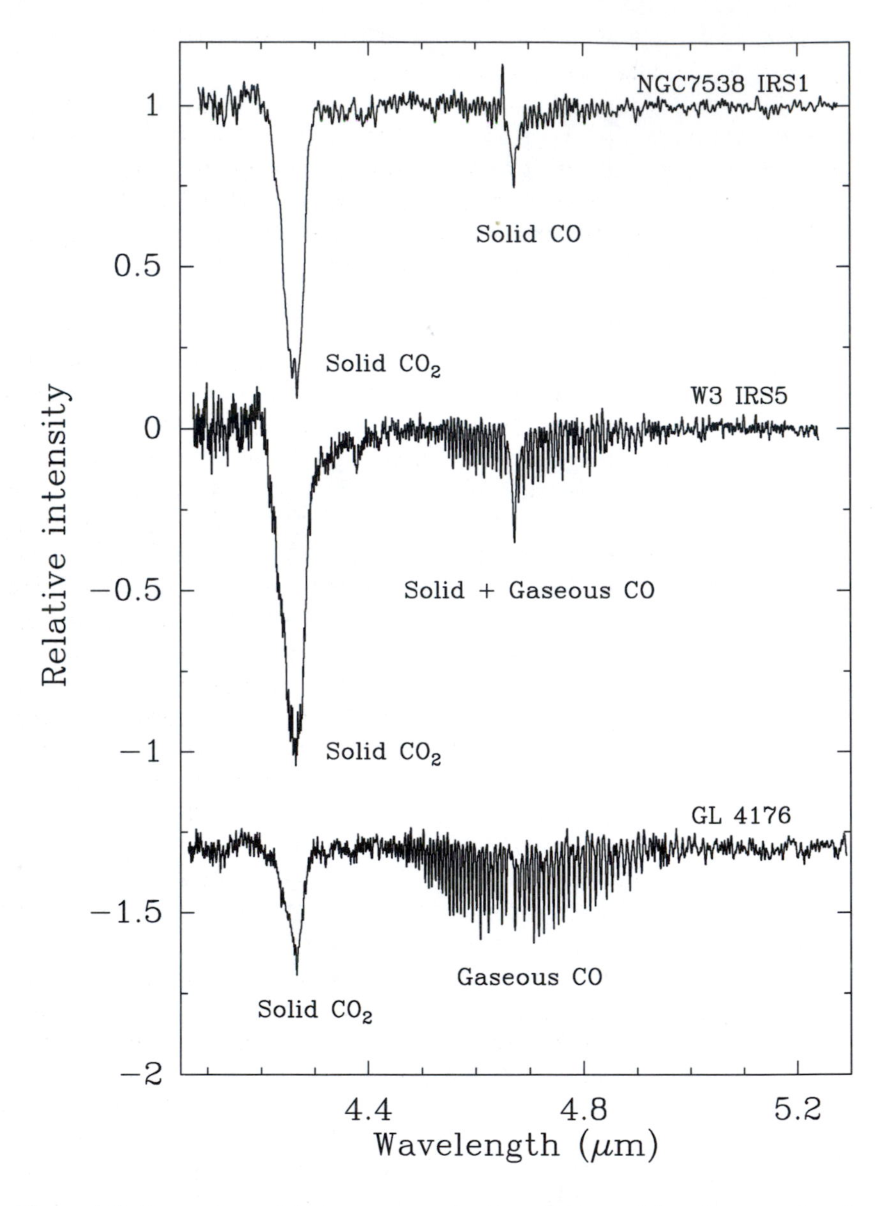

Figure 8.7. Comparison of 4–5 μm spectra for three YSOs at different stages of cycling of their circumstellar envelopes. Features of solid and gaseous CO centred near 4.67 μm (figure 5.1) are present with differing relative strengths: toward NGC 7538 IRS1, most of the CO remains condensed on the grains; toward W3 IRS5 and GL 4176, it has been partially and completely sublimed, respectively. CO_2, which is less volatile than CO, produces absorption at 4.27 μm in all three spectra. Data from ISO SWS observations; figure courtesy of Ewine van Dishoeck.

have weak or absent solid CO features, indicating that annealing of the polar ices is accompanied by sublimation of the apolar ices.

Although it is convenient to characterize the ices in terms of distinct polar and apolar components deposited sequentially, as depicted in figure 5.14 and discussed in section 8.4.1, the boundary between them may not, in fact, be sharp. One may envisage an intermediate stage of mantle growth in which H_2O and CO accumulate at comparable rates. Competing oxidation and hydrogenation reactions involving CO may lead to the production of CO_2 and CH_3OH (section 8.4.1) and an amorphous mixture containing CO_2, CH_3OH and H_2O in similar proportions may thus be formed. If this mixture is subsequently heated, the resulting structural changes lead to the formation of complexes, in which the O atoms of CH_3OH link with the C atoms of CO_2 to form strong intermolecular bonds (Lewis acid–base pairs). These complexes are stable and have distinctive spectral properties that are most clearly seen in the region of the CO_2 bending-mode feature near 15.3 μm (Ehrenfreund *et al* 1998, 1999), discussed previously in chapter 5 (see section 5.3.5 and figure 5.15(*b*)). Figure 8.8 compares 15 μm profiles for several sources arranged in an evolutionary sequence. The O=C=O bend is degenerate and naturally has a double-peaked structure in pure CO_2 ices. The laboratory experiments of Ehrenfreund *et al* show that this structure is weak or absent in amorphous H_2O:CH_3OH:CO_2 mixtures but becomes prominent upon heating: it effectively provides a measure of the ice temperature. In addition to the relatively sharp CO_2 peaks at 15.15 and 15.25 μm, broader structure identified with the $CH_3OH \cdots CO_2$ complexes is seen at 15.4–15.5 μm. This structure is absent in highly polar (H_2O-dominated) ices and its presence in the observed spectra indicates that H_2O cannot be the majority species in this component of the ices. Thus, the observations provide strong evidence not only for thermal processing but also for segregation of $CH_3OH \cdots CO_2$ complexes from the polar ice layer. This phase of the ice is most probably located on the surface of the mantle after the CO-dominated apolar phase has sublimed.

Thermal evolution results primarily from absorption of infrared radiation emitted by stars that remain deeply embedded in placental gas and dust (figure 8.6). As circumstellar material gradually disperses, however, more energetic radiation and particle winds may permeate the surrounding medium. Photon energies at the level of a few eV may break chemical bonds and convert saturated molecules such as H_2O, NH_3 and CH_4 into radicals. If irradiation is accompanied by warming, the radicals may be free to migrate through the mantles and react with other species. The production of CO_2 in this manner (equation (8.26)) has already been noted. Laboratory simulations show that other possible products include both kerogen-like organic polymers and prebiotic molecules such as amino acids (e.g. Agarwal *et al* 1985, Briggs *et al* 1992, Bernstein *et al* 2002). The more exotic species are not produced in sufficient numbers to be detectable in the ices by infrared techniques (a more fruitful approach is to study their sublimation products in warm gas; see section 9.1.4). However, as a generic class, the C≡N-bearing molecules are a helpful exception:

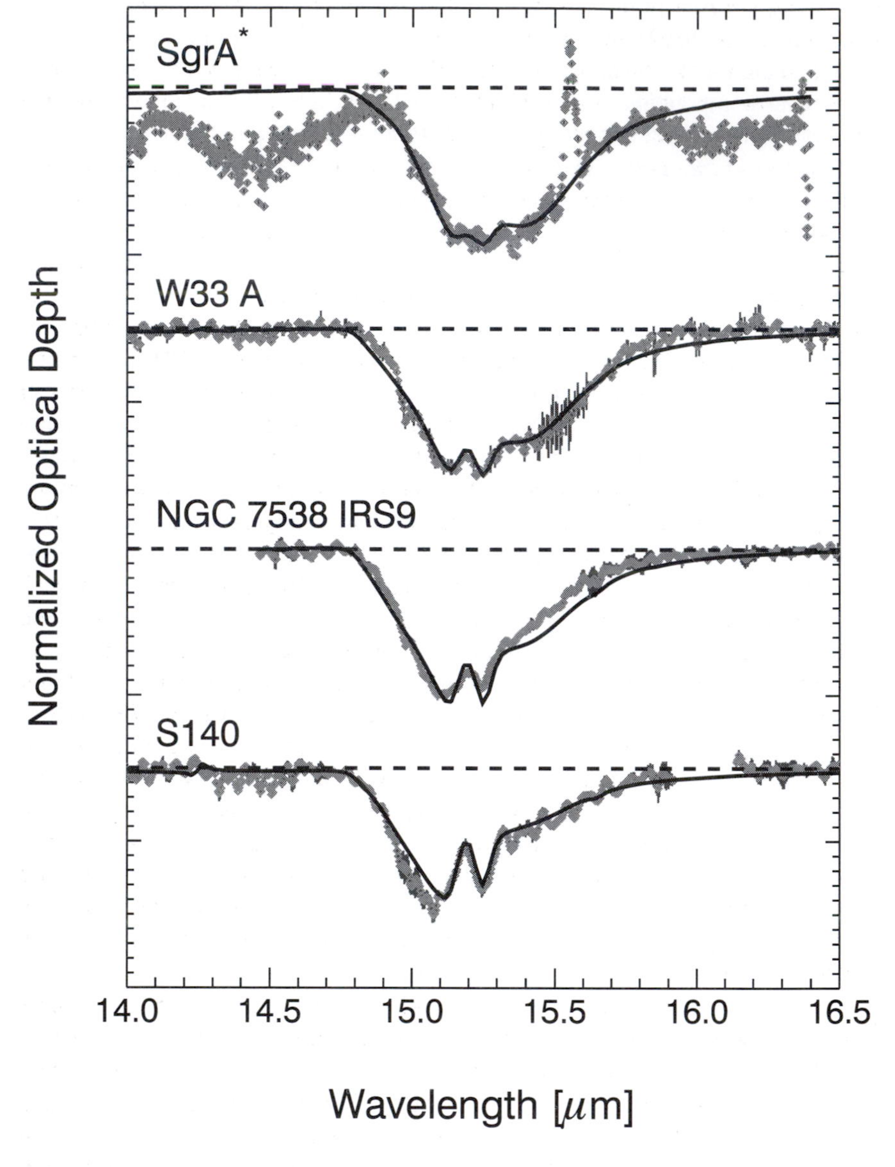

Figure 8.8. Comparison of the CO_2 bending mode near 15 μm in several lines of sight, illustrating thermal evolution of the ices. The sources are displayed in order of increasing mean gas temperature (T_g = 16 K, 23 K, 26 K and 28 K for SgrA*, W33A, NGC7538 IRS9 and S140, respectively). Full curves are fits based on laboratory data for an ice mixture (H_2O:CH_3OH:CO_2 = 1:1:1) at various temperatures (Gerakines *et al* 1999). The strengths of the narrow features near 15.15 and 15.25 μm increase systematically with temperature. Data from ISO SWS observations; figure courtesy of Perry Gerakines.

observations reviewed in section 5.3.6 show that they are present toward certain luminous YSOs (and generally absent elsewhere). The profile of the relevant spectral feature at 4.62 μm is well matched by laboratory data for ices subjected to UV photolysis or ion irradiation (Pendleton *et al* 1999; see figure 5.16), providing corroborative evidence for energetic processing of the ices.

One of the principal models for interstellar dust (section 1.6) assumes that grains retain mantles acquired within molecular clouds when they return to the diffuse ISM. The mantles are assumed to be heavily processed by photochemical reactions that convert volatile ices into organic refractory matter (ORM), a scenario that gains credence from laboratory simulations (e.g. Briggs *et al* 1992, Strazzulla and Baratta 1992, Jenniskens *et al* 1993, Greenberg *et al* 1995). If ORM is produced efficiently in the environments of embedded YSOs, we should expect to see evidence for this in their spectra. The most prominent features in ORM occur at ~3 and 6 μm (see figure 5.9). Of these, the first overlaps the strong H_2O-ice stretching mode and the latter the rather weaker bending mode. It was noted in section 5.3.3 that H_2O-ice column densities calculated from the strength of the bending-mode feature appear systematically overestimated in some YSOs compared with results from other H_2O features, suggesting the presence of blending at 6 μm. Could this be evidence for ORM? The case of W33A is illustrated in figure 8.9: $N(H_2O)$ is well constrained by observations of the stretching, combination and libration mode features, such that the bending mode accounts for only 25–30% of the optical depth at 6.0 μm (Gibb and Whittet 2002). Other ices that have resonances in this spectral region, such as NH_3, H_2CO and HCOOH, cannot explain the shortfall (although they may explain some of the structure in the profile). A substantial contribution from ORM is, however, consistent with the data, as shown in figure 8.9. The asymmetric shape of the ORM feature, in particular, seems compatible with the structure of the observed 6 μm profile. If ORM is, indeed, present in this line of sight, it will also absorb at other wavelengths, notably in the 2.8–3.5 μm region, but its contribution to the observed 3 μm profile (figure 5.5) will be overwhelmed by the very deep H_2O-ice stretching mode.

Comparing other sources with W33A, it is found that the 6 μm excess relative to the predicted H_2O-ice bending mode correlates with the strength of the 4.62 μm XCN feature. In sources where no XCN is detected, no significant excess is found. As the XCN carrier is evidently formed by energetic processing, as discussed earlier, this correlation supports an origin for the 6 μm excess in ORM, likewise a product of energetic processing. It should not surprise us that organic material is produced more efficiently toward some embedded stars than others. The radiative environment to which ices are exposed along the line of sight to a particular YSO will depend on a number of factors, including not only the mass and age of the star but also the distribution of dust and the orientation of the circumstellar disc (see figure 8.6).

Although these results make a case for synthesis of organic refractory matter around some YSOs, evidence for widespread cycling of the products into the

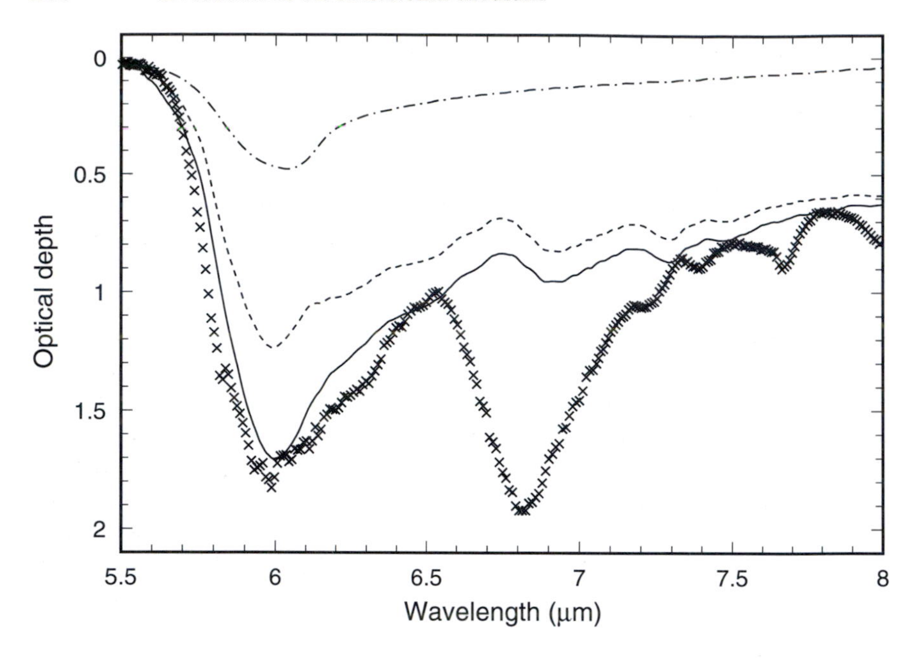

Figure 8.9. Spectrum of W33A illustrating the possible contribution of organic refractory matter (ORM) to the absorption profile near 6 μm (Gibb and Whittet 2002). The observations (crosses) are compared with a model (full curve) that combines H_2O-ice and ORM absorption. The contribution of H_2O (dot–dash curve) is based on the mean column density estimated from features at other wavelengths (stretching, combination and libration modes). The ORM contribution (broken curve) is scaled so that the total matches the depth of the 6.0 μm feature. Note that the feature at 6.8 μm is not accounted for.

general ISM is not compelling. Comparisons of laboratory data for ORM with observations of the diffuse ISM have generally focused on the C–H stretch feature at 3.4 μm (e.g. Pendleton 1997). However, the interstellar feature now seems securely identified with hydrogenated amorphous carbon nanoparticles rather than with mantles on larger grains that contribute to visual extinction (section 5.2.4), contrary to the predictions of the core–mantle model for interstellar dust (Li and Greenberg 1997). More seriously, diffuse-ISM spectra (e.g. figures 5.3 and 5.4) show no hint of the 6.0 μm feature, yet it should be considerably stronger than that at 3.4 μm according to data for laboratory analogues. Long-term exposure to the interstellar radiation field may reduce organic matter to amorphous carbon, thus suppressing features associated with bonds such as O–H and C–O.

8.5 Refractory dust

8.5.1 Destruction

The refractory grains that provide substrates for mantle growth and dissipation in molecular clouds are themselves subject to disruptive forces in harsher environments. Important destruction mechanisms include sputtering and grain–grain collisions (Barlow 1978a, b, Draine and Salpeter 1979, Seab 1988, Draine 1989b, McKee 1989, Tielens *et al* 1994). Sputtering occurs when grain surfaces are eroded by impacting gas-phase atoms or ions, a process that may be chemical or physical. Chemical sputtering arises when the formation of chemical bonds between the surface and impinging particles of relatively low kinetic energy leads to the desorption of molecules containing atoms that were originally part of the surface. Physical sputtering involves the removal of surface atoms by energetic impact: this may be thermal, resulting from kinetic motion of high-temperature gas, or non-thermal in cases where gas and dust are in rapid relative motion (see Dwek and Arendt 1992 for a review of dust–gas interactions in hot plasmas). If the grains also have high velocities relative to one another, collisions between them can lead to evaporation of grain material.

The dominant destruction mechanism and the rate of destruction are sensitive to environment. Chemical sputtering is likely to be efficient only in regions that are both warm and dense, such as compact envelopes surrounding young stars (Barlow 1978b, Draine 1989b, Lenzuni *et al* 1995). In the diffuse ISM, destruction is dominated by physical mechanisms driven by supernova-generated shock waves (Seab and Shull 1983, Seab 1988, McKee 1989, Jones *et al* 1994). A shock wave is an irreversible, pressure-driven disturbance that leads to impulsive heating of the shocked gas (see Draine and McKee (1993) for detailed discussion of the theory of interstellar shocks). Thermal sputtering results from immersion of the dust in high-temperature gas: the atoms within the lattice structure of a refractory grain typically have binding energies $\sim$5 eV, and temperatures in excess of $\sim 10^5$ K are thus required for this process to be efficient. As a grain crosses a shock front, its velocity does not change as quickly as that of the gas, resulting in differential motion. If the grain bears electric charge, it will gyrate about magnetic field lines and may undergo betatron acceleration (Spitzer 1976). Either of these effects may lead to non-thermal sputtering and enhanced rates of grain–grain collisions. Calculations by Jones *et al* (1994) show that thermal sputtering is important at the highest shock speeds ($v > 150$ km s^{-1}), whilst non-thermal sputtering is dominant for the range 50–150 km s^{-1}. Results are similar for both carbon and silicate grains. Grain–grain collisions appear to play a somewhat lesser role in grain destruction but will, of course, modify the size distribution by shattering at impact speeds above $\sim$1 km s^{-1} (Barlow 1978b, Tielens *et al* 1994, Jones *et al* 1994, 1996, Borkowski and Dwek 1995).

Supernova explosions generally occur in low-density phases of the ISM, resulting in shock waves that propagate through large volumes of low-density gas.

Suppose a shock wave of velocity v_0 travelling in a medium of number density n_0 encounters a cloud of number density $n_c \gg n_0$. The cloud is subjected to a sudden increase in ambient pressure that drives the shock into the cloud with velocity

$$v_c = (n_0/n_c)^{\frac{1}{2}} v_0 \tag{8.30}$$

(Draine 1989b). This reduction in shock speed as the square root of the relative density implies that shocks capable of destroying grains in the intercloud medium are decelerated to such a degree that they do not generally destroy grains in clouds. As an illustrative example, let us assume $v_0 = 500$ km s^{-1}; adopting values of $n_0 = 5 \times 10^3$ m^{-3} and $n_c = 3 \times 10^7$ m^{-3}, typical of the intercloud medium and a diffuse H I cloud, respectively (table 1.1), we deduce $v_c = 6$ km s^{-1} (equation (8.30)), with correspondingly lower values in denser clouds. But shock speeds ~50 km s^{-1} and above are needed for effective destruction of refractory dust. Thus, destruction by shocks is efficient in the intercloud medium (where the grain number density is low) and generally inoperative in clouds (where the grain number density is high). On this basis, it seems reasonable to assume that destruction of refractory dust occurs predominantly in the intermediate warm phase of the ISM (McKee 1989, Jones *et al* 1994). Note that ices, which have binding energies $\lesssim$0.5 eV, are removed easily by low velocity shocks (as well as by photodesorption and heating) in diffuse phases of the ISM; they survive only in molecular clouds (Seab 1988, Barlow 1978b, c).

Observational evidence for shock destruction of dust is provided by studies of element depletions in high-velocity clouds. These diffuse clouds are presumed to have been accelerated by supernova explosions or other energetic events, such as winds from luminous stars. The degree of depletion shows a tendency to correlate with cloud velocity, in the sense that fewer atoms are in the dust in the fastest-moving clouds. This is illustrated in figure 8.10 for the case of silicon. The data are consistent with Si being virtually undepleted (almost the full solar abundance in the gas) in clouds with $v > 50$ km s^{-1}, whereas <10% remains in the gas in clouds with $v < 10$ km s^{-1}. A model in which grains are thermally sputtered at 100–200 km s^{-1} is consistent with this trend, assuming that the shocked gas is subsequently decelerated as it sweeps up matter with 'normal' depletions (Barlow and Silk 1977a, Cowie 1978). It appears that silicate grains are being destroyed with an overall efficiency of ~50% in high velocity clouds.

The correlation between depletion and mean gas density illustrates the synergy between grain destruction and renewal. This is shown in figure 2.8 for titanium (see Jenkins 1987 and references therein for other examples). The correlation embraces a wide range of environments in which different mechanisms control the exchange of matter between the dust and the gas. Seab (1988) estimates that the mean time interval between destructive shocks in the warm phase of the ISM is ~7 Myr, compared with ~3 Gyr for a diffuse cloud. The corresponding timescales for adsorption (equation (8.28)) are ~400 Myr and ~50 Myr, respectively. Thus, in clouds, adsorption dominates and very high

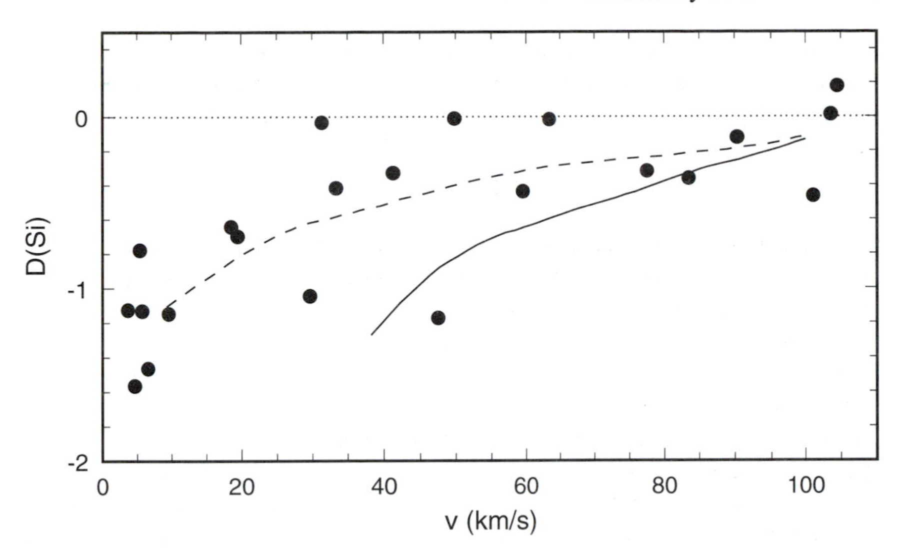

Figure 8.10. Plot of silicon depletion index against velocity relative to the local standard of rest for various discrete clouds (points). The horizontal dotted line indicates zero depletion for solar reference abundances. The data are compared with a model for grain destruction by sputtering in shocks (full curve), the broken curve representing shocked material that is subsequently decelerated by sweeping up unshocked gas. (Adapted from Cowie 1978.)

levels of depletion are maintained for most of the condensible heavy elements. In tenuous regions, the situation is reversed and grain material is eroded more rapidly than atoms can re-adsorb, leading to reduced levels of depletion. It should be noted that evaporation of relatively small quantities of dust can lead to dramatic increases in gas-phase abundances for the most highly depleted (group III) elements such as Fe, Ti, Ca and Al (section 2.4.4). Mixing of dense and diffuse phases of the ISM is estimated to occur on a timescale of $\sim$100 Myr (Seab 1988), which is sufficiently rapid to account for the rather high levels of depletion observed for these refractory elements even in low density regions. The behaviour of the somewhat more volatile group II elements Si and Mg suggests that they are being removed from the grains on timescales comparable with the mixing time (Tielens 1998).

Average lifetimes for refractory grains are notoriously difficult to estimate, which is unfortunate as the results are vitally important for an understanding of grain ecology in the galactic environment. Given that supernova-driven shocks are the dominant destructive agent, the destruction rate will depend on the supernova rate and the galactic distribution of their progenitors, the mean expansion rate, the phase structure of the ISM and, of course, the properties of the dust grains themselves. Current estimates of the destruction timescale lie typically in the range 0.2–0.6 Gyr (Jones *et al* 1994, 1996). Comparison with injection timescales for stardust (~2.5 Gyr; section 7.3.5) leads to the conclusion that stars are not the primary source of new grain material. Some possible alternatives (re-accretion of gas-phase atoms; formation of organic refractory mantles) have been discussed in this chapter, but it is highly questionable whether such processes can create grain materials that satisfy observational constraints, either qualitatively or quantitatively. Of greatest concern is the fact that observations require virtually all of the available Si to be in silicates (section 5.2.2), yet there is no known interstellar process that can form them. Perhaps the estimated destruction rates contain systematic errors, such as might arise if unrealistic assumptions are made regarding dust properties (e.g. structure, porosity) or the strength of the ambient magnetic field.

8.5.2 Size distribution

The size distribution is governed by the balance of destruction and re-growth. A power-law (MRN) distribution of the form $n(a) \propto a^{-q}$ (typically with $q \approx 3.5$) is broadly consistent with observed properties of the dust, including extinction (section 3.7), polarization (section 4.4) and infrared continuum emission (section 6.2). Indeed, Kim *et al* (1994) have *derived* size distributions from extinction curves that approximate to the MRN formulation. The significance of this general result is that it concurs with the physics of grain–grain collisions: it is predicted in situations where the size distribution is regulated by shattering and coagulation (Biermann and Harwit 1980, Jones *et al* 1996). An alternative scenario is one in which sizes are regulated primarily by gas–grain interactions, i.e. by evaporation and condensation. In this case, the predicted form is $n(a) \propto \exp(-Ka^3)$ where K is a constant (Greenberg 1968). This form is capable of fitting the visual extinction curve, as originally shown by Oort and van de Hulst (1946), but it seriously underestimates the number of small grains needed to explain the level of UV extinction relative to that in the visual.

Although grain–grain collisions are evidently powerful moderators of the size distribution, gas–grain interactions are clearly important as well – we have identified sputtering as the most effective destruction mechanism for refractory dust and there is observational evidence for re-accretion and mantle growth. Sputtering leads to *size-independent* rates of erosion because the sputtering rate is proportional to the cross-sectional area of the grain: the situation is analogous to the treatment of mantle growth in section 8.3.2, in the sense that sputtering can be considered as negative growth. McKee (1989) estimates that a $v \sim 300$ km s^{-1} shock will typically remove a surface layer ~0.02 μm thick, leading to annihilation of small grains whilst leaving the cores of larger grains intact. The model of Oort and van de Hulst (1946) assumes that grains condense directly from interstellar gas; but suppose instead that mantles condense onto and sputter from pre-existing grain cores. If the core-size distribution is determined by grain–grain collisions, then it will naturally have a power-law form and this will tend to be conserved as mantles are deposited and removed.

8.5.3 Metamorphosis

Energetic processes may lead to internal structural changes in the grains as well as to vaporization and shattering. Propagation of shock waves through the grain material or heating induced by photon absorption may result in solid-state phase transitions. A problem of great interest in this context is the nature of solid carbon in interstellar dust. In the presence of a suitable energy source, carbon is expected to evolve toward a more highly ordered state in the sequence amorphous carbon $\rightarrow$ graphite $\rightarrow$ diamond. Observations of evolved stars indicate a preponderance of amorphous carbon in stardust (section 7.2) but graphite is considered an essential ingredient of interstellar dust (section 3.7) as the presumed carrier of the 2175 Å extinction bump. To add another twist, diamond is the most abundant form of stardust in meteorites (section 7.2.4). Can interstellar processes account for these differences?

The amorphous carbon $\rightarrow$ graphite transition may be driven by UV photon absorption, as proposed by Sorrell (1990, 1991). Absorption of an incident photon will induce heating and, if the resulting increase in temperature is of sufficient amplitude, a phase change will occur. If the grain is sufficiently small, the entire particle will be affected; more generally, because amorphous carbon is highly porous, annealing may tend to be localized in larger particles to regions close to the point of incidence (see section 6.1.4). Irradiation will also tend to desorb hydrogen from the grains, as required by the graphite model for the bump (section 3.5.2). Sorrell (1990) estimates that significant graphitization can occur on timescales $\sim$100 Myr, shorter than the predicted lifetimes of the grains (Jones *et al* 1994). Laboratory simulations reported by Mennella *et al* (1996, 1998) provide empirical verification of the feasibility of this proposal.

Grain–grain collisions may also lead to structural evolution of the grain material. Tielens *et al* (1987) discuss the conversion of graphitic carbon to diamond by collisionally induced shock processing. Diamond is the thermodynamically favoured form of solid carbon at high temperature and pressure. The graphite–diamond phase change has a high activation energy (7.5 eV per C atom) because complete rearrangement of the crystal lattice is required. Tielens *et al* (1987) argue that diamond nanocrystals ranging up to $\sim$10 nm in size may be formed from graphite with an efficiency of $\sim$5%. Once created, the diamonds would be hard to destroy and could survive for long periods in the ISM. This process has been proposed as a source of the nanodiamonds in meteorites. However, an objection arises from the fact that diamonds are overwhelmingly more abundant than graphite in the pre-solar fraction of the meteorites (table 7.2): as interstellar graphite $\rightarrow$ diamond conversion is inefficient, we should expect the opposite to be true. It seems unlikely, if not impossible, that such a large discrepancy could result from selection effects (section 7.2.4). An alternative and perhaps more likely explanation is that the nanodiamonds form by vapour deposition in C-rich stellar ejecta rather than by metamorphosis in the solid phase, and do not therefore need the prior existence of graphite.

Finally, we consider exposure of the dust to bombardment by cosmic-ray particles. A cosmic ray is an atomic nucleus, most typically a proton, travelling at relativistic speed. When it strikes a grain, it will strip atoms from the lattice. Because the kinetic energy is many orders of magnitude greater than that of an atom or ion travelling at thermal speeds, it will penetrate into the grain and cause internal damage as well as surface sputtering. Exposure to cosmic rays thus tends to destroy long-range order within the grain material, reversing the crystallization process.

8.5.4 Dust in galactic nuclei

So far, we have considered dust in 'normal' interstellar environments – clouds and tenuous intercloud gas. We conclude this chapter with a brief discussion of dust in the extreme conditions of galactic nuclear zones. At the heart of our own Galaxy is a massive, compact object (coincident with the radio source Sgr A*), surrounded by a circumnuclear disc or torus of radius $\sim$5 pc. The nucleus and disc are embedded in a molecular zone some 200 pc in radius, in which high densities ($n_{\mathrm{H}} > 10^{10}$ m^{-3}) are coupled with warmer temperatures ($\sim$100 K) than are typical of molecular gas in the spiral arms (see Morris and Serabyn 1996 for a review). Other differences include much higher degrees of turbulence and stronger magnetic fields. The overall trend in gas motion is inward, toward the nucleus, and this matter must ultimately either form stars or accrete onto the nucleus itself. The observed presence of young stars does, indeed, attest to recent star formation within the molecular zone. In the prevailing physical conditions, star formation is most likely triggered by energetic events, such as shocks associated with supernova explosions, cloud–cloud collisions or nuclear activity. In any case, star formation activity appears to be episodic. Thus, our Galaxy displays evidence for processes analogous to those seen in major classes of 'active' galaxy: starbursts and accretion onto a compact nucleus.

It seems clear from this discussion that the evolution of dust will tend to be greatly accelerated in such regions. High densities and temperatures, the prevalence of shocks and the strong magnetic fields that control grain dynamics will lead to accelerated rates of sputtering, both physical and chemical, and more frequent grain–grain collisions. The harsh radiative environment will also lead to enhanced rates of photoprocessing and photoevaporation. FUV fields are typically higher by factors $\sim 10^4$ in starburst regions compared with the local diffuse ISM (Wolfire *et al* 1990, Carral *et al* 1994). Indeed, physical conditions within starburst nuclei appear to be generally similar (on a different size scale) to those prevailing in the photodissociation zones surrounding massive H II regions such as the Orion nebula. Accretion within active galactic nuclei will also generate copious FUV and x-ray fields that will ablate circumnuclear clouds (Pier and Voit 1995). Timescales for grain destruction may be as short as 10^3–10^4 years

(Villar-Martin *et al* 2001). Unless production timescales are similarly shortened, dust seems likely to be a scarce commodity in the centres of even mildly active galaxies.

Nevertheless, some dust survives. This is demonstrated by observations of various relevant phenomena, including infrared line and continuum emission, infrared absorption features and optical polarization (e.g. Roche 1989b, Roche *et al* 1986a, 1991, Cimatti *et al* 1993, Hough 1996, Dudley and Wynn-Williams 1997, Ivison *et al* 1998, Genzel and Cesarsky 2000). For some systems, such as starbursts and type 2 Seyferts, dust-related phenomena are amongst their defining characteristics (see chapter 6, figures 6.5 and 6.9). The jet-emitting radio galaxies present a particularly perplexing problem (De Young 1998, Villar-Martin *et al* 2001). Observations show a close spatial correspondence between optical polarization (by scattering from dust) and radio continuum (by synchrotron emission from gas) in jets that extend outward for several tens of kiloparsecs from the nucleus. Whereas dust in the nucleus itself might be replenished (e.g. by stellar mass-loss or local influx of interstellar matter), it is not clear how this could happen in the extended jets.

Recommended reading

- *Dust Metamorphosis in the Galaxy*, by J Dorschner and T Henning, in Astronomy and Astrophysics Reviews, vol 6, pp 271–333 (1995).
- *Chemical Evolution of Protostellar Matter*, by William D Langer *et al*, in *Protostars and Planets IV*, ed V Mannings, A P Boss and S Russell (University of Arizona Press, Tucson), pp 29–57 (2000).
- *Theory of Interstellar Shocks*, by B T Draine and C F McKee, in Annual Reviews of Astronomy and Astrophysics, vol 31, pp 373–432 (1993).

Problems

1. In ion–molecule chemistry, why is H_2O formed by recombination of an electron with H_3O^+ (rather than H_2O^+)? What is the expected product when H_2O^+ recombines?
2. Two dark interstellar clouds (clouds A and B) in the solar neighbourhood of the Galaxy are similar in size, mass, age and structure and each produces visual extinction $A_V = 3$ mag through its centre. However, the dust in cloud A is characterized by a 'normal' extinction curve with $R_V = 3.0$, whereas that in cloud B has $R_V = 4.5$. Assuming that the 'CCM' empirical law (section 3.4.3) applies in each case, discuss with reasoning what differences (if any) you would expect between the abundances of simple gas-phase molecules in the two clouds.
3. The binding energy of an H_2O molecule in bulk H_2O-ice (0.52 eV) is much greater than the adsorption (attachment) energy of an isolated H_2O

molecule on a grain surface, which is typically ~0.1 eV. Compare the range of photon wavelengths capable of desorbing ices from grains for the bulk and isolated cases. Does this difference help to explain the 'threshold effect' (section 5.3.2) for detection of ice in dark clouds?

4. (a) Show that thermal evaporation of *isolated* H_2O molecules from a grain surface is negligible within cloud lifetimes for dust temperatures $T_d <$ 19 K.
   (b) Estimate the timescales for evaporation of an H_2O molecule from the surface of a *bulk* H_2O-ice mantle, considering dust temperatures of (i) 19 K, (ii) 100 K and (iii) 200 K. Briefly comment on the significance of the results.
5. Distinguish between thermal processing and UV photolysis as mechanisms leading to grain mantle evolution in regions of active star formation within molecular clouds. What observational evidence do we have for each of these processes?
6. Compare with reasoning the composition and structure of the ice mantles you would expect to find on grains in (a) a cold ($T_d \sim 10$ K), quiescent molecular cloud lacking internal sources of energy; (b) a cold ($T_d \sim 10$ K) region of a molecular cloud that is subject to an intense local source of UV photons; and (c) a warm ($T_d \sim 75$ K) protostellar condensation within a molecular cloud that remains shielded from energetic radiation.
7. A shock wave propagating at a speed of 600 km s^{-1} through the intercloud medium (density $n_0 \approx 5 \times 10^3$ m^{-3}) encounters a molecular cloud of mean internal density $n_c \approx 2 \times 10^8$ m^{-3}. Estimate the speed of the shock inside the cloud. Is the shock likely to be capable of removing ice mantles from grain surfaces within the cloud?

Chapter 9

Dust in the envelopes of young stars

"At one time, it was thought that an X-solar-mass star resulted from the collapse of an X-solar-mass cloud..."

F H Shu *et al* (1989)

New stars are born when regions of an interstellar cloud fragment and collapse under their own gravity. Until the advent of infrared astronomy this process was largely hidden from view, veiled by layers of obscuring dust. The parent molecular clouds are much more massive than the stellar populations they spawn and, during each generation of starbirth, only a minor fraction of associated interstellar material is converted into stars. Newly born stars disrupt and dissipate the placental material by means of their radiative energy, winds and shocks. Some effects of star formation on molecular clouds were reviewed in the previous chapter. In this chapter, we focus on the fate of material contained within the circumstellar envelopes of young stars themselves. Some of this material may be returned to the ISM in vigorous stellar winds, some may ultimately be incorporated into planets.

We begin with a brief overview of the star formation process and discuss some observed properties of dust around young stars (section 9.1). Protoplanetary discs are reviewed in section 9.2, with emphasis on stars of low and intermediate mass that seem most likely to be realistic analogues of the early Solar System. Most of the chemical elements needed to form planetesimals were carried into the solar nebula by interstellar dust. Comparisons between interstellar dust and primitive bodies such as comets and asteroids in the present-day Solar System thus provide insight into the modification processes that operated in the solar nebula, a topic discussed in section 9.3. In the final section, we consider the possible relevance of such bodies as reservoirs of the organics and volatiles needed to form planetary biospheres and life.

9.1 The early phases of stellar evolution

9.1.1 Overview

Densities required for the collapse of interstellar matter into stars are expected to arise only in molecular clouds (e.g. Elmegreen 1985). Millimetre-wave observations of CO and other tracers of molecular gas, coupled with surveys at infrared wavelengths, demonstrate that stars in the earliest phases of their evolution are, indeed, embedded within the cold, dense cores of molecular clouds (figure 9.1). These clouds behave as compressible, turbulent magnetohydrodynamic fluids (Vázquez-Semadeni *et al* 2000). The fragmentation of a cloud into individual prestellar condensations depends on the detailed interplay of gravity, magnetic fields and turbulence, and is probably the least well understood phase of the entire star formation process (e.g. Mestel 1985, Shu *et al* 1987, Larson 1989, Mouschovias and Ciolek 1999, Myers *et al* 2000, Balsara *et al* 2001). In general, high-mass stars form with much lower frequency than low-mass stars, as first discussed in detail by Salpeter (1955). Two fundamentally different modes of star formation appear to exist in the disc of our Galaxy, characterized as 'high mass' and 'low mass' (Shu *et al* 1989; Lada *et al* 1993; Wilking 1997). Young high-mass stars tend to be found within the dense cores of giant molecular clouds (Churchwell 1990), many of which are likely to evolve to become gravitationally bound OB associations or open clusters. In its most profligate form, this is the 'starburst' mode of star formation. Whilst some low-mass stars form together with high-mass stars in giant molecular clouds, others form in dark clouds within much looser groupings that lack OB stars. The latter seem destined to become part of the field star population rather than members of bound clusters (Wilking 1989). Regions of high- and low-mass star formation in the solar neighbourhood are typified by the Orion (M42/OMC–1) and Taurus clouds, respectively.

The early evolution of a star may be characterized in terms of distinct phases, illustrated in figure 9.1. We may describe these as the collapse phase, the embedded-YSO phase and the T Tauri phase. During the collapse phase, luminosity is derived primarily from gravitational energy released by infall of accreting material (Wynn-Williams 1982, Beichman *et al* 1986, André *et al* 2000, Myers *et al* 2000). If the original condensation has some net rotation, conservation of angular momentum causes the development of a rotating disc as collapse proceeds. Meanwhile, the temperature of the nucleus rises until thermonuclear reactions are ignited and the young star begins to drive a wind. The wind is rapidly decelerated when it encounters the equatorial disc but travels more freely in other directions, blowing cavities in the polar regions of the envelope and establishing a bipolar outflow pattern (Lada 1988, Edwards *et al* 1993). During both collapse and embedded-YSO phases, the young star remains deeply embedded in and hidden by dense molecular gas and dust, such that most of its luminous energy emerges in the mid- to far infrared. Eventually,

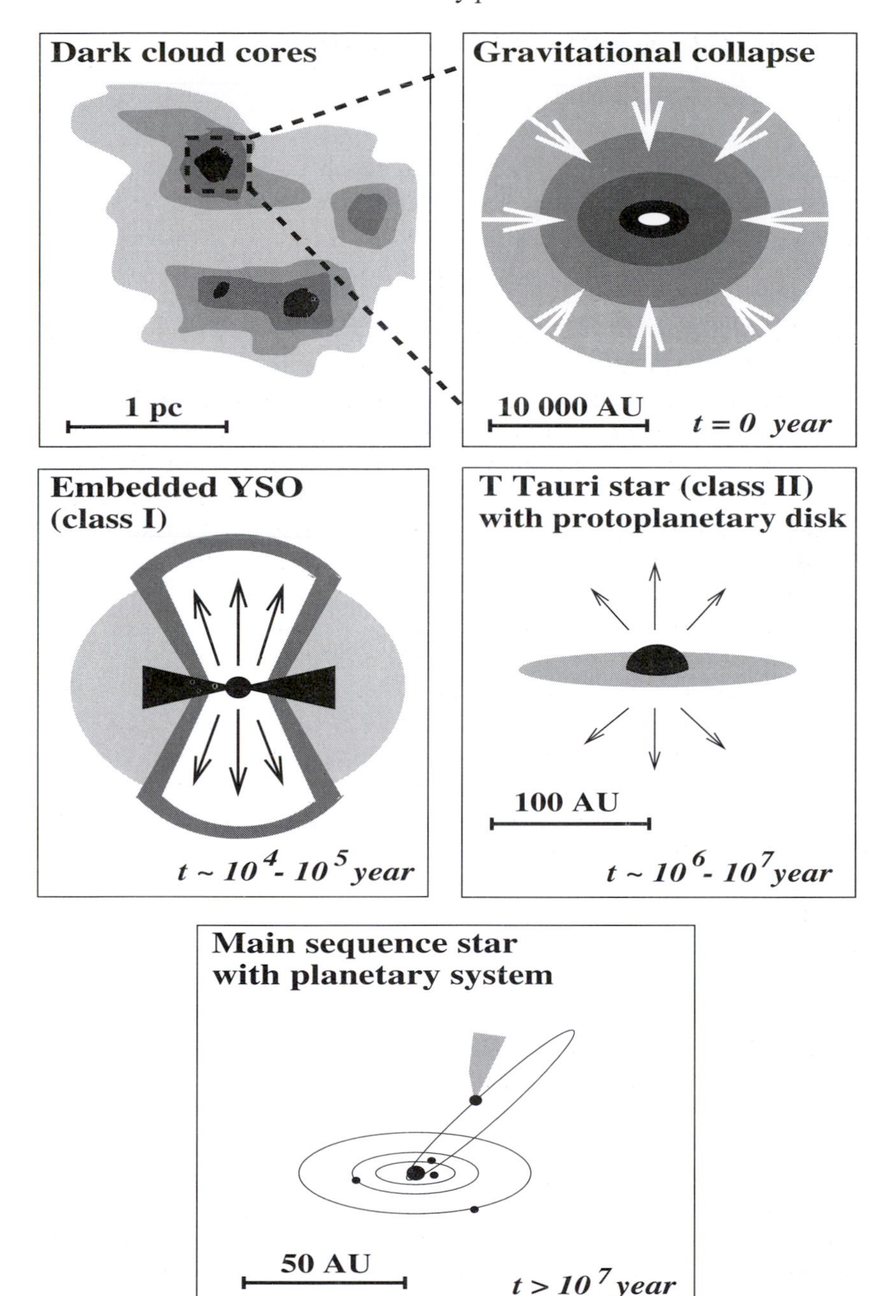

Figure 9.1. Schematic illustration of the early evolution of an intermediate-mass star from initial collapse to arrival on the main sequence. Adapted from van Dishoeck and Blake (1998), courtesy of Ewine van Dishoeck.

the material in the extended envelope either falls into the disc or is swept up and ejected by the outflow. The T Tauri phase is then reached, in which the object may become a visible pre-main-sequence star accompanied by an optically thick accretion disc (Basri and Bertout 1993). Note, however, that the visibility of the central object is highly dependent on viewing angle. For lines of sight away from the plane of the disc, circumstellar extinction is greatly diminished; but in cases where the disc is viewed approximately edge on, the star will remain hidden at wavelengths below about 1 μm. The visible light from the star may, nevertheless, sometimes be detected indirectly, via reflection in which starlight is scattered into our line of sight by concentrations of dust lying out of the plane of the disc (e.g. Weintraub *et al* 1995). Many Herbig–Haro nebulae are thought to arise in this way (Schwartz 1983). Finally, the T Tauri star evolves toward the main sequence along predictable evolutionary tracks in the HR diagram (e.g. Cohen and Kuhi 1979, Bodenheimer 1989). This is illustrated in figure 9.2, in which models for the pre-main-sequence evolution of stars of different mass are compared with the distribution of T Tauri stars. The entire formation process from initial condensation to arrival on the main sequence takes approximately 40 Myr for a 1 $M_{\odot}$ star.

This discussion assumes that initial collapse leads to the formation of a single star. However, more than 50% of all stars in the solar neighbourhood are members of binary or multiple systems. The crucial parameter appears to be the angular momentum of the system, which may determine whether collapse results in a stable disc around a single protostar or a fragmented disc with more than one major mass concentration (Bodenheimer *et al* 1993). This issue will not be considered further here: as one of the goals of this chapter is to examine what can be learned about the early evolution of the Solar System, we will naturally focus on single stars.

9.1.2 Infrared emission from dusty envelopes

The spectral energy distribution of a young star changes dramatically during the course of its evolution through the various phases described earlier. Models have been constructed that predict the shape of the continuous spectrum as the source evolves from a prestellar core (detected by emission from dust) to a young star with a visible photosphere and greatly reduced dust emission (Adams *et al* 1987). Comparisons of observed and predicted fluxes provide a valuable diagnostic technique for classifying infrared sources, identifying young stars still hidden by foreground extinction and characterizing the nature of their circumstellar envelopes (Wilking 1989, Wilking *et al* 1989, André *et al* 1993, Kenyon *et al* 1993). Some examples are shown in figures 9.3 and 9.4.

Prestellar cores and protostars undergoing gravitational collapse (figure 9.1) emit appreciable flux only at the longest infrared wavelengths ($\lambda > 20$ μm). Their flux distributions (figure 9.3) are consistent with a simple isothermal model of the

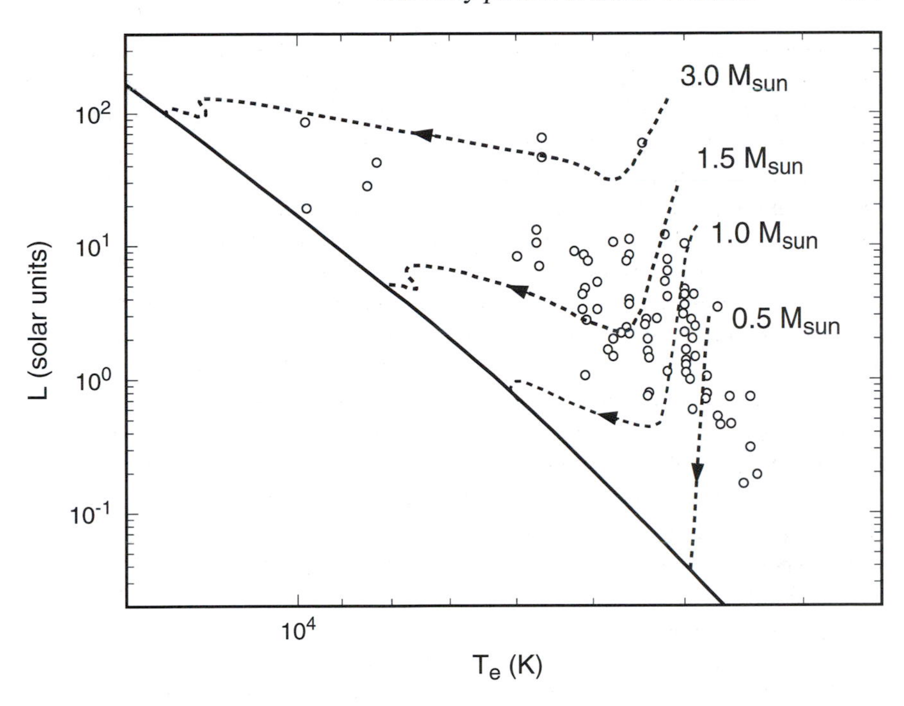

Figure 9.2. Hertzsprung–Russell diagram showing the pre-main-sequence evolution of low-mass stars. Evolutionary tracks are shown for stars with masses in the range 0.5–3.0 $M_\odot$ (broken curves). The change in slope arises because of a transition from fully convective to partially radiative transport of energy within the star for masses $>0.5\ M_\odot$. The full diagonal line indicates the loci of stars that have just reached a stable hydrogen-burning state (the 'zero-age' main sequence). T Tauri stars in the Orion region (Cohen and Kuhi 1979) are plotted for comparison.

form

$$F_\nu = B_\nu(T_d)[1 - \exp(-\tau_\nu)]\Omega \tag{9.1}$$

(André *et al* 2000), where Ω is the solid angle subtended by the source and $B_\nu(T_d)$ is the Planck function for dust at temperature T_d. The optical depth is assumed to vary with frequency as $\tau_\nu \propto \nu^\beta$ (where β is the emissivity spectral index, defined in section 6.1.1). Fits to the L1544 prestellar core (figure 9.3) indicate temperatures $T_d \approx 13$ K, i.e. not significantly different from those prevailing in cold molecular clouds, whereas dust in the collapse-phase protostar IRAS 16293-2422 appears to have been warmed appreciably ($T_d \approx 30$ K).

A classification scheme for more evolved YSOs is based on the slope of the spectral energy distribution, given by the mean value of

$$\alpha_{IR} = \frac{d(\log \lambda F_\lambda)}{d(\log \lambda)} \tag{9.2}$$

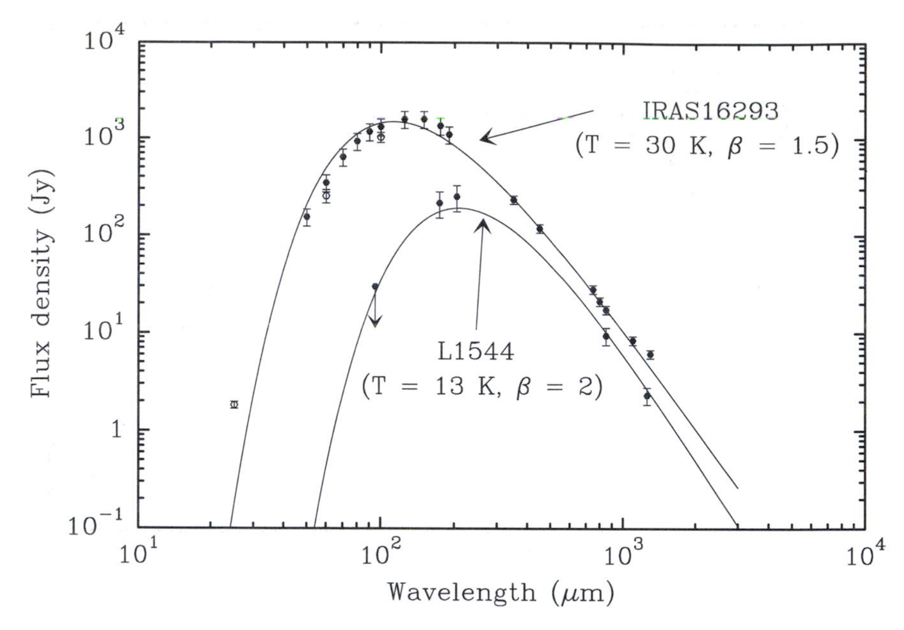

Figure 9.3. Spectral flux distributions of the prestellar core L1544 and the collapse-phase ('class 0') protostar IRAS 16293-2422 (André *et al* 2000). The curves represent models based on equation (9.1), with values for dust temperature and emissivity spectral index as indicated. Figure courtesy of Derek Ward-Thompson.

from near- to mid-infrared wavelengths (2–25 μm; Wilking 1989). The youngest objects detected in the near infrared, designated class I, are characterized by spectra that increase toward longer wavelength ($\alpha_{IR} > 0$)[1]. As an embedded YSO evolves, the photospheric contribution increases, dust emission becomes less prominent and the slope turns negative. Class II sources are associated optically with T Tauri stars (figure 9.1) or their more massive counterparts, the Herbig Ae/Be stars, and have slopes in the range $0 > \alpha_{IR} > -2$. Finally, stars close to or on the zero-age main sequence (figure 9.2) have spectral energy distributions resembling blackbodies at photospheric temperatures, for which the infrared slope approaches the value expected in the Rayleigh–Jeans limit ($\alpha_{IR} = -3$). These objects, designated Class III, have active chromospheres and are often strong x-ray emitters.

The infrared spectra of young stars frequently exhibit structure arising from dust-related emission or absorption features, as discussed in detail elsewhere (sections 5.1.3, 5.3, 6.3.1 and 8.4.3). In general, the emergent spectrum may contain contributions from dust both within the circumstellar envelope and in the foreground molecular cloud. The 9.7 μm silicate feature is commonly detected

[1] Protostars in the earlier collapse phase, for which α_{IR} is indeterminate, were subsequently designated 'class 0' (André *et al* 1993, 2000).

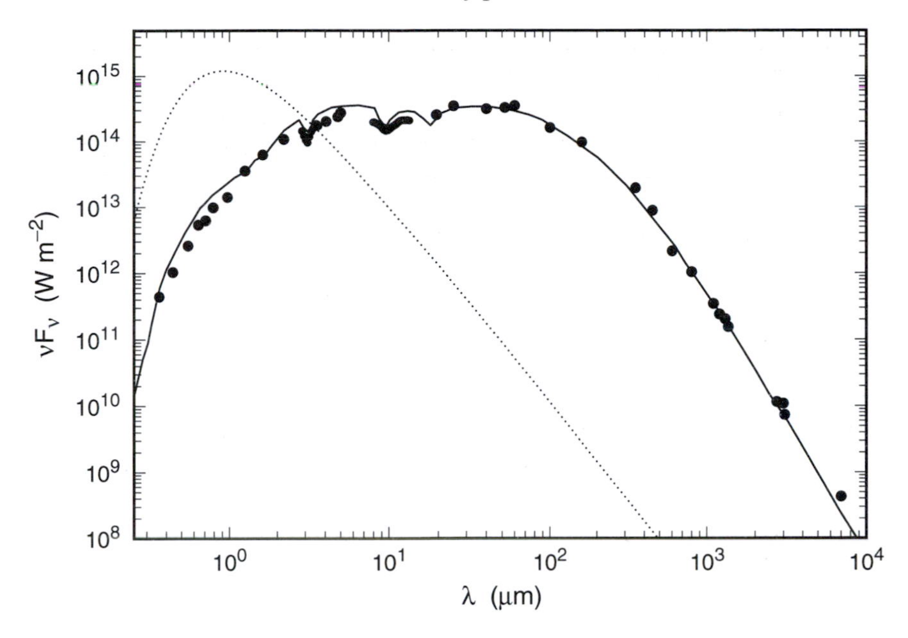

Figure 9.4. Spectral energy distribution of the YSO HL Tauri. Observational data (points) are from the compilation of previous literature by Men'shchikov *et al* (1999). The dotted line represents the expected stellar continuum (a blackbody with $T = 4000$ K) if no dust were present. The continuous line represents a model fitted to the observations, assuming a toroidal circumstellar envelope of mass 0.11 $M_\odot$, viewed at an angle 43° from its equatorial plane and containing dust composed of silicates, ices, metal oxides and amorphous carbon (see Men'shchikov *et al* 1999 for full details of the model and section 9.2.1 for further discussion).

and its profile is generally consistent with models based on the Trapezium emissivity curve (section 6.3.1; Hanner *et al* 1998). Crystalline subfeatures are sometimes present (section 9.3.1). Whether the feature is seen in net absorption or net emission in a particular line of sight may depend on the viewing geometry (Cohen and Witteborn 1985). In cases where a circumstellar disc is viewed edge on, it will generally appear in absorption, whereas emission from warm silicate grains may dominate the spectrum for other orientations. Stars with silicate absorption have a high incidence of associated Herbig–Haro nebulae, suggesting common constraints on viewing geometry (section 9.1.1). Some objects with silicate absorption also show ice absorption features (Willner *et al* 1982, Whittet *et al* 1988, Sato *et al* 1990, Boogert *et al* 2000b). In many lines of sight, the ice may be located in the foreground molecular cloud but in some cases, of which HL Tauri is considered a good example, ice and silicate dust appear to co-exist in the disc (Cohen 1983, Whittet *et al* 1988, Bowey and Adamson 2001).

The envelopes of young stars may be mapped spatially in both far infrared continuum emission and millimetre-wave spectral lines. Results show a structure generally consistent with the evolutionary state of the YSO as indicated by its spectral energy distribution: large, spheroidal envelopes around the youngest objects and flattened discs or toroids around more mature pre-main-sequence stars (Beckwith and Sargent 1993, Chandler and Richer 2000, Mundy *et al* 2000). Whilst the infrared and millimetre observations map the distribution of dust and gas, respectively, the latter also provide velocity information. This shows that, in many cases, the disc is in Keplerian rotation about the stellar nucleus (i.e. with tangential speed $v(r) \propto r^{-1/2}$) on size scales extending over several hundred AU (Sargent and Beckwith 1987, Beckwith and Sargent 1993, Thi *et al* 2001). The available resolution is not yet good enough to explore the inner regions of the envelopes on scales comparable with our planetary system, but this situation should change within the next few years (van Dishoeck 2002).

9.1.3 Polarization and scattering

Observations of polarized radiation from young stars are important for several reasons: they may help to constrain the properties of the dust grains, the orientation and strength of the local magnetic field and the geometry of the YSO environment (see Weintraub *et al* 2000 for an extensive review). Polarization may be introduced in two ways. First, if the line of sight contains aspherical grains that are being systematically aligned (e.g. by a magnetic field), then linear polarization will result from dichroic extinction in the transmitted beam at visible or near infrared wavelengths, as discussed in detail in chapter 4. Corresponding polarized emission will be seen in the far infrared (section 6.2.5). Second, if starlight is scattered by circumstellar dust, that light will be polarized even if the grains are spherical or have no net alignment. Scattered light from a star embedded in a uniform dusty medium will produce a centrosymmetric pattern of linear polarization vectors, as commonly observed in reflection nebulae; in the case of a YSO, this will be moderated by the disc structure of the circumstellar envelope (Weintraub *et al* 2000). Polarizations produced by dichroism and by scattering will have somewhat different spatial distributions, arising from their different dependences on magnetic field direction and the direction of incident radiation. Results of model calculations by Whitney and Wolff (2002) are illustrated in figure 9.5. The primary effect of the aligned grains on the linear polarization is to enhance its degree, whilst preserving the basic centrosymmetric distribution (compare upper frames in figure 9.5). Such patterns are commonly observed in many YSOs at near infrared wavelengths (e.g. Gledhill and Scarrott 1989, Whitney *et al* 1997, Lucas and Roche 1998).

If the dust grains are spherical, only linear polarization is produced when unpolarized radiation is scattered. However, circular polarization will result if light that is already linearly polarized is scattered or if unpolarized light is scattered by aspherical particles that are aligned. Either of these situations is quite likely to arise in YSO envelopes, e.g. if light that has already been linearly

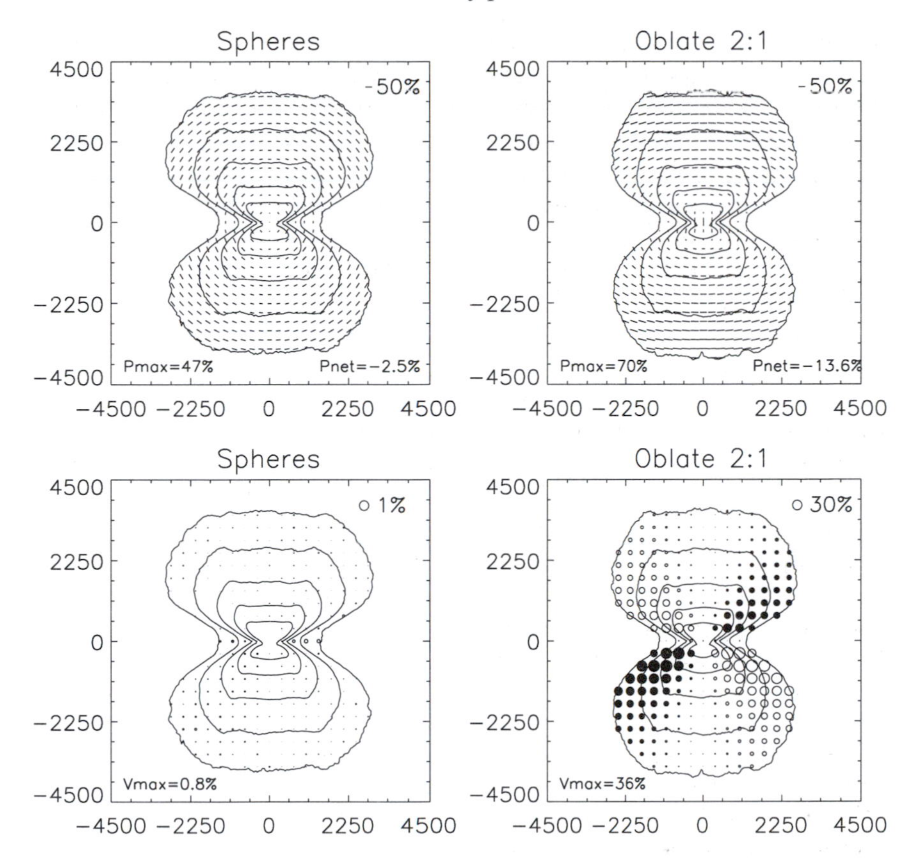

Figure 9.5. Models for the distribution of linear and circular polarization in a protostellar envelope. The x and y axes represent distance in AU, with the protostar centred at (0, 0) and with the circumstellar disc along the $y = 0$ axis, inclined at an angle of 84° to the plane of the page. Contours indicate the distribution of intensity. The upper frames show linear polarization for spheres (left) and oblate spheroids (right); in the latter case, grains of axial ratio 2:1 are assumed to be aligned by a magnetic field perpendicular to the plane of the disc. The lower frames show the corresponding circular polarization (filled and open circles distinguish left- and right-handed cases). Note that the maximum degree of circular polarization is dramatically increased in the case of aligned spheroids. Figure courtesy of Barbara Whitney, adapted from Whitney and Wolff (2002).

polarized by a scattering event is scattered again (multiple scattering) or if target grains similar to those in the ISM are being aligned. Observations have revealed remarkably high degrees of near infrared circular polarization in some YSOs (Londsdale *et al* 1980, Bailey *et al* 1998, Chrysostomou *et al* 2000, Clark *et al* 2000). In the case of OMC-1, levels ~20% have been detected, vastly greater than those measured in the ISM (section 4.3.5). Models indicate that such high degrees of circular polarization can be reproduced only if aspherical grains are being aligned (Chrysostomou *et al* 2000, Gledhill and McCall 2000, Whitney and Wolff 2002; compare lower frames in figure 9.5). The observed distribution of circularly polarized light displays a quadrupolar structure, consistent with model predictions (figure 9.5, lower right). Finally, the degree of ellipticity (the ratio of circular-to-linear polarization) places constraints on the composition and size of the grains. Results for OMC–1 are consistent with dielectrics such as silicates provided that sufficiently large grains are included in the size distribution. Very small grains and highly absorbing grains can be ruled out.

9.1.4 Ice sublimation in hot cores

The embedded-YSO phase of a massive star may last some 10^5 years, during which the density and temperature of the molecular envelope may reach levels $\sim 10^{12}$–10^{14} m^{-3} and 100–300 K, much elevated compared with the surrounding molecular cloud. These so-called 'hot cores' are typically $\sim$0.1 pc in size, comparable with the dimensions of the Solar System's Oort cloud of comets. The combination of high density and high temperature is unusual in terms of interstellar environments and presents opportunities for chemical evolution not possible in colder clouds (Blake *et al* 1987, Walmsley and Schilke 1993, Millar 1993, van Dishoeck and Blake 1998, Lahuis and van Dishoeck 2000, Gibb *et al* 2000b). Dust temperatures may rise above the sublimation point for most or all of the ice mantle constituents (see figure 8.6). Species commonly observed by radio techniques in hot cores include some quite complex organic molecules, such as CH_2CHCN (vinyl cyanide) and CH_3CH_2CN (ethyl cyanide). It is of great interest to understand how these species form and whether they are present in the mantles prior to evaporation. The possibility thus exists to use radio astronomy as an indirect method of studying grain-mantle composition, and of identifying mantle constituents with abundances too low for direct detection in the solid phase by the techniques described in chapter 5.

The abundances of some gaseous molecules are selectively enhanced in hot cores compared with those in cold molecular clouds, as was first demonstrated by Sweitzer (1978). Figure 9.6 compares abundances for the Orion hot core with those for the surrounding Orion ridge. Physical conditions in the ridge appear to be less extreme than in the hot core and more representative of normal molecular gas. The plot shows that many N-bearing and S-bearing species are substantially more abundant in the hot core compared with the ridge, often by factors of more than 10 and sometimes by factors of more than 100. This might be taken simply as evidence for the evaporation of icy mantles in the hot core, leading to enhanced molecular abundances in the gas, but in reality the situation is more complex.

A molecule observed in a hot core may originate in three ways (Ehrenfreund and Charnley 2000). First, gas-phase reactions in the parent molecular cloud (section 8.2) may be important for the production of some species, notably CO, which might then either persist in the gas or condense onto and evaporate from the

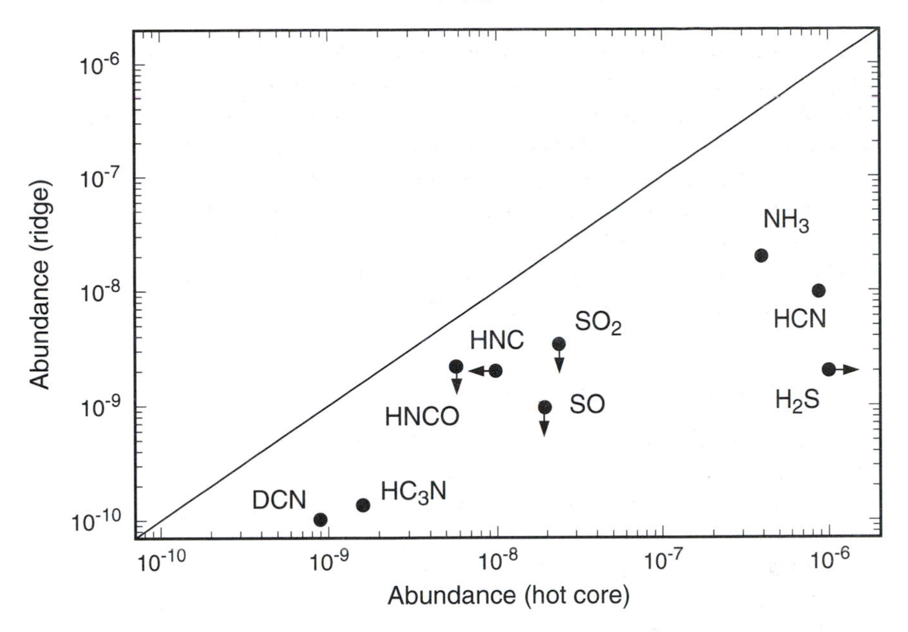

Figure 9.6. Correlation of gas-phase abundances relative to hydrogen for various interstellar molecules observed in the Orion ridge and Orion hot-core regions. The diagonal line indicates exact agreement; arrows indicate limiting values. Systematically higher abundances in the hot-core region are attributed to differences in the chemical and thermal histories of the regions and the different contributions of ice evaporation products. Data from Walmsley and Schilke (1993) and references therein.

dust without undergoing further chemical change. Second, grain surface reactions accompanying the growth of mantles (section 8.4) are the most likely source of saturated molecules such as H_2O and NH_3. The third possibility is that molecules are generated in the hot core itself, by chemical reactions acting on the ambient gas, including the products of mantle evaporation (Charnley *et al* 1992, Caselli *et al* 1993, Charnley 1997). Whereas ion–molecule reactions tend to dominate gas-phase chemistry at low temperature, many neutral–neutral reactions may become important in hot cores. Enhanced temperatures and the presence of shocks (Viti *et al* 2001) will provide the means to overcome activation energies and drive endothermic reactions.

Do the observations enable us to distinguish between these possibilities? If most of the species present in a hot core formed at low temperature, either in the gas or on the grains in the parent cloud, we would expect them to exhibit large degrees of deuteration (Tielens 1983) and this is highly consistent with observations in many cases (Turner 1990, Rodgers and Millar 1996). But it is not possible to explain abundances in hot cores purely in terms of low-temperature chemistry and mantle evaporation. Indeed, many of the molecules with enhanced abundances in hot cores remain undetected in the mantles (section 5.3.1): examples include HCN and the higher nitriles and several S-bearing compounds (SO, SO_2, H_2S). In some cases, this may simply reflect sensitivity limits – species present at below ~1% abundance relative to H_2O in the solid phase are typically difficult or impossible to detect. However, a clear difference between ice mantles and hot-core gas has been established in the case of CN-bearing species (Whittet *et al* 2001b). Infrared observations (section 5.3.6) support the presence of cyanate (–OCN) groups, formed by energetic processing of the ices in the vicinities of some massive YSOs (section 8.4.3) but radio observations of the gas indicate a preponderance of nitriles (HCN, CH_3CN, CH_2CHCN, etc, lacking the adjacent O atom). HCN most probably forms by gas-phase reactions fuelled by ammonia released from grains in hot cores:

$$\begin{aligned} NH_3 + H_3^+ &\rightarrow NH_4^+ + H_2 \\ NH_4^+ + e^- &\rightarrow NH_2 + H_2 \\ NH_2 + C &\rightarrow HNC + H \\ HNC + H &\rightarrow HCN + H \end{aligned} \tag{9.3}$$

(Charnley *et al* 1992, Millar 1993), a sequence inhibited in cold clouds not only by a relative dearth of gaseous NH_3 but also by the high activation energy of the final step. The cyanogen (CN) radical can then form by H abstraction reactions involving HCN or HNC, and CN may react with various hydrocarbons to produce the higher nitriles.

Gaseous CH_2CHCN and CH_3CH_2CN have been shown to concentrate within a region small compared with the Oort cloud in the envelope of a protostar in the Sgr B2 molecular cloud (Liu and Snyder 1999). Understanding the chemistry of hot cores around YSOs may thus be an important step toward understanding the evolution of protoplanetary matter. Although the examples studied to date are massive compared with the Sun, some interesting chemistry doubtless occurs on a more modest scale in warm gas associated with less massive YSOs that might ultimately form solar systems such as our own.

9.2 Protoplanetary discs

Both theoretical models and observational results suggest that the formation process for a single star of low-to-intermediate mass ($0.1 < M < 5\ M_\odot$) results naturally in the development of a circumstellar disc that might subsequently become a planetary system. Dusty discs appear to be ubiquitous around pre-main-sequence stars (Beckwith and Sargent 1993) and may sometimes persist around more mature stars that have reached the main sequence (Backman and Paresce 1993). The discs are detected not only by their far-infrared emission (section 9.1.2) but also, in some instances, by absorption and scattering in the visible and near infrared. Well known examples include the 'proplyds' seen in silhouette toward the Orion nebula (O'Dell *et al* 1993, McCaughrean and O'Dell 1996). The study of such discs is an area of enormous current activity and growth, matched by the parallel discovery of extrasolar planets in ever increasing numbers (Marcy and Butler 1998). There is thus growing evidence that planetary systems are commonplace, at least around single stars (although it remains to be seen how many contain *habitable* planets). I will not attempt to review this entire field here; in this section, I focus on the evolution of dust in the discs of $\sim$1 $M_\odot$ stars. Several examples are known that appear likely to be reasonable analogues of the early Solar System (Koerner 1997).

9.2.1 T Tauri discs

Between 25 and 50% of pre-main-sequence stars in nearby dark clouds have detectable circumstellar discs (Skrutskie *et al* 1990, Beckwith and Sargent 1993, 1996). Their masses lie in the range 0.001–0.5 $M_\odot$, contributing up to about 10% of the total mass of the system. They are typically several hundred AU in extent, with temperatures ranging from $\sim$1000 K near the inner boundary to $\sim$30 K in the outermost regions. Many display evidence for Keplerian rotation about the central star (section 9.1.2). The discs are thought to be flared rather than strictly planar, i.e. the thickness increases with radial distance from the centre (Kenyon and Hartmann 1987), as depicted in figures 9.1 (frame 3) and 9.7. This geometry is needed to explain the fact that the surface layers of the outer discs are often warmer than would be expected for flat-disc models: flaring exposes material above and below the midplane to light from the central star, whereas material at the midplane is well shielded. One consequence of this is that gas-phase molecules are often present that might otherwise be frozen onto the dust (Willacy and Langer 2000, Boogert *et al* 2000b, Aikawa *et al* 2002).

Detailed modelling of their spectral energy distributions provide important constraints on the properties of the discs. As the discs around class I and class II objects remain optically thick, detailed radiative transfer calculations are required for realistic modelling. As a specific example, consider the case of HL Tau, an embedded YSO of mass $\sim$1 $M_\odot$, often regarded as a good analogue of the Sun at the corresponding point in its evolution (Cohen 1983, Sargent and Beckwith 1987, Stapelfeldt *et al* 1995, Close *et al* 1997, Koerner 1997). With an estimated age of $\sim$0.1 Myr, HL Tau is younger than most objects identified as T Tauri stars[2]. Its spectral energy distribution (figure 9.4) is well sampled from the near infrared to millimetre wavelengths. Models described by Men'shchikov *et al* (1999) require two distinct populations of dust grains in the disc: (i) an inner torus of radius $\sim$100 AU, containing predominantly very large dust grains ($a \geq 100$ μm); and (ii) an outer torus containing much smaller ($a \leq 1$ μm) grains. The density in the disc varies with radial distance as $\rho(r) \propto r^{-q}$ where q increases from 1.25 in the

[2] HL Tau was originally classified as a T Tauri star on the basis of its optical spectrum, but the visible 'star' was later shown to be a reflection nebula; the star itself remains hidden by over 20 magnitudes of visual extinction (Stapelfeldt *et al* 1995).

inner torus to 2 in the outer torus. Opacity toward the central star is provided by a combination of grey (approximately wavelength-independent) extinction from the large grains and normal wavelength-dependent extinction from the small grains. The assumed masses of the two dust components are comparable. The need for a population of large grains to match the spectral energy distribution of the star (figure 9.4) hints that in HL Tau we might be observing the early stages of particle accumulation that could lead to the formation of planets.

Evolution of a YSO through the various stages described in section 9.1.1 is accompanied by a progressive thinning of the circumstellar environment. This is manifested by changes in the spectral energy distribution, with declining infrared emission and emergence of a photosphere as the visual attenuation is reduced. During evolution from the embedded YSO phase to the T Tauri phase, mass is lost from the system in bipolar flows, whilst material may continue to fall into outer regions of the disc from the envelope; net mass-loss rates $\sim 10^{-7}$ to 10^{-8} $M_{\odot}$ yr^{-1} appear to be typical (Edwards *et al* 1993). Observations show that the discs themselves become progressively thinner during the T Tauri phase: whereas $\sim$50% of those less than 3 Myr in age have optically thick discs, this has dropped to $<$10% at age 10 Myr (Skrutskie *et al* 1990). How does this occur? Physical processes leading to their dissipation include radiation pressure and the Poynting–Robertson effect; in the latter case, absorption and re-emission of radiation leads to the decay of a particle's orbit such that it spirals into the star (Hodge 1981). Both of these mechanisms tend selectively to remove the smaller particles. Detailed calculations by Artymowicz (1988) suggest that radiation pressure is generally the dominant process once the discs become optically thin: particles of a given composition and with dimensions less than some critical size will then be efficiently expelled. For any realistic grain composition, this critical size is $a \sim 1$ μm. If the initial size distribution were typical of interstellar grains, radiation pressure could thus lead to almost complete dissipation of the disc in the absence of a competing process. However, particle growth by coagulation may operate efficiently in the midplane regions of the discs (Beckwith *et al* 2000, Suttner and Yorke 2001, Wood *et al* 2002), producing clusters with dimensions $>$10 μm, i.e. above the critical size, on timescales of 10^3–10^4 years. The optical properties of the disc evolve as the particles grow: for a given volume of dust, coagulation diminishes the total surface area, and both the visual optical depth and the infrared luminosity are therefore reduced.

9.2.2 Vega discs

The discovery of optically thin dusty discs around main sequence stars (Aumann *et al* 1984), the so-called 'Vega phenomenon', was unexpected: models suggest that virtually all circumstellar matter should have either dispersed or accreted onto larger bodies by the time this stage of a star's evolution is reached. The stars concerned are predominantly single, with spectral types in the range A–K and estimated ages in the range $\sim$50 Myr to 5 Gyr; the incidence of discs around such stars may be as high as 15% in the solar neighbourhood (Backman and Paresce 1993, Lagrange *et al* 2000). The primary mode of detection is by observations of FIR emission far above that predicted for a stellar photosphere. The emission is generally spatially extended and, in the case of β Pictoris, the presence of a resolved circumstellar disc is confirmed optically (Smith and Terrile 1984). The radii of the discs are typically up to a few hundred AU and dust temperatures in the range 50–125 K are indicated by the spectral energy distributions. The total mass of dust needed to account for the FIR emission is typically $\sim 10^{-7}$ $M_\odot$, or 0.03 Earth masses, which exceeds by a large factor ($\sim 10^7$) the estimated total mass of interplanetary dust in our Solar System (Millman 1975). Observations of CO line emission suggest that the discs have a low gas content, but this inference may be affected by systematic errors in the CO/H_2 conversion introduced by selective photodissociation of CO (Thi *et al* 2001): direct detections of H_2 by means of rotational line emission in the discs of several Vega systems suggest dust-to-gas ratios more in line with the average for the ISM. If this is the case, Vega discs may range up to 10^{-3} $M_\odot$ in mass, comparable with the lowest-mass T Tauri discs. Vega systems thus fit into the general pattern of decreasing circumstellar mass with age for young stars.

An observational constraint on particle size arises from the equation describing the balance of energy absorbed and re-emitted in an optically thin circumstellar disc. Considering a grain of radius a at a distance r from a star of radius R_s and surface temperature T_s, we have

$$\left(\frac{\pi a^2}{4\pi r^2}\right) 4\pi R_s^2 \sigma T_s^4 \langle Q_V \rangle = 4\pi a^2 \sigma T_d^4 \langle Q_{FIR} \rangle \tag{9.4}$$

which reduces to

$$\frac{r}{R_s} = 0.5 \left(\frac{T_s}{T_d}\right)^2 \left\langle \frac{Q_V}{Q_{FIR}} \right\rangle^{\frac{1}{2}} \tag{9.5}$$

(Walker and Wolstencroft 1988), where $\langle Q_V \rangle$ and $\langle Q_{FIR} \rangle$ are the mean absorptivity and emissivity of the dust grains evaluated at the appropriate wavelengths (see section 6.1.1). A self-consistent model for the discs of Vega-like stars that accounts simultaneously for the dust temperature (estimated from the spectral energy distribution) and the disc radius (estimated from the angular size) requires $\langle Q_V / Q_{FIR} \rangle \sim 1$ (Aumann *et al* 1984, Gillett 1986). As $Q_V \sim 1$ for most grain materials, Q_{FIR} must also be approximately unity, i.e. the particles behave as blackbodies, and the emitted flux is given by equation (6.8) with Q_λ set to a constant. This may be understood if the grains are large compared with the wavelength emitted[3]. Quantitatively, we require $2\pi a > \lambda_{peak}$, where λ_{peak} is the wavelength of peak emission, and for $\lambda_{peak} \sim 60$ μm, $a > 10$ μm. Thus, the grains are general larger than those in the ISM. This result is highly consistent

[3] For comparison, classical silicate grains with $a \sim 0.1$ μm have FIR emissivities that are small ($Q_{FIR} \ll 1$) and strongly wavelength dependent ($Q_{FIR} \propto \lambda^{-\beta}$ where $\beta \sim 1$–2; see section 6.1.1).

with expectations based on the effect of radiation pressure (section 9.2.1), which will selectively eject the smaller grains from the disc.

Matter may be lost from Vega discs via the combined effects of collisions, Poynting–Robertson drag and stellar winds as well as radiation pressure. The estimated timescale for dissipation of a disc is typically 10–100 Myr, much shorter than the age of the star in most cases (Backman and Paresce 1993, Lagrange *et al* 2000). Hence, the discs cannot represent a permanent population of unaccreted remnants of pre-main-sequence evolution: they are now generally accepted to be 'debris discs' that are being continuously replenished by collisions between larger bodies. Images of some Vega systems indicate the presence of clear inner regions on size scales of several AU, together with a structure that might result from the gravitational influence of planets. The debris discs may thus correspond to a period of late planetary accretion. An intriguing possibility, in the case of our Solar System, is that this might correspond to the period of intense bombardment, during which most of the impact craters were formed.

9.2.3 The solar nebula

Although many details remain to be clarified, a general model for the origin of the Solar System has emerged that is now widely accepted (Cameron 1988, Wetherill 1989, Lunine 1997). This model attempts to explain the composition and dynamics of the present-day Solar System in terms of what we have learned about circumstellar discs around young Sun-like stars. The planets, their moons and the numerous smaller bodies are presumed to have formed some 4.6 Gyr ago in a circumstellar disc referred to as the solar nebula. Simulations of the formation and evolution of such discs have reached a level of sophistication where it is possible to draw conclusions about the physical and chemical processes that control them.

A schematic cross section of the solar nebula is shown in figure 9.7. The young Sun at its heart drives photon and particle winds that flow freely away from the disc, whilst cold matter from the parent cloud continues to fall into its outer regions. The heat balance of the disc is governed by solar heating, infall and thermal re-radiation. The predicted temperature in the midplane declines with solar radial distance from >1000 K at 1 AU to ~ 160 K at 5 AU and ~ 20 K at 100 AU (Wood and Morfill 1988, Lunine 1997, Bell *et al* 1997, Boss 1998). The 'snowline' (figure 9.7) is the approximate distance beyond which H_2O-ice is expected to be stable (Stevenson and Lunine 1988), which occurs at ~ 5 AU, i.e. just inside the orbit of Jupiter. Planetesimals are thus expected to be volatile poor within the snowline and volatile rich beyond it, in qualitative agreement with the distribution of rocky and icy bodies in the present-day Solar System.

The fate of dust falling into the solar nebula will depend on its location. In the hot inner zone of the disc, all solids will be vaporized, whereas at large distances from the Sun, interstellar dust may survive essentially unaltered (Chick and Cassen 1997). Between these extremes, substantial modification is to be

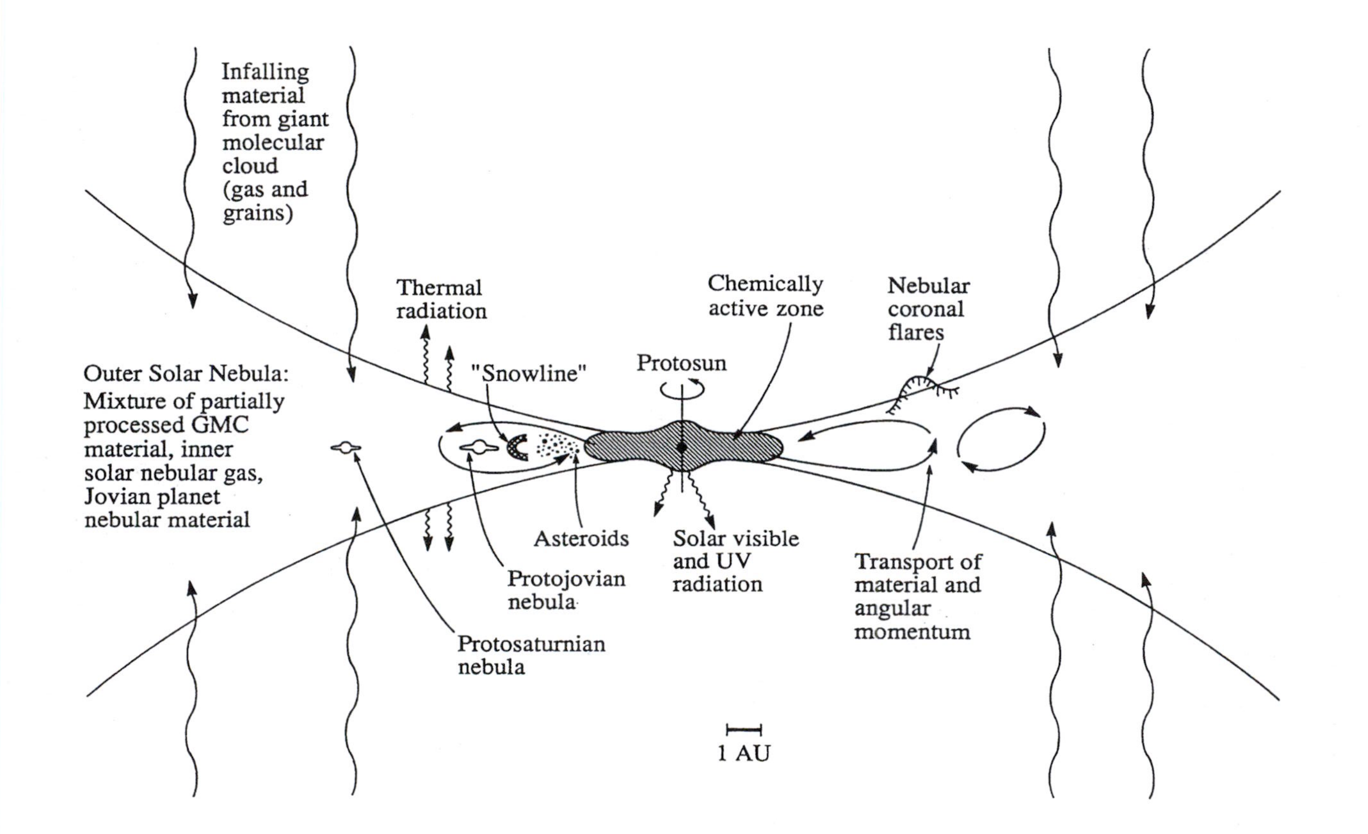

Figure 9.7. Schematic cross section of the solar nebula model discussed in section 9.2.3 (Lunine 1997). The thickness of the flared disc has been somewhat exaggerated for clarity. Figure courtesy of Jonathan Lunine.

expected. Infall creates an accretion shock that will compress the gas and result in frictional heating of the dust. This will lead to efficient vaporization of icy mantles 5–40 AU from the Sun and may be sufficient to anneal silicates at 5–10 AU (Lunine *et al* 1991, Neufeld and Hollenbach 1994, Harker and Desch 2002). As the midplane temperatures remain low, the evaporated mantles are expected to subsequently recondense, but in an altered state, probably with much higher degrees of crystallinity compared with unprocessed interstellar ices. As crystalline ices are less able to accommodate impurities compared with amorphous ices, this may have important implications for the composition of comets and other icy bodies originating in the giant-planet region of the solar nebula.

The formation of planets is a two-step process. It begins with an aggregational phase, in which stochastic grain–grain collisions lead to growth by coagulation; this is followed by an accretional phase in which the interactions of large aggregates (planetesimals) are dominated by gravity. Observations of solar analogues (section 9.2.1) indicate that the first step must be accomplished within a timescale of $\sim$10 Myr. Growth from micron-sized grains to kilometre-sized bodies is plausible but not yet well understood (e.g. Weidenschilling 1997, Beckwith *et al* 2000). Whether a grain–grain collision is constructive or destructive (section 8.3) depends not only on the impact speed but also on the detailed properties of the grains (density, structure, rigidity, stickiness). For example, the presence or absence of mantles may affect the sticking probability when grains collide (Suttner and Yorke 2001). Both model calculations (Weidenschilling and Ruzmaikina 1994, Suttner and Yorke 2001) and laboratory experiments (Blum and Wurm 2000, Poppe *et al* 2000, Kouchi *et al* 2002) show that growth can be quite rapid, subject to reasonable assumptions.

9.3 Clues from the early Solar System

Comets and asteroids are believed to be surviving remnants of planetary formation in our Solar System and, as such, they may bear a chemical memory of past events. The material they contain may include virtually unaltered pre-solar dust and ices as well as solids condensing in the solar nebula. Comparisons between the constituents of such primitive bodies and interstellar dust thus provide insight into the physical conditions in the nebula and the chemical processes that were occurring. Information is gathered from astronomical observations, space probes and laboratory analysis of material that falls to Earth. One example, the identification of isotopically distinct pre-solar grains in meteorites, was previously discussed in section 7.2.4.

No permanent population of diffuse matter exists in the inner Solar System. Much of the debris now in Earth-crossing orbits is likely to have been dispersed within the past 10 000 years by collisions between asteroids or by ablation of comets near perihelion. The fate of such debris upon entering the

Earth's atmosphere is a function of size: the smallest (micron-sized) grains are collisionally decelerated without significant melting and are thus available for collection, whereas millimetre-sized grains are completely ablated as 'shooting stars'. A meteorite fall results from the arrival of a larger fragment, some fraction of which survives passage through the atmosphere. Spectroscopic evidence suggests that asteroids are the parent bodies of most classes of meteorite, whereas interplanetary dust may include both asteroidal and cometary components.

9.3.1 Comets

Comets are undoubtedly the most volatile-rich of all surviving remnants from the early Solar System. Their basic composition is a mixture of molecular ices and dust (the 'dirty snowball' model of Whipple 1950, 1951), perhaps concealing a rocky core. Two contrasting scenarios have been proposed for their origin (see Irvine *et al* 2000 for a review). In the classical view of planetesimal formation, the solar nebula became sufficiently hot that all pre-existing solids were vaporized and homogenized. Cometary ices are then simply nebular condensates forming at the lowest temperatures and greatest solar distances (Lewis 1972), the more refractory dust being added through a radial mixing of condensates. At the opposite extreme is the proposal that comets originate almost entirely from aggregation of pre-solar ice and dust from the parent molecular cloud, without passing through a vapour phase (Greenberg 1982, 1998, Greenberg and Hage 1990; see also Whipple 1987). I refer to these as the 'nebular' and 'pre-solar' models for cometary volatiles, respectively. According to the pre-solar model, comets should have essentially the same composition as the solids in molecular clouds, i.e. unannealed silicates, ices and carbon; we shall discuss the extent to which this is true later. A nebular model in which the gas was fully homogenized no longer seems viable, most crucially because it fails to explain the deuterium content of comets (section 9.4.2); a more plausible scenario is that they formed in the wake of an accretion shock that temporarily vaporized the ices (section 9.2.3).

Different methods are used to investigate the refractory and volatile components of comets (e.g. Mumma *et al* 1993). As a comet sweeps past the Sun, refractory dust is released as the ices sublimate. This dust may be observed remotely, by infrared spectroscopy of solid-state emission features; in the case of comet Halley, we also have data from *in situ* experiments flown during the 1985/6 apparition. Although H_2O-ice is also sometimes detectable via its infrared spectral features (e.g. Lellouch *et al* 1998), the primary method of determining the volatile content is indirect, by spectroscopic analysis of the gas-phase products of sublimation.

The presence of silicate dust is confirmed by detection of the 10 μm Si–O stretch feature, which may become prominent in emission in comets close to perihelion. Evidence for the presence of profile structure consistent with partial annealing of the silicates has been found in Halley (Bregman *et al* 1987) and subsequently in other comets (Hanner *et al* 1994, 1997); data for comet

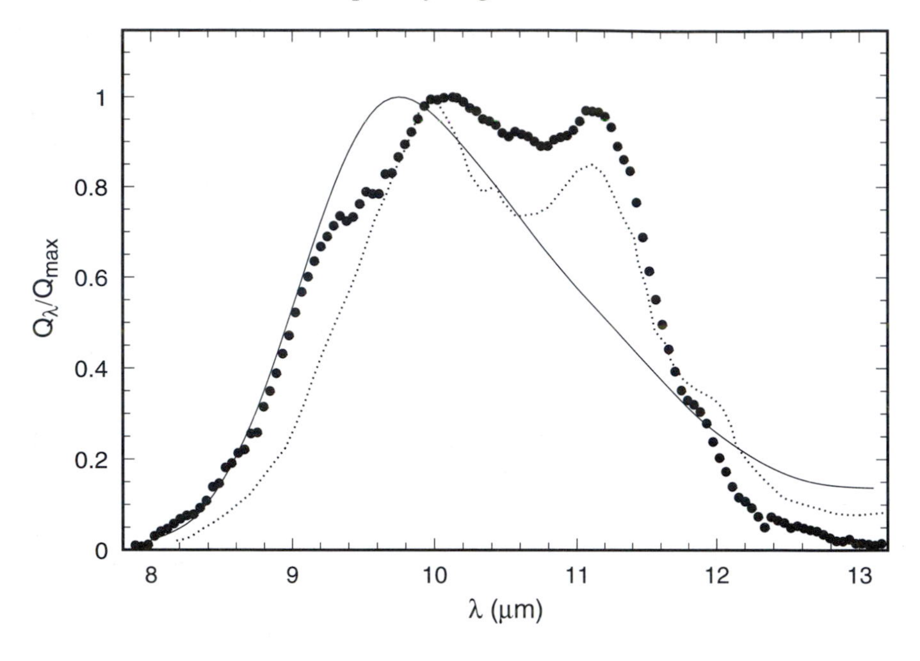

Figure 9.8. The profile of the 10 μm silicate feature in the comet Hale–Bopp at distance $r = 0.79$ AU from the Sun (Hanner *et al* 1997, points), compared with laboratory data for amorphous olivine ($MgFeSiO_4$, continuous curve) and annealed fosterite (Mg_2SiO_4, dotted curve). Each curve is normalized to unity at the peak. Silicate absorption in the diffuse ISM is well fitted by amorphous olivine (see figure 5.6). The silicates in Hale–Bopp are at least partially annealed.

Hale–Bopp are illustrated in figure 9.8. The peak at 11.2 μm is attributed to annealed Mg_2SiO_4. It might be argued that annealing is a recent event associated with heating of the grains near perihelion, but this is implausible given the occurrence of similar features in Halley and Hale–Bopp, comets with quite different orbital characteristics and recent thermal histories. At least some of the silicates incorporated into comets seem likely to have been annealed in the solar nebula; clearly they are not, for the most part, unaltered pre-solar grains, which are expected to be amorphous (section 5.2.2). A hint that annealing may be commonplace in protoplanetary discs is provided by the remarkable similarity between mid-infrared features attributed to annealed silicates in the spectra from Hale–Bopp and the Herbig-type pre-main-sequence star HD 100546 (figure 9.9).

The mass spectrometers carried by the Giotto and Vega missions to Halley produced the first *in situ* analyses of the composition of cometary dust (see Brownlee and Kissel 1990 and Schulze *et al* 1997 for reviews). Abundances were determined for particles with diameters typically in the range 0.1–1 μm. Averaged over many particles, the abundances of common mineral-forming

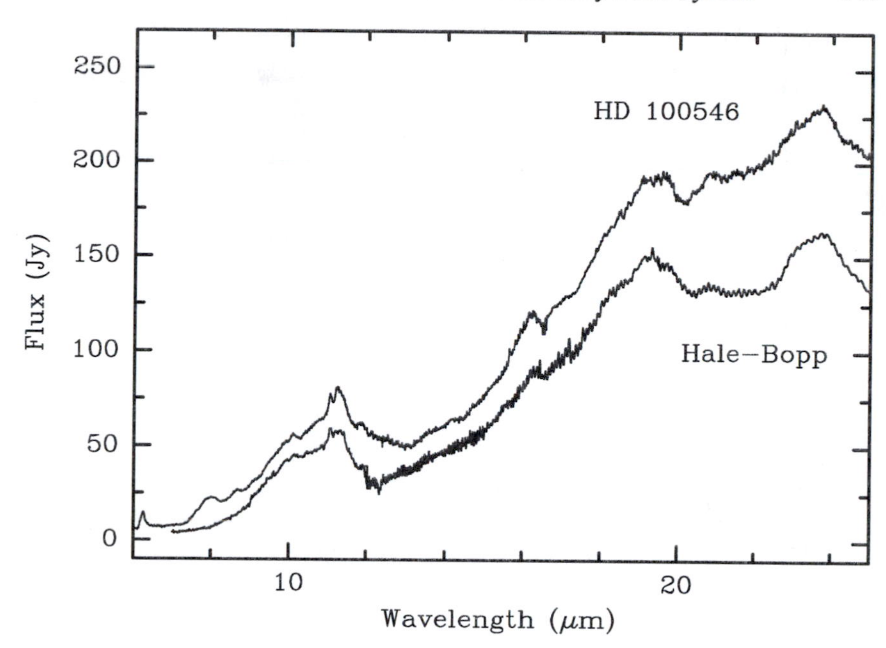

Figure 9.9. Mid-infrared spectra of comet Hale–Bopp and the young star HD 100546 compared. Data are from Crovisier *et al* (1997) and Malfait *et al* (1998), respectively. The comet was observed at $r = 2.9$ AU. Emission features common to both objects at 11.2, 19.5 and 23.7 μm are attributed to annealed silicates (compare figure 7.7). Weak PAH features at 6.2, 7.7, 8.6 and 11.3 μm are also present in the HD 100546 spectrum.

elements (O, Mg, Al, Si, Ca, Fe) were found to be chondritic, i.e. they match data for carbonaceous chondrites to within a factor of two. However, C and N show a significant excess compared with chondrites, their abundances more closely matching those in the solar atmosphere (see figure 2.2). Halley dust may contain as much as 20% by mass of carbon. Mass spectra for individual impacts indicate two major classes of grain material: refractory organics ('CHON') and magnesium silicates (Langevin *et al* 1987, Lawler and Brownlee 1992). Both materials are usually present in any given particle but in varying proportions from one to another. Statistical investigations of ionic molecular lines in the mass spectra allow some characterization of the organic fraction of the dust, and results suggest that it is composed primarily of unsaturated hydrocarbons of low oxygen content (Kissel and Krueger 1987).

Emission structure in the 3.1–3.6 μm (C–H stretch) region of cometary spectra has been detected in several comets (Baas *et al* 1986, Bockelée-Morvan *et al* 1995). The profile observed in Halley (figure 9.10) bears some resemblance to an inversion of the 3.4 μm absorption feature seen in absorption in the diffuse ISM (figure 5.10), suggesting that it might be a signature of organic refractory matter

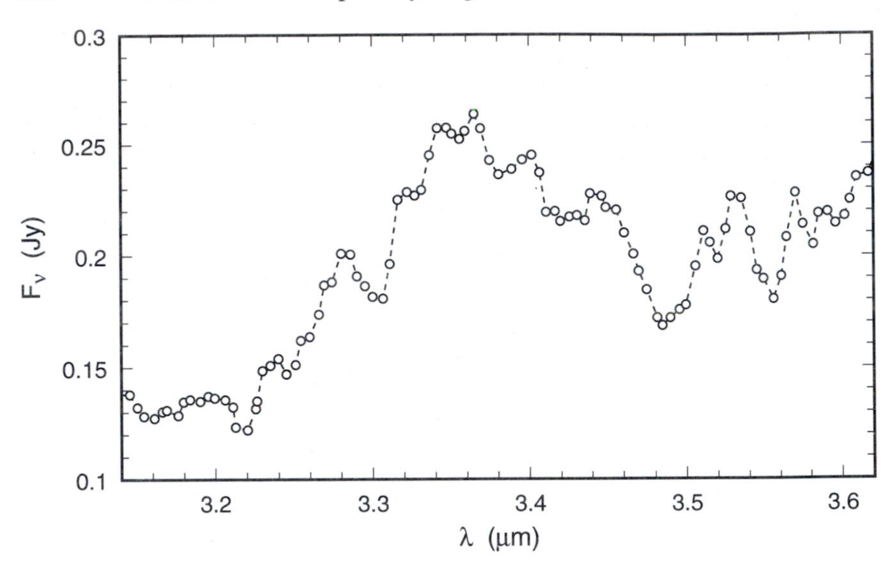

Figure 9.10. Spectrum of comet Halley from 3.1 to 3.6 μm (Baas *et al* 1986), illustrating the presence of a broad emission attributed to methanol and hydrocarbons.

(Chyba *et al* 1989). However, it is now clear that the feature arises primarily in gaseous molecules released by sublimation of the ices, of which methanol (CH_3OH) is the main contributer (Bockelée-Morvan *et al* 1995, Mumma 1997). CH_3OH abundances in the range 0.5–5% relative to H_2O are implied, with significant variations between different comets. Aromatic hydrocarbons may be responsible for the peak at 3.28 μm.

Sublimation of ices from the surface of a comet as it passes the Sun produces a stream of gaseous molecules, many of which are quickly ionized and/or dissociated by solar radiation. Determination of abundances in the original ices is not straightforward (Irvine *et al* 2000). However, great advances were made during the 1990s, thanks to improved models for cometary ablation, together with better observational facilities and the fortuitous arrival of two bright comets (Hyakutake and Hale–Bopp). Comet Hale–Bopp, in particular, has been used as a test case for the pre-solar model for cometary volatiles (Bockelée-Morvan *et al* 2000). A summary comparison of abundance data for ices in comets and those observed toward YSOs is given in table 9.1. The YSO sample is dominated by class I objects (section 9.1): note that the line-of-sight average will generally include a contribution from ices in the foreground molecular cloud, but only rarely will it include a significant contribution from the protoplanetary disc. Thus, cometary and YSO ices should be similar if little or no thermal processing occurs during accretion into the disc, as predicted by the pre-solar model.

The results in table 9.1 suggest some similarities but also some differences. Oxidized forms of carbon (CO, CO_2) tend to dominate over hydrogenated forms

Table 9.1. An inventory of cometary ices. Abundances are expressed as percentages of the H_2O abundance. Values for Hale–Bopp (Bockelée-Morvan *et al* 2000) and typical average values for other comets (Cottin *et al* 1999, Irvine *et al* 2000 and references therein) are compared with those for low- and high-mass YSOs (section 5.3.1). Cometary data are determined for a solar distance $r \sim 1$ AU, with the exception of CO_2 in Hale–Bopp, which was observed at 2.9 AU. A range of values generally indicates real variation. Values followed by a colon are particularly uncertain. Entries for HCN in YSOs are actually 'XCN' (see section 5.3.6). A dash indicates that no data are currently available.

	Comets		YSOs	
Species	Hale–Bopp	Others	Low-mass	High-mass
H_2O	100	100	100	100
CO	23	1–20	0–60	0–25
CO_2	6:	3–10	20–30	10–35
CH_3OH	2	0.5–5	≤ 5	3–30
H_2CO	1	0.2–1	<2	2–6:
HCOOH	0.1	—	<1	2–6:
CH_4	0.6	0.7	<2	2
C_2H_6	0.3	0.8	—	<0.6
NH_3	0.7	0.1–1	≤ 8	2–15
HCN	0.3	0.05–0.2	<0.5	0.2
H_2S	1.5	0.2–1.5	—	<0.4
OCS	0.5	—	<0.5	0.2

(CH_4, C_2H_6) in both comets and YSOs. However, the abundance of CO_2 in comets appears to be significantly lower than that expected for a pre-solar model. As CO_2 and H_2O are the least volatile constituents of the ices, redistribution of their abundances seems likely to require sublimation. CO_2 may be sublimated and then destroyed by gas-phase reactions during passage through shocked gas (Charnley and Kaufman 2000), such that the recondensed ices could be deficient in CO_2. Another notable discrepancy is the surprisingly low ammonia abundance in comets ($<1\%$). This is in contrast with values ranging up to $\sim 15\%$ in YSOs (table 9.1; see section 5.3.6). NH_3 is expected to form simultaneously with H_2O in interstellar clouds, with which it may join to form ammonium hydrate groups in the polar ice layer (sections 5.3.3 and 8.4.1). The concentration of other molecules that can be accommodated into H_2O-ice depends on the degree of order: an amorphous ice can readily absorb large concentrations of NH_3 but a crystalline ice can accommodate, at most, a few per cent. Hence, the divergence between protostellar and cometary NH_3 abundances might be understood if the latter have vaporized and recondensed in a more ordered state (section 9.2.3).

Cometary ices are quite highly deuterated, with D/H ratios typically a factor of ~20 greater than the mean value for atomic gas in the local ISM (Irvine *et al* 2000). This result provides an important clue because chemical reactions operating at interstellar temperatures tend to promote efficient deuterium fractionation (section 9.4.2). A direct comparison between cometary D/H ratios and those in interstellar ices is impracticable, as the techniques used to study the latter are not sufficiently sensitive to detect deuterated species (with the possible exception of HDO; Teixeira *et al* 1999). However, cometary values are well within the range observed in interstellar gas phase molecules (e.g. Turner 2001), including those in hot cores where the ices have sublimed (section 9.1.4). Comets are thus inferred to have formed, at least in part, from fractionated interstellar material, either by direct accumulation of deuterated ices or by freeze-out of deuterated water vapour and other interstellar molecules that became concentrated in cool regions of the solar nebula.

Abundances of volatile noble gases in comets provide a powerful means of investigating their thermal histories. Because they play no part in chemical processes, their concentrations should be determined purely by temperature: if a noble gas with a given sublimation temperature is present with an abundance that approaches its solar value, the comet must have formed and been maintained below that temperature. Argon, which sublimes at 35–40 K, was detected in Hale–Bopp (Stern *et al* 2000), whereas neon, which sublimes at 15–20 K, appears to be highly depleted in both Hale–Bopp and Hyakutake (Krasnopolsky *et al* 1997, Krasnopolsky and Mumma 2001). These results imply that Hale–Bopp, at least, has undergone only mild heating. It will be important to apply this technique to other comets.

Shock processing of ices falling into the nebular disc would have been most efficient in the giant-planet region, i.e. at distances 5–40 AU from the Sun. This is precisely the location in which Oort-cloud comets such as Hyakutake and Hale–Bopp are thought to have formed (e.g. Mumma *et al* 1993); gravitational interactions with the giant planets led to their subsequent ejection into the farthest reaches of the Solar System, from which they may intermittently return. Compositional differences amongst Oort-cloud comets may be related to their point of origin (see Mumma *et al* 2001): Hale–Bopp seems likely to have formed near the outer limits of the giant-planet zone. Comets originating beyond 40 AU would have formed from material that suffered even lower degrees of processing, preserving a greater fraction of pre-solar volatiles. Such bodies are most likely to be found in the outer disc, in the region now known as the Kuiper Belt (Cruikshank 1997).

9.3.2 Interplanetary dust

The collection and study of interplanetary dust particles (IDPs) was pioneered by Brownlee (1978). Particles with diameters ranging from 1 μm up to about 100 μm are sufficiently decelerated by the Earth's atmosphere to allow non-destructive collection by aircraft in the stratosphere. IDPs at the upper end of this size range have also been recovered from Antarctic ice (Engrand and Maurette 1998). Their extraterrestrial origin is confirmed by evidence for cosmic-ray and solar-wind exposure and by the presence of isotopic anomalies, notably excess deuterium. The particles are composed principally of silicates, together with about 10% by mass of carbon and smaller quantities of various other minerals, including magnetite and iron sulphide. The carbon, which is predominantly amorphous, frequently coats the mineral grains and is responsible for the dark visual appearance of IDPs. Abundances of the mineral-forming elements are approximately chondritic. Individual IDPs may be grouped into two general classes according to the nature of the silicates they contain (Sandford 1989): those dominated by hydrous (layer-lattice) silicates and those dominated by anhydrous silicates (principally olivine and pyroxene). In mineralogical terms, the hydrous IDPs are similar to carbonaceous chondrites and seem likely to have formed on asteroidal parent bodies (section 9.3.3). Some anhydrous IDPs may also come from asteroids. Others, termed 'cluster' IDPs, are so porous and fragile that they seem unlikely to have been subjected to the compaction that would tend to occur on asteroids; they are presumed to be of cometary origin.

The anhydrous cluster IDPs are of greatest interest because they are evidently the most primitive survivors of the early Solar System currently available for laboratory analysis. They consist of loose aggregates of much smaller particles, typically $\sim$0.1 μm in size, some of which are monolithic, whilst others are themselves aggregates of smaller particles $\sim$0.01 μm or less in size. A variety of chemically and structurally distinct mineral grains generally coexist in the same aggregate, implying diverse origins. The macroscopic particles evidently formed by aggregation of a heterogeneous mixture of grains with sizes in the range characteristic of interstellar dust. As such small grains would be swept away by solar radiation pressure as the nebular disc became optically thin, aggregation must have occurred early, within the first 10 Myr. Does this imply that anhydrous IDPs are, in fact, aggregates of pre-solar interstellar grains? Isotopic studies can provide the answer. Recent work by Messenger *et al* (2002) shows that at least some of the particles in the aggregates exhibit oxygen isotope ratios that deviate dramatically from solar values, indicating a pre-solar origin.

If anhydrous IDPs contain pristine interstellar silicates, we should expect the profile of the 10 μm silicate feature to be smooth and devoid of substructure associated with crystallization. Studies by Sandford and Walker (1985) show that IDPs contain annealed silicates with average 10 μm profiles similar to those in comets (Hanner *et al* 1994, 1997), exhibiting the distinctive crystallization feature at 11.2 μm (see figure 9.8). However, the experimental technique used by Sandford and Walker did not distinguish individual grains within the IDPs and so their results do not exclude the possibility that some are entirely amorphous.

Included in the matrices of many anhydrous IDPs are glassy, submicron-sized silicate grains known as GEMS (glasses with embedded metal and sulphides). GEMS are proposed to be pre-solar interstellar grains that have been subjected to minimal alteration (Bradley 1994, Bradley *et al* 1999). They contain Fe-rich inclusions that have been linked to the superparamagnetic model for interstellar grain alignment (section 4.5.3). Infrared spectra of some GEMS show evidence of mild annealing, whilst others show smooth 10 μm profiles characteristic of amorphous silicates and similar to those observed in interstellar clouds (Bradley *et al* 1999). One possible objection to the pre-solar model concerns the high sulphur content of the GEMS, which conflicts with the low depletion of S into grains in the diffuse ISM (sections 2.4–2.5).

The carbonaceous material in IDPs contains both aromatic and aliphatic hydrocarbons (Clemett *et al* 1993, Flynn *et al* 2002). Their spectra appear to be quite similar to those of meteoritic kerogens (see section 9.3.3 and figure 5.9), displaying features at 3.4 and 5.85 μm associated with C–H and C=O groups. Deuterium enrichment supports a pre-solar source for the organics in IDPs (Keller *et al* 2000, Messenger 2000, Aléon *et al* 2001). Isotopic imaging shows that dramatic D/H enhancements (by factors of up to 1000 relative to the mean local ISM) occur in submicron-sized zones within some IDPs, presumably corresponding to individual grains that carry an isotopic fingerprint of interstellar chemistry.

In summary, anhydrous cluster IDPs appear to be carriers of pre-solar interstellar grains that survived the formation of the Solar System with minimal alteration. They may perhaps represent the refractory mix of dust in the outer giant-planet region of the solar nebula, where they were accreted into comets and subsequently preserved, deep-frozen, until their recent release into the interplanetary environment.

9.3.3 Meteorites

Several classes of meteorite have been identified, but discussion here will be limited to the carbonaceous chondrites as these are clearly the most primitive (Cronin and Chang 1993). Because they come from asteroidal parent bodies on which aqueous processes can occur (Grimm and McSween 1989, Gaffey *et al* 1993), carbonaceous chondrites provide a different window on chemical evolution in the solar nebula compared with cometary IDPs. They are composed of a fine-grained aggregate of micron-sized particles (the matrix), in which much larger (millimetre-sized) particles are embedded. The latter include glassy spheres (chondrules) and Ca, Al-rich mineral phases, both of which appear to have condensed at relatively high temperature. The matrix is composed predominantly of magnesium silicates with varying degrees of hydration; other ingredients include magnesium sulphate ($MgSO_4$), magnetite (Fe_3O_4) and troilite (FeS). The carbonaceous fraction amounts to a few per cent by mass. Most of the minerals present appear likely to have condensed in the solar nebula (Barshay and Lewis 1976) in a manner analogous to that outlined in section 7.1.3. A small but highly significant fraction of the matrix grains display isotopic anomalies that point to a pre-solar origin, as discussed in detail in chapter 7 (section 7.2.4).

Matrix silicates in carbonaceous chondrites might have formed in a hydrous state in the solar nebula (Barshay and Lewis 1976) but it is more probable that they were initially anhydrous and subsequently hydrated (Browning and Bourcier 1998, Bischoff 1998). Hydration is attributed to aqueous alteration: ices and silicates co-accreted into asteroidal parent bodies, where they were subsequently heated, principally by decay of short-lived radioactive elements such as ^{26}Al; the ices melted and liquid water reacted with the silicates to form hydrated phases.

The C-rich component of the meteorites appears to have had a varied and complex history. It includes organics and carbonates, together with more refractory phases (diamond, graphite and SiC). The latter are identified as pre-solar but account for only a few per cent of the mass (Alexander *et al* 1998), whilst the carbonates are presumed to have formed by aqueous processes on asteroidal parent bodies. The bulk of the carbonaceous matter is organic, only a small fraction of which consists of well characterized molecules; the rest is composed of kerogen-like material, i.e. complex, insoluble organic matter of high molecular weight that contains both aromatic groups and aliphatic chains (de Vries *et al* 1993). Are the organics in meteorites pre-solar, or did they form in the solar nebula or on the parent bodies? In common with comets and IDPs, the presence of deuterium enrichment, typically by factors of 10–50 relative to the ISM (Kerridge 1989), supports a contribution from pre-solar material. Detailed analyses of H, C and N isotopes (Alexander *et al* 1998) suggest that the pre-solar fraction could be as high as 60%.

A number of processes may have led to synthesis of organic molecules in the early Solar System. Surface reactions on warm ($T_d > 60$ K) pre-solar carbon grains, in which absorbed H and O atoms bond with lattice C atoms, may convert the grains to simple hydrocarbons and aldehydes (Barlow and Silk 1977b); this might provide an explanation for the scarcity in meteorites of phases such as graphite and amorphous carbon thought to be common in the ISM. Photolytic processing of simple organics and ices by solar UV radiation (Sagan and Khare 1979) might have driven the formation of kerogen-like organics in unshielded regions of the solar nebula. However, a more abundant potential source of carbon is locked up in gas-phase CO. We noted in chapter 7 (section 7.1) that the stability of CO normally prevents the condensation of C-rich solids in a stellar atmosphere of solar composition. However, this reservoir might be tapped by Fischer–Tropsch-type (FTT) reactions that involve hydrogenation of CO on warm (300–400 K) catalytic surfaces (Anders 1971, Hayatsu and Anders 1981, Kress and Tielens 2001). Suitable catalysts include magnetite, layer-lattice silicates and metallic iron – solids known to exist on meteorite parent bodies. Generic FTT reactions that produce hydrocarbons of the form C_nH_{2n+2} may be written

$$n\mathrm{CO} + (2n+1)\mathrm{H_2} \rightarrow \mathrm{C}_n\mathrm{H}_{2n+2} + n\mathrm{H_2O} \tag{9.6}$$

$$2n\mathrm{CO} + (n+1)\mathrm{H_2} \rightarrow \mathrm{C}_n\mathrm{H}_{2n+2} + n\mathrm{CO_2}. \tag{9.7}$$

Such reactions can readily account for the presence of aliphatic hydrocarbons. However, synthesis of aromatic hydrocarbons and kerogen requires *sustained* heating of the primary aliphatics (higher temperatures and/or longer heating times over periods of up to 10^5 years), consistent with current models for the thermal evolution of the asteroid belt (Gaffey 1997). Aqueous processes on asteroids may have produced important prebiotic molecules such as amino acids as well (Shock and Schulte 1990).

9.4 Ingredients for life

9.4.1 Motivation

Interstellar dust grains are carriers of carbon and other biologically important elements formed by nucleosynthesis in stars (chapter 2). Evidence is presented elsewhere in this book for the production of organic molecules in a variety of astrophysical environments, including the atmospheres of C-rich red giants, dense interstellar clouds, protostellar condensations and protoplanetary discs. Those present in or on dust grains include aliphatic and aromatic hydrocarbons, kerogens and organic ices. A variety of organic molecules are also present in the gas phase: simple prebiotic species such as hydrogen cyanide (HCN) and formaldehyde (H_2CO) are quite widespread; and somewhat more complex ones such as methanimine (CH_2NH) and formic acid (HCOOH) have been detected in the denser regions. Although amino acids have yet to be detected in the ISM, they are clearly present in meteorites (section 9.4.3). If such molecules were delivered to the atmosphere and oceans of the primitive Earth in sufficient quantities, they could have played a significant part in the early stages of prebiotic evolution that led to the emergence of life some 3.8 or more billion years ago (Whittet 1997).

The potential importance of extraterrestrial organic matter is emphasized by developments in our understanding of the nature of the Earth's early atmosphere. For many years, it was assumed that the atmosphere was originally highly *reducing*, resembling the current atmospheres of the giant planets, i.e. rich in H_2 and other hydrogenated gases such as CH_4, NH_3 and H_2O. The classic Miller–Urey experiment (Miller 1953, Miller and Urey 1959; see Miller 1992 for a modern account) showed that amino acids and other biologically significant molecules are produced efficiently when an energy source such as an electric discharge is available in such an atmosphere. However, there is now general agreement that the Earth's atmosphere originated by accumulation of gases released from the surface and that its composition is constrained by the previous history of volcanic emissions. Present-day volcanic emissions are non-reducing (mostly CO_2, H_2O and N_2) and geochemical analysis of ancient magmas suggests that they were no more reducing 3.9 Gyr ago than they are today (Delano 2001). The nature of the atmosphere prior to 3.9 Gyr is much less certain but seems likely to have been dominated by CO_2 (Walker 1985); in any case, life may not have been possible at much earlier times because of the frequency of planet-sterilizing impacts (Maher and Stevenson 1988, Sleep *et al* 1989). Miller–Urey-type experiments in H-poor, CO_2-rich atmospheres give orders-of-magnitude lower yields of prebiotic molecules compared with reducing (hydrogenated) atmospheres (Schlessinger and Miller 1983). This arises because CO_2 and N_2 are more tightly bound molecules than CH_4 and NH_3 and it is thus more difficult to produce radicals that will react with each other to form molecules of greater complexity. Astrophysical environments are, in contrast, highly reducing. Of course, this finding does not necessarily imply that amino acids and other prebiotic molecules implicated in the origin of terrestrial life were of cosmic origin, but it illustrates the importance of exploring that possibility.

9.4.2 The deuterium diagnostic

A recurrent theme in earlier sections of this chapter is the occurrence of deuterium enrichment as an isotopic fingerprint of interstellar chemistry. Deuterium and hydrogen are easily fractionated because of the large mass ratio. Whilst modest levels of fractionation are thus readily obtainable by a variety of processes, chemical reactions operating at interstellar temperatures can lead to orders-of-magnitude increases in D/H (Watson 1976, Tielens 1983, Herbst 1987). Molecular hydrogen may become deuterated in the gas phase by the ion–molecule reaction

$$D^+ + H_2 \rightarrow HD + H^+ \qquad (9.8)$$

and the important and highly reactive ion H_3^+ (section 8.2) then becomes deuterated via the exothermic reaction

$$H_3^+ + HD \rightarrow H_2D^+ + H_2. \qquad (9.9)$$

As the reverse of reaction (9.9) is endothermic, the abundance ratio H_2D^+/H_3^+ becomes systematically enhanced at low temperatures compared with HD/H_2. Carbon-bearing molecules formed from reaction sequences that begin with H_3^+ will inherit its deuterium endowment. Grain-surface chemistry also leads to fractionation, arising from the fact that catalytic HD formation on grains is less efficient than catalytic H_2 formation (section 8.1; Tielens 1983). Because of this, the D/HD ratio is higher than the H/H_2 ratio in molecular gas, and molecules such as H_2O, NH_3 and CH_4 formed by reactions between atomic hydrogen and heavier elements on grain surfaces (section 8.4.1) are thus systematically deuterated.

Figure 9.11 provides an overview of deuterium fractionation with respect to temperature from interstellar clouds to the Earth's oceans. The baseline of 2×10^{-5} (broken line) is set by the average 'cosmic' value of D/H in unfractionated interstellar gas; we expect this value to represent the solar nebula as a whole. Interstellar chemistry allows enhancements by factors of up to about 10^4 at temperatures $\sim$20 K, with rapid decline toward higher temperatures (broken curve). Partitioning by exchange reactions such as

$$H_2O + HD \rightarrow HDO + H_2$$

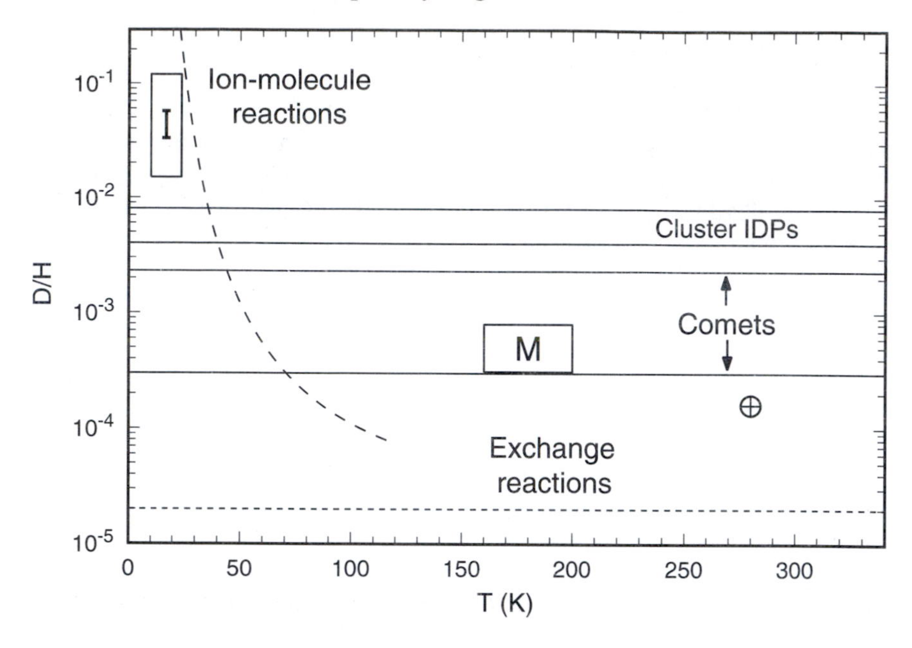

Figure 9.11. Plot of D/H ratio versus temperature for interstellar molecular clouds (box labelled I), meteorite parent bodies (box labelled M) and terrestrial ocean water (⊕). The broken curve is a prediction for molecules formed by ion–molecule chemistry as a function of temperature according to the model of Geiss and Reeves (1981). The range of D/H ratios measured in cluster IDPs and comets is indicated by horizontal lines. The horizontal broken line indicates the D/H ratio in unfractionated interstellar matter (determined principally by the primordial abundance ratio). Data are from Irvine *et al* (2000) and references therein.

and

$$CH_4 + HD \rightarrow CH_3D + H_2$$

can fractionate warmer gas to more modest levels (Geiss and Reeves 1981, Kerridge 1989, Lecluse and Robert 1994): factors of 10 are possible but not factors of 10^3–10^4. The very high D/H values found in cluster IDPs, in particular, cannot be readily explained without invoking interstellar chemistry.

The mean D/H value in the Earth's oceans ($\sim 1.6 \times 10^{-4}$) is enhanced by a factor of about 10 over the baseline value (figure 9.11). Whilst this does not prove that the Earth's water and volatiles came from material bearing the imprint of interstellar processes, it is highly consistent with the hypothesis (Oro 1961, Delsemme 2000) that at least some were delivered by cometary impact. In fact, the ocean D/H value can be explained if approximately half of the water came from cometary ices, the remainder from rocky material formed in the inner solar

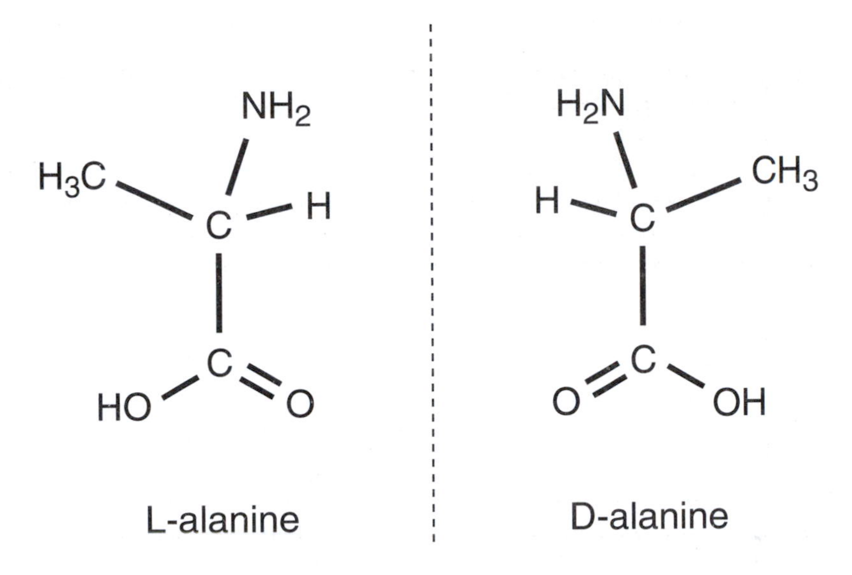

Figure 9.12. Structure of the amino acid alanine, comparing left-handed (laevo, L) and right-handed (dextro, D) enantiomers.

system at temperatures high enough to equilibriate H_2O and H_2 (Laufer *et al* 1999, Owen and Bar-Nun 2001).

9.4.3 Amino acids and chirality

Amino acids are fundamental to terrestrial life as they are the building blocks of peptides, proteins and enzymes. Each amino acid consists of an amino group (NH_2) and a carboxyl group (COOH) joined by a central C atom, to which a peripheral H atom and side chain (denoted by R) are attached. Thus, the simplest amino acid, glycine (R = H) may be written NH_2CH_2COOH. The two chiral forms of alanine (R = CH_3) are illustrated in figure 9.12. Many different amino acids are possible, each distinguished by a unique side chain, but terrestrial biology utilizes only 20.

It may be noted that none of the individual units that make up amino acids such as glycine or alanine is particularly complex or exotic: lists of detected interstellar molecules (e.g. van Dishoeck *et al* 1993, Snyder 1997) include several obvious precursors such as HCN, CH_2NH and HCOOH. The possibility that glycine might also be an interstellar molecule has prompted observers to search for its rotational spectral lines at radio wavelengths (see Snyder 1997 for a review) but, despite some tantalizing hints, no confirmed detection has been made at the time of writing.

Whilst the quest for amino acids has proved frustrating for astronomers, meteoriticists have met with notable success: the soluble organic fraction of carbonaceous chondrites has been shown to contain them in appreciable quantities and with great diversity (Kvenvolden *et al* 1970, Cronin and Chang 1993). Several lines of evidence demonstrate that they are intrinsic to the meteorites and not terrestrial contaminants. The two strongest arguments may be summarized as follows: (i) over 70 different amino acids have been identified in carbonaceous chondrites, with no particular preference for those utilized by terrestrial biology; and (ii) the meteoritic amino acids have D/H ratios higher than those typically present in any terrestrial equivalent. These results show that molecules as complex as amino acids are, indeed, synthesized in interstellar and/or interplanetary environments. They also illustrate the viability of delivery to Earth.

A third argument that has sometimes been used to support the extraterrestrial origin of meteoritic amino acids is the fact that they are approximately *racemic* (e.g. Chyba 1990). With the exception of glycine, all amino acids are chiral, i.e. they have left-handed and right-handed forms (L and D enantiomers) that are chemically identical but physically they are mirror images of one another. This is shown in figure 9.12 in the case of alanine. A racemic mixture contains L and D enantiomers in equal numbers, whereas biological systems appear be made exclusively from one or the other: terrestrial proteins are composed only of L amino acids (and nucleic acids contain only D sugars)[4]. The lack of a substantial L excess in meteorites is thus consistent with an abiotic (and presumed extraterrestrial) origin. However, as experimental techniques have improved, it has become clear that the meteoritic amino acids do contain a small but significant excess of the L enantiomers, at least in the two best studied cases (the Murchison and Murray meteorites; see Cronin and Pizzarello 1997, 2000, Pizzarello and Cronin 2000). As contamination may be excluded on the basis of other lines of evidence, as mentioned above, it is possible to turn the argument around and hypothesize that the data support an intrinsic excess of L enantiomers in the meteorites. The origin of biological homochirality on Earth is unknown, and the meteoritic data open up the intriguing possibility that matter accreted onto the early Earth might have contained an asymmetry that tipped the balance in favour of L amino acids. Experimental work has shown that it is possible for small enantiomeric excesses to be amplified by autocatalytic activity (Shibata *et al* 1998).

What could cause enantiomeric excesses in meteoritic amino acids? Catalytic reactions on claylike minerals (e.g. Ferris *et al* 1996) might lead to some selectivity at the molecular level within parent bodies, but the net result of such reactions on large enough size scales seems likely to be a racemic mixture in the absence of any pre-existing bias. What is needed is an asymmetry on the scale of the solar nebula itself. A hypothesis that has been widely discussed in recent years is exposure to circularly polarized radiation (CPR) (e.g. Bonner 1991, Bailey 2001). The fact that ultraviolet CPR of a given handedness can selectively destroy one enantiomeric form of a chiral molecule was first demonstrated in the

[4] Even tiny concentrations of monomers with the 'wrong' handedness in crucial biopolymers such as peptides and RNA strands greatly reduce their stability and biological functionality.

laboratory many years ago (Kuhn and Braun 1929). The yield depends on both the degree of circular polarization and the spectrum of the source; and is generally highest for radiation in the 2000–2300 Å region: enantiomeric excesses of 10%, comparable with the highest meteoritic values, are routinely obtainable in the laboratory and much higher levels are possible (Flores *et al* 1977, Norden 1977, Bailey 2001). Three extrasolar sources of CPR have been proposed that might lead to enantiomeric selection in the early Solar System: neutron stars, magnetic white dwarfs and YSOs. Note that, in each case, a chance encounter between the source and the solar nebula is required. Whereas this has a very low probability in the first two cases, an encounter between the young Sun and another, perhaps more luminous, YSO within the parent cloud is natural and logical.

Observations reviewed in section 9.1.3 indicate that very high degrees of circular polarization are present at infrared wavelengths in regions containing massive YSOs. The physical process involved appears to be scattering of stellar radiation by aligned grains located in the circumstellar envelope and/or the surrounding molecular cloud. This discovery led Bailey *et al* (1998) to propose a model in which the solar nebula was irradiated by light scattered from a nearby, but obscured, young star. The direct line of sight should contain sufficient opacity to effectively block unpolarized UV radiation from the star (which would lead to photodissociation without enantiomeric selection). Such a geometric configuration is entirely feasible, e.g. for a circumstellar disc orientated roughly 'edge on'. Figure 9.5 (lower right frame) illustrates a relevant model: note that both the strength and the handedness of the CPR depends on the orientation. A further possibility is that linearly polarized radiation from the star may undergo further scattering to generate CPR in the solar nebula itself.

Because massive YSOs are embedded in dense molecular clouds, the CPR that they produce is observable only in the infrared. There is, of course, no guarantee that ultraviolet CPR will be present in sources where infrared CPR is observed. However, scattering calculations indicate that the proposed mechanism should be effective if a UV flux is present (Bailey *et al* 1998). There is, indeed, ample evidence that energetic radiation from massive stars has a dramatic effect on the surrounding environment, generating compact H II regions encased in photodissociation regions and hot cores (sections 1.4.4 and 9.1.4). Moreover, the Orion nebula contains examples of evident interaction between the UV field from massive stars and the envelopes of lower-mass stars nearby (Bally *et al* 1998, Störzer and Hollenbach 1999). The timescale for production of a significant enantiomeric excess under such circumstances is estimated to be quite short, ~2000 years (Bailey *et al* 1998), well within the probable lifetime of a photodissociation region. An obvious caveat is that the mechanism is viable only if the Sun formed in a massive star-formation region like Orion rather than in a low-mass region like Taurus (section 9.1.1). Independent support for this view is provided by evidence for high abundances of short-lived radioactive isotopes such as ^{26}Al in the solar nebula (Goswami and Vanhala 2000), explicable in terms of contamination from a nearby massive star.

9.4.4 Did life start with RNA?

Amino acids and nucleic acids have very different but complementary roles in modern terrestrial biology. In polymerized form, they make proteins and genetic codes, analogous to the hardware and software of a computer. As life cannot function without either one, we are faced with a 'chicken and egg' dilemma regarding which was the first true biomolecule. The current paradigm is the 'RNA world'. It is proposed that because RNA (ribonucleic acid) can both store genetic information and act as a catalyst, it could have served dual roles, simulating the functions of DNA (deoxyribonucleic acid) and protein enzymes in modern biology. The discussion here will be limited to some concerns regarding the origin of RNA. The reader is referred to chapters 12 and 13 of the text by Lunine (1999) for an overview of current concepts in origins of life, and to Chyba and McDonald (1995) for an in-depth review of key issues.

Whereas amino acids are relatively easy to synthesize abiotically, nucleic acids are far more challenging because each monomer (nucleotide) contains three diverse units, a base, a sugar and a phosphate. Of these, the bases are the least problematic. They have been found in trace amounts in the Murchison meteorite (Cronin *et al* 1988) and may be formed in the laboratory by aqueous polymerization of HCN (Oro 1960). Sugars with the generic formula $C_nH_{2n}O_n$ may be synthesized by polymerization of H_2CO (the formose reaction) but ribose (the sugar in RNA) is not particularly preferred over the many other possible forms (Chyba and McDonald 1995). Moreover, ribose is chiral and only D enantiomers are present in biotic RNA. As Lunine (1999) has remarked, it is a daunting task to create a properly functioning RNA polymer out of a random mix of L and D sugars. Sugar-like compounds have been detected in carbonaceous chondrites (Cooper *et al* 2001) but there is as yet no evidence for an extraterrestrial source of chiral asymmetry in sugars.

Phosphates present another set of problems. P is the least abundant of all the chemical elements that are essential to biology (table 2.1), suggesting that its availability could limit nucleotide synthesis in any relevant environment. Inorganic phosphate (apatite) was no doubt present on the early Earth but its availability for prebiotic chemistry would have been further limited by its very low solubility in water. Moreover, apatite is an orthophosphate, lacking the high-energy phosphodiester bonds that drive polymerization in modern biological systems. Energetic pyrophosphates may condense from orthophosphates in aqueous environments at temperatures >300 K (Chyba and McDonald 1995), conditions that might arise, for example, in hydrothermal systems where ocean water is in contact with hot igneous rock.

Given these problems with RNA, it is logical to question whether a structure containing amino acids might be a better choice as the first biomolecule. A plausible candidate is peptide nucleic acid (PNA). Peptides are formed by polymerization of amino acids, in the case of glycine by linkage of $-NHCH_2CO-$ units. In PNA, a peptide dimer substitutes for the phosphate–sugar 'backbone', to which a normal RNA base becomes attached. Nelson *et al* (2000) show that PNA may be synthesized abiotically and appears to polymerize efficiently at temperatures ~373 K. It will be important to explore the properties of PNA further to determine its suitability as a primitive biomolecule fulfilling both catalytic and genetic roles.

9.4.5 Delivery to Earth

The current model for the origin of the Earth (Wetherill 1990) proposes that it formed in a hot, largely molten state by coalescence of giant planetary embryos, ranging in size up to that of present-day Mars. The last of these giant collisions is thought to have led to the formation of the Moon some 4.5 Gyr ago. As a result of that event, the Earth would have lost any prior atmosphere it might have had and would have been essentially devoid of water and organics.

Elemental carbon and organic matter are delivered to the present-day Earth by meteorites and interplanetary dust, of which the latter dominates in terms of mass. The current accretion rate of unablated carbonaceous matter is estimated to be $\sim 3 \times 10^5$ kg/yr (Anders 1989). If this rate were to remain constant, the timescale to accrete a biomass (i.e. the total mass of organic carbon in the current biosphere, estimated by Chyba *et al* 1990 to be $\sim 6 \times 10^{14}$ kg) would be ~2 Gyr. However, it seems reasonable to suppose that the accretion rate would have been much higher at earlier times, during the final stages of planetary formation. Adopting an accretion model based on the lunar cratering record (Chyba and Sagan 1992), the rate 4 Gyr ago could have been $\sim 5 \times 10^7$ kg/yr, or a biomass in only ~10 Myr. Although subject to considerable uncertainty, these simple calculations illustrate that the contribution of exogenous organic matter to the early Earth could well have been substantial.

Another probable difference, comparing the Earth 4 Gyr ago with today, is the density of the atmosphere. A 10 bar CO_2 atmosphere has been proposed (Walker 1985, Chyba *et al* 1990), consistent with geochemical evidence and with the need for a substantial greenhouse effect to offset lower solar luminosity at that epoch. Although it is difficult to quantify the effect this would have had on the survival of incoming material, it seems likely that it would have broadened the size range of particles sufficiently air braked to preserve their organic content, especially for those approaching at grazing incidence.

Whilst the principle of delivery is demonstrated for IDPs and meteorites, a case for cometary accretion is more difficult to make. Water and organic molecules will not survive impacts at speeds above about 5–10 km s^{-1} (Chyba *et al* 1990). The impact speed depends on the entry speed and the degree of deceleration in the atmosphere. The orbits of Oort-cloud comets are highly eccentric, with no preferred direction of orbital motion or inclination relative to the ecliptic plane. Because of this, they may enter the Earth's atmosphere at speeds ranging from 11 to 80 km s^{-1}, the lower and upper limits being set by co-orbital and head-on collisions, respectively (e.g. Steel 1992). In comparison, because their orbits are approximately co-orbital with that of the Earth, the range for asteroids is much less, 11–25 km s^{-1}. The degree of deceleration depends on the size and mass of the projectile and the density of the atmosphere. Calculations by Chyba *et al* (1990) show that for a 10 bar atmosphere, a small (100 m) comet with an entry speed of 25 km s^{-1} may be sufficiently decelerated to avoid complete destruction. Comets of this size order often explode in the atmosphere prior to reaching the surface but this may not necessarily be detrimental.

Impactors above about 1 km in size will probably not break up in the atmosphere and will experience little deceleration, hitting the surface at a speed close to their entry speeds. On average, such events may reduce rather than increase the hydrospheric mass through impact erosion. Such impacts have been associated with mass extinctions during the Earth's more recent history, most notably the disappearance of the dinosaurs some 65 Myr ago at the Cretaceous–Tertiary (K/T) boundary. An interesting result that has emerged from study of the geological record is the occurrence of significant abiotic amino acid deposits in the sediments associated with this event. It seems unlikely that these molecules are survivors of the impact itself, however. They might have been carried to Earth by interplanetary dust released from the projectile during prior ablation in the solar wind, or they might have been synthesized on Earth by shock-driven chemical reactions in the post-impact fireball (see Chyba and Sagan 1997 for further discussion and references). Impact shocks could have been important instigators of organic synthesis at earlier epochs as well.

Could accretion of interplanetary debris onto the Earth have delivered life itself, rather than merely the raw materials? The concept of panspermia, i.e. the seeding of the Earth with life from space, was first discussed a century ago by Arrhenius (1908) and has since reappeared in a number of forms (e.g. Hoyle and Wickramasinghe 1979, Crick 1981). Of course, to invoke an external source of Earth-based life does not solve the problem of its origin but merely displaces it. Davies (1988) describes panspermia as 'Unlikely, unsupported but just possible', a view with which this author concurs. There is no *a priori* objection to the concept of interstellar organisms (Sagan 1973) but the densities prevailing in molecular clouds are insufficient to allow appreciable evolution within cloud lifetimes, and organisms exposed to the harsh radiation field permeating regions outside molecular clouds are rapidly destroyed by UV irradiation (Weber and Greenberg 1985). If biological organisms exist in non-planetary environments, the most viable location is probably comets. However, comets merely provide a possible means of cold storage and transport, not a medium for formation and growth. Excitement has been generated by the proposed presence of microbial fossils in a meteorite originating from Mars (McKay *et al* 1996). If confirmed, this will be a momentous discovery; but implications in terms of panspermia are limited to the possibility that life originating in the putative Martian biosphere might have been delivered to Earth (or conversely; see Melosh 1988). No credible evidence has yet been presented for the existence of extraterrestrial organisms of wider provenance, either within or beyond the Solar System.

Recommended reading

- *Chemical Evolution of Star-Forming Regions*, by Ewine F van Dishoeck and Geoffrey A Blake, in Annual Reviews of Astronomy and Astrophysics, vol 36, pp 317–63 (1998).
- *From prestellar cores to protostars: The initial conditions of star formation*, by Philippe André, Derek Ward-Thompson and Mary Barsony, in *Protostars and Planets IV*, ed V Mannings, A P Boss and S Russell (University of Arizona Press, Tucson), pp 59–96 (2000).
- *The structure and evolution of envelopes and discs in young stellar systems*, by Lee G Mundy, Leslie W Looney and William J Welch, in *Protostars and Planets IV*, ed V Mannings, A P Boss and S Russell (University of Arizona Press, Tucson), pp 355–76 (2000).
- *Comets: A link between interstellar and nebular chemistry*, by William M Irvine *et al*, in *Protostars and Planets IV*, ed V Mannings, A P Boss and S Russell (University of Arizona Press, Tucson), pp 1159–200 (2000).
- *Planetary and Interstellar Processes Relevant to the Origins of Life*, ed Douglas C B Whittet (Kluwer, Dordrecht, 1997).
- *Earth: Evolution of a Habitable World*, by Jonathan I Lunine (Cambridge University Press, 1999).

Problems

1. What YSO class should be assigned to the YSO HL Tauri on the basis of its spectral energy distribution shown in figure 9.4?
2. Show that $\alpha_{IR} = -3$ describes the slope of a spectral energy distribution in the Rayleigh–Jeans limit.
3. Explain how spectroscopic analysis of the gases released by a comet as it passes by the Sun allows us to place constraints on its point of origin in the solar nebula. What differences would you expect to see between an Oort-cloud comet and a comet that originated in the Kuiper belt?
4. Why are cometary IDPs thought to be anhydrous, given that comets are composed largely of H_2O?
5. Given that meteorites and interplanetary dust particles deliver significant quantities of organic matter to the surface of the Earth, is it surprising that samples of the lunar regolith contain almost no trace of carbonaceous material?
6. Suppose that the circular polarization model for the origin of terrestrial homochirality is correct. If primitive life should prove to exist elsewhere in our Solar System (e.g. on Mars or Europa), would you expect it to display the same chiral configuration as terrestrial life? What about life in other planetary systems? Explain your reasoning.

Chapter 10

Toward a unified model for interstellar dust

> *"This phenomenon is so puzzling... one is almost tempted to give up a satisfactory explanation. However, the man of science ought not to recoil, either before the obscurity of a phenomenon or before the difficulty of a research. Whether he is in possession of earlier work, or whether he tries to increase the knowledge of a phenomenon by new, precise observations, he can be sure of a certain success in his studies, if he employs calm speculation, without abandoning himself to an excited and preoccupied imagination. However little ground he gains, he will always enlarge it by returning to his problem with that persistence which is the indispensable condition of study. It is thus, guided by analysis and calculation, that he can even derive unexpected results which, however, enjoy considerable certainty."*
>
> F G W Struve (1847)

A wide variety of observational and empirical constraints on the properties of dust in galactic environments have been described in previous chapters. The key results are the abundances of the condensible elements and their observed depletions (sections 2.2–2.5), the wavelength-dependence of interstellar extinction and scattering (sections 3.3–3.4), the presence of a strong mid-ultraviolet absorption feature (section 3.5), the polarization and alignment properties of the grains (sections 4.2–4.5), their infrared absorption features (sections 5.2–5.3) and their continuum and spectral line emissions (sections 6.2–6.3). Studies of dusty envelopes around mass-ejecting stars give insight into the circumstellar origins of refractory interstellar grains (chapter 7), whilst variations in all of these phenomena as functions of environment help us to understand how the grains evolve as they cycle between diffuse and dense phases of the ISM (chapter 8) and become incorporated into new stars and planetary systems (chapter 9). Finally, laboratory isolation of pre-solar inclusions in meteorites and interplanetary dust provides us with tangible examples of some interstellar grains against which to test our models (sections 7.2.4 and 9.3.2).

The observations imply a need for multi-component models in which different grain populations account for different aspects of the data. These populations may differ from each other in terms of size distribution, composition and physical structure (i.e. monolithic or composite, mantled or unmantled, oblate or prolate, etc.). A unified model for interstellar dust must be capable of matching all known constraints unambiguously and in a self-consistent way, invoking substances and size distributions that are physically reasonable and have an obvious source in terms of stellar mass-loss processes or production in the ISM itself. None of the existing models (section 1.6) is entirely successful in achieving this ideal. It is notable, however, that diverse observational results can place common or similar constraints on grain properties. Consequently, a number of models have common factors, suggesting a degree of convergence. Rather than present a detailed assessment of the merits and failures of previous proposals, the goal of this chapter is to attempt an overview that focuses on common factors and highlights the main areas of uncertainty, with a view to future progress toward a truly unified and rigorous model.

The discussion that follows is based entirely on evidence and arguments presented in the preceding chapters. I have not included cross-references to the many relevant sections as this would have interrupted the flow of the text to an unreasonable degree. The contents and index pages should provide the reader with easy access as required.

10.1 Areas of consensus

10.1.1 A generic grain model

There is general agreement on the need for both O-rich and C-rich grain materials in any successful model, and on the need for grain sizes that span the range from macromolecular dimensions up to about 1 μm. The presence of O-rich and C-rich grains is naturally explained by nucleation and growth in the atmospheres of late-type stars of differing C/O ratio. The requirement for a grain component that scatters efficiently supports dielectric compounds such as silicates for the O-rich component. A substantial contribution from carbon is needed to account for the opacity of the dust per H atom as well as to explain certain spectral features. Large (‘classical’) grains with dimensions approaching the wavelength of visible light produce the λ^{-1} extinction law in this spectral region. Evidence for distinct populations of much smaller ($a < 0.02$ μm) grains arises from the shape of the extinction curve and the behaviour of the phase function in the ultraviolet, and from the detection of near and mid-infrared emission greatly in excess of that expected from classical-sized grains in thermal equilibrium with the interstellar radiation field. Mathematical description of the size distribution in terms of a power law of the form $n(a) \propto a^{-q}$ ($q \approx 3.5$) seems well established and

Table 10.1. An overview of grain components in a general model for dust in the diffuse interstellar medium.

	Silicate cores	Sooty mantles	Bump grains	Carbon VSGs
Composition	amorphous $MgSiO_3$, etc with Fe-rich inclusions	amorphous carbon	partially graphitic carbon	PAHs diamond
Origin	O-rich stardust	ISM	C-rich stardust (modified)	C-rich stardust or ISM
Size (μm)	≤ 1	(on cores)	≤ 0.02	~ 0.001
Elements depleted	O, Mg, Al, Si, Fe, etc.	C	C	C
Extinction	← UV–visible–IR →		Mid UV	Far UV
Alignment?	Yes	Yes	No	No
Polarization	← UV–visible–IR →		None	None
Absorption features	9.7, 18.5 μm	—	2175 Å 3.4 μm?	—
Emission	← FIR continuum →		← MIR continuum →	
Emission features	9.7, 18.5 μm	—	—	3.3, 6.2, 7.7, 8.6, 11.3 μm

physically reasonable as the expected outcome of grain–grain collisions in stellar winds and interstellar clouds. Star-to-star variations in the extinction curve are correlated with respect to wavelength across the available spectral range, and this behaviour may be characterized in terms of the parameter $R_V = A_V/E_{B-V}$, taken as a measure of growth by coagulation.

Table 10.1 presents an overview of grain components in a general model for dust in the diffuse ISM that is consistent with current observational constraints. The nature and observed properties of each component are summarized. I have chosen to represent the 'big' grains as silicate cores with sooty amorphous-carbon mantles. However, they should not be envisioned as idealized core–mantle grains but rather as silicate clusters coated with soot, resembling anhydrous interplanetary dust particles. A composite structure in which silicate and carbon grains are intermixed is equally feasible. Various aspects of this general model are discussed in the following paragraphs.

10.1.2 Silicates

As a generic class, silicates are identified spectroscopically in virtually all of the environments discussed in this book, from O-rich red giants to interstellar clouds, from protostellar envelopes to comets. The chemical composition of interstellar silicates is not uniquely defined. However, abundances dictate that only magnesium and iron silicates need be considered, and both depletion data and spectral profile matching favour Mg-bearing forms – $MgSiO_3$, $(Mg,Fe)SiO_3$, Mg_2SiO_4, $(Mg,Fe)_2SiO_4$ – over pure iron silicates. In the diffuse ISM, silicates appear to be essentially amorphous and anhydrous. Some crystallization is observed in circumstellar envelopes, both young and old, and hydration may occur in protoplanetary discs.

10.1.3 Carbon

The carbon component has traditionally been the point of departure for most grain models. However, despite some unease with graphite, there is still no viable alternative as the carrier of the 2175 Å mid-ultraviolet feature after some 40 years of study. Graphite and amorphous carbon seem likely to coexist and to undergo phase changes with respect to one another when appropriate conditions prevail. Very small grains containing PAHs are well established by their association with infrared continuum and line emission; as is the case with silicates, they are identified as a generic class rather then as specific molecules. The FUV extinction rise is tentatively attributed to PAHs here (table 10.1) but other assignments are not excluded. Nanodiamonds are gaining ground as a potentially important component of the dust, given their ubiquitous presence in the pre-solar fraction of meteorites. Whilst all of these forms of carbon might originate as stardust in stellar winds, it is probable that some are formed in the ISM. Organic refractory mantles, synthesized by photolysis of ices, have long been discussed as viable candidates for the C-rich component of large grains. However, sooty amorphous-carbon mantles now seem more plausible, given the lack of spectroscopic evidence for organic refractories in the diffuse ISM. Sooty material may be the product of long-term exposure of organics to the interstellar radiation field. Soot-coated silicate grains are consistent with the structure of interplanetary dust particles, for which there is growing evidence of a direct link to interstellar dust.

10.1.4 Ices

There is general consensus that refractory dust grains (table 10.1) acquire icy mantles whilst resident in dense molecular clouds. The ices are transient and are quickly dissipated when the grains return to the diffuse ISM. Mantling (as distinct from condensation of separate ice grains) is physically reasonable and predicted by theory; indirect support is provided by spectropolarimetric observations showing that ices and silicates display the same alignment properties in lines of sight where both are seen. Although many details remain to be clarified, the nature of the mantles seems reasonably well understood, thanks principally to observations made with the Infrared Space Observatory. They are composed primarily ($\sim$60%) of H_2O, together with CO, CO_2, NH_3, CH_3OH, CH_4 and a number of other species in varying concentrations. They form in an amorphous state in cold clouds and contain at least two distinct phases, polar and apolar, dominated by H_2O and CO. Many of the molecules present form by surface catalysis rather than by adsorption of molecules formed in the gas (CO being the most important exception). As they are warmed, the ices undergo selective sublimation, segregation and annealing. Photolysis and exposure to stellar winds drives reactions that produce new species, notably CN-bearing molecules. Evaporated mantles contribute to the inventory of gas-phase molecules in hot cores surrounding luminous protostars. Some interstellar ices may accrete directly into icy planetesimals in protoplanetary discs, others may evaporate and recondense.

10.1.5 Alignment

Grains are optically anisotropic and aligned by the interstellar magnetic field. The alignment efficiency depends on grain size (big grains are better aligned than small grains) and on environment (grains are less well aligned in dark clouds compared with the diffuse ISM). The wavelength of maximum polarization is a measure of the mean size of the aligned grains; this parameter characterizes the observed linear polarization curves in an analogous way to R_V for extinction. Spectropolarimetry of relevant spectral features indicates that the aligned component includes silicates and (where present) ices but probably does not include the carriers of the 3.4 μm 'hydrocarbon' feature. Similarly, the carriers of the 2175 Å 'graphite' feature are poorly aligned at best. The general principles of magnetic alignment seem well understood, although many details remain to be clarified. In general, oblate grains with modest axial ratios seem to give better fits to relevant data than prolate grains. The polarization vectors map the average magnetic field direction on the sky; but toward dark clouds they are dominated by the surface rather than the internal field, because the alignment efficiency is so much lower within the clouds.

10.2 Open questions

Having reviewed areas of broad agreement, let us now consider some key questions. What follows is a personal selection of important issues and is not necessarily exhaustive.

What is the preferred structure of the grains responsible for visual extinction? Models based on MRN that assume separate populations of carbon and silicate grains are unphysical, yet they have wide currency in the literature. We have ample evidence that grains grow by coagulation and accretion in interstellar clouds and there is no basis for supposing that the carbon and silicates somehow avoid each other during this process. If the carbonaceous component of big grains is largely synthesized in the ISM, then sooty mantles on silicate clusters are to be expected, as suggested in section 10.1.1; if it is largely stardust, then both carbon grains and silicate grains should be present in the clusters.

What quantity and quality of dust is formed in supernova explosions? It is widely assumed that supernovae make a substantial, perhaps dominant, contribution to the total mass of dust injected by stars into the ISM, but observational corroboration is limited at best. This is a question of great significance, with implications for our understanding of dust production not only in recent times but also in the early Universe. The observed presence of dust in high redshift galaxies is hard to explain without a major contribution from massive stars that end their lives as supernovae; low- and intermediate-mass stars that produce stardust today would not have reached the high-mass-loss phase of their life cycle in the age of the Universe at redshifts $z \geq 5$. This leads us to another question: how has the average composition of injected dust evolved over galactic history as the lower-mass stars became active sources?

How can estimates of the stardust injection rate be reconciled with those of grain destruction in shocks? If the current estimates are of the right order, then an interstellar source of grain material is essential to account for the opacity per unit H atom of the ISM. Whilst this seems plausible for organic or amorphous carbon, we know of no process that can form silicates at the temperatures and pressures prevailing in the ISM. The calculations are thus in conflict with the requirement that essentially the full interstellar abundance of Si must be in silicates to explain the strength of the 9.7 μm feature. The problem is compounded by the fact that some stellar winds will inject Si in other forms, such as SiC dust and, in the case of hot stars, as pure atomic Si. It is possible that the presence of mantles on silicate grains provides some degree of protection from sputtering. However, if silicates are, indeed, much more resistant to destruction than current estimates suggest, it becomes hard to account for the correlations of Si and Mg depletions with cloud density.

Why is Si bonded almost exclusively with O in silicates, whilst silicon carbide is rare? If carbon stars contribute 20–50% of the dust injected by stars into the ISM, as is generally assumed, a corresponding profusion of SiC is to be expected. However, the mid-infrared spectrum of interstellar dust indicates that less than 5%

of the available Si is in SiC, a result corroborated by the scarcity of pre-solar SiC in meteorites. It is not simply a case of finding a way to destroy SiC to explain the observational limit on its interstellar abundance, but also of somehow reaccreting the Si into silicates.

How do silicates evolve in the ISM and in stellar envelopes? We have seen that significant differences exist in the spectral profiles of silicates in diffuse and dense clouds – how may these be understood? Is the absence of detectable signatures of crystallinity in interstellar silicates consistent with the existence of crystalline forms in circumstellar shells around both young and old stars? Might *young* stars be important sources of interstellar silicate dust? This possibility was suggested some years ago (Herbig 1970; see also Burke and Silk 1976) but has received little attention. Given the high pre-existing depletions of the heavy elements in dense clouds, it is, perhaps, unlikely that star formation can produce more dust than it consumes, but it may recycle it in interesting ways. For example, if SiC dust is destroyed by oxidation in warm regions of protoplanetary discs, the products might then recondense into silicates that might subsequently be swept away by the stellar outflow and returned to the ISM.

What are the dominant forms of carbon in interstellar dust, and what qualitative and quantitative differences exist between C-rich stardust and carbonaceous material synthesized in the ISM? Are PAHs and C-rich small grains formed exclusively in C-rich environments (the atmospheres of carbon stars) or can they be made in the ISM (e.g. by shattering of photolysed organic mantles)? What specific PAHs are present in the ISM and how do they contribute to interstellar extinction? Can they account for the diffuse interstellar bands? How are pre-solar meteoritic nanodiamonds formed and what contribution do they make to the extinction curve? Why is the carrier of the 3.4 μm diffuse-ISM feature absent in dense molecular clouds? What are the carriers of the 3.47 and 6.85 μm features in dense clouds? Is the assignment of the 2175 Å 'bump' to small, graphitized carbon grains correct? Can the process of graphitization be simulated in the laboratory and, if so, can the position-independent variations in the width of the feature be reproduced?

What is the role of iron in interstellar dust? Fe is the most abundant heavy metal. It appears to be highly depleted into dust over a wide range of interstellar environments, making a contribution to the total grain mass comparable with those of Mg and Si (table 2.2). Yet there seems to be no consensus on what form it takes. Interstellar silicates will account for some, but the bulk must be in other forms such as oxides, sulphides, carbides and metal alloys. Are the Fe-rich phases in interplanetary particles (GEMS) representative of Fe-rich phases in interstellar dust? Do such inclusions render the grains superparamagnetic and thus contribute to their alignment?

How big are the biggest grains in the ISM and how common are they? Studies of dust currently entering the Solar System and of pre-solar meteoritic grains, together with observations of x-ray scattering halos, suggest the existence of grains with radii in the range $0.25 < a < 4$ μm, larger than the maximum size generally adopted for models of interstellar extinction. Can such big grains be incorporated into grain models without violating other constraints, such as element abundances?

What contribution does interstellar dust make to the organic inventories of early solar systems? Studies of isotope ratios in general, and D/H ratios in particular, make a clear case for survival of carbonaceous interstellar material in the solar nebula. The products of interstellar chemistry are seen in interplanetary dust grains and meteorites that rain down on us today. Was this the dominant source of elemental carbon and prebiotic compounds on the early Earth? Does the chiral selectivity of modern terrestrial biology have its roots in the radiative environment of the solar nebula and its parent cloud? A study of chiral molecules in samples returned by the Stardust mission might be particularly informative.

There seems to be no shortage of problems to engage the interested student or more seasoned campaigner. If I should ever be foolhardy enough to attempt a third edition of *'Dust in the Galactic Environment'*, perhaps the answers to some of these questions will adorn its pages.

Appendix A

Glossary

The aim here is to provide a convenient list of definitions, covering units, terms and acronyms used frequently in this book or in cited literature that might not be familiar to the reader. The units list should be regarded as a supplement to standard Système Internationale (SI) units. Completeness is not guaranteed. Readers unfamiliar with basic astrophysical terms and concepts are recommended to consult an introductory text such as *The Physical Universe* (Shu 1982) or *Foundations of Astronomy* (Seeds 1997).

A.1 Units and constants

Ångstrom (Å) Unit of length commonly used to denote wavelength in astronomical spectra, especially in the visible and ultraviolet; 1 Å $= 10^{-10}$ m $= 0.1$ nm.

Astronomical unit (AU) Mean Sun–Earth distance; 1 AU $= 1.496 \times 10^{11}$ m.

Dex Quasi-unit of logarithm to base 10: log 10 = 1 dex, log 100 = 2 dex, etc.

Gauss (G) CGS unit of magnetic flux density, hence 1 μG $= 10^{-6}$ G, the unit of choice for interstellar magnetic fields. The equivalent SI unit is the Tesla (1 T $= 10^4$ G).

Jansky (Jy) Unit of luminous flux density; 1 Jy $= 10^{-26}$ W m^{-2} Hz^{-1}.

Magnitude (mag) Logarithmic unit of stellar brightness based on an ancient visual ranking system, such that two stars of intensities I_1 and I_2 differ in magnitude by $m_2 - m_1 = -2.5 \log(I_2/I_1)$. Hence, an intensity ratio of 100 corresponds to a magnitude difference of exactly 5. Note that magnitudes are numerically greater for dimmer stars.

Micron (equivalent to SI micrometre, μm) Unit of length, commonly used to denote particle size; also used to denote wavelength in astronomical spectra, especially in the infrared; 1 μm $= 10^{-6}$ m.

Parsec (pc) Unit of distance based on stellar annual parallax; 1 pc $= 3.086 \times 10^{16}$ m. Similarly, kiloparsec (1 kpc $= 10^3$ pc), megaparsec (1 Mpc $= 10^6$ pc).

Solar luminosity $1\ L_\odot = 3.826 \times 10^{26}$ W.

Solar mass $1\ M_\odot = 1.989 \times 10^{30}$ kg.

Solar radius $1\ R_\odot = 6.960 \times 10^8$ m.

A.2 Physical, chemical and astrophysical terms

Adsorption Adhesion to a solid surface; commonly used to describe the attachment of gaseous atoms and molecules to the surface of a dust grain.

Aliphatic hydrocarbons Organic molecules composed of linear carbon chains with hydrogen attachments, typically with the generic formula $CH_3(CH_2)_nCH_3$.

Amorphous carbon Sootlike carbon, composed of clusters of hexagonal rings assembled randomly.

Apatite Commonest mineral form of phosphate, $Ca_5(F, Cl)(PO_4)_3$.

Aromatic hydrocarbons Planar organic molecules structurally related to benzene, i.e. composed of linked hexagonal carbon rings with hydrogen attachments.

Aromatic infrared spectrum Collective term for the family of infrared emission features attributed to polycyclic aromatic hydrocarbons (see table 6.1).

Asymptotic giant branch (AGB) Region of the Hertzsprung–Russell diagram (figure 7.1) occupied by intermediate-mass stars in their final phase of red-giant evolution, in which energy is generated primarily by helium-shell-burning. AGB stars undergo copious mass-loss and generally evolve to become planetary nebulae.

Bipolar outflow Stellar wind that is collimated to flow outward in directions close to the poles of a rotating star with an equatorial circumstellar disc.

Carbonaceous chondrite A broad class of meteorite, characterized by the presence of chondrules (glassy spherules), together with up to $\sim$3% by weight of carbonaceous matter, in a fine-grained matrix. The matrix contains hydrous silicates believed to have formed on asteroidal parent bodies in the presence of liquid water. The carbon is mostly in the form of complex organic polymers and amorphous carbon but also includes isolated grains of graphite, diamond and silicon carbide that are believed to pre-date the

Solar System. Amino acids and other prebiotic molecules are present in trace amounts.

Carbon star A red giant that has a carbon-enriched atmosphere ($C/O > 1$).

Chiral molecule An organic molecule (such as an amino acid or sugar) that has right-handed and left-handed structural forms (termed dextro, D and laevo, L). The two forms (enantiomers) are chemically identical but physically they are mirror images of one another. Ensembles of chiral molecules produced abiotically are usually racemic, i.e. they contain approximately equal numbers of D and L. Terrestrial life is homochiral, i.e. composed exclusively from L amino acids and D sugars.

Classical grain An interstellar dust grain that contributes to the λ^{-1} dependence of interstellar extinction at visual wavelengths, as determined in early work by Trumpler and others. Such a grain has a diameter typically in the range 0.1–1 μm.

Cosmic ray Highly energetic particle travelling through space at relativistic speed. Most cosmic rays are protons but all common atomic nuclei are represented. Most low-energy cosmic rays ($<10^{15}$ eV) probably originate in energetic stellar events such as supernova explosions in our Galaxy; those of the highest energies probably originate outside the Galaxy.

Depletion Shortfall in the abundances of chemical elements measured in interstellar gas, attributed to their assumed presence in dust grains.

Diamond Crystalline form of solid carbon in which C atoms are arranged tetrahedrally in a cubic structure.

Early-type stars Stars characterized by hot ($T > 10\,000$ K) photospheres that emit intense ultraviolet radiation, including stars of spectral types O and B and also the Wolf–Rayet stars. Their distribution in the Milky Way closely follows the galactic disc. As they are relatively short lived, early-type stars are very young compared with stars like the Sun.

Emission nebula Generic term for a cloud of hot, ionized gas, such as an H II region or a planetary nebula, maintained in an ionized state by radiation from nearby star(s). Their spectra are characterized by emission lines.

Enantiomer One of the two possible forms (L or D) of a chiral molecule.

Flux density Measure of irradiance per unit wavelength or frequency interval (F_λ or F_ν, respectively). These forms are related by the equation $F_\lambda\, d\lambda = F_\nu\, d\nu$, hence $F_\lambda = (-c/\lambda^2) F_\nu$. The SI unit of F_ν is $W\, m^{-2}\, Hz^{-1}$; for convenience the Jansky (see earlier) is generally used. The unit of choice for F_λ is $W\, m^{-2}\, \mu m^{-1}$.

Graphite Highly ordered form of solid carbon, in which platelets composed of linked hexagonal rings are regularly stacked.

Herbig Ae/Be stars Young (pre-main-sequence) stars of intermediate mass (2–10 $M_\odot$), higher mass counterparts to T Tauri stars.

Herbig–Haro objects Bright, irregular nebulosities associated with young stars, especially those with bipolar outflows. Some may be knots of emission from ionized gas; others are reflection nebulae.

H II region Interstellar matter maintained in an ionized state by stellar ultraviolet radiation. Compact H II regions form in the dense gas surrounding embedded OB stars. Diffuse H II regions are more extensive and lower in density (see section 1.4.4).

Hydrogenated amorphous carbon Amorphous (sootlike) carbon with hydrogen atoms attached to surface sites.

Infrared carbon star C-rich evolved star that has developed an optically thick dust shell as a consequence of rapid mass-loss on the asymptotic giant branch; characterized by intense infrared continuum emission from carbonaceous circumstellar dust.

Kerogen Complex, non-volatile organic material with tarlike properties. Present in carbonaceous chondrites and similar to organic refractory matter synthesized in the laboratory by energetic processing of ices.

Late-type stars Stars characterized by cool ($T < 4000$ K) photospheres (spectral types K, M, N and S). This term is most commonly used to refer to those stars (red giants and supergiants) that have evolved beyond the main sequence.

Luminescence Visible or infrared radiation emitted by a solid as the result of prior excitation. In the astronomical context, excitation generally arises from absorption of shorter-wavelength (UV–visible) radiation (photoluminescence), but luminescence can also be caused by particle bombardment, mechanical strain or chemical reaction. Fluorescence and phosphorescence are forms of luminescence, distinguished by the timescale for emission (fluorescence decays rapidly if the energy source is removed; phosphorescence decays more slowly, resulting in ‘afterglow’).

Luminosity class Ranking system based on roman numerals, used to denote luminosities of main-sequence and post-main-sequence stars. I: supergiants; II: bright giants; III: giants; IV: subgiants; V: main sequence.

Main sequence The most stable phase of a star’s evolution, in which energy is produced by core hydrogen burning.

Metallicity A measure of the heavy-element content of an astrophysical system such as a cloud, star or galaxy. All elements other than H and He are generally included.

Nanoparticles Particles with diameters of order 1 nm.

Nova (classical) Stellar outburst caused by explosive thermonuclear ignition of matter in the outer layers of a white dwarf in a close binary system, triggered by mass transfer from its companion star. Unlike a type Ia supernova (which also involves a white dwarf in a close binary system), a nova event does not destroy the progenitor and can recur periodically. The explosion ejects matter from the system. Evidence for dust condensation in the expanding envelopes of some novae is provided by the onset of a period of intense infrared emission, accompanied by a drop in visual brightness.

OB stars Generic term for stars of spectral classes O and B (see early-type stars).

OH-IR star O-rich evolved star that has developed an optically thick dust shell as a consequence of rapid mass-loss on the asymptotic giant branch; characterized by OH maser line emission at radio wavelengths and intense continuum emission from circumstellar dust in the infrared.

Organic refractory matter (ORM) Complex, non-volatile organic material synthesized in the laboratory by ultraviolet photolysis or energetic ion bombardment of ices containing molecules thought to be present in molecular clouds and comets.

Planetary nebula (PN) Expanding shell of hot ($\sim$10 000 K) ionized gas ejected from a central star. The precursor of a PN is a red giant of intermediate mass (typically 1–5 $M_\odot$) on the asymptotic giant branch of the HR diagram (see figure 7.1). The expanding envelope of the star forms the nebula, whilst the hot, compact core becomes a white dwarf.

Planetesimal Solid body of cometary or asteroidal dimensions, formed by accretion of dust and ice in a protoplanetary disc.

Platt particles Very small (nano-sized) particles originally proposed by Platt (1956), postulated to produce visual extinction by continuous absorption. The term is occasionally applied more generally to very small absorbing particles.

Polycyclic aromatic hydrocarbons (PAHs) Class of organic molecules composed of planar sheets of regularly linked hexagonal carbon rings with peripheral hydrogen attachments.

Pre-main-sequence star A YSO with a visible photosphere that has not yet reached a stable state of hydrogen-burning on the main sequence of the Hertzsprung–Russell diagram.

Protoplanetary disc Disc of material (gas and dust) in Keplerian rotation around a newly formed star. Low-speed collisions between co-rotating particles in the disc may lead to accretion, forming planetesimals, and ultimately, planets.

Protoplanetary nebula (PPN) This term has two possible meanings, generally distinguishable from the context: (i) the circumstellar nebula (or disc) around a pre-main-sequence star, from which planets may form; or (ii) a post-main-sequence star in the process of evolving from the asymptotic giant branch to become a planetary nebula.

Protostar A YSO powered mainly by gravitational energy released by infalling matter. Sometimes used more broadly to mean any young star that remains embedded in dust and emits most of its observable flux in the infrared.

R Coronae Borealis (RCB) stars A class of variable star characterized by dramatic, irregular declines in apparent visual brightness attributed to episodes of circumstellar dust condensation. Spectra indicate that their atmospheres are hydrogen deficient and C rich; they are believed to be in the post-asymptotic-giant-branch phase of their evolution.

Red giant Generic term for a luminous, late-type star of intermediate mass, occupying the mid- to upper right-hand region of the Hertzsprung–Russell diagram (figure 7.1), including both 'first assent' (luminosity class III) and AGB stars. Stars of higher mass become red supergiants.

Reflection nebula A cloud or region that reflects starlight, i.e. one that is visible because starlight is scattered by dust grains within it. Reflection nebulae are distinguishable from emission nebulae by their blue appearance and the general weakness or absence of emission lines in their spectra.

Silicates Class of minerals based on linked tetrahedral SiO_4 units. Forms thought to be common in the interstellar medium have generic formulae $MSiO_3$ (pyroxene) and M_2SiO_4 (olivine), where M can be Mg or Fe.

Solar nebula The protoplanetary nebula from which our Solar System formed.

Starburst An episode of intense star formation activity, especially that occurring in the nuclear regions of certain galaxies.

Supernova (SN) Catastrophic explosion of a highly evolved star at the end of its life. Supernovae are classified type I (hydrogen-deficient) and type II (hydrogen-rich) on the basis of their spectra, with type I divided into three subclasses. A type Ia supernova occurs when a white dwarf in a close binary system accretes enough mass from its companion to initiate a runaway thermonuclear explosion in its core. Types Ib, Ic and II are all associated with gravitational core collapse in a single, massive star but differ in the evolutionary state of the progenitor when the crisis point is reached. If the progenitor has retained a hydrogen-rich envelope, a type II SN will be the outcome. If hydrogen has been lost but helium retained, the event is classified type Ib. Finally, if both hydrogen and helium are deficient, we have a type Ic supernova.

Supernova remnant (SNR) Expanding cloud of hot gas emanating from a supernova explosion.

Surface brightness or intensity Measure of intensity per unit wavelength or frequency interval per unit solid angle of the radiation received from an extended object such as a nebula. The flux density received in a beam of solid angle Ω in a direction (θ, ϕ) is

$$F_\nu = \int I_\nu(\theta, \phi)\, \mathrm{d}\Omega$$

where I_ν is the surface brightness in W m^{-2} Hz^{-1} sr^{-1}. If the beam is smaller than the scale for variations in I_ν, then $F_\nu \approx I_\nu \Omega$. The wavelength form I_λ (W m^{-2} μm^{-1} sr^{-1}) is related to the frequency form I_ν in the same way as F_λ to F_ν (see earlier).

T Tauri stars Young (pre-main-sequence) stars of low to moderate mass. The Sun is presumed to have passed through a T Tauri phase prior to becoming a main-sequence star.

Very small grains (VSGs) Generic term for grains much smaller than classical grains, most commonly used to describe particles small enough ($a < 0.02\ \mu$m) to undergo substantial temperature increases upon absorption of individual ultraviolet photons (section 6.1). See also nanoparticles, PAHs, Platt particles.

WC stars C-rich Wolf–Rayet stars; some members of this class have evidence for dust formation in their circumstellar envelopes.

Wolf–Rayet stars Massive, post-main-sequence stars characterized by very hot ($\sim$50 000 K) surface temperatures and spectra with broad emission lines of He, C and O (WC stars) or He and N (WN stars). Some are also central stars of planetary nebulae. (See also early-type stars.)

A.3 Acronyms

CCM	Cardelli, Clayton and Mathis (formulation of extinction curve).
COBE	Cosmic Background Explorer (satellite).
CPR	Circularly polarized radiation.

DCM	Dark cloud medium.
DDA	Discrete dipole approximation.
DGL	Diffuse galactic light.
EMT	Effective medium theory.
ERE	Extended red emission.
EUV	Extreme ultraviolet.
FIR	Far infrared.
FUV	Far ultraviolet.
FWHM	Full-width at half-maximum intensity (of a spectral feature).
GEMS	Glasses with embedded metal and sulphides.
GL	Geophysical Laboratory (prefix for numerical entry in IRS catalogue).
HAC	Hydrogenated amorphous carbon
HD	Henry Draper (prefix for numerical entry in star catalogue).
HR	Hertzsprung–Russell (diagram).
HST	Hubble Space Telescope.
IDP	Interplanetary dust particle.
IRAS	Infrared Astronomical Satellite.
IRC	Infrared catalogue (prefix for numerical entry).
IRS	Infrared source.
ISM	Interstellar medium.
ISRF	Interstellar radiation field.
ISO	Infrared Space Observatory.
IUE	International Ultraviolet Explorer (satellite).
LMC	Large Magellanic Cloud.
MIR	Mid-infrared.
MRN	Mathis, Rumpl and Nordsieck (grain-size distribution).
NGC	New General Catalog (of star clusters, nebulae and galaxies).
NIR	Near infrared.
ORM	Organic refractory matter (or mantles).
PAH	Polycyclic aromatic hydrocarbon.
PN	Planetary nebula (plural PNe).
RCB	R Coronae Borealis (star).
SMC	Small Magellanic Cloud.
SN	Supernova (plural SNe).
SPM	Superparamagnetic.
UV	Ultraviolet.
VSG	Very small grain.
WR	Wolf–Rayet (star).
YSO	Young stellar object.

References

Aannestad P A and Greenberg J M 1983 *Astrophys. J.* **272** 551

Aannestad P A and Kenyon S J 1979 *Astrophys. Space Sci.* **65** 155

Aannestad P A and Purcell E M 1973 *Annu. Rev. Astron. Astrophys.* **11** 309

Abbott D C and Conti P S 1987 *Annu. Rev. Astron. Astrophys.* **25** 113

Acosta-Pulido J A *et al* 1996 *Astron. Astrophys.* **315** L121

Adams F C, Lada C J and Shu F H 1987 *Astrophys. J.* **312** 788

Adamson A J and Whittet D C B 1992 *Astrophys. J.* **398** L69

——1995 *Astrophys. J.* **448** L49

Adamson A J, Whittet D C B, Chrysostomou A, Hough J H, Aitken D K, Wright G and Roche P F 1999 *Astrophys. J.* **512** 224

Adamson A J, Whittet D C B and Duley W W 1990 *Mon. Not. R. Astron. Soc.* **243** 400

——1991 *Mon. Not. R. Astron. Soc.* **252** 234

Agarwal V K *et al* 1985 *Origins of Life* **16** 21

Aikawa Y, van Zadelhoff G J, van Dishoeck E F and Herbst E 2002 *Astron. Astrophys.* **386** 622

Aitken D K 1989 *22nd ESLAB Symposium, Infrared Spectroscopy in Astronomy* ed B H Kaldeich (ESA publication SP-290) p 99

——1996 *Polarimetry of the Interstellar Medium* ed W G Roberge and D C B Whittet (ASP Conference Series 97, San Francisco, CA) p 225

Aitken D K and Roche P F 1985 *Mon. Not. R. Astron. Soc.* **213** 777

Aitken D K, Roche P F, Smith C H, James S D and Hough J H 1988 *Mon. Not. R. Astron. Soc.* **230** 629

Aitken D K, Roche P F, Spenser P and Jones B 1979 *Astrophys. J.* **233** 925

Aitken D K, Smith C H and Roche P F 1989 *Mon. Not. R. Astron. Soc.* **236** 919

Aléon J, Engrand C, Robert F and Chaussidon M 2001 *Geochim. Cosmochim. Acta* **65** 4399

Alexander C M O'D 1997 *Astrophysical Implications of the Laboratory Study of Presolar Materials (AIP Conference Series)* ed T Bernatowicz and E Zinner (New York: Springer) p 567

Alexander C M O'D, Russell S S, Arden J W, Ash R D, Grady M M and Pillinger C T 1998 *Meteoritics Planetary Sci.* **33** 603

Allamandola L J, Sandford S A, Tielens A G G M and Herbst T M 1992 *Astrophys. J.* **399** 134

Allamandola L J, Tielens A G G M and Barker J R 1985 *Astrophys. J.* **290** L25

——1987 *Interstellar Processes* ed D J Hollenbach and H A Thronson (Dordrecht: Reidel) p 471

——1989 *Astrophys. J. Suppl.* **71** 733

Allen D A, Baines D W T, Blades J C and Whittet D C B 1982 *Mon. Not. R. Astron. Soc.* **199** 1017

Alton P B, Stockdale D P, Scarrott S M and Wolstencroft R D 2000b *Astron. Astrophys.* **357** 443

Alton P B, Xilouris E M, Bianchi S, Davies J and Kylafis N 2000a *Astron. Astrophys.* **356** 795

Amari S, Anders E, Virag A and Zinner E 1990 *Nature* **345** 238

Anders E 1971 *Annu. Rev. Astron. Astrophys.* **9** 1

——1989 *Nature* **342** 255

Anders E and Grevesse N 1989 *Geochim. Cosmochim. Acta* **53** 197

Anders E and Zinner E 1993 *Meteoritics* **28** 490

André P, Ward-Thompson D and Barsony M 1993 *Astrophys. J.* **406** 122

——2000 *Protostars & Planets* vol IV, ed V Mannings *et al* (Tucson, AZ: University of Arizona Press) p 59

Arendt R G, Dwek E and Moseley S H 1999 *Astrophys. J.* **521** 234

Arnett D 1996 *Supernovae and Nucleosynthesis* (Princeton, NJ: Princeton University Press)

Arnoult K M, Wdowiak T J and Beegle L W 2000 *Astrophys. J.* **535** 815

Arrhenius S 1908 *Worlds in the Making* (New York: Harper)

Artymowicz P 1988 *Astrophys. J.* **335** L79

Aumann H H *et al* 1984 *Astrophys. J.* **278** L23

Avery R W, Stokes R A, Michalsky J J and Ekstrom P A 1975 *Astron. J.* **80** 1026

Axon D J and Ellis R S 1976 *Mon. Not. R. Astron. Soc.* **177** 499

Baade W and Minkowski R 1937 *Astrophys. J.* **86** 123

Baas F, Allamandola L J, Geballe T R, Persson S E and Lacy J H 1983 *Astrophys. J.* **265** 290

Baas F, Geballe T R and Walther D M 1986 *Astrophys. J.* **311** L97

Baas F, Grim R J A, Geballe T R, Schutte W and Greenberg J M 1988 *Dust in the Universe* ed M E Bailey and D A Williams (Cambridge: Cambridge University Press) p 55

Backman D E and Paresce F 1993 *Protostars & Planets* vol III, ed E H Levy *et al* (Tucson, AZ: University of Arizona Press) p 1253

Bailey J 2001 *Origins Life Evol. Biosphere* **31** 167

Bailey J *et al* 1998 *Science* **281** 627

Bakes E L O and Tielens A G G M 1994 *Astrophys. J.* **427** 822

Bakes E L O, Tielens A G G M and Bauschlicher C W 2001 *Astrophys. J.* **556** 501

Bally J, Langer W D and Liu W 1991 *Astrophys. J.* **383** 645

Bally J, Sutherland R S, Devine D and Johnstone D 1998 *Astron. J.* **116** 293

Balsara D, Ward-Thompson D and Crutcher R M 2001 *Mon. Not. R. Astron. Soc.* **327** 715

Banwell C N and McCash E M 1994 *Fundamentals of Molecular Spectroscopy* 4th edn (London: McGraw-Hill)

Barker J R, Allamandola L J and Tielens A G G M 1987 *Astrophys. J.* **315** L61

Barlow M J 1978a *Mon. Not. R. Astron. Soc.* **183** 367

——1978b *Mon. Not. R. Astron. Soc.* **183** 397

——1978c *Mon. Not. R. Astron. Soc.* **183** 417

Barlow M J and Silk J 1977a *Astrophys. J.* **211** L83

——1977b *Astrophys. J.* **215** 800

Barnard E E 1910 *Astrophys. J.* **31** 8

——1913 *Astrophys. J.* **38** 496

——1919 *Astrophys. J.* **49** 1

——1927 *Atlas of Selected Regions of the Milky Way* ed E B Frost and M R Calvert (Washington, DC: Carnegie Institute)

Barshay S S and Lewis J S 1976 *Annu. Rev. Astron. Astrophys.* **14** 81

Basri G and Bertout C 1993 *Protostars & Planets* vol III, ed E H Levy *et al* (Tucson, AZ: University of Arizona Press) p 543

Bastiaansen P A 1992 *Astron. Astrophys. Suppl.* **93** 449

Batten A H 1988 *The Lives of Wilhelm and Otto Struve* (Dordrecht: Reidel) p 149

Bazell D and Dwek E 1990 *Astrophys. J.* **360** 142

Beck R 1996 *Polarimetry of the Interstellar Medium* ed W G Roberge and D C B Whittet (ASP Conference Series 97, San Francisco, CA) p 475

Becklin E E and Neugebauer G 1975 *Astrophys. J.* **200** L71

Beckwith S V W, Henning T and Nakagawa Y 2000 *Protostars & Planets* vol IV, ed V Mannings *et al* (Tucson, AZ: University of Arizona Press) p 533

Beckwith S V W and Sargent A I 1993 *Protostars & Planets* vol III, ed E H Levy *et al* (Tucson, AZ: University of Arizona Press) p 521

——1996 *Nature* **383** 139

Begemann B, Dorschner J, Henning T, Mutschke H, Gürtler J, Kömpe C and Nass R 1997 *Astrophys. J.* **476** 199

Beichman C A 1987 *Annu. Rev. Astron. Astrophys.* **25** 521

Beichman C A, Myers P C, Emerson J P, Harris S, Mathieu R, Benson P J and Jennings R E 1986 *Astrophys. J.* **307** 337

Beintema D A *et al* 1996 *Astron. Astrophys.* **315** L369

Bell K R, Cassen P M, Klahr H H and Henning T 1997 *Astrophys. J.* **486** 372

Benvenuti P and Porceddu I 1989 *Astron. Astrophys.* **223** 329

Bergin E A, Langer W D and Goldsmith P F 1995 *Astrophys. J.* **441** 222

Bernatowicz T J, Amari S, Zinner E K and Lewis R S 1991 *Astrophys. J.* **373** L73

Bernatowicz T J, Cowsik R, Gibbons P C, Lodders K, Fegley B, Amari S and Lewis R S 1996 *Astrophys. J.* **472** 760

Bernatowicz T J, Fraundorf G, Tang M, Anders E, Wopenka B, Zinner E and Fraundorf P 1987 *Nature* **330** 728

Bernatowicz T J and Walker R M 1997 *Phys. Today* **50** (no. 12) 26

Bernstein M P, Dworkin J P, Sandford S A, Cooper G W and Allamandola L J 2002 *Nature* **416** 401

Bernstein M P, Sandford S A and Allamandola L J 1997 *Astrophys. J.* **476** 932

Bernstein M P, Sandford S A, Allamandola L J, Chang S and Scharberg M A 1995 *Astrophys. J.* **454** 327

Bertoldi F and Cox P 2002 *Astron. Astrophys.* **384** L11

Biermann P and Harwit M 1980 *Astrophys. J.* **241** L105

Bischoff A 1998 *Meteoritics Planetary Sci.* **33** 1113

Blades J C and Whittet D C B 1980 *Mon. Not. R. Astron. Soc.* **191** 701

Blake G A, Sutton E C, Masson C R and Phillips T G 1987 *Astrophys. J.* **315** 621

Blum J and Wurm G 2000 *Icarus* **143** 138

Bockelée-Morvan D, Brooke T Y and Crovisier J 1995 *Icarus* **116** 18

Bockelée-Morvan D *et al* 2000 *Astron. Astrophys.* **353** 1101

Bode M F 1988 *Dust in the Universe* ed M E Bailey and D A Williams (Cambridge: Cambridge University Press) p 73

——1989 *22nd ESLAB Symposium, Infrared Spectroscopy in Astronomy* ed B H Kaldeich (ESA publication SP-290) p 317

Bode M F and Evans A 1983 *Mon. Not. R. Astron. Soc.* **203** 285

Bode M F, Evans A, Whittet D C B, Aitken D K, Roche P F and Whitmore B 1984 *Mon. Not. R. Astron. Soc.* **207** 897

Bodenheimer P 1989 *The Formation and Evolution of Planetary Systems* ed H A Weaver and L Danly (Cambridge: Cambridge University Press) p 243

Bodenheimer P, Ruzmaikina T and Mathieu R D 1993 *Protostars & Planets* vol III, ed E H Levy *et al* (Tucson, AZ: University of Arizona Press) p 367

Boesgaard A M and Steigman G 1985 *Annu. Rev. Astron. Astrophys.* **23** 319

Bohlin R C, Savage B D and Drake J F 1978 *Astrophys. J.* **224** 132

Bohren C F and Huffman D R 1983 *Absorption and Scattering of Light by Small Particles* (New York: Wiley)

Bok B J 1956 *Astron. J.* **61** 309

Bonner W A 1991 *Origins Life Evol. Biosphere* **21** 59

Boogert A C A, Helmich F P, van Dishoeck E F, Schutte W A, Tielens A G G M and Whittet D C B 1998 *Astron. Astrophys.* **336** 352

Boogert A C A, Schutte W A, Helmich F P, Tielens A G G M and Wooden D H 1997 *Astron. Astrophys.* **317** 929

Boogert A C A *et al* 2000a *Astron. Astrophys.* **353** 349

——2000b *Astron. Astrophys.* **360** 683

Borkowski K J and Dwek E 1995 *Astrophys. J.* **454** 254

Boss A P 1998 *Annu. Rev. Earth Planet. Sci.* **26** 53

Boulanger F, Boissel P, Cesarsky D and Ryter C 1998 *Astron. Astrophys.* **339** 194

Boulanger F and Pérault M 1988 *Astrophys. J.* **330** 964

Boulanger F *et al* 1996 *Astron. Astrophys.* **312** 256

Boulares A and Cox D P 1990 *Astrophys. J.* **365** 544

Bowey J E and Adamson A J 2001 *Mon. Not. R. Astron. Soc.* **320** 131

Bowey J E, Adamson A J and Whittet D C B 1998 *Mon. Not. R. Astron. Soc.* **298** 131

Bradley J P 1994 *Science* **265** 925

Bradley J P *et al* 1999 *Science* **285** 1716

Breger M, Gehrz R D and Hackwell J A 1981 *Astrophys. J.* **248** 963

Bregman J 1989 *IAU Symp. 135, Interstellar Dust* ed L J Allamandola and A G G M Tielens (Dordrecht: Kluwer) p 109

Bregman J D, Campins H, Witteborn F C, Wooden D H, Rank D, Allamandola L J, Cohen M and Tielens A G G M 1987 *Astron. Astrophys.* **187** 616

Briggs R, Ertem G, Ferris J P, Greenberg J M, McCain P J, Mendoza-Gomez C X and Schutte W 1992 *Origins Life Evol. Biosphere* **22** 287

Brooke T Y, Sellgren K and Geballe T R 1999 *Astrophys. J.* **517** 883

Brooke T Y, Sellgren K and Smith R G 1996 *Astrophys. J.* **459** 209

Browning L and Bourcier W 1998 *Meteoritics Planetary Sci.* **33** 1213

Brownlee D E 1978 *Protostars & Planets* ed T Gehrels (Tucson, AZ: University of Arizona Press) p 134

Brownlee D E and Kissel J 1990 *Comet Halley: Investigations, Results and Interpretations, Vol 2: Dust, Nucleus, Evolution* ed J W Mason (Chichester: Ellis Horwood) p 89

Burke J R and Silk J 1976 *Astrophys. J.* **210** 341

Burki G, Cramer N, Burnet M, Rufener F, Pernier B and Richard C 1989 *Astron. Astrophys.* **213** L26

Burstein D and Heiles C 1982 *Astron. J.* **87** 1165

Buss R H, Lamers H J G L M and Snow T P 1989 *Astrophys. J.* **347** 977

Buss R H and Snow T P 1988 *Astrophys. J.* **335** 331

Buss R H, Tielens A G G M, Cohen M, Werner M W, Bregman J D and Witteborn F C 1993 *Astrophys. J.* **415** 250

Butchart I and Whittet D C B 1983 *Mon. Not. R. Astron. Soc.* **202** 971

Butchart I, McFadzean A D, Whittet D C B, Geballe T R and Greenberg J M 1986 *Astron. Astrophys.* **154** L5

Calzetti D, Bohlin R C, Gordon K D, Witt A N and Bianchi L 1995 *Astrophys. J.* **446** L97

Cameron A G W 1988 *Annu. Rev. Astron. Astrophys.* **26** 441

Campbell M F *et al* 1976 *Astrophys. J.* **208** 396

Cardelli J A and Savage B D 1988 *Astrophys. J.* **325** 864

Cardelli J A, Clayton G C and Mathis J S 1989 *Astrophys. J.* **345** 245

Cardelli J A, Meyer D M, Jura M and Savage B D 1996 *Astrophys. J.* **467** 334

Carnochan D J 1989 *Interstellar Dust* ed L J Allamandola and A G G M Tielens (NASA Conference Publication 3036) p 11

Carral P, Hollenbach D J, Lord S D, Colgan S W J, Haas M R, Rubin R H and Erickson E F 1994 *Astrophys. J.* **423** 223

Caselli P, Hasegawa T I and Herbst E 1993 *Astrophys. J.* **408** 548

Caselli P, Walmsley C M, Tafalla M, Dore L and Myers P C 1999 *Astrophys. J.* **523** L165

Cecci-Pestellina C, Aiello S and Barsella B 1995 *Astrophys. J. Suppl.* **100** 187

Chandler C J and Richer J S 2000 *Astrophys. J.* **530** 851

Charnley S B 1997 *Astrophys. J.* **481** 396

Charnley S B and Kaufman M J 2000 *Astrophys. J.* **529** L111

Charnley S B, Tielens A G G M and Millar T J 1992 *Astrophys. J.* **399** L71

Chiar J E, Adamson A J, Kerr T H and Whittet D C B 1995 *Astrophys. J.* **455** 234

Chiar J E, Adamson A J, Pendleton Y J, Whittet D C B, Caldwell D A and Gibb E L 2002 *Astrophys. J.* **507** 198

Chiar J E, Adamson A J and Whittet D C B 1996 *Astrophys. J.* **472** 665

Chiar J E, Gerakines P A, Whittet D C B, Pendleton Y J, Tielens A G G M and Adamson A J 1998a *Astrophys. J.* **498** 716

Chiar J E, Pendleton Y J, Geballe T R and Tielens A G G M 1998b *Astrophys. J.* **507** 281

Chiar J E and Tielens A G G M 2001 *Astrophys. J.* **550** L207

Chiar J E, Tielens A G G M, Whittet D C B, Schutte W, Boogert A, Lutz D, van Dishoeck E F and Bernstein M P 2000 *Astrophys. J.* **537** 749

Chick K M and Cassen P 1997 *Astrophys. J.* **477** 398

Chlewicki G and Greenberg J M 1990 *Astrophys. J.* **365** 230

Chrysostomou A, Gledhill T M, Ménard F, Hough J H, Tamura M and Bailey J 2000 *Mon. Not. R. Astron. Soc.* **312** 103

Chrysostomou A, Hough J H, Whittet D C B, Aitken D K, Roche P F and Lazarian A 1996 *Astrophys. J.* **465** L61

Churchwell E 1990 *Astron. Astrophys. Rev.* **2** 79

Chyba C F 1990 *Nature* **348** 113

Chyba C F and McDonald G D 1995 *Annu. Rev. Earth Planet. Sci.* **23** 215

Chyba C F and Sagan C 1992 *Nature* **355** 125

——1997 *Comets and the Origin and Evolution of Life* ed P J Thomas *et al* (New York: Springer) p 147

Chyba C F, Sagan C and Mumma M J 1989 *Icarus* **79** 362

Chyba C F, Thomas P J, Brookshaw L and Sagan C 1990 *Science* **249** 366
Cimatti A, Alighieri S, Fosbury R, Salvati M and Taylor D 1993 *Mon. Not. R. Astron. Soc.* **264** 421
Clark D H and Stephenson F R 1977 *The Historical Supernovae* (Oxford: Pergamon)
Clark S, McCall A, Chrysostomou A, Gledhill T, Yates J and Hough J H 2000 *Mon. Not. R. Astron. Soc.* **319** 337
Clayton D D 1975 *Astrophys. J.* **199** 765
——1981 *Astrophys. J.* **251** 374
——1982 *Quart. J. R. Astron. Soc.* **23** 174
——1988 *Dust in the Universe* ed M E Bailey and D A Williams (Cambridge: Cambridge University Press) p 145
Clayton G C, Gordon K D and Wolff M J 2000 *Astrophys. J. Suppl.* **129** 147
Clayton D D and Hoyle F 1976 *Astrophys. J.* **203** 490
Clayton D D, Liu W and Dalgarno A 1999 *Science* **283** 1290
Clayton G C and Mathis J S 1988 *Astrophys. J.* **327** 911
Clayton G C, Martin P G and Thompson I 1983 *Astrophys. J.* **265** 194
Clayton G C, Wolff M J, Allen R G and Lupie O L 1995 *Astrophys. J.* **445** 947
Clayton G C *et al* 1996 *Astrophys. J.* **460** 313
Clemett S J, Maechling C R, Zare R N, Swan P D and Walker R M 1993 *Science* **262** 721
Close L M, Roddier F, Northcott M J, Roddier C and Graves J E 1997 *Astrophys. J.* **478** 766
Cohen M 1983 *Astrophys. J.* **270** L69
——1984 *Mon. Not. R. Astron. Soc.* **206** 137
Cohen M and Kuhi L V 1979 *Astrophys. J. Suppl.* **41** 743
Cohen M and Witteborn F C 1985 *Astrophys. J.* **294** 345
Cohen M *et al* 1986 *Astrophys. J.* **302** 737
——1989 *Astrophys. J.* **341** 246
Cook D J and Saykally R J 1998 *Astrophys. J.* **493** 793
Cooper G, Kimmich N, Belisle W, Sarinana J, Brabham K and Garrel L 2001 *Nature* **414** 879
Cottin H, Gazeau M C and Raulin F 1999 *Planetary Space Sci.* **47** 1141
Cowie L L 1978 *Astrophys. J.* **225** 887
Cowie L L and Songaila A 1986 *Annu. Rev. Astron. Astrophys.* **24** 499
Cox D P 1995 *The Physics of the Interstellar Medium and Intergalactic Medium* ed A Ferrera *et al* (ASP Conference Series 80, San Francisco, CA) p 317
Cox D P and Reynolds R J 1987 *Annu. Rev. Astron. Astrophys.* **25** 303
Cox P, Krugel E and Mezger P G 1986 *Astron. Astrophys.* **155** 380
Cox P and Mezger P G 1987 *Star Formation in Galaxies* ed C J Lonsdale Persson (NASA Conference Publication 2466) p 23
——1989 *Astron. Astrophys. Rev.* **1** 49
Coyne G V, Gehrels T and Serkowski K 1974 *Astron. J.* **79** 581
Coyne G V, Tapia S and Vrba F J 1979 *Astron. J.* **84** 356
Crawford D L 1975 *Publ. Astron. Soc. Pacific* **87** 481
Creese M, Jones T J and Kobulnicky H A 1995 *Astron. J.* **110** 268
Crick F 1981 *Life Itself: Its Origin and Nature* (New York: Simon and Schuster)
Cronin J R and Chang S 1993 *The Chemistry of Life's Origins* ed J M Greenberg *et al* (Dordrecht: Kluwer) p 209
Cronin J R and Pizzarello S 1997 *Science* **275** 951

——2000 *Adv. Space Res.* **23** 293

Cronin J R, Pizzarello S and Cruikshank D P 1988 *Meteorites and the Early Solar System* ed J F Kerridge and M S Matthews (Tucson, AZ: University of Arizona Press) p 819

Crovisier J *et al* 1997 *Science* **275** 1904

Cruikshank 1997 *From Stardust to Planetesimals* ed Y J Pendleton and A G G M Tielens (ASP Conference Series 122, San Francisco, CA) p 315

Dai Z R, Bradley J P, Joswiak D J, Brownlee D E, Hills H G M and Genge M J 2002 *Nature* **418** 157

Dale D A, Helou G, Contursi A, Silbermann N A and Kolhatkar S 2001 *Astrophys. J.* **549** 215

Dame T M *et al* 1987 *Astrophys. J.* **322** 706

Danziger I J, Lucy L B, Gouiffes C and Bouchet P 1991 *Supernovae* ed S E Woosley (New York: Springer) p 69

Dartois E, Schutte W A, Geballe T R, Demyk K, Ehrenfreund P and d'Hendecourt L 1999 *Astron. Astrophys.* **342** L32

Davies R E 1986 *Nature* **324** 10

——1988 *Acta Astronautica* **17** 129

Davies R E, Delluva A M and Kock R H 1984 *Nature* **311** 748

Davis L and Greenstein J L 1951 *Astrophys. J.* **114** 206

Day K L 1979 *Astrophys. J.* **234** 158

——1981 *Astrophys. J.* **246** 110

Day K L and Huffman D R 1973 *Nature Phys. Sci.* **243** 50

Debye P 1909 *Ann. Phys., Lpz.* **30** 59

——1912 *Ann. Phys., Lpz.* **39** 789

Delano J W 2001 *Origins Life Evol. Biosphere* **31** 311

Delsemme A H 2000 *Icarus* **146** 313

de Muizon M, Geballe T R and d'Hendecourt L B 1986 *Astrophys. J.* **306** L105

Désert F X and Dennefeld M 1988 *Astron. Astrophys.* **206** 227

Désert F X, Boulanger F, Léger A, Puget J L and Sellgren K 1986 *Astron. Astrophys.* **159** 328

Désert F X, Bazel D and Boulanger F 1988 *Astrophys. J.* **334** 815

Désert F X, Boulanger F and Puget J L 1990 *Astron. Astrophys.* **237** 215

Désert F X, Jenniskens P and Dennefeld M 1995 *Astron. Astrophys.* **303** 223

de Vaucouleurs G and Buta R 1983 *Astron. J.* **88** 939

de Vries M S, Reihs K, Wendt H R, Golden W G, Hunziker H E, Fleming R, Peterson E and Chang S 1993 *Geochim. Cosmochim. Acta* **57** 933

De Young D S 1998 *Astrophys. J.* **507** 161

d'Hendecourt L B 1997 *From Stardust to Planetesimals* ed Y J Pendleton and A G G M Tielens (ASP Conference Series 122, San Francisco, CA) p 129

d'Hendecourt L B, Allamandola L J and Greenberg J M 1985 *Astron. Astrophys.* **152** 130

d'Hendecourt L B, Allamandola L J, Grim R J A and Greenberg J M 1986 *Astron. Astrophys.* **158** 119

Disney M 1990 *Nature* **346** 105

Dolginov A Z and Mytrophanov I G 1976 *Astrophys. Space Sci.* **43** 291

Dominik C, Gail H P, Sedlmayr E and Winters J M 1990 *Astron. Astrophys.* **240** 365

Dominik C and Tielens A G G M 1997 *Astrophys. J.* **480** 647

Donn B 1968 *Astrophys. J.* **152** L129

Donn B and Nuth J A 1985 *Astrophys. J.* **288** 187

Dorschner J and Henning T 1986 *Astrophys. Space Sci.* **128** 47
——1995 *Astron. Astrophys. Rev.* **6** 271
Dorschner J, Friedemann C, Gürtler J and Henning T 1988 *Astron. Astrophys.* **198** 223
Douvion T, Lagage P O, Cesarcky C J and Dwek E 2001 *Astron. Astrophys.* **373** 281
Dragovan M 1986 *Astrophys. J.* **308** 270
Draine B T 1984 *Astrophys. J.* **277** L71
——1988 *Astrophys. J.* **333** 848
——1989a *IAU Symp. 135, Interstellar Dust* ed L J Allamandola and A G G M Tielens (Dordrecht: Kluwer) p 313
——1989b *Evolution of Interstellar Dust and Related Topics* ed A Bonetti *et al* (Amsterdam: North-Holland) p 103
——1990 *The Interstellar Medium in Galaxies* ed H A Thronson and J M Shull (Dordrecht: Kluwer) p 483
——1996 *Polarimetry of the Interstellar Medium* ed W G Roberge and D C B Whittet (ASP Conference Series 97, San Francisco, CA) p 16
Draine B T and Anderson N 1985 *Astrophys. J.* **292** 494
Draine B T and Lee H M 1984 *Astrophys. J.* **285** 89
Draine B T and Malhotra S 1993 *Astrophys. J.* **414** 632
Draine B T and McKee C F 1993 *Annu. Rev. Astron. Astrophys.* **31** 373
Draine B T and Salpeter E E 1979 *Astrophys. J.* **231** 438
Draine B T and Weingartner J C 1996 *Astrophys. J.* **470** 551
——1997 *Astrophys. J.* **480** 633
Drilling J S, Hecht J H, Clayton G C, Mattei J A, Landolt A U and Whitney B A 1997 *Astrophys. J.* **476** 865
Drilling J S and Schönberner D 1989 *Astrophys. J.* **343** L45
Dudley C C and Wynn-Williams C G 1997 *Astrophys. J.* **488** 720
Duley W W 1973 *Astrophys. Space Sci.* **23** 43
——1974 *Astrophys. Space Sci.* **26** 199
——1984 *Quart. J. R. Astron. Soc.* **25** 109
——1988 *Dust in the Universe* ed M E Bailey and D A Williams (Cambridge: Cambridge University Press) p 209
——1993 *Dust and Chemistry in Astronomy* ed T J Millar and D A Williams (Bristol: IOP Publishing) p 71
——2001 *Astrophys. J.* **553** 575
Duley W W, Jones A P, Whittet D C B and Williams D A 1989b *Mon. Not. R. Astron. Soc.* **241** 697
Duley W W, Jones A P and Williams D A 1989a *Mon. Not. R. Astron. Soc.* **236** 709
Duley W W, Scott A D, Seahra S and Dadswell G 1998 *Astrophys. J.* **503** L183
Duley W W and Seahra S 1998 *Astrophys. J.* **507** 874
Duley W W and Whittet D C B 1992 *Mon. Not. R. Astron. Soc.* **255** 243
Duley W W and Williams D A 1981 *Mon. Not. R. Astron. Soc.* **196** 269
——1983 *Mon. Not. R. Astron. Soc.* **205** 67P
——1984 *Interstellar Chemistry* (London: Academic)
——1988 *Mon. Not. R. Astron. Soc.* **231** 969
——1990 *Mon. Not. R. Astron. Soc.* **247** 647
Dupree A K 1986 *Annu. Rev. Astron. Astrophys.* **24** 377
Dwek E 1998 *Astrophys. J.* **501** 643
Dwek E and Arendt R G 1992 *Annu. Rev. Astron. Astrophys.* **30** 11

Dwek E and Scalo J M 1980 *Astrophys. J.* **239** 193

Dwek E *et al* 1997 *Astrophys. J.* **475** 565

Dyck H M and Lonsdale C J 1981 *IAU Symp. 96, Infrared Astronomy* ed C G Wynn-Williams and D P Cruikshank (Dordrecht: Reidel) p 223

Dyson J E and Williams D A 1997 *The Physics of the Interstellar Medium* 2nd edn (Bristol: IOP Publishing)

Eales S A, Wynn-Williams C G and Duncan W 1989 *Astrophys. J.* **339** 859

Eddington A S 1926 *Proc. R. Soc.* A **111** 423

Edvardsson B, Andersen J, Gustafsson B, Lambert D L, Nissen P E and Tomkin J 1993 *Astron. Astrophys.* **275** 101

Edwards S Ray T and Mundt R 1993 *Protostars & Planets* vol III, ed E H Levy *et al* (Tucson, AZ: University of Arizona Press) p 567

Ehrenfreund P and Charnley S B 2000 *Annu. Rev. Astron. Astrophys.* **38** 427

Ehrenfreund P, Breukers R, d'Hendecourt L and Greenberg J M 1992 *Astron. Astrophys.* **260** 431

Ehrenfreund P, Boogert A C A, Gerakines P A, Jansen D, Schutte W, Tielens A G G M and van Dishoeck E F 1996 *Astron. Astrophys.* **315** L341

Ehrenfreund P, Dartois E, Demyk K and d'Hendecourt L 1998 *Astron. Astrophys.* **339** L17

Ehrenfreund P *et al* 1999 *Astron. Astrophys.* **350** 240

Eiroa C and Hodapp K W 1989 *Astron. Astrophys.* **210** 345

Elmegreen B G 1985 *Protostars & Planets* vol II, ed D C Black and M S Matthews (Tucson, AZ: University of Arizona Press) p 33

Elmegreen B G 1997 *Astrophys. J.* **477** 196

——2002 *Astrophys. J.* **564** 773

Elsila J, Allamandola L J and Sandford S A 1997 *Astrophys. J.* **479** 818

Engrand C and Maurette M 1998 *Meteoritics Planetary Sci.* **33** 565

Epchtein N, Le Bertre T and Lépine J R D 1990 *Astron. Astrophys.* **227** 82

Evans A and Rawlings J M C 1994 *Mon. Not. R. Astron. Soc.* **269** 427

Evans A, Whittet D C B, Davies J K, Kilkenny D and Bode M F 1985 *Mon. Not. R. Astron. Soc.* **217** 767

Evans A, Geballe T R, Rawlings J M C, Eyres S P S and Davies J K 1997 *Mon. Not. R. Astron. Soc.* **292** 192

Fabian D, Jäger C, Henning T, Dorschner J and Mutschke H 2000 *Astron. Astrophys.* **364** 282

Fahlman G G and Walker G A H 1975 *Astrophys. J.* **200** 22

Feitzinger J V and Stüwe J A 1986 *Astrophys. J.* **305** 534

Ferlet R 1999 *Astron. Astrophys. Rev.* **9** 153

Ferris J P, Hill A R, Liu R and Orgel L E 1996 *Nature* **381** 59

Field G B 1974 *Astrophys. J.* **187** 453

Field G B, Partridge R B and Sobel H 1967 *Interstellar Grains* ed J M Greenberg and T P Roark (Washington, DC: NASA) p 207

FitzGerald M P 1968 *Astron. J.* **73** 983

Fitzpatrick E L 1985 *Astrophys. J.* **299** 219

——1986 *Astron. J.* **92** 1068

——1989 *IAU Symp. 135, Interstellar Dust* ed L J Allamandola and A G G M Tielens (Dordrecht: Kluwer) p 37

——1996 *Astrophys. J.* **473** L55

——1999 *Publ. Astron. Soc. Pacific* **111** 63

Fitzpatrick E L and Massa D 1986 *Astrophys. J.* **307** 286
——1988 *Astrophys. J.* **328** 734
——1990 *Astrophys. J. Suppl.* **72** 163
Fitzpatrick E L and Spitzer L 1997 *Astrophys. J.* **475** 623
Flores J J, Bonner W A and Massey G A 1977 *J. Am. Chem. Soc.* **99** 3622
Flynn G J, Keller L P, Joswiak D and Brownlee D E 2002 *33rd Lunar Planetary Sci. Conf.* abstract no 1320
Fogel M E and Leung C M 1998 *Astrophys. J.* **501** 175
Foing B H and Ehrenfreund P 1997 *Astron. Astrophys.* **317** L59
Forrest W J, Houck J R and McCarthy J F 1981 *Astrophys. J.* **248** 195
Forrest W J, McCarthy J F and Houck J R 1979 *Astrophys. J.* **233** 611
Fosalba P, Lazarian A, Prunet S and Tauber J 2002 *Astrophys. J.* **564** 762
Frenklach M, Carmer C S and Feigelson E D 1989 *Nature* **339** 196
Frenklach M and Feigelson E D 1989 *Astrophys. J.* **341** 372
——1997 *From Stardust to Planetesimals* ed Y J Pendleton and A G G M Tielens (ASP Conference Series 122, San Francisco, CA) p 107
Frerking M A, Langer W D and Wilson R W 1982 *Astrophys. J.* **262** 590
Frisch P C *et al* 1999 *Astrophys. J.* **525** 492
Furton D G, Laiho J W and Witt A N 1999 *Astrophys. J.* **526** 752
Furton D G and Witt A N 1992 *Astrophys. J.* **386** 587
Gaffey M J 1997 *Origins Life Evol. Biosphere* **27** 185
Gaffey M J, Burbine T H and Binzel R P 1993 *Meteoritics* **28** 161
Gail H P and Sedlmayr E 1986 *Astron. Astrophys.* **166** 225
——1987 *Physical Processes in Interstellar Clouds* ed G E Morfill and M Scholer (Dordrecht: Reidel) p 275
Galazutdinov G A, Krelowski J, Musaev F A, Ehrenfreund P and Foing B H 2000 *Mon. Not. R. Astron. Soc.* **317** 750
Gaustad J E 1971 *Dark Nebulae, Globules and Protostars* ed B T Lynds (Tucson, AZ: University of Arizona Press) p 91
Geballe T R 1991 *Mon. Not. R. Astron. Soc.* **251** 24P
——1997 *From Stardust to Planetesimals* ed Y J Pendleton and A G G M Tielens (ASP Conference Series 122, San Francisco, CA) p 119
Geballe T R, Baas F, Greenberg J M and Schutte W 1985 *Astron. Astrophys.* **146** L6
Geballe T R, Joblin C, d'Hendecourt L B, Jourdain de Muizon M, Tielens A G G M and Léger A 1994 *Astrophys. J.* **434** L15
Geballe T R, Kim Y H, Knacke R F and Noll K S 1988 *Astrophys. J.* **326** L65
Geballe T R, Noll K S, Whittet D C B and Waters L B F M 1989a *Astrophys. J.* **340** L29
Geballe T R, Tielens A G G M, Allamandola L J, Moorhouse A and Brand P W J L 1989b *Astrophys. J.* **341** 278
Gehrz R D 1988 *Annu. Rev. Astron. Astrophys.* **26** 377
Gehrz R D, Grasdalen G L, Greenhouse M, Hackwell J A, Heyward T and Bentley A F 1986 *Astrophys. J.* **308** L63
Gehrz R D, Greenhouse M A, Hayward T L, Houck J R, Mason C G and Woodward C E 1995 *Astrophys. J.* **448** L119
Gehrz R D and Ney E P 1987 *Proc. Natl Acad. Sci.* **84** 6961
——1990 *Proc. Natl Acad. Sci.* **87** 4354
Geiss J and Reeves H 1981 *Astron. Astrophys.* **93** 189
Genzel R and Cesarsky C J 2000 *Annu. Rev. Astron. Astrophys.* **38** 761

Gerakines P A, Schutte W A, Greenberg J M and van Dishoeck E F 1995a *Astron. Astrophys.* **296** 810
Gerakines P A, Whittet D C B and Lazarian A 1995b *Astrophys. J.* **455** L171
Gerakines P A *et al* 1999 *Astrophys. J.* **522** 357
Giard M, Lamarre J, Pajot F and Serra G 1994 *Astron. Astrophys.* **286** 203
Gibb E L *et al* 2000a *Astrophys. J.* **536** 347
Gibb E L, Nummelin A, Irvine W M, Whittet D C B and Bergman P 2000b *Astrophys. J.* **545** 309
Gibb E L and Whittet D C B 2002 *Astrophys. J.* **566** L113
Gibb E L, Whittet D C B and Chiar J E 2001 *Astrophys. J.* **558** 702
Gies D R and Lambert D L 1992 *Astrophys. J.* **387** 673
Gillett F C 1986 *Light on Dark Matter* ed F P Israel (Dordrecht: Reidel) p 61
Gillett F C and Soifer B T 1976 *Astrophys. J.* **207** 780
Gillett F C, Forrest W J and Merrill K M 1973 *Astrophys. J.* **183** 87
Gillett F C, Jones T W, Merrill K M and Stein W A 1975a *Astron. Astrophys.* **45** 77
Gillett F C, Forrest W J, Merrill K M, Capps R W and Soifer B T 1975b *Astrophys. J.* **200** 609
Gilman R C 1969 *Astrophys. J.* **155** L185
Glasse A C H, Towlson W A, Aitken D K and Roche P F 1986 *Mon. Not. R. Astron. Soc.* **220** 185
Glassgold A E 1996 *Annu. Rev. Astron. Astrophys.* **34** 241
Gledhill T M and McCall A 2000 *Mon. Not. R. Astron. Soc.* **314** 123
Gledhill T M and Scarrott S M 1989 *Mon. Not. R. Astron. Soc.* **236** 139
Goebel J H and Moseley S H 1985 *Astrophys. J.* **290** L35
Gold T 1952 *Mon. Not. R. Astron. Soc.* **112** 215
Gonzalez G 1999 *Mon. Not. R. Astron. Soc.* **308** 447
Goodman A A and Whittet D C B 1995 *Astrophys. J.* **455** L181
Goodman A A, Jones T J, Lada E A and Myers P C 1995 *Astrophys. J.* **448** 748
Gordon K D, Calzetti D and Witt A N 1997 *Astrophys. J.* **487** 625
Gordon K D, Witt A N, Carruthers G R, Christensen S A, Dohne B C and Hulburt E O 1994 *Astrophys. J.* **432** 641
Gordon K D *et al* 2000 *Astrophys. J.* **544** 859
Gordon M A 1995 *Astron. Astrophys.* **301** 853
Goswami J N and Vanhala H A T 2000 *Protostars & Planets* vol IV, ed V Mannings *et al* (Tucson, AZ: University of Arizona Press) p 963
Greenberg J M 1968 *Nebulae and Interstellar Matter* ed B M Middlehurst and L H Aller (Chicago, IL: University of Chicago Press) p 221
——1971 *Astron. Astrophys.* **12** 240
——1978 *Cosmic Dust* ed J A M McDonnell (New York: Wiley) p 187
——1982 *Comets* ed L L Wilkening (Tucson, AZ: University of Arizona Press) p 131
——1998 *Astron. Astrophys.* **330** 375
Greenberg J M and Hage J I 1990 *Astrophys. J.* **361** 260
Greenberg J M and Hong S S 1974 *IAU Symp. 60, Galactic Radio Astronomy* ed F J Kerr and S C Simonson (Dordrecht: Reidel) p 155
Greenberg J M and Meltzer A S 1960 *Mon. Not. R. Astron. Soc.* **132** 667
Greenberg J M and Shah G A 1971 *Astron. Astrophys.* **12** 250
Greenberg J M, Li A, Mendoza-Gomez C X, Schutte W A, Gerakines P A and de Groot M 1995 *Astrophys. J.* **455** L177

Greenberg J M *et al* 2000 *Astrophys. J.* **531** L71
Grevesse N and Noels A 1993 *Origin and Evolution of the Elements* ed N Prantzos *et al* (Cambridge: Cambridge University Press) p 15
Grevesse N and Sauval A J 1998 *Space Sci. Rev.* **85** 161
Grim R J A and d'Hendecourt L B 1986 *Astron. Astrophys.* **167** 161
Grim R J A, Baas F, Geballe T R, Greenberg J M and Schutte W 1991 *Astron. Astrophys.* **243** 473
Grimm R E and McSween H Y 1989 *Icarus* **82** 244
Grishko V I and Duley W W 2000 *Astrophys. J.* **543** L85
Groenewegen M A T 1997 *Astron. Astrophys.* **317** 503
Grün E *et al* 1993 *Nature* **362** 428
——1994 *Astron. Astrophys.* **286** 915
Guglielmo F, Epchtein N and Le Bertre T 1998 *Astron. Astrophys.* **334** 609
Gummersbach C A, Kaufer A, Schäfer D R, Szeifert T and Wolf B 1998 *Astron. Astrophys.* **338** 881
Hall J S 1949 *Science* **109** 166
Hall J S and Serkowski K 1963 *Basic Astronomical Data* ed K Aa Strand (Chicago, IL: University of Chicago Press) p 293
Hallenbeck S L, Nuth J A and Nelson R N 2000 *Astrophys. J.* **535** 247
Hanner M S, Lynch D K and Russell R W 1994 *Astrophys. J.* **425** 274
Hanner M S *et al* 1997 *Earth, Moon and Planets* **79** 247
Hanner M S, Brooke T Y and Tokunaga A T 1998 *Astrophys. J.* **502** 871
Harker D E and Desch S J 2002 *Astrophys. J.* **565** 109
Harris A W, Gry C and Bromage G E 1984 *Astrophys. J.* **284** 157
Harris D H 1973 *IAU Symp. 52, Interstellar Dust and Related Topics* ed J M Greenberg and H C van de Hulst (Dordrecht: Reidel) p 31
Harrison T E and Stringfellow G S 1994 *Astrophys. J.* **437** 827
Hartquist T W and Williams D A 1990 *Mon. Not. R. Astron. Soc.* **247** 343
Harwit M 1970 *Nature* **226** 61
Hayatsu R and Anders E 1981 *Topics Current Chem.* **99** 1
He L, Whittet D C B, Kilkenny D and Spencer Jones J H 1995 *Astrophys. J. Suppl.* **101** 335
Hecht J H 1986 *Astrophys. J.* **305** 817
——1991 *Astrophys. J.* **367** 635
Hecht J H, Holm A V, Donn B and Wu C C 1984 *Astrophys. J.* **280** 228
Heck A, Egret D, Jaschek M and Jaschek C 1984 *IUE Low-Dispersion Spectra Reference Atlas – Part 1: Normal Stars* (ESA publication SP-1052)
Heiles C 1987 *Interstellar Processes* ed D J Hollenbach and H A Thronson (Dordrecht: Reidel) p 171
——1996 *Polarimetry of the Interstellar Medium* ed W G Roberge and D C B Whittet (ASP Conference Series 97, San Francisco, CA) p 457
——2000 *Astron. J.* **119** 923
Helmich F P 1996 *PhD Thesis* (University of Leiden)
Helou G 1989 *IAU Symp. 135, Interstellar Dust* ed L J Allamandola and A G G M Tielens (Dordrecht: Kluwer) p 285
Helou G, Lu N Y, Werner M W, Malhotra S and Silbermann N 2000 *Astrophys. J.* **532** L21
Henning T and Stognienko R 1993 *Astron. Astrophys.* **280** 609
Herbig G H 1970 *Mém. Soc. R. Sci. Liège* 5th series **19** 13

——1975 *Astrophys. J.* **196** 129
——1995 *Annu. Rev. Astron. Astrophys.* **33** 19
Herbst E 1987 *Interstellar Processes* ed D J Hollenbach and H A Thronson (Dordrecht: Reidel) p 611
Herbst E and Klemperer W 1973 *Astrophys. J.* **185** 505
Herbst E and Leung C M 1986 *Mon. Not. R. Astron. Soc.* **222** 689
Hildebrand R H 1983 *Quart. J. R. Astron. Soc.* **24** 267
——1988a *Quart. J. R. Astron. Soc.* **29** 327
——1988b *Astrophys. Lett. Commun.* **26** 263
Hildebrand R H, Dotson J L, Dowell C D, Schleuning D A and Vaillancourt J E 1999 *Astrophys. J.* **516** 834
Hildebrand R H and Dragovan M 1995 *Astrophys. J.* **450** 663
Hildebrand R H, Dragovan M and Novak G 1984 *Astrophys. J.* **284** L51
Hill H G M, Jones A P and d'Hendecourt L B 1998 *Astron. Astrophys.* **336** L41
Hiltner W A 1949 *Science* **109** 165
Hobbs L M, Blitz L and Magnani L 1986 *Astrophys. J.* **306** L109
Hobbs L M, York D G and Oegerle W 1982 *Astrophys. J.* **252** L21
Hodge P W 1981 *Interplanetary Dust* (New York: Gordon and Breach) p 201
Hollenbach D J and Salpeter E E 1971 *Astrophys. J.* **163** 155
Hollenbach D J and Tielens A G G M 1997 *Annu. Rev. Astron. Astrophys.* **35** 179
Holweger H 2001 *Solar and Galactic Composition (AIP Conference Series)* ed R F Wimmer-Schweingruber (New York: Springer) p 23
Hong S S and Greenberg J M 1980 *Astron. Astrophys.* **88** 194
Hough J H 1996 *Polarimetry of the Interstellar Medium* ed W G Roberge and D C B Whittet (ASP Conference Series 97, San Francisco, CA) p 569
Hough J H, Bailey J A, Rouse M F and Whittet D C B 1987 *Mon. Not. R. Astron. Soc.* **227** 1P
Hough J H *et al* 1988 *Mon. Not. R. Astron. Soc.* **230** 107
Hough J H, Whittet D C B, Sato S, Yamashita T, Tamura M, Nagata T, Aitken D K and Roche P F 1989 *Mon. Not. R. Astron. Soc.* **241** 71
Hough J H, Peacock T and Bailey J A 1991 *Mon. Not. R. Astron. Soc.* **248** 74
Hough J H, Chrysostomou A, Messinger D W, Whittet D C B, Aitken D K and Roche P F 1996 *Astrophys. J.* **461** 902
Howk J C and Savage B D 1997 *Astron. J.* **114** 2463
Hoyle F and Wickramasinghe N C 1962 *Mon. Not. R. Astron. Soc.* **124** 417
——1979 *Diseases from Space* (London: Dent)
——1986 *Quart. J. R. Astron. Soc.* **27** 21
Hrivnak B J, Volk K and Kwok S 2000 *Astrophys. J.* **535** 275
Huffman D R 1977 *Adv. Phys.* **26** 129
——1989 *IAU Symp. 135, Interstellar Dust* ed L J Allamandola and A G G M Tielens (Dordrecht: Kluwer) p 329
Huffman D R and Stapp J L 1971 *Nature Phys. Sci.* **229** 45
Iben I 1967 *Annu. Rev. Astron. Astrophys.* **5** 571
Iben I and Renzini A 1983 *Annu. Rev. Astron. Astrophys.* **21** 271
Irvine W M, Schloerb F P, Crovisier J, Fegley B and Mumma M J 2000 *Protostars & Planets* vol IV, ed V Mannings *et al* (Tucson, AZ: University of Arizona Press) p 1159
Issa M R, MacLaren I and Wolfendale A W 1990 *Astron. Astrophys.* **236** 237
Itoh Y, Tamura M and Gatley I 1996 *Astrophys. J.* **465** L129

Ivison R J *et al* 1998 *Astrophys. J.* **494** 211

Jäger C, Molster F J, Dorschner J, Henning T, Mutschke H and Waters L B F M 1998 *Astron. Astrophys.* **339** 904

Jenkins E B 1987 *Interstellar Processes* ed D J Hollenbach and H A Thronson (Dordrecht: Reidel) p 533

——1989 *IAU Symp. 135, Interstellar Dust* ed L J Allamandola and A G G M Tielens (Dordrecht: Kluwer) p 23

Jenkins E B, Jura M and Loewenstein M 1983 *Astrophys. J.* **270** 88

Jenniskens P 1994 *Astron. Astrophys.* **284** 227

Jenniskens P and Greenberg J M 1993 *Astron. Astrophys.* **274** 439

Jenniskens P, Baratta G A, Kouchi A, de Groot M S, Greenberg J M and Strazzulla G 1993 *Astron. Astrophys.* **273** 583

Joblin C, Léger A and Martin P 1992 *Astrophys. J.* **393** L79

Joblin C, Tielens A G G M, Allamandola L J and Geballe T R 1996 *Astrophys. J.* **458** 610

Johnson H L 1963 *Basic Astronomical Data* ed K Aa Strand (Chicago, IL: University of Chicago Press) p 204

——1968 *Nebulae and Interstellar Matter* ed B M Middlehurst and L H Aller (Chicago, IL: University of Chicago Press) p 167

Johnson P E 1982 *Nature* **295** 371

Jones A P 1990 *Mon. Not. R. Astron. Soc.* **245** 331

Jones A P and Williams D A 1984 *Mon. Not. R. Astron. Soc.* **209** 955

Jones A P, Tielens A G G M and Hollenbach D J 1996 *Astrophys. J.* **469** 740

Jones A P, Tielens A G G M, Hollenbach D J and McKee C F 1994 *Astrophys. J.* **433** 797

Jones R V and Spitzer L 1967 *Astrophys. J.* **147** 943

Jones T J 1989a *Astron. J.* **98** 2062

——1989b *Astrophys. J.* **346** 728

——1996 *Polarimetry of the Interstellar Medium* ed W G Roberge and D C B Whittet (ASP Conference Series 97, San Francisco, CA) p 381

Jones T J, Klebe D and Dickey J M 1992 *Astrophys. J.* **389** 602

Jourdain de Muizon M, d'Hendecourt L B and Geballe T R 1990 *Astron. Astrophys.* **235** 367

Jura M 1980 *Astrophys. J.* **235** 63

——1986 *Astrophys. J.* **303** 327

——1987 *Interstellar Processes* ed D J Hollenbach and H A Thronson (Dordrecht: Reidel) p 3

——1989 *Evolution of Interstellar Dust and Related Topics* ed A Bonetti *et al* (Amsterdam: North-Holland) p 143

——1994 *Astrophys. J.* **434** 713

——1996 *Astrophys. J.* **472** 806

——1999 *Astrophys. J.* **515** 706

Jura M and Kleinmann S G 1989 *Astrophys. J.* **341** 359

——1990 *Astrophys. J. Suppl.* **73** 769

Kamijo F 1963 *Publ. Astron. Soc. Japan* **15** 440

Kapteyn J C 1909 *Astrophys. J.* **29** 46

Katz N, Furman I, Biham O, Pirronello V and Vidali G 1999 *Astrophys. J.* **522** 305

Keane J V, Tielens A G G M, Boogert A C A, Schutte W A and Whittet D C B 2001 *Astron. Astrophys.* **376** 254

Keller L P, Messenger S and Bradley J P 2000 *J. Geophys. Res.* **105** 10 397

Kemp J C and Wolstencroft R D 1972 *Astrophys. J.* **176** L115
Kenyon S J, Calvet N and Hartmann L 1993 *Astrophys. J.* **414** 676
Kenyon S J and Hartmann L 1987 *Astrophys. J.* **323** 714
Kenyon S J, Lada E A and Barsony M 1998 *Astron. J.* **115** 252
Kerridge J F 1989 *IAU Symp. 135, Interstellar Dust* ed L J Allamandola and A G G M Tielens (Dordrecht: Kluwer) p 383
Kim S H and Martin P G 1994 *Astrophys. J.* **431** 783
——1995a *Astrophys. J.* **442** 172
——1995b *Astrophys. J.* **444** 293
——1996 *Astrophys. J.* **462** 296
Kim S H, Martin P G and Hendry P D 1994 *Astrophys. J.* **422** 164
Kirsten T 1978 *The Origin of the Solar System* ed S F Dermott (New York: Wiley) p 267
Kissel J and Krueger F R 1987 *Nature* **326** 755
Kitta K and Krätschmer W 1983 *Astron. Astrophys.* **122** 105
Knacke R F and Krätschmer W 1980 *Astron. Astrophys.* **92** 281
Knacke R F, Cudaback D D and Gaustad J E 1969 *Astrophys. J.* **158** 151
Knapp G R 1985 *Astrophys. J.* **293** 273
——1986 *Astrophys. J.* **311** 731
Knapp G R and Morris M 1985 *Astrophys. J.* **292** 640
Knuth D E 1986 *The TEXbook* (Reading, MA: Addison-Wesley)
Kobayashi N, Nagata T, Tamura M, Takeuchi T, Takami H, Kobayashi Y and Sato S 1999 *Astrophys. J.* **517** 256
Koerner D W 1997 *Origins Life Evol. Biosphere* **27** 157
Koike C, Hasegawa H and Hattori T 1987 *Astrophys. Space Sci.* **134** 95
Koornneef J 1982 *Astron. Astrophys.* **107** 247
Kouchi A *et al* 2002 *Astrophys. J.* **566** L121
Kramer C, Alves J, Lada C J, Lada E A, Sievers A, Ungerechts H and Walmsley C M 1999 *Astron. Astrophys.* **342** 257
Krasnopolsky V A and Mumma M J 2001 *Astrophys. J.* **549** 629
Krasnopolsky V A *et al* 1997 *Science* **277** 1488
Krätschmer W and Huffman D R 1979 *Astrophys. Space Sci.* **61** 195
Krätschmer W, Lamb L D, Fostiropoulos K and Huffman D R 1990 *Nature* **347** 354
Krelowski J and Walker G A H 1987 *Astrophys. J.* **312** 860
Kress M E and Tielens A G G M 2001 *Meteoritics Planetary Sci.* **36** 75
Krügel E, Siebenmorgen R, Zota V and Chini R 1998 *Astron. Astrophys.* **331** L9
Kuhn W and Braun E 1929 *Naturwissenschaften* **17** 227
Kuijken K and Gilmore G 1989 *Mon. Not. R. Astron. Soc.* **239** 605
Kulkarni S R and Heiles C 1987 *Interstellar Processes* ed D J Hollenbach and H A Thronson (Dordrecht: Reidel) p 87
Kvenvolden K, Lawless J, Pering K, Peterson E, Flores J, Ponnamperuma C, Kaplan I R and Moore C 1970 *Nature* **228** 923
Kwok S, Volk K M and Hrivnak B J 1989 *Astrophys. J.* **345** L51
Lacy J H, Baas F, Allamandola L J, Persson S E, McGregor P J, Lonsdale C J, Geballe T R and van de Bult C E P 1984 *Astrophys. J.* **276** 533
Lacy J H, Faraji H, Sandford S A and Allamandola L J 1998 *Astrophys. J.* **501** L105
Lada C J 1988 *Galactic and Extragalactic Star Formation* ed R E Pudritz and M Fich (Dordrecht: Kluwer) p 5

Lada E A, Strom K M and Myers P C 1993 *Protostars & Planets* vol III, ed E H Levy *et al* (Tucson, AZ: University of Arizona Press) p 245
Lagache G, Abergel A, Boulanger F and Puget J L 1998 *Astron. Astrophys.* **333** 709
Lagage P O, Claret A, Ballet J, Boulanger F, Césarsky C J, Fransson C and Pollock A 1996 *Astron. Astrophys.* **315** L273
Lagrange A M, Backman D E and Artymowicz P 2000 *Protostars & Planets* vol IV, ed V Mannings *et al* (Tucson, AZ: University of Arizona Press) p 639
Lahuis F and van Dishoeck E F 2000 *Astron. Astrophys.* **355** 699
Lallement R, Bertin P, Ferlet R, Vidal-Madjar A and Bertaux J L 1994 *Astron. Astrophys.* **286** 898
Lambert D L, Smith V V and Hinkle K H 1990 *Astron. J.* **99** 1612
Langevin Y, Kissel J, Bertaux J L and Chassefiere E 1987 *Astron. Astrophys.* **187** 761
Larson K A, Whittet D C B and Hough J H 1996 *Astrophys. J.* **472** 755
Larson R B 1989 *Structure and Dynamics of the Interstellar Medium* ed G Tenorio-Tagle *et al* (Berlin: Springer) p 44
Lattimer J M, Schramm D N and Grossman L 1978 *Astrophys. J.* **219** 230
Laufer D, Notesco G and Bar-Nun A 1999 *Icarus* **140** 446
Laureijs R J, Mattila K and Schnur G 1987 *Astron. Astrophys.* **184** 269
Lawler M E and Brownlee D E 1992 *Nature* **359** 810
Lazarian A 1995a *Mon. Not. R. Astron. Soc.* **274** 679
——1995b *Mon. Not. R. Astron. Soc.* **277** 1235
——1995c *Astrophys. J.* **451** 660
——1997 *Astrophys. J.* **483** 296
Lazarian A, Goodman A A and Myers P C 1997 *Astrophys. J.* **490** 273
Lazarian A and Roberge W G 1997a *Astrophys. J.* **484** 230
——1997b *Mon. Not. R. Astron. Soc.* **287** 941
Le Bertre T 1987 *Astron. Astrophys.* **176** 107
——1997 *Astron. Astrophys.* **324** 1059
Lecluse C and Robert F 1994 *Geochim. Cosmochim. Acta* **58** 2927
Lee H H, Bettens R P A and Herbst E 1996 *Astron. Astrophys. Suppl.* **119** 111
Léger A 1983 *Astron. Astrophys.* **123** 271
Léger A and d'Hendecourt L 1988 *Dust in the Universe* ed M E Bailey and D A Williams (Cambridge: Cambridge University Press) p 219
Léger A, Gauthier S, Defourneau D and Rouan D 1983 *Astron. Astrophys.* **117** 164
Léger A, Jura M and Omont A 1985 *Astron. Astrophys.* **144** 147
Léger A and Puget J L 1984 *Astron. Astrophys.* **137** L5
Léger A, Verstraete L, d'Hendecourt L, Défourneau D, Dutuit O, Schmidt W and Lauer J C 1989 *IAU Symp. 135, Interstellar Dust* ed L J Allamandola and A G G M Tielens (Dordrecht: Kluwer) p 173
Lehtinen K and Mattila K 1996 *Astron. Astrophys.* **309** 570
Leitch-Devlin M A and Williams D A 1985 *Mon. Not. R. Astron. Soc.* **213** 295
Lellouch E *et al* 1998 *Astron. Astrophys.* **339** 9L
Lemke D, Mattila K, Lehtinen K, Laureijs R J, Liljeström T, Léger A and Herbstmeier U 1998 *Astron. Astrophys.* **331** 742
Lenzuni P, Gail H P and Henning T 1995 *Astrophys. J.* **447** 848
Lenzuni P, Natta A and Panagia N 1989 *Astrophys. J.* **345** 306
Leung C M 1975 *Astrophys. J.* **199** 340
LeVan P D, Sloan G C, Little-Marenin I R and Grasdalen G L 1992 *Astrophys. J.* **392** 702

Lewis J S 1972 *Icarus* **16** 241
Lewis R S, Anders E and Draine B T 1989 *Nature* **339** 117
Lewis R S, Tang M, Wacker J F, Anders E and Steel E 1987 *Nature* **326** 160
Li A and Draine B T 2001 *Astrophys. J.* **550** L213
——2002 *Astrophys. J.* **564** 803
Li A and Greenberg J M 1997 *Astron. Astrophys.* **323** 566
Lillie C F and Witt A N 1976 *Astrophys. J.* **208** 64
Lindblad B 1935 *Nature* **135** 133
Linsky J L *et al* 1993 *Astrophys. J.* **402** 694
Little-Marenin I R 1986 *Astrophys. J.* **307** L15
Liu S Y and Snyder L E 1999 *Astrophys. J.* **523** 683
Longo R, Stalio R, Polidan R S and Rossi L 1989 *Astrophys. J.* **339** 474
Londsdale C J, Dyck H M, Capps R W and Wolstencroft R D 1980 *Astrophys. J.* **238** L31
Lorenz-Martins S and Lefèvre J 1993 *Astron. Astrophys.* **280** 567
Low F J *et al* 1984 *Astrophys. J.* **278** L19
Lucas P W and Roche P F 1998 *Mon. Not. R. Astron. Soc.* **299** 699
Lucke P B 1978 *Astron. Astrophys.* **64** 367
Lucy L B, Danziger I J, Gouiffes C and Bouchet P 1991 *Supernovae* ed S E Woosley (New York: Springer) p 82
Lunine J I 1997 *Origins Life Evol. Biosphere* **27** 205
——1999 *Earth: Evolution of a Habitable World* (Cambridge University Press)
Lunine J I, Engel S, Rizk B and Horanyi M 1991 *Icarus* **94** 333
Maciel W J 1981 *Astron. Astrophys.* **98** 406
MacLean S, Duley W W and Millar T J 1982 *Astrophys. J.* **256** L61
Magnani L, Blitz L and Mundy L 1985 *Astrophys. J.* **295** 402
Maher K A and Stevenson D J 1988 *Nature* **331** 612
Malfait K, Waelkens C, Waters L B F M, Vandenbussche B, Huygen E and de Graauw M S 1998 *Astron. Astrophys.* **332** L25
Marcy G W and Butler R P 1998 *Annu. Rev. Astron. Astrophys.* **36** 57
Martin P G 1971 *Mon. Not. R. Astron. Soc.* **152** 279
——1972 *Mon. Not. R. Astron. Soc.* **158** 63
——1974 *Astrophys. J.* **187** 461
——1975 *Astrophys. J.* **202** 393
——1978 *Cosmic Dust, its Impact on Astronomy* (Oxford: Oxford University Press)
——1989 *IAU Symp. 135, Interstellar Dust* ed L J Allamandola and A G G M Tielens (Dordrecht: Kluwer) p 55
Martin P G and Angel J R P 1974 *Astrophys. J.* **188** 517
——1975 *Astrophys. J.* **195** 379
——1976 *Astrophys. J.* **207** 126
Martin P G, Clayton G C and Wolff M J 1999 *Astrophys. J.* **510** 905
Martin P G and Rogers C 1987 *Astrophys. J.* **322** 374
Martin P G and Shawl S J 1982 *Astrophys. J.* **253** 86
Martin P G and Whittet D C B 1990 *Astrophys. J.* **357** 113
Martin P G *et al* 1992 *Astrophys. J.* **392** 691
Massa D, Savage B D and Fitzpatrick E L 1983 *Astrophys. J.* **266** 662
Mathewson D S and Ford V L 1970 *Mem. R. Astron. Soc.* **74** 139
Mathis J S 1979 *Astrophys. J.* **232** 747
——1986 *Astrophys. J.* **308** 281

——1994 *Astrophys. J.* **422** 176
——1996a *Astrophys. J.* **472** 643
——1996b *Polarimetry of the Interstellar Medium* ed W G Roberge and D C B Whittet (ASP Conference Series 97, San Francisco, CA) p 3
——1998 *Astrophys. J.* **497** 824
——2000 *Allen's Astrophysical Quantities* 4th edn, ed A N Cox (New York: Springer) p 523
Mathis J S, Cohen D, Finley J P and Krautter J 1995 *Astrophys. J.* **449** 320
Mathis J S and Lee C W 1991 *Astrophys. J.* **376** 490
Mathis J S, Mezger P G and Panagia N 1983 *Astron. Astrophys.* **128** 212
Mathis J S, Rumpl W and Nordsieck K H 1977 *Astrophys. J.* **217** 425
Mathis J S and Wallenhorst S G 1981 *Astrophys. J.* **244** 483
Mathis J S and Whiffen G 1989 *Astrophys. J.* **341** 808
Mattila K *et al* 1996 *Astron. Astrophys.* **315** L353
McCaughrean M J and O'Dell C R 1996 *Astron. J.* **111** 1977
McClure R D and Crawford D L 1971 *Astron. J.* **76** 31
McKay D S *et al* 1996 *Science* **273** 924
McKee C F 1989 *IAU Symp. 135, Interstellar Dust* ed L J Allamandola and A G G M Tielens (Dordrecht: Kluwer) p 431
McKee C F and Ostriker J P 1977 *Astrophys. J.* **218** 148
McKellar A 1940 *Publ. Astron. Soc. Pacific* **52** 187
McLachlan A and Nandy K 1984 *Observatory* **104** 29
McMillan R S 1976 *Astron. J.* **81** 970
——1978 *Astrophys. J.* **225** 880
McMillan R S and Tapia S 1977 *Astrophys. J.* **212** 714
Meikle W P S, Spyromilio J, Allen D A, Varani G F and Cumming R J 1993 *Mon. Not. R. Astron. Soc.* **261** 535
Melosh H J 1988 *Nature* **332** 687
Mennella V, Brucato J R, Colangeli L and Palumbo P 1999 *Astrophys. J.* **524** L71
Mennella V, Colangeli L, Bussoletti E, Palumbo P and Rotundi A 1998 *Astrophys. J.* **507** L177
Mennella V, Colangeli L, Palumbo P, Rotundi A, Schutte W and Bussoletti E 1996 *Astrophys. J.* **464** L191
Mennella V, Muñoz Caro G M, Ruiterkamp R, Schutte W A, Greenberg J M, Brucato J R and Colangeli L 2001 *Astron. Astrophys.* **367** 355
Men'shchikov A B, Henning T and Fischer O 1999 *Astrophys. J.* **519** 257
Merrill K M and Stein W A 1976a *Publ. Astron. Soc. Pacific* **88** 285
——1976b *Publ. Astron. Soc. Pacific* **88** 294
Messenger S 2000 *Nature* **404** 968
Messenger S, Keller L P and Malker R M 2002 *33rd Lunar Planetary Sci. Conf.* abstract no 1887
Messinger D W, Whittet D C B and Roberge W G 1997 *Astrophys. J.* **487** 314
Mestel L 1985 *Protostars & Planets* vol II, ed D C Black and M S Matthews (Tucson, AZ: University of Arizona Press) p 320
Meyer A W, Smith R G, Charnley S B and Pendleton Y J 1998a *Astron. J.* **115** 2509
Meyer D M and Savage B D 1981 *Astrophys. J.* **248** 545
Meyer D M, Cardelli J A and Sofia U J 1997 *Astrophys. J.* **490** L103
Meyer D M, Jura M and Cardelli J A 1998b *Astrophys. J.* **493** 222

Michalsky J J and Schuster G J 1979 *Astrophys. J.* **231** 73
Mie G 1908 *Ann. Phys., Lpz.* **25** 377
Mihalas D and Binney J 1981 *Galactic Astronomy* (San Francisco, CA: Freeman)
Millar T J 1982 *Mon. Not. R. Astron. Soc.* **200** 527
——1993 *Dust and Chemistry in Astronomy* ed T J Millar and D A Williams (Bristol: IOP Publishing) p 249
Miller S L 1953 *Science* **117** 528
——1992 *Major Events in the History of Life* ed J W Schopf (London: Jones and Bartlett) p 1
Miller S L and Urey H C 1959 *Science* **130** 245
Millman P M 1975 *The Dusty Universe* ed G B Field and A G W Cameron (New York: Neale Watson) p 185
Mirabel I F *et al* 1998 *Astron. Astrophys.* **333** L1
Misselt K A, Clayton G C and Gordon K D 1999 *Astrophys. J.* **515** 128
Mitchell G F, Ginsburg J L and Kuntz P J 1978 *Astrophys. J. Suppl.* **38** 39
Mitchell G F, Allen M and Maillard J P 1988 *Astrophys. J.* **333** L55
Mitchell G F, Maillard J P, Allen M, Beer R and Belcourt K 1990 *Astrophys. J.* **363** 554
Mizutani K, Suto H and Maihara T 1989 *Astrophys. J.* **346** 675
——1994 *Astrophys. J.* **421** 475
Molster F J *et al* 1999 *Astron. Astrophys.* **350** 163
——2001 *Astron. Astrophys.* **372** 165
Moneti A, Pipher J L, Helfer H L, McMillan R S and Perry M L 1984 *Astrophys. J.* **282** 508
Moore M H and Donn B 1982 *Astrophys. J.* **257** L47
Morgan D H, Nandy K and Thompson G I 1978 *Mon. Not. R. Astron. Soc.* **185** 371
Morris M and Serabyn G 1996 *Annu. Rev. Astron. Astrophys.* **34** 645
Moseley S H, Dwek E, Glaccum W, Graham J R and Loewenstein R F 1989 *Nature* **340** 697
Mouschovias T Ch 1987 *Physical Processes in Interstellar Clouds* ed G E Morfill and M Scholer (Dordrecht: Reidel) p 453
Mouschovias T Ch and Ciolek G E 1999 *The Origin of Stars and Planetary Systems* ed C J Lada and N D Kylafis (Dordrecht: Kluwer) p 305
Moutou C, Krelowski J, d'Hendecourt L and Jamroszczak J 1999 *Astron. Astrophys.* **351** 680
Muci A M, Blanco A, Fonti S and Orofino V 1994 *Astrophys. J.* **436** 831
Mumma M J 1997 *From Stardust to Planetesimals* ed Y J Pendleton and A G G M Tielens (ASP Conference Series 122, San Francisco, CA) p 369
Mumma M J, DiSanti M A, Dello Russo N, Fomenkova M, Magee-Sauer K, Kaminski C D and Xie D X 1996 *Science* **272** 1310
Mumma M J, Weissman P R and Stern S A 1993 *Protostars & Planets* vol III, ed E H Levy *et al* (Tucson, AZ: University of Arizona Press) p 1177
Mumma M J *et al* 2001 *Astrophys. J.* **546** 1183
Mundy L G, Looney L W and Welch W J 2000 *Protostars & Planets* vol IV, ed V Mannings *et al* (Tucson, AZ: University of Arizona Press) p 355
Myers P C, Evans N J and Ohashi N 2000 *Protostars & Planets* vol IV, ed V Mannings *et al* (Tucson, AZ: University of Arizona Press) p 217
Nagata T 1990 *Astrophys. J.* **348** L13
Nagata T, Kobayashi N and Sato S 1994 *Astrophys. J.* **423** L113

Nandy K 1984 *IAU Symp. 108, Structure and Evolution of the Magellanic Clouds* ed S van den Bergh and K de Boer (Dordrecht: Reidel) p 341

Nandy K and Wickramasinghe N C 1971 *Mon. Not. R. Astron. Soc.* **154** 255

Natta A and Panagia N 1981 *Astrophys. J.* **248** 189

Neckel T and Klare G 1980 *Astron. Astrophys. Suppl.* **42** 251

Nelson K E, Levy M and Miller S L 2000 *Proc. Natl Acad. Sci.* **97** 3868

Neufeld D A and Hollenbach D J 1994 *Astrophys. J.* **428** 170

Ney E P and Hatfield B F 1978 *Astrophys. J.* **219** L111

Nittler L R *et al* 1995 *Astrophys. J.* **453** L25

Nittler L R, Amari S, Zinner E, Woosley S E and Lewis R S 1996 *Astrophys. J.* **462** L31

Nittler L R, Alexander C M O'D, Gao X, Walker R M and Zinner E 1997 *Astrophys. J.* **483** 475

Norden B 1977 *Nature* **266** 567

Nummelin A, Whittet D C B, Gibb E L, Gerakines P A and Chiar J E 2001 *Astrophys. J.* **558** 185

Nuth J A 1990 *Nature* **345** 207

——1996 *The Cosmic Dust Connection* ed J M Greenberg (Dordrecht: Kluwer) p 205

Nuth J A, Hallenbeck S L and Rietmeijer F J M 2000 *J. Geophys. Res.* **105** 10 387

Nuth J A, Moseley S H, Silverberg R F, Goebel J H and Moore W J 1985 *Astrophys. J.* **290** L41

O'Dell C R, Wen Z and Hu X 1993 *Astrophys. J.* **410** 696

O'Donnell J E 1994a *Astrophys. J.* **422** 158

——1994b *Astrophys. J.* **437** 262

O'Donnell J E and Mathis J S 1997 *Astrophys. J.* **479** 806

Olofsson H, Eriksson K, Gustafsson B and Carltröm U 1993 *Astrophys. J. Suppl.* **87** 267

Omont A 1986 *Astron. Astrophys.* **164** 159

Omont A, Moseley S H, Forveille T, Glaccum W J, Harvey P M, Likkel L, Lowenstein R F and Lisse C M 1990 *Astrophys. J.* **355** L27

Omont A *et al* 1999 *Astron. Astrophys.* **348** 755

Onaka T, de Jong T and Willems F J 1989 *Astron. Astrophys.* **218** 169

Oort J H 1932 *Bull. Astron. Inst. Netherlands* **6** 249

Oort J H and van de Hulst H C 1946 *Bull. Astron. Inst. Netherlands* **10** 187

Oro J 1960 *Biochem. Biophys. Res. Commun.* **2** 407

——1961 *Nature* **190** 389

Ostriker J P and Heisler J 1984 *Astrophys. J.* **278** 1

Owen T C and Bar-Nun A 2001 *Origins Life Evol. Biosphere* **31** 435

Pagel B E J and Edmunds M G 1981 *Annu. Rev. Astron. Astrophys.* **19** 77

Palumbo M E, Geballe T R and Tielens A G G M 1997 *Astrophys. J.* **479** 839

Palumbo M E and Strazzulla G 1993 *Astron. Astrophys.* **269** 568

Palumbo M E, Strazzulla G, Pendleton Y J and Tielens A G G M 2000 *Astrophys. J.* **534** 801

Peimbert M 1992 *Observational Astrophysics* ed R E White (Bristol: IOP Publishing) p 1

Pendleton Y J 1997 *Origins Life Evol. Biosphere* **27** 53

Pendleton Y J and Allamandola L J 2002 *Astrophys. J. Suppl.* **138** 75

Pendleton Y J, Sandford S A, Allamandola L J, Tielens A G G M and Sellgren K 1994 *Astrophys. J.* **437** 683

Pendleton Y J, Tielens A G G M, Tokunaga A T and Bernstein M P 1999 *Astrophys. J.* **513** 294

Pendleton Y J, Tielens A G G M and Werner M W 1990 *Astrophys. J.* **349** 107

Penprase B E 1992 *Astrophys. J. Suppl.* **83** 273

Perrin J M and Sivan J P 1990 *Astron. Astrophys.* **228** 238

Perry C L and Johnston L 1982 *Astrophys. J. Suppl.* **50** 451

Pier E A and Voit G M 1995 *Astrophys. J.* **450** 628

Pipher J L 1973 *IAU Symp. 52, Interstellar Dust and Related Topics* ed J M Greenberg and H C van de Hulst (Dordrecht: Reidel) p 559

Pizzarello S and Cronin J R 2000 *Geochim. Cosmochim. Acta* **64** 329

Platt J R 1956 *Astrophys. J.* **123** 486

Poppe T, Blum J and Henning T 2000 *Astrophys. J.* **533** 454

Pottasch S R, Baud B, Beintema D, Emerson J, Habing H J, Harris S, Houck J, Jennings R and Marsden P 1984 *Astron. Astrophys.* **138** 10

Prasad S S and Huntress W T 1980 *Astrophys. J. Suppl.* **43** 1

Prasad S S and Tarafdar S P 1983 *Astrophys. J.* **267** 603

Prévot M L, Lequeux J, Maurice E, Prévot L and Rocca-Volmerange B 1984 *Astron. Astrophys.* **132** 389

Puget J L and Léger A 1989 *Annu. Rev. Astron. Astrophys.* **27** 161

Purcell E M 1969 *Astrophys. J.* **158** 433

——1975 *The Dusty Universe* ed G B Field and A G W Cameron (New York: Neale Watson) p 155

——1976 *Astrophys. J.* **206** 685

——1979 *Astrophys. J.* **231** 404

Purcell E M and Pennypacker C R 1973 *Astrophys. J.* **186** 705

Purcell E M and Spitzer L 1971 *Astrophys. J.* **167** 31

Rawlings J M C and Williams D A 1989 *Mon. Not. R. Astron. Soc.* **240** 729

Rayleigh Lord 1871 *Phil. Mag.* **41** 107

Reach W T, Pound M W, Wilner D J and Lee Y 1995a *Astrophys. J.* **441** 244

Reach W T *et al* 1995b *Astrophys. J.* **451** 188

Reid M J 1989 *IAU Symposium 136, The Center of the Galaxy* ed M Morris (Dordrecht: Reidel) p 37

Reimann H G and Friedemann C 1991 *Astron. Astrophys.* **242** 474

Reimers D 1975 *Mém. Soc. R. Sci. Liège* 6th series **8** 369

Rice W, Boulanger F, Viallefond F, Soifer B T and Freedman W L 1990 *Astrophys. J.* **358** 418

Rieke G H and Lebofsky M J 1985 *Astrophys. J.* **288** 618

Rietmeijer F J M, Nuth J A and Karner J M 1999 *Astrophys. J.* **527** 395

Roberge W G 1996 *Polarimetry of the Interstellar Medium* ed W G Roberge and D C B Whittet (ASP Conference Series 97, San Francisco, CA) p 401

Roberge W G and Lazarian A 1999 *Mon. Not. R. Astron. Soc.* **305** 615

Roberge W G, Hanany S and Messinger D W 1995 *Astrophys. J.* **453** 238

Robertson J 1996 *Phys. Rev.* B **53** 16 302

Robertson J and O'Reilly E P 1987 *Phys. Rev.* B **35** 2946

Roche P F 1988 *Dust in the Universe* ed M E Bailey and D A Williams (Cambridge: Cambridge University Press) p 415

——1989a *IAU Symposium 131, Planetary Nebulae* ed S Torres-Peimbert (Dordrecht: Reidel) p 117

——1989b *IAU Symp. 135, Interstellar Dust* ed L J Allamandola and A G G M Tielens (Dordrecht: Kluwer) p 303

——1996 *Polarimetry of the Interstellar Medium* ed W G Roberge and D C B Whittet (ASP Conference Series 97, San Francisco, CA) p 551

Roche P F and Aitken D K 1984a *Mon. Not. R. Astron. Soc.* **208** 481

——1984b *Mon. Not. R. Astron. Soc.* **209** 33P

——1985 *Mon. Not. R. Astron. Soc.* **215** 425

Roche P F, Aitken D K and Smith C H 1989a *Mon. Not. R. Astron. Soc.* **236** 485

——1993 *Mon. Not. R. Astron. Soc.* **261** 522

Roche P F, Aitken D K, Smith C H and James S D 1986a *Mon. Not. R. Astron. Soc.* **218** 19P

——1989b *Nature* **337** 533

Roche P F, Aitken D K, Smith C H and Ward M J 1991 *Mon. Not. R. Astron. Soc.* **248** 606

Roche P F, Allen D A and Bailey J A 1986b *Mon. Not. R. Astron. Soc.* **220** 7P

Roche P F, Lucas P W, Hoare M G, Aitken D K and Smith C H 1996 *Mon. Not. R. Astron. Soc.* **280** 924

Rodgers S D and Millar T J 1996 *Mon. Not. R. Astron. Soc.* **280** 1046

Rodrigues C V, Magalhães A M, Coyne G V and Piirola V 1997 *Astrophys. J.* **485** 618

Rodriguez-Espinosa J M, Pérez Garcia A M, Lemke D and Meisenheimer K 1996 *Astron. Astrophys.* **315** L129

Roelfsema P R *et al* 1996 *Astron. Astrophys.* **315** L289

Rogers C and Martin P G 1979 *Astrophys. J.* **228** 450

Roser J E, Vidali G, Manico G and Pirronello V 2001 *Astrophys. J.* **555** L61

Rouleau F, Henning T and Stognienko R 1997 *Astron. Astrophys.* **322** 633

Routly P M and Spitzer L 1952 *Astrophys. J.* **115** 227

Rowan-Robinson M, Lock T D, Walker D W and Harris S 1986 *Mon. Not. R. Astron. Soc.* **222** 273

Ruffle D P and Herbst E 2001 *Mon. Not. R. Astron. Soc.* **324** 1054

Russell R W, Soifer B T and Merrill K M 1977 *Astrophys. J.* **213** 66

Sagan C 1973 *Molecules in the Galactic Environment* ed M A Gordon and L E Snyder (New York: Wiley) p 451

Sagan C and Khare B N 1979 *Nature* **277** 102

Sakata A, Nakagawa N, Iguchi T, Isobe S, Morimoto M, Hoyle F and Wickramasinghe N C 1977 *Nature* **266** 241

Sakata A, Wada S, Okutsu Y, Shintani H and Nakata Y 1983 *Nature* **301** 493

Salama F 1999 *Solid Interstellar Matter: The ISO Revolution* ed L d'Hendecourt *et al* (Les Ulis: EDP) p 65

Salama F, Galazutdinov G A, Krelowski J, Allamandola L J and Musaev F A 1999 *Astrophys. J.* **526** 265

Salpeter E E 1955 *Astrophys. J.* **121** 161

——1974 *Astrophys. J.* **193** 579

——1977 *Annu. Rev. Astron. Astrophys.* **15** 267

Sandage A 1976 *Astron. J.* **81** 954

Sanders D B and Mirabel I F 1996 *Annu. Rev. Astron. Astrophys.* **34** 749

Sandford S A 1989 *IAU Symp. 135, Interstellar Dust* ed L J Allamandola and A G G M Tielens (Dordrecht: Kluwer) p 403

Sandford S A, Allamandola L J, Tielens A G G M and Valero G J 1988 *Astrophys. J.* **329** 498

Sandford S A, Allamandola L J, Tielens A G G M, Sellgren K, Tapia M and Pendleton Y 1991 *Astrophys. J.* **371** 607

Sofue Y, Fujimoto M and Wielebinski R 1986 *Annu. Rev. Astron. Astrophys.* **24** 459
Soifer B T, Houck J R and Neugebauer G 1987 *Annu. Rev. Astron. Astrophys.* **25** 187
Soifer B T, Neugebauer G, Franx M, Matthews K and Illingworth G D 1998 *Astrophys. J.* **501** L171
Soifer B T, Russell R W and Merrill K M 1976 *Astrophys. J.* **210** 334
Soifer B T, Willner S P, Capps R W and Rudy R J 1981 *Astrophys. J.* **250** 631
Somerville W B *et al* 1994 *Astrophys. J.* **427** L47
Sonnentrucker P, Cami J, Ehrenfreund P and Foing B H 1997 *Astron. Astrophys.* **327** 1215
Sopka R J *et al* 1985 *Astrophys. J.* **294** 242
Sorrell W H 1990 *Mon. Not. R. Astron. Soc.* **243** 570
——1991 *Mon. Not. R. Astron. Soc.* **248** 439
——1995 *Mon. Not. R. Astron. Soc.* **273** 169
Speck A K, Barlow M J and Skinner C J 1997 *Meteoritics Planetary Sci.* **32** 703
Spitzer L 1976 *Comments Astrophys.* **6** 177
——1978 *Physical Processes in the Interstellar Medium* (New York: Wiley)
——1982 *Searching Between the Stars* (New Haven, CT: Yale University Press)
——1985 *Astrophys. J.* **290** L21
Spitzer L and Fitzpatrick E L 1993 *Astrophys. J.* **409** 299
Spitzer L and Jenkins E B 1975 *Annu. Rev. Astron. Astrophys.* **13** 133
Spitzer L and McGlynn T A 1979 *Astrophys. J.* **231** 417
Spitzer L and Tukey J W 1951 *Astrophys. J.* **114** 187
Stapelfeldt K R *et al* 1995 *Astrophys. J.* **449** 888
Stark R 1995 *Astron. Astrophys.* **301** 873
Stasinska G and Szczerba R 1999 *Astron. Astrophys.* **352** 297
Stebbins J, Huffer C M and Whitford A E 1939 *Astrophys. J.* **90** 209
Stecher T P 1965 *Astrophys. J.* **142** 1683
Stecher T P and Donn B 1965 *Astrophys. J.* **142** 1681
Steel D 1992 *Origins Life Evol. Biosphere* **21** 339
Steel T M and Duley W W 1987 *Astrophys. J.* **315** 337
Stephens J R 1980 *Astrophys. J.* **237** 450
Stern S A *et al* 2000 *Astrophys. J.* **544** L169
Sternberg A, Dalgarno A and Lepp S 1987 *Astrophys. J.* **320** 676
Stevenson D J and Lunine J I 1988 *Icarus* **75** 146
Stokes G M 1978 *Astrophys. J. Suppl.* **36** 115
Störzer H and Hollenbach D 1999 *Astrophys. J.* **515** 669
Strazzulla G and Baratta G A 1992 *Astron. Astrophys.* **266** 434
Strömgren B 1966 *Annu. Rev. Astron. Astrophys.* **4** 433
——1987 *The Galaxy* ed G Gilmore and R Carswell (Dordrecht: Reidel) p 229
Struve F G W 1847 *Etudes d'Astronomie Stellaire*
Suttner G and Yorke H W 2001 *Astrophys. J.* **551** 461
Sweitzer J S 1978 *Astrophys. J.* **225** 116
Sylvester R J, Kemper F, Barlow M J, de Jong T, Waters L B F M, Tielens A G G M and Omont A 1999 *Astron. Astrophys.* **352** 587
Szczerba R, Henning T, Volk K, Kwok S and Cox P 1999 *Astron. Astrophys.* **345** L39
Tanabé T, Nakada Y, Kamijo F and Sakata A 1983 *Publ. Astron. Soc. Japan* **35** 397
Tanaka M, Sato S, Nagata T and Yamamoto T 1990 *Astrophys. J.* **352** 724
Tang M and Anders E 1988 *Astrophys. J.* **335** L31
Tang M, Anders E, Hoppe P and Zinner E 1989 *Nature* **339** 351

Tayler R J 1975 *The Origin of the Chemical Elements* (London: Wykeham)
Teixeira T C, Devlin J P, Buch V and Emerson J P 1999 *Astron. Astrophys.* **347** 19
Terebey S and Fich M 1986 *Astrophys. J.* **309** L73
Thi W F *et al* 2001 *Astrophys. J.* **561** 1074
Thronson H A 1988 *Galactic and Extragalactic Star Formation* ed R E Pudritz and M Fich (Dordrecht: Kluwer) p 621
Thronson H A, Latter W B, Black J H, Bally J and Hacking P 1987 *Astrophys. J.* **322** 770
Tielens A G G M 1983 *Astron. Astrophys.* **119** 177
——1990 *Carbon in the Galaxy: Studies from Earth and Space* ed J Tarter *et al* (Washington, DC: NASA) p 59
——1993 *Dust and Chemistry in Astronomy* ed T J Millar and D A Williams (Bristol: IOP Publishing) p 103
——1998 *Astrophys. J.* **499** 267
Tielens A G G M and Allamandola L J 1987a *Interstellar Processes* ed D J Hollenbach and H A Thronson (Dordrecht: Reidel) p 397
——1987b *Physical Processes in Interstellar Clouds* ed G E Morfill and M Scholer (Dordrecht: Reidel) p 333
Tielens A G G M and Charnley S B 1997 *Origins Life Evol. Biosphere* **27** 23
Tielens A G G M and Hagen W 1982 *Astron. Astrophys.* **114** 245
Tielens A G G M, Hony S, van Kerckhoven C and Peeters E 1999 *The Universe as Seen by ISO* ed P Cox and M F Kessler (ESA publication SP-427) p 579
Tielens A G G M, McKee C F, Seab C G and Hollenbach D J 1994 *Astrophys. J.* **431** 321
Tielens A G G M, Seab C G, Hollenbach D J and McKee C F 1987 *Astrophys. J.* **319** L109
Tielens A G G M and Snow T P (ed) 1995 *The Diffuse Interstellar Bands* (Dordrecht: Kluwer)
Tielens A G G M, Wooden D H, Allamandola L J, Bregman J and Witteborn F C 1996 *Astrophys. J.* **461** 210
Toller G N 1981 *Thesis* (State University of New York at Stony Brook)
Treffers R R and Cohen M 1974 *Astrophys. J.* **188** 545
Trimble V 1975 *Rev. Mod. Phys.* **47** 877
——1991 *Astron. Astrophys. Rev.* **3** 1
——1997 *Origins Life Evol. Biosphere* **27** 3
Trumpler R J 1930a *Lick Obs. Bull.* **14** 154
——1930b *Publ. Astron. Soc. Pacific* **42** 214
——1930c *Publ. Astron. Soc. Pacific* **42** 267
Turner B E 1990 *Astrophys. J.* **362** L29
——2001 *Astrophys. J. Suppl.* **136** 579
Turon P and Mennessier M O 1975 *Astron. Astrophys.* **44** 209
Twarog B A 1980 *Astrophys. J.* **242** 242
van Breda I G and Whittet D C B 1981 *Mon. Not. R. Astron. Soc.* **195** 79
van de Bult C E P M, Greenberg J M and Whittet D C B 1985 *Mon. Not. R. Astron. Soc.* **214** 289
van de Hulst H C 1946 *Recherches Astronomiques de l'Observatoire d'Utrecht* **11** 1
——1957 *Light Scattering by Small Particles* (New York: Wiley)
——1989 *Evolution of Interstellar Dust and Related Topics* ed A Bonetti *et al* (Amsterdam: North-Holland) p 1
van den Bergh S and Tammann G A 1991 *Annu. Rev. Astron. Astrophys.* **29** 363
Vandenbussche B *et al* 1999 *Astron. Astrophys.* **346** L57

van der Tak F, van Dishoeck E F and Caselli P 2000 *Astron. Astrophys.* **361** 327
van Dishoeck E F 2002 *The Origins of Stars and Planets: The VLT View* ed J Alves and M McCaughrean (New York: Springer) in press
van Dishoeck E F and Black J H 1987 *Physical Processes in Interstellar Clouds* ed G E Morfill and M Scholer (Dordrecht: Reidel) p 241
van Dishoeck E F and Blake G A 1998 *Annu. Rev. Astron. Astrophys.* **36** 317
van Dishoeck E F, Blake G A, Draine B T and Lunine J I 1993 *Protostars & Planets* vol III, ed E H Levy *et al* (Tucson, AZ: University of Arizona Press) p 163
van Dishoeck E F *et al* 1996 *Astron. Astrophys.* **315** L349
Vázquez-Semadeni E, Ostriker E C, Passot T, Gammie C F and Stone J M 2000 *Protostars & Planets* vol IV, ed V Mannings *et al* (Tucson, AZ: University of Arizona Press) p 3
Villar-Martin M, De Young D, Alonso-Herrero A, Allen M and Binette L 2001 *Mon. Not. R. Astron. Soc.* **328** 848
Viola V E and Mathews G J 1987 *Sci. Am.* **255** 35
Viti S, Caselli P, Hartquist T W and Williams D A 2001 *Astron. Astrophys.* **370** 1017
Voit G M 1992 *Mon. Not. R. Astron. Soc.* **258** 841
von Helden G, Tielens A G G M, van Heijnsbergen D, Duncan M A, Honey S, Waters L B F M and Meijer G 2000 *Science* **288** 313
Vrba F J, Coyne G V and Tapia S 1981 *Astrophys. J.* **243** 489
——1993 *Astron. J.* **105** 1010
Vrba F J, Strom S E and Strom K M 1976 *Astron. J.* **81** 958
Vriend W J 2000 *Masters Thesis* (University of Groningen)
Wagner D R, Kim H S and Saykally R J 2000 *Astrophys. J.* **545** 854
Walker H J and Wolstencroft R D 1988 *Publ. Astron. Soc. Pacific* **100** 1509
Walker J C G 1985 *Origins of Life* **16** 117
Wall W F *et al* 1996 *Astrophys. J.* **456** 566
Wallerstein G and Knapp G R 1998 *Annu. Rev. Astron. Astrophys.* **36** 369
Walmsley C M and Schilke P 1993 *Dust and Chemistry in Astronomy* ed T J Millar and D A Williams (Bristol: IOP Publishing) p 37
Wannier P G 1989 *IAU Symposium 136, The Center of the Galaxy* ed M Morris (Dordrecht: Reidel) p 107
Wannier P G, Sahai R, Andersson B G and Johnson H R 1990 *Astrophys. J.* **358** 251
Watanabe I, Hasegawa S and Kurata Y 1982 *Japanese J. Appl. Phys.* **21** 856
Waters L B F M *et al* 1989 *Astron. Astrophys.* **211** 208
——1996 *Astron. Astrophys.* **315** L361
——1998 *Astron. Astrophys.* **331** L61
Watson W D 1976 *Rev. Mod. Phys.* **48** 513
Weber P and Greenberg J M 1985 *Nature* **316** 403
Weidenschilling S J 1997 *From Stardust to Planetesimals* ed Y J Pendleton and A G G M Tielens (ASP Conference Series 122, San Francisco, CA) p 281
Weidenschilling S J and Ruzmaikina T V 1994 *Astrophys. J.* **430** 713
Weiland J L, Blitz L, Dwek E, Hauser M G, Magnani L and Rickard L J 1986 *Astrophys. J.* **306** L101
Weingartner J C and Draine B T 1999 *Astrophys. J.* **517** 292
——2001 *Astrophys. J.* **548** 296
Weintraub D A, Goodman A A and Akeson R L 2000 *Protostars & Planets* vol IV, ed V Mannings *et al* (Tucson, AZ: University of Arizona Press) p 247
Weintraub D A, Kastner J H and Whitney B A 1995 *Astrophys. J.* **452** L141

Welty D E and Fowler J R 1992 *Astrophys. J.* **393** 193
Westerlund B E 1997 *The Magellanic Clouds* (Cambridge: Cambridge University Press)
Wetherill G W 1989 *The Formation and Evolution of Planetary Systems* ed H A Weaver and L Danly (Cambridge: Cambridge University Press) p 1
——1990 *Annu. Rev. Earth Planet. Sci.* **18** 205
Whipple F L 1950 *Astrophys. J.* **111** 375
——1951 *Astrophys. J.* **113** 464
——1987 *Astron. Astrophys.* **187** 852
Whiteoak J B 1966 *Astrophys. J.* **144** 305
Whitford A E 1958 *Astron. J.* **63** 201
Whitney B A, Kenyon S J and Gomez M 1997 *Astrophys. J.* **485** 703
Whitney B A and Wolff M J 2002 *Astrophys. J.* **574** 205
Whittet D C B 1977 *Mon. Not. R. Astron. Soc.* **180** 29
——1979 *Astron. Astrophys.* **72** 370
——1984a *Observatory* **104** 159
——1984b *Mon. Not. R. Astron. Soc.* **210** 479
——1988 *Dust in the Universe* ed M E Bailey and D A Williams (Cambridge: Cambridge University Press) p 25
——1993 *Dust and Chemistry in Astronomy* ed T J Millar and D A Williams (Bristol: IOP Publishing) p 9
——1997 *Origins Life Evol. Biosphere* **27** 249
Whittet D C B, Adamson A J, Duley W W, Geballe T R and McFadzean A D 1989 *Mon. Not. R. Astron. Soc.* **241** 707
Whittet D C B and Blades J C 1980 *Mon. Not. R. Astron. Soc.* **191** 309
Whittet D C B, Bode M F, Longmore A J, Adamson A J, McFadzean A D, Aitken D K and Roche P F 1988 *Mon. Not. R. Astron. Soc.* **233** 321
Whittet D C B, Bode M F, Longmore A J, Baines D W T and Evans A 1983 *Nature* **303** 218
Whittet D C B and Duley W W 1991 *Astron. Astrophys. Rev.* **2** 167
Whittet D C B, Duley W W and Martin P G 1990 *Mon. Not. R. Astron. Soc.* **244** 427
Whittet D C B, Gerakines P A, Carkner A L, Hough J H, Martin P G, Prusti T and Kilkenny D 1994 *Mon. Not. R. Astron. Soc.* **268** 1
Whittet D C B, Martin P G, Hough J H, Rouse M F, Bailey J A and Axon D J 1992 *Astrophys. J.* **386** 562
Whittet D C B, McFadzean A D and Geballe T R 1984 *Mon. Not. R. Astron. Soc.* **211** 29P
Whittet D C B and van Breda I G 1978 *Astron. Astrophys.* **66** 57
Whittet D C B, van Breda I G and Glass I S 1976 *Mon. Not. R. Astron. Soc.* **177** 625
Whittet D C B *et al* 1996 *Astrophys. J.* **458** 363
——1997 *Astrophys. J.* **490** 729
——1998 *Astrophys. J.* **498** L159
Whittet D C B, Gerakines P A, Hough J H and Shenoy S S 2001a *Astrophys. J.* **547** 872
Whittet D C B, Pendleton Y J, Gibb E L, Boogert A C A, Chiar J E and Nummelin A 2001b *Astrophys. J.* **550** 793
Wickramasinghe N C and Nandy K 1971 *Mon. Not. R. Astron. Soc.* **153** 205
Wilking B A 1989 *Publ. Astron. Soc. Pacific* **101** 229
——1997 *Origins Life Evol. Biosphere* **27** 135
Wilking B A, Lada C J and Young E T 1989 *Astrophys. J.* **340** 823

Wilking B A, Lebofsky M J, Martin P G, Rieke G H and Kemp J C 1980 *Astrophys. J.* **235** 905

Wilking B A, Lebofsky M J and Rieke G H 1982 *Astron. J.* **87** 695

Willacy K and Langer W D 2000 *Astrophys. J.* **544** 903

Willacy K, Langer W D and Velusamy T 1998 *Astrophys. J.* **507** L171

Willacy K and Williams D A 1993 *Mon. Not. R. Astron. Soc.* **260** 635

Willems F J and de Jong T 1986 *Astron. Astrophys.* **309** L39

Williams D A 1985 *Quart. J. R. Astron. Soc.* **26** 463

——1993 *Dust and Chemistry in Astronomy* ed T J Millar and D A Williams (Bristol: IOP Publishing) p 143

Williams D A, Hartquist T W and Whittet D C B 1992 *Mon. Not. R. Astron. Soc.* **258** 599

Willner S P *et al* 1982 *Astrophys. J.* **253** 174

Wilson R 1960 *Mon. Not. R. Astron. Soc.* **120** 51

Wilson W L, Szajowski P F and Brus L E 1993 *Science* **262** 1242

Witt A N 1988 *Dust in the Universe* ed M E Bailey and D A Williams (Cambridge: Cambridge University Press) p 1

——1989 *IAU Symp. 135, Interstellar Dust* ed L J Allamandola and A G G M Tielens (Dordrecht: Kluwer) p 87

Witt A N, Bohlin R C and Stecher T P 1983 *Astrophys. J.* **267** L47

——1984 *Astrophys. J.* **279** 698

Witt A N and Boroson T A 1990 *Astrophys. J.* **355** 182

Witt A N, Gordon K D and Furton D G 1998 *Astrophys. J.* **501** L111

Witt A N, Petersohn J K, Bohlin R C, O'Connell R W, Roberts M S, Smith A M and Stecher T P 1992 *Astrophys. J.* **395** L5

Witt A N, Petersohn J K, Holberg J B, Murthy J, Dring A and Henry R C 1993 *Astrophys. J.* **410** 714

Witt A N and Schild R E 1986 *Astrophys. J. Suppl.* **62** 839

——1988 *Astrophys. J.* **325** 837

Witt A N, Smith R K and Dwek E 2001 *Astrophys. J.* **550** L201

Witt A N, Walker G A H, Bohlin R C and Stecher T P 1982 *Astrophys. J.* **261** 492

Witte M, Rosenbauer H, Banaszkiewicz M and Fahr H 1993 *Adv. Space Rev.* **13** 121

Wolff M J, Clayton G C and Gibson S J 1998 *Astrophys. J.* **503** 815

Wolff M J, Clayton G C, Kim S H, Martin P G and Anderson C M 1997 *Astrophys. J.* **478** 395

Wolff M J, Clayton G C, Martin P G and Schulte-Ladbeck R E 1994 *Astrophys. J.* **423** 412

Wolff M J, Clayton G C and Meade M R 1993 *Astrophys. J.* **403** 722

Wolfire M G, Tielens A G G M and Hollenbach D 1990 *Astrophys. J.* **358** 116

Wood J A and Morfill G E 1988 *Meteorites and the Early Solar System* ed J F Kerridge and M S Matthews (Tucson, AZ: University of Arizona Press) p 329

Wood K, Wolff M J, Bjorkman J E and Whitney B 2002 *Astrophys. J.* **564** 887

Wooden D H *et al* 1993 *Astrophys. J. Suppl.* **88** 477

Woolf N J 1973 *IAU Symp. 52, Interstellar Dust and Related Topics* ed J M Greenberg and H C van de Hulst (Dordrecht: Reidel) p 485

Wright C M, Aitken D K, Smith C H, Roche P F and Laureijs R J 2002 *The Origins of Stars and Planets: The VLT View* ed J Alves and M McCaughrean (New York: Springer) in press

Wurm G and Schnaiter M 2002 *Astrophys. J.* **567** 370

Wynn-Williams G C 1982 *Annu. Rev. Astron. Astrophys.* **20** 587

Yamamura I, Dominik C, de Jong T, Waters L B F M and Molster F J 2000 *Astron. Astrophys.* **363** 629
York D G 1971 *Astrophys. J.* **166** 65
Zaikowski A, Knacke R F and Porco C C 1975 *Astrophys. Space Sci.* **35** 97
Zijlstra A A, Loup C, Waters L B F M and de Jong T 1992 *Astron. Astrophys.* **265** L5
Zinner E 1998 *Annu. Rev. Earth Planet. Sci.* **26** 147

Index